Lecture Notes in Computer Scien

Edited by G. Goos, J. Hartmanis and J. van L

T0237806

Springer
Berlin
Heidelberg
New York
Barcelona
Hong Kong
London
Milan
Paris
Singapore
Tokyo

Mogens Nielsen   Dan Simpson (Eds.)

Application and Theory of Petri Nets 2000

21st International Conference, ICATPN 2000
Aarhus, Denmark, June 26-30, 2000
Proceedings

Springer

Series Editors

Gerhard Goos, Karlsruhe University, Germany
Juris Hartmanis, Cornell University, NY, USA
Jan van Leeuwen, Utrecht University, The Netherlands

Volume Editors

Mogens Nielsen
University of Aarhus, Department of Computer Science
Ny Munkegade bldg 540, 8000 Aarhus C, Denmark
E-mail: mn@brics.dk

Dan Simpson
University of Brighton, Watts Building
Lewes Road, Brighton BN2 4GJ, East Sussex, UK
E-mail: Dan.Simpson@brighton.ac.uk

Cataloging-in-Publication Data applied for

Die Deutsche Bibliothek - CIP-Einheitsaufnahme

Application and theory of Petri nets 2000 : 21st international
conference ; proceedings / ICATPN 2000, Aarhus, Denmark, June 26 - 30,
2000. Morgens Nielsen ; Dan Simpson (ed.). - Berlin ; Heidelberg ; New
York ; Barcelona ; Hong Kong ; London ; Milan ; Paris ; Singapore ;
Tokyo : Springer, 2000
    (Lecture notes in computer science ; Vol. 1825)
    ISBN 3-540-67693-7

CR Subject Classification (1998): F.1-3, C.1-2, G.2.2, D.4, D.2, J.4

ISSN 0302-9743
ISBN 3-540-67693-7 Springer-Verlag Berlin Heidelberg New York

Springer-Verlag is a company in the BertelsmannSpringer publishing group.
© Springer-Verlag Berlin Heidelberg 2000
Printed in Germany

Typesetting: Camera-ready by author, data conversion by Boller Mediendesign
Printed on acid-free paper   SPIN: 10721145   06/3142   5 4 3 2 1 0

Preface

This volume contains the proceedings of the 21st International Conference on Application and Theory of Petri Nets. The aim of the Petri net conferences is to create a forum for discussing progress in the application and theory of Petri nets. Typically, the conferences have 100–150 participants – one third of these coming from industry while the rest are from universities and research institutions. The conferences always take place in the last week of June.

The conference and a number of other activities are co-ordinated by a steering committee with the following members: G. Balbo (Italy), J. Billlington (Australia), G. De Michelis (Italy), C. Girault (France), K. Jensen (Denmark), S. Kumagai (Japan), T. Murata (USA), C. A. Petri (Germany; honorary member), W. Reisig (Germany), G. Roucairol (France), G. Rozenberg (The Netherlands; chairman), and M. Silva (Spain).

Other activities before and during the 2000 conference included tool presentations and demonstrations, a meeting on "Interchange Formats", extensive introductory tutorials, two advanced tutorials on "Hardware Design" and "Timed and Hybrid Automata", and two workshops on "Software Engineering" and "Practical Use of High-Level Nets". The tutorial notes and workshop proceedings are not published in these proceedings, but copies are available from the organisers.

The 2000 conference was organised by the CPN Group at the University of Aarhus, Denmark. We would like to thank the organisers (see next page) and their teams.

We would like to thank very much all those who submitted papers to the Petri net conference. We received a total of 57 submissions from 20 different countries. This volume comprises the papers that were accepted for presentation. Invited lectures were given by Jordi Cortadella, Philippe Darondeau, Gregor Engels, Serge Haddad, Kim Guldstrand Larsen, and Ole Lehrmann Madsen.

The submitted papers were evaluated by a programme committee. The programme committee meeting took place in Aarhus, Denmark. We would like to express our gratitude to the members of the programme committee, and to all the referees who assisted them. The names of these are listed on the following pages.

Finally, we would like to mention the excellent co-operation with Springer-Verlag, and to thank Janne Christensen and Teresa Hampton who handled all the submissions, and Uffe H. Engberg for preparing this volume.

April 2000   Mogens Nielsen   Dan Simpson

Organising Committee

Kurt Jensen (Chair)
Søren Christensen (Chair)
Helle Holm-Nielsen (Secretary)

Tools Demonstration

Kjeld Høyer Mortensen (Chair)

Programme Committee

Wil van der Aalst (The Netherlands)
Gianfranco Balbo (Italy)
Luca Bernardinello (Italy)
Eike Best (Germany)
Jonathan Billington (Australia)
Wilfried Brauer (Germany)
Ginafranco Ciardo (USA)
José-Manuel Colom (Spain)
Luis Gomes (Portugal)
Serge Haddad (France)
Kunihiko Hiraishi (Japan)
Mogens Nielsen (Denmark;
        co-chair; theory)

Leo Ojala (Finland)
Wojciech Penczek (Poland)
Karsten Schmidt (Germany)
Sol Shatz (USA)
Dan Simpson (UK;
        co-chair; applications)
P. S. Thiagarajan (India)
Robert Valette (France)
Antti Valmari (Finland)
Alex Yakovlev (UK)
Wlodek Zuberek (Canada)

Referees

Alessandra Agostini
Marco Ajmone Marsan
Stanislaw Ambroszkiewicz
S. Arun-Kumar
Pierre Azema
Joao-Paulo Barros
Marek Bednarczyk
Marco Bernardo
Andrea Bobbio
Andrzej Borzyszkowski
Marc Boyer
Gary Bundell
Nadia Busi
Javier Campos
Giovanni Chiola
Armando-Walter Colombo
Giorgio De Michelis
Susanna Donatelli
Javier Esparza
Joaquin Ezpeleta

Carlo Ferigato
Hans Fleischhack
Adrianna Foremniak
Giuliana Franceschinis
Jörn Freiheit
Rossano Gaeta
Maike Gajewski
Fernando Garcia-Vallés
Qi-Wei Ge
Jaco Geldenhuys
Claude Girault
Steven Gordon
Marco Gribaudo
Keijo Heljanko
Nisse Husberg
Jean-Michel Ilie
Ryszard Janicki
Pawel Janowski
Guy Juanole
Tommi Junttila

Table of Contents

Invited Papers

Full Papers

Tools Presentations

Author Index   485

Hardware and Petri Nets:
Application to Asynchronous Circuit Design

Jordi Cortadella[1], Michael Kishinevsky[2], Alex Kondratyev[3],
Luciano Lavagno[4], and Alex Yakovlev[5]

[1] Universitat Politècnica de Catalunya, Department of Software,
Campus Nord, Mòdul C6, 08034 Barcelona, Spain.
jordic@lsi.upc.es
[2] Intel Corporation, JFT-104, 2111 N.E. 25th Ave., Hillsboro, OR 97124-5961, USA.
mkishine@ichips.intel.com
[3] Theseus Logic, 710 Lakeway Drive, suite 230, Sunnyvale, CA 94087, USA.
alex.kondratyev@theseus.com
[4] Universitá di Udine, DIEGM, Via delle Scienze 208, I-33100 Udine, Italy.
lavagno@uniud.it
[5] University of Newcastle upon Tyne, Department of Computing Science,
Claremont Tower, Claremont Road, Newcastle upon Tyne, NE1 7RU, UK.
Alex.Yakovlev@newcastle.ac.uk

Abstract. Asynchronous circuits is a discipline in which the theory
of concurrency is applied to hardware design. This paper presents an
overview of a design framework in which Petri nets are used as the main
behavioral model for specification. Techniques for synthesis, analysis and
formal verification of asynchronous circuits are reviewed and discussed.

1 Introduction

Finite State Machines has been the most traditional model of computation for
sequential circuits [25, 26]. It is a state-based model in which the system, being
in a state, reads some inputs, writes some outputs and moves to another state.
Time is discretized by the notion of cycle, which is the time that takes the system
to move from one state to another. This model is appropriate to derive circuit
implementations with a periodic signal, the *clock*, that dictates the time instants
in which the system changes state. The cycle is the finest degree of granularity
at which operations are scheduled. Thus, two operations are concurrent if they
are scheduled at the same cycle. The cycle delay is determined by the worst-case
delay the circuit takes to perform the operations scheduled at any of the cycles.
Synchronization among operations is implicit, i.e. the initiation of an operation
always starts at a clock edge and completes after a fixed quantity of cycles at
another clock edge. Thus, clock edges indicate the *initiation and completion of
several actions* simultaneously.

After a long period of hibernation, asynchronous circuits woke up fifteen
years ago as a potential solution to some of the design problems posed by VLSI
technologies [10]. In asynchronous circuits, the sequencing of operations is no

M. Nielsen, D. Simpson (Eds.): ICATPN 2000, LNCS 1825, pp. 1 15, 2000.

longer dictated by a clock, but by *events* that indicate the *initiation and completion of individual actions*. The correctness of an asynchronous circuit not only depends on its structure, but also on the timing behavior of the individual gates and their interaction. Thus, a circuit can be modeled as a set of processes (gates) that communicate through channels (wires) and modify the state of the system represented by a set of Boolean variables (signals). A gate is *enabled* when the value of its output is different from the value calculated by its logic function. An enabled gate can produce a transition (event) on the output signal by changing its value.

The view of an asynchronous circuit as a concurrent system makes event-based models of computation more appropriate for analysis and synthesis. For this reason, process algebras, such as CSP [16], and Petri nets [34] have raised the interest of the researchers in this discipline.

This paper presents an overview of a design methodology for asynchronous circuits that uses Petri nets as the underlying model for specification, synthesis and verification. Section 2 explains how the behavior of an asynchronous circuit can be specified by using Petri nets. Section 3 presents a set of sufficient properties for a specification to be implementable as a speed-independent circuit. The techniques to derive an implementation with logic gates are described in Sect. 4. The retrieval of the actual circuit behavior as a Petri net is known as back-annotation and is presented in Sect. 5. Finally, different strategies to fight against the state explosion problem in analysis and formal verification are reviewed in Sect. 6.

2   Timing Diagrams, Petri Nets, and Signal Transition Graphs

For most circuit designers, Petri nets resemble timing diagrams, a model to specify asynchronous interfaces as signal waveforms that explictly indicate the causality and concurrency relations among signal transitions.

Figure 1(a) depicts the block diagram of a VME bus controller. According to its functionality, the controller has three sets of "handshake" signals: those interacting with the bus (DSr, DSw and $DTACK$), those interacting with the device connected to the bus (LDS and $LDTACK$), and that controlling the transceiver that connects the bus with the device (D).

The behavior of the controller is as follows: a request to read from or write into the device is received by one of the signals DSr or DSw respectively. In a read cycle, a request to read is done through signal LDS. When the device has the data ready ($LDTACK$), the controller must open the transceiver to transfer data to the bus (signal D). In the write cycle, data is first transferred to the device. Next, a request to write is done (LDS). Once the device acknowledges the reception of the data ($LDTACK$) the transceiver must be closed to isolate the device from the bus. Each transaction must be completed by a return-to-zero of all interface signals, seeking for a maximum parallelism between the bus and the device operations. Figure 1(b) shows the timing diagram of the READ cycle.

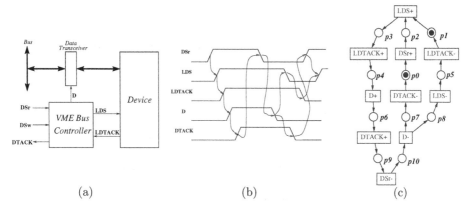

(a)   (b)   (c)

Fig. 1. (a) VME bus controller, (b) READ cycle, (c) Signal Transition Graph.

Figure 1(c) is a Petri net describing the same behavior. The events are interpreted as rising (+) and falling (−) signal transitions. Petri Nets with such interpretation are called *Signal Transition Graphs* (STG) [5, 38]. Three types of signals can be distinguished in an STG: input, output and internal. The behavior of the input signals (*DSr*, *DSw* and *LDTACK*) is determined by the environment. The behavior of the output (*D*, *DTACK* and *LDS*) and internal signals is determined by the system and is the one that must be implemented by the circuit. Typically, internal signals are incorporated during the synthesis of the circuit to solve some implementation problems (encoding, decomposition) and do not appear in the original specification of the system.

Figure 2 depicts the STG that describes the complete behavior of the controller[1]. Unlike timing diagrams, STGs can specify choice and non-determinism. In this example, the initial marking models the situation in which the environment can non-deterministically choose to initiate a read cycle by firing *DSr*+, or a write cycle by firing *DSw*+.

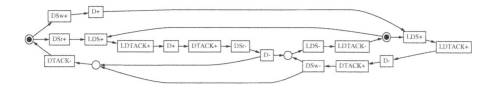

Fig. 2. STG for the READ and WRITE cycles.

---

[1] Usually, places with only one predecessor and one successor transition are not explicitly drawn in STGs

This paper presents a methodology to synthesize asynchronous circuits from STGs. This methodology has been completely automated and implemented in a synthesis tool called `petrify` [35]. The specification of the READ cycle shown in Fig. 1(c) will be used as an example to illustrate this methodology along the paper. For the sake of simplicity, the WRITE cycle will be ignored.

3 Implementability Properties

The goal of the synthesis methodology is to derive a *speed-independent* circuit that realizes the specified behavior. Speed independence is a property that guarantees a correct behavior under the assumption that all gates have an unbounded delay and all wires have a negligible delay [28].

A specification must fulfil certain properties to be implementable as a speed-independent circuit. These properties can be better described on the state graph of the specification.

3.1 State Graph

The state graph (SG) of a specification is the transition system obtained from the reachability analysis of an STG. Each state corresponds to a marking of the STG and each arc corresponds to the firing of a signal transition. Figure 3 shows the SG of the read cycle specified in Fig. 1(c).

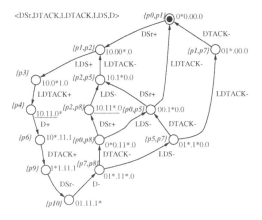

Fig. 3. State graph for the READ cycle.

In the SG, each state is assigned a binary vector with the value of all signals at that state. For the sake of readability, the control signals corresponding to the left handshake, right handshake and data transceiver are separated by dots. Enabled signals are marked with an asterisk. For example, the state corresponding to the marking $\{p_7\ p_8\}$ has the code 01* 11* 0, indicating that DSr and D are at 0, and

$DTACK$, $LDTACK$ and LDS are at 1. Moreover, signals $DTACK$ and LDS are both enabled, i.e. ready to change their value. The initial state corresponds to the marking $\{p_0 \ p_1\}$ and all signals at 0.

3.2   Properties for Implementability

The properties required for the specification to be implementable as a speed-independent circuit are the following:

Boundedness of the STG that guarantees the SG to be finite.

Consistency of the STG, that consists in ensuring that the rising and falling transitions of each signal alternate in all possible runs of the specification.

Completeness of state encoding that ensures that there are no two different states with the same signal encoding and different behavior of the output or internal signals.

Persistency of signal transitions in such a way that no signal transition can be disabled by another signal transition, unless both signals are inputs. This property ensures that no short glitches, known as *hazards*, will appear at the disabled signals.

The SG of Fig. 3 fulfils boundedness, consistency and persistency. However, it does not have completeness of state encoding. The states corresponding to the markings $\{p_4\}$ and $\{p_2 \ p_8\}$ have the same code. Moreover, the behavior of the output signals in those states is different. In the state $\{p_4\}$, the event $D+$ is enabled, whereas in the state $\{p_2 \ p_8\}$, the event $LDS-$ is enabled. Intuitively, this means that the information provided by the value of the signals is not enough to determine the future behavior of the system. This will result in an ambiguity in the definition of the next-state logic functions.

4   Logic Synthesis

The goal of logic synthesis is to derive a gate netlist that implements the behavior defined by the specification. For simplicity, we will illustrate this step by synthesizing a speed-independent circuit for the read cycle of the VME bus (see Fig. 3).

The main steps in logic synthesis are the following:

Encode the SG in such a way that the complete state coding property holds. This may require the addition of internal signals.

Derive the *next-state* functions for each output and internal signal of the circuit.

Map the functions onto a netlist of gates.

4.1   Complete State Coding

As mentioned in Sect. 3.2, the SG of Fig. 3 has state conflicts. A possible method to solve this problem is to insert new state signals that disambiguate the encoding conflicts. Figure 4 depicts a new SG in which a new signal, $csc0$, has been inserted. Now, the next-state functions for signals LDS and D can be uniquely defined. The insertion of new signals must be done in such a way that the resulting SG preserves the properties for implementability.

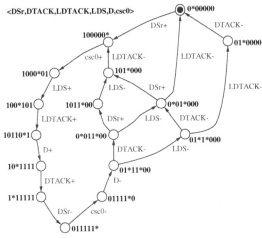

Fig. 4. SG for the READ cycle with complete state coding.

4.2   Next-State Functions

When an SG fulfills all the implementability properties, a next-state function can be derived for each non-input signal.

Given a signal z, we can classify the states of the SG into four sets: positive and negative *excitation regions* ($\mathsf{ER}(z+)$ and $\mathsf{ER}(z-)$) and *quiescent regions* ($\mathsf{QR}(z+)$ and $\mathsf{QR}(z-)$).

A state belongs to $\mathsf{ER}(z+)$ if $z = 0$ and $z+$ is enabled in that state. In this situation, the value of the signal is denoted by 0^* in the SG. A state belongs to $\mathsf{QR}(z+)$ if s is in stable 1 state. These definitions are analogous for $\mathsf{ER}(z-)$ and $\mathsf{QR}(z-)$.

The next-state function for a signal z is defined as follows:

$$f_z(s) = \begin{array}{ll} 1 & \text{if } s \in \mathsf{ER}(z+) \cup \mathsf{QR}(z+) \\ 0 & \text{if } s \in \mathsf{ER}(z-) \cup \mathsf{QR}(z-) \\ - & \text{otherwise} \end{array}$$

where s denotes the binary code of a state. The fact that $f_z(s) = -$ indicates that there is no state with such code in the SG and, thus, s can be considered as a *don't care* condition for Boolean minimization.

Once the next-state function has been derived, Boolean minimization can be performed to obtain a logic equation that implements the behavior of the signal. In this step it is crucial to make an efficient use of the don't care conditions derived from those binary codes not corresponding to any state of the SG. For the example of Fig. 4, the following equations can be obtained:

$$D = LDTACK \land csc0; \qquad LDS = D \lor csc0$$
$$DTACK = D; \qquad csc0 = DSr \land (csc0 \lor \neg LDTACK)$$

A well known result in the theory of asynchronous circuits is that any circuit implementing the next-state function of each signal with only one atomic complex gate is speed independent. By atomic gate we mean a gate without internal hazardous behavior [18, 22]. Two possible hazard-free gate mappings for the next-state function of the READ cycle example are shown in Fig. 5(a) and 5(b).

However, there could be two obstacles in the actual implementation of the next state functions:

a logic function can be too complex to be mapped into one gate available in the library,

the solution requires the use of gates which are not typically present in standard synchronous libraries

The second is the case with the solution in Fig. 5(a). A gate drawn as a circle with "C" is the so-called C-element [27]: a popular asynchronous latch with the next state function $c = (a \land b) \lor (c \land (a \lor b))$. Its output, c, goes high (low) if both inputs, a and b, go high (low), otherwise it keeps the previous value.

Decomposing the circuit into a set of simpler gates that can be implemented in a given technology is a problem that has been studied by several authors [2, 4, 7, 20].

4.3   Size of the State Space

The derivation of Boolean equations from a specification requires to calculate the encoding of all the states of the system. Unfortunately, the size of the state space of a concurrent system can be exponential on the size of the specification.

The existing tools for synthesis of asynchronous circuits use different methods to fight against the state explosion problem. In *Burst-mode automata* concurrency is restricted in such a way that bursts of input and output events are serialized [31]. Thus, the behavior can be represented by a Mealy-like automata with a manageable number of states in which the concurrency is annotated as input/output bursts on the arcs of the automata.

When no constraints are imposed on the type of concurrency manifested by the system, the knowledge and manipulation of the state space usually becomes the dominant part on the complexity of the synthesis algorithms. This is the case when Petri nets are used as the specification formalism. Different strategies have been used to calculate the state space:

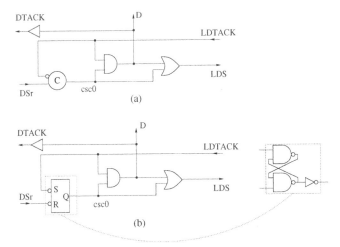

Fig. 5. Implementations with 2-input gates

Restrict the specification to a certain class of Petri nets, e.g. Free-choice Petri nets [11], and use structural methods that can manipulate the state space by only analyzing the underlying graph of the net and without explicitly generating the states [33]. Techniques for the efficient calculation of concurrency relations are crucial in this context [21].

Annotate timing on the events, e.g. min/max firing delays, and calculate the reduced state space reachable only under the specified timing constraints. This is the approach used for the synthesis of timed asynchronous circuits [30]. Use symbolic methods, such as Binary Decision Diagrams [3], to implicitly represent and manipulate the complete state space. This is the approach used in the synthesis tool `petrify` [35].

5   Back-Annotation

As important as the structure of the circuit resulting from the synthesis of the specification is the actual behavior of the circuit. During synthesis, internal signals might have been added to encode the states and decompose complex gates.

On the other hand, and although paradoxical, reducing concurrency [8] is one of the proposed approaches to improve the efficiency of the final circuit. Reducing concurrency directly results in a reduction of the state space and, thus, in an increase of the don't care conditions for logic minimization. In general, concurrency reduction produces smaller circuits, but it may also produce faster circuits: the system manifests less concurrency but the events take less time to fire.

In the synthesis flow, signal insertion and concurrency reduction are usually performed at the level of SG. Providing a behavioral description of the synthesized circuit with the same formalism used for specification, e.g. Petri nets, allows the designer to easily interact with the synthesis framework and manually introduce those optimizations that automatic tools cannot find.

The problem of deriving a Petri net from a transition system was first tackled in [14] and studied by other authors [1, 12]. In these works, the theory of regions was developed to characterize the class of transition systems which correspond to elementary Petri nets. That work was extended in [9] to propose algorithms for the synthesis of safe Petri nets from any finite transition system. This work was crucial to provide the synthesis tool `petrify` [35] the capability of deriving a succint behavioral description of the synthesized circuits.

An example of back-annotation is presented in Fig. 6. The circuit is an implementation of the READ cycle specified in Fig. 1(a). In this implementation, only combinational 2-input gates have been used. With respect to the original specification, two new signals have been incorporated: *csc0* to uniquely encode the states, as shown in Fig. 4, and *map0* to decompose a complex gate into smaller gates. By using the theory of regions, the Petri net of Fig. 6(b) is automatically synthesized and shown to the designer for analysis of the circuit's behavior.

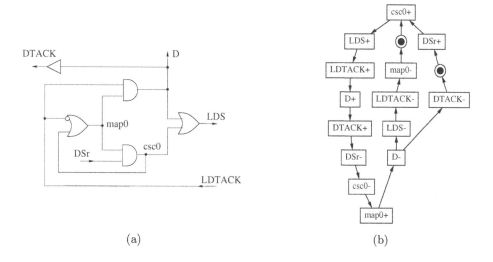

(a)   (b)

Fig. 6. (a) A 2-input gate circuit with no latches, (b) circuit behavior.

6   Analysis and Formal Veri cation

Analysis and formal verification are used at different stages of the design flow of asynchronous circuits. In particular for:

Property veri cation. After specifying the design it is required to check implementability properties to answer the following question: "Can the specification be implemented with an asynchronous circuit?" [18, 19]. Other properties of the specification can be of interest as well, e.g., absence of deadlocks, fairness in serving requests, etc. General purpose verification techniques can be employed for this analysis [23].

Implementation veri cation. After the design has been done fully automatically or with some manual intervention it is often desirable to check that the implementation conforms the given specification [13, 37].

Performance analysis and separation between events is required (a) for determining the latency and throughput of the circuit and (b) for logic optimization based on timing information [17, 30].

6.1 Techniques

As mentioned in Sect. 4.3, the state space of a concurrent specification is one of the major bottlenecks for the analysis of this type of systems. Here we present some techniques that have been succesfully applied in the area of formal verification of asynchronous circuits.

Symbolic Binary Decision Diagram-based (BDD) [3] traversal of a reachability graph allows its implicit representation, which is generally much more compact than an explicit enumeration of states [37].

Partial order reductions ([15], stubborn sets [39], identification method [18]) can abstract and ignore many of the irrelevant states for the verification of certain properties.

Structural properties of Petri nets (e.g., place invariants) can provide fast upper approximation of the reachability space [6, 11, 29] and can be also used for dense variable encoding of states in the reachability graph. Structural reductions are useful as a preprocessing step in order to simplify the structure of the net before traversal or analysis, keeping all important properties.

Unfoldings [19, 24] are finite *acyclic* prefixes of the Petri net behavior, representing all reachable markings. They are often more compact than the reachability graph and well-suited for extracting ordering relations between places and transitions (concurrency, conflict and precedence). Different types of unfoldings are also used for performance analysis [17].

As an example, we next illustrate how structural properties of a Petri net and BDD-based representations can be combined for an efficient analysis of the state space. Figure 7 is the result of applying linear reductions to the STG from Fig. 2. Using more elaborate reductions (place and transition fusions) it is possible to reduce the whole Petri net from Fig. 1(c) to a single self-loop transition [29].

The BDD-based method used for deriving the transition function and calculating the reachable markings of a Petri net are similar to those used for reachability analysis and equivalence checking of finite state machines [23]: starting from the initial marking and by iteratively applying the transition relation until a fix point is reached, the characteristic function of the reachability set is

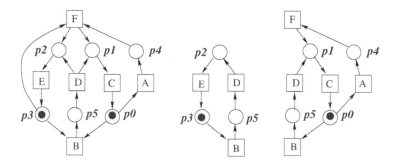

Fig. 7. STG after linear reduction and two state machine components

calculated. Under the assumption that the Petri net is safe, one could use a naive encoding, such as one Boolean variable per place, to encode the markings. However, this may be too costly for large specifications.

The following observation can be made: the sets of places $P_0 = \{p_2\ p_3\ p_5\}$ and $P_1 = \{p_0\ p_1\ p_4\ p_5\}$ of the Petri net in Fig. 7 define two state machines [11, 29] with the sets of transitions $T_0 = \{B\ D\ E\}$ and $T_1 = \{A\ B\ C\ D\ F\}$, respectively. This information can be structurally obtained by using algebraic methods. State machines (see Fig. 7) correspond to place-invariants of the Petri net and preserve their token count in all reachable markings. Therefore, the following are two invariants for the net:

$$I_1(p_2\ p_3\ p_5): \qquad\qquad M(p_2) + M(p_3) + M(p_5) = 1$$
$$I_2(p_0\ p_1\ p_4\ p_5): \qquad M(p_0) + M(p_1) + M(p_4) + M(p_5) = 1$$

If invariants $I_1(p_2\ p_3\ p_5)$ and $I_2(p_0\ p_1\ p_4\ p_5)$ are represented as Boolean functions (e.g., by using BDDs), then the conjunction on these two functions give, for this example, the characteristic function of the exact reachability set of markings. In general a conjunction of any set of invariants gives an *upper approximation* of the reachability set, which is useful for conservative verification.

On the other hand, and by using the previous invariants, a dense encoding for places with the vector of Boolean variables $V = (v_0\ v_1\ v_2\ v_3)$ can be proposed (see Table 1).

$_p$ is the characteristic function of a place and represents the those markings that have a token in place p, whereas $_M$ is the characteristic function of marking M. The set of variables v_0 and v_1 is used to encode the places involved in the invariant I_1. Similarly, v_2 and v_3 are used to encode the places involved in invariant I_2. Place p_5 supports both invariants and is encoded with variables from both sets. The characteristic function of all reachable markings, $R(V)$, can be represented by a single Boolean equation obtained as the disjunction of the characteristic functions of the markings. When simplified, the equation can be represented as follows:

$$R(V) \;=\; \overline{v_0}(\overline{v_2} \vee \overline{v_3}) \vee v_0 v_2 v_3 \;=\; v_0 \Leftrightarrow v_2 v_3$$

Table 1. Encoding for the places and reachable markings of the Petri net in Fig. 7.

place	v_0	v_1	v_2	v_3	p
p_2	0	0	-	-	$\overline{v_0}\,\overline{v_1}$
p_3	0	1	-	-	$\overline{v_0}v_1$
p_5	1	-	1	1	$v_0 v_2 v_3$
p_0	-	-	0	0	$\overline{v_2}\,\overline{v_3}$
p_1	-	-	0	1	$\overline{v_2}v_3$
p_4	-	-	1	0	$v_2\overline{v_3}$

marking	M
$p_0\ p_3$	$\overline{v_0}\,v_1\overline{v_2}\,v_3$
p_5	$v_0 v_2 v_3$
$p_1\ p_2$	$\overline{v_0}\,\overline{v_1}\,\overline{v_2}v_3$
$p_1\ p_3$	$\overline{v_0}v_1\overline{v_2}v_3$
$p_0\ p_2$	$\overline{v_0}\,\overline{v_1}\,\overline{v_2}\,\overline{v_3}$
$p_3\ p_4$	$\overline{v_0}v_1 v_2\overline{v_3}$
$p_2\ p_4$	$\overline{v_0}\,\overline{v_1}v_2\overline{v_3}$

With $R(V)$ it is possible to verify properties of the net. For example, one might what to verify that transitions D and F are not concurrent[2]. This can be verified by proving that there is no reachable marking in which both transitions are enabled. The characteristic function of D and F being enabled can be represented as:

$$\text{Enabled}(D) \wedge \text{Enabled}(F) = \quad p_5 \wedge (\quad p_3 \wedge \quad p_4)$$

Therefore, the characteristic function of the markings in which both transitions are enabled is

$$R(V) \wedge \text{Enabled}(D) \wedge \text{Enabled}(F)$$

It can be easily observed that this logic proposition is a contradiction, thus proving that there is no reachable marking in which D and F are enabled.

This type of techniques has been succesfully applied to verify concurrent systems specified with Petri nets [32].

7 Conclusions

Event-based models for concurrency are being used for the specification, synthesis, analysis and verification of asynchronous circuits. Among them, Petri nets seem to be the most appropriate model that offers the following features:

A succint representation of the behavior of concurrent systems with no restriction on the type of allowed concurrency among events.

A easy calculation and manipulation of the state space, thus enabling the use of efficient techniques for state encoding and logic synthesis.

An intermediate representation for higher-level formalisms like process algebras.

---

[2] This property can be easily verified by using algebraic methods. Here, we only want to illustrate the mechanics of formal verification when using symbolic encoding techniques.

Asynchronous circuit design is still in its prehistory. In the future, we foresee an increasing interest in such type of circuits and design methodologies. High-level synthesis tools will be constructed and used in a broader set of applications. For this reason, techniques to synthesize and verify highly-complex concurrent systems will be required. This is an area where theoreticians on concurrency have a chance to apply their findings and collect new challenging problems to solve.

Acknowledgements. This work has been partially funded by the ACiD-WG (ESPRIT 21949), a grant by Intel Corporation, CICYT TIC 98-0410, CICYT TIC 98-0949 and CIRIT 1999SGR-150.

References

[1] E. Badouel and Ph. Darondeau. Theory of regions. In *[36]*, pages 529–586. Springer-Verlag, 1998.

[2] P. Beerel and T.H.-Y. Meng. Automatic gate-level synthesis of speed-independent circuits. In *Proc. International Conf. Computer-Aided Design (ICCAD)*, pages 581–587. IEEE Computer Society Press, November 1992.

[3] R. Bryant. Symbolic boolean manipulation with ordered binary-decision diagrams. *ACM Computing Surveys*, 24(3):293–318, September 1992.

[4] S. M. Burns. General condition for the decomposition of state holding elements. In *Proc. International Symposium on Advanced Research in Asynchronous Circuits and Systems*. IEEE Computer Society Press, March 1996.

[5] T.-A. Chu and L. A. Glasser. Synthesis of self-timed control circuits form graphs: An example. In *Proc. International Conf. Computer Design (ICCD)*, pages 565–571. IEEE Computer Society Press, 1986.

[6] J. Cortadella. Combining structural and symbolic methods for the verification of concurrent systems. In *Proc. of the International Conference on Application of Concurrency to System Design*, pages 2–7, March 1998.

[7] Jordi Cortadella, Michael Kishinevsky, Alex Kondratyev, Luciano Lavagno, Enric Pastor, and Alexandre Yakovlev. Decomposition and technology mapping of speed-independent circuits using Boolean relations. *IEEE Transactions on Computer-Aided Design*, 18(9), September 1999.

[8] Jordi Cortadella, Michael Kishinevsky, Alex Kondratyev, Luciano Lavagno, and Alexandre Yakovlev. Automatic handshake expansion and reshuffling using concurrency reduction. In *Proc. of the Workshop Hardware Design and Petri Nets (within the International Conference on Application and Theory of Petri Nets)*, pages 86–110, June 1998.

[9] Jordi Cortadella, Michael Kishinevsky, Luciano Lavagno, and Alexandre Yakovlev. Deriving Petri nets from finite transition systems. *IEEE Transactions on Computers*, 47(8):859–882, August 1998.

[10] Al Davis and Steven M. Nowick. An introduction to asynchronous circuit design. In A. Kent and J. G. Williams, editors, *The Encyclopedia of Computer Science and Technology*, volume 38. Marcel Dekker, New York, February 1998.

[11] J. Desel and J. Esparza. *Free-choice Petri Nets*, volume 40 of *Cambridge Tracts in Theoretical Computer Science*. Cambridge University Press, 1995.

[12] J. Desel and W. Reisig. The synthesis problem of Petri nets. *Acta Informatica*, 33(4):297–315, 1996.

[13] David L. Dill. *Trace Theory for Automatic Hierarchical Veri cation of Speed-Independent Circuits*. ACM Distinguished Dissertations. MIT Press, 1989.

[14] A. Ehrenfeucht and G. Rozenberg. Partial (Set) 2-Structures. Part I, II. *Acta Informatica*, 27:315–368, 1990.

[15] P. Godefroid. Using partial orders to improve automatic verification methods. In E.M Clarke and R.P. Kurshan, editors, *Proc. International Workshop on Computer Aided Veri cation*, 1990. DIMACS Series in Discrete Mathematica and Theoretical Computer Science, 1991, pages 321-340.

[16] C. A. R. Hoare. *Communicating Sequential Processes*. Prentice-Hall, 1985.

[17] H. Hulgaard, S. M. Burns, T. Amon, and G. Borriello. An algorithm for exact bounds on the time separation of events in concurrent systems. *IEEE Transactions on Computers*, 44(11):1306–1317, November 1995.

[18] Michael Kishinevsky, Alex Kondratyev, Alexander Taubin, and Victor Varshavsky. *Concurrent Hardware: The Theory and Practice of Self-Timed Design*. Series in Parallel Computing. John Wiley & Sons, 1994.

[19] Alex Kondratyev, Michael Kishinevsky, Alexander Taubin, and Sergei Ten. Analysis of Petri nets by ordering relations in reduced unfoldings. *Formal Methods in System Design*, 12(1):5–38, January 1998.

[20] Alex Kondratyev, Michael Kishinevsky, and Alex Yakovlev. Hazard-free implementation of speed-independent circuits. *IEEE Transactions on Computer-Aided Design*, 17(9):749–771, September 1998.

[21] A. Kovalyov. A Polynomial Algorithm to Compute the Concurrency Relation of a Regular STG. In A. Yakovlev, L. Gomesa, and L. Lavagno, editors, *Hardware Design and Petri Nets*, pages 107–126. Kluwer Academic Publishers, March 2000.

[22] Luciano Lavagno and Alberto Sangiovanni-Vincentelli. *Algorithms for Synthesis and Testing of Asynchronous Circuits*. Kluwer Academic Publishers, 1993.

[23] K. L. McMillan. *Symbolic Model Checking*. Kluwer Academic Publishers, 1993.

[24] K. L. McMillan. Trace theoretic verification of asynchronous circuits using unfoldings. In *Proc. International Workshop on Computer Aided Veri cation*, 1995.

[25] G.H. Mealy. A method for synthesizing sequential circuits. *Bell System Technical J.*, 34(5):1045–1079, 1955.

[26] E.F. Moore. Gedanken experiments on sequential machines. *Automata Studies*, pages 129–153, 1956.

[27] David E. Muller. Asynchronous logics and application to information processing. In *Symposium on the Application of Switching Theory to Space Technology*, pages 289–297. Stanford University Press, 1962.

[28] David E. Muller and W. S. Bartky. A theory of asynchronous circuits. In *Proceedings of an International Symposium on the Theory of Switching*, pages 204–243. Harvard University Press, April 1959.

[29] T. Murata. Petri Nets: Properties, analysis and applications. *Proceedings of the IEEE*, pages 541–580, April 1989.

[30] Chris J. Myers and Teresa H.-Y. Meng. Synthesis of timed asynchronous circuits. *IEEE Transactions on VLSI Systems*, 1(2):106–119, June 1993.

[31] S. M. Nowick and B. Coates. Automated design of high-performance asynchronous state machines. In *Proc. International Conf. Computer Design (ICCD)*. IEEE Computer Society Press, October 1994.

[32] E. Pastor, J. Cortadella, and M.A. Peña. Structural methods to improve the symbolic analysis of Petri nets. In *Application and Theory of Petri Nets 1999*, Lecture Notes in Computer Science, June 1999.

[33] Enric Pastor, Jordi Cortadella, Alex Kondratyev, and Oriol Roig. Structural methods for the synthesis of speed-independent circuits. *IEEE Transactions on Computer-Aided Design*, 17(11):1108–1129, November 1998.

[34] C. A. Petri. *Kommunikation mit Automaten*. PhD thesis, Bonn, Institut für Instrumentelle Mathematik, 1962. (technical report Schriften des IIM Nr. 3).

[35] petrify: a tool for the synthesis of Petri nets and asynchronous controllers. http://www.lsi.upc.es/~jordic/petrify.

[36] W. Reisig and G. Rozenberg, editors. *Lectures on Petri Nets I: Basic Models*, volume 1491 of *Lecture Notes in Computer Science*. Springer-Verlag, 1998.

[37] Oriol Roig, Jordi Cortadella, and Enric Pastor. Verification of asynchronous circuits by BDD-based model checking of Petri nets. In *16th International Conference on the Application and Theory of Petri Nets*, volume 815 of *Lecture Notes in Computer Science*, pages 374–391, 1995.

[38] L. Y. Rosenblum and A. V. Yakovlev. Signal graphs: from self-timed to timed ones. In *Proceedings of International Workshop on Timed Petri Nets*, pages 199–207, Torino, Italy, July 1985. IEEE Computer Society Press.

[39] A. Valmari. Stubborn sets for reduced state space generation. In *Lecture Notes in Computer Science, Advances in Petri Nets 1990*, volume 483, pages 491–515. Springer Verlag, 1991.

Region Based Synthesis of P/T-Nets and Its Potential Applications

Philippe Darondeau

IRISA, campus de Beaulieu, F35042 Rennes Cedex
darondeau@irisa.fr

Abstract. This talk is an informal presentation of ideas put forward by Badouel, Bernardinello, Caillaud and me for solving various types of P/T-net synthesis problems, with hints at the potential role of net synthesis in distributed software and distributed control. The ideas are theirs as much as mine. The lead is to start from Ehrenfeucht and Rozenberg's axiomatic characterization of behaviours of elementary nets, based on regions, to adapt the characterization to P/T-nets in line with Mukund's extended regions with integer values, and to pro t from algebraic properties of graphs and languages for converting decision problems about regions to linear algebra.

1 Nets and Regions

Petri nets may be presented as matrices: rows are places, columns are events, and entries define relations between places and events. A simple net N with set of places P and set of events E is a matrix $N : P \times E \to \mathcal{E}$ where $\mathcal{E}$ enumerates all possible relations between places and events in some fixed class of Petri nets.

Thus for the P/T-nets, $\mathcal{E} = \mathbb{N} \times \mathbb{N}$ and $N(p, e) = (w_1, w_2)$ represents two arcs: one arc weighted w_1 from p to e, and one arc weighted w_2 from e to p.

On this basis, the dynamics of Petri nets may be described in a uniform way. Given a class of nets, let $\mathcal{N}$ be the net in this class with a single place π and with set of events $\mathcal{E}$ such that $\mathcal{N}(\pi, \varepsilon) = \varepsilon$ for every $\varepsilon \in \mathcal{E}$ (thus $\mathcal{N} : \{\pi\} \times \mathcal{E} \to \mathcal{E}$). The state graph of $\mathcal{N}$, let $\mathcal{T} = (\mathcal{S}, \mathcal{E}, T)$ with $T \subseteq \mathcal{S} \times \mathcal{E} \times \mathcal{S}$, determines the state graph of any other net in the class. First, it defines the set $\mathcal{S}$ of all possible values for places. Second, an event $e \in E$ of a net $N : P \times E \to \mathcal{E}$ has concession at marking $M : P \to \mathcal{S}$ if and only if, for each $p \in P$, the event $N(p, e) \in \mathcal{E}$ has concession at the marking of the net $\mathcal{N}$ such that place π holds $M(p)$; and then $M \xrightarrow{e} M$ where M' is the marking of N such that, for each $p \in P$, $M(p) \xrightarrow{N(p\,e)} M(p)$ is a transition in T.

Thus for the P/T-nets, $\mathcal{T} = (\mathcal{S}, \mathcal{E}, T)$ is the infinite graph defined with $\mathcal{S} = \mathbb{N}$, $\mathcal{E} = \mathbb{N} \times \mathbb{N}$, and $T = \{m \xrightarrow{(w_1\, w_2)} m \mid m \geq w_1 \wedge m' = m - w_1 + w_2\}$.

M. Nielsen, D. Simpson (Eds.): ICATPN 2000, LNCS 1825, pp. 16 23, 2000.

The state graph of a net $N : P \times E \to \mathcal{E}$ embeds into the synchronous product $\mathcal{T}^P$ of $|P|$ copies of $\mathcal{T}$: given $e \in E$ and markings M and M', $M \xrightarrow{e} M$ if and only if this transition projects for each place $p \in P$ to a transition $M(p) \xrightarrow{N(p\ e)} M(p)$ in $\mathcal{T}$; and since e is the (column) vector with entries $N(p, e)$, the transition $M \xrightarrow{e} M$ is precisely the synchronous product of its projections. Conversely, if for some vector $\varepsilon : P \to \mathcal{E}$, $M(p) \xrightarrow{(p)} M(p)$ for all $p \in P$, then $M \longrightarrow M$ is a transition of N if and only if ε is a column of the matrix N, i.e. if $\varepsilon \in E$. To sum up, the state graph of N is the Arnold-Nivat product of $|P|$ copies of $\mathcal{T}$, using the columns of N as the synchronization vectors.

Let $T = (S, E, T)$ be the state graph of N. By construction of state graphs, each place p of N induces a map from T to $\mathcal{T}$ projecting transitions $M \longrightarrow M$ to transitions $M(p) \xrightarrow{(p)} M(p)$. This map from T to $\mathcal{T}$ is determined by two maps operating respectively on states and on events, namely $\sigma_p : S \to \mathcal{S}$ with $\sigma_p(M) = M(p)$ and $\eta_p : E \to \mathcal{E}$ with $\eta_p(e) = e(p) = N(p, e)$, hence it is a morphism of transition systems.

The net synthesis problem for transition systems consists in approximating at best a reachable transition system T (taken as input) by the reachable state graph of some initialized net in a specified class of Petri nets. Equivalently, the problem consists in approximating at best a transition system by the reachable restriction of some Arnold-Nivat product of copies of the state graph of the representative net $\mathcal{N}$. Approximations are from above, with $T \leq T'$ if there exists a morphism of transition systems from T to T' that acts bijectively on events and that preserves the initial state. The best approximation (the least one) may be seen as a closure. A crucial problem is to decide when the approximation is exact, that is when T coincides (up to isomorphism) with the reachable state graph of some net.

Morphisms of transition sytems from $T = (S, E, T)$ to $\mathcal{T} = (\mathcal{S}, \mathcal{E}, \mathcal{T})$ are privileged tools for solving the above described problems. A morphism from T to $\mathcal{T}$ is a pair (σ, η) of maps $\sigma : S \to \mathcal{S}$ and $\eta : E \to \mathcal{E}$ such that $(s) \xrightarrow{(e)} (s)$ whenever $s \xrightarrow{e} s$. Each morphism (σ, η) determines an initialized net N whose state graph approximates T: this net has a unique place p, defined with $N(p, e) = \eta(e)$ for all $e \in E$, and its initial marking $M_0(p) = \sigma(s_0)$ is the image of the initial state of T. As a matter of fact, the state graph of a net N with a single place p approximates T if and only if this place derives from a morphism $(\sigma, \eta) : T \to \mathcal{T}$ as indicated. For the elementary nets, these morphisms are in bijective correspondence with Ehrenfeucht and Rozenberg's regions. For the P/T-nets, these morphisms are essentially Mukund's regions (although we do not consider here step transition systems): a morphism (σ, η) sends each state $s \in S$ to a non negative integer $\sigma(s)$ and each event $e \in E$ to a pair of weights (w_1, w_2) such that $\sigma(s) \geq w_1$ and $\sigma(s') = \sigma(s) - w_1 + w_2$ for every transition $s \xrightarrow{e} s$ in T.

Ehrenfeucht and Rozenberg's regions allowed an axiomatic characterization of the transition systems that coincide (up to isomorphism) with reachable state graphs of elementary net systems. Mukund's regions allowed a similar characterization of the (step) transition systems which represent behaviours of P/T-nets. The regions as morphisms analogy permits to extend uniformly this characterization to an arbitrary class of nets. Two axioms must be satisfied. One axiom, called state separation, requires that there always exists for distinct states s, s' of the transition system some region (σ, η) such that $\sigma(s) \neq \sigma(s')$, explaining why they differ in terms of nets. The second axiom, called event-state separation, requires that whenever an event e is not enabled at $\sigma(s)$ in the transition system, there exists a region (σ, η) such that $\eta(e)$ is not enabled at $\sigma(s)$ in T. When both axioms are satisfied, the transition system is isomorphic to the reachable state graph of the *saturated* net with regions as places and with M_0 as the initial marking defined as follows: for every place p produced from a corresponding region (σ, η), let $N(p, e) = \eta(e)$ for all $e \in E$ and let $M_0(p) = \sigma(s_0)$ where s_0 is the initial state of T.

When both separation axioms are satisfied, one may construct smaller nets with an identical state graph, also isomorphic to the given transition system. Desel and Reisig pointed out that this is the case of all admissible subnets of the saturated net defined as its induced restrictions on subsets of regions large enough to witness the satisfaction of the separation axioms. More generally, the regions as morphisms analogy allows to show that the state graph of the saturated net is the best approximation of a transition system by the state graph of a net, independently of the separation axioms.

Let us now focus on P/T-nets. As the P/T-regions of a finite transition system form already an infinite set, it is unclear whether one can ever decide that the separation axioms are satisfied w.r.t. P/T-regions. Nor is it obvious, unless restricting nets to bounded P/T-nets, that the saturated net with all regions as places has always a finite subnet with an isomorphic state graph. Both questions may in fact be answered positively relying on linear algebra. It takes time polynomial in the size of a transition system to check the separation axioms and to produce when they are satisfied a minimal P/T-net realization of the transition system. I shall not describe here the polynomial time synthesis method but a simpler method, efficient in practice, implemented in the second version of the tool SYNET.

2   A Method for Synthesizing P/T-Nets from Finite Automata, with Potential Applications to Distribution

In order to help the intuition, P/T-regions may be given a more convenient presentation. Let $T = (S, E, T, s_o)$ be a reachable transition system, with a set of events $E = \{e_1, \ldots, e_n\}$ and with the initial state s_0. A P/T-region (σ, η) of T may be identified with a vector of $2n + 1$ non negative integers, let

$$p = (M_0(p), p^\bullet e_1, e_1{}^\bullet p, \ldots, p^\bullet e_n, e_n{}^\bullet p)$$

where $M_0(p) = \sigma(s_0)$ and $(p^\bullet e_i, e_i{}^\bullet p) = \eta(e_i)$ for all i $(1 \leq i \leq n)$. Under this form, a region appears more clearly as an initialized P/T-net with a single place. Conversely, an initialized P/T-net with a single place defined by a vector as above yields an upper approximation of T if and only if the relations $\sigma(s_0) = M_0(p)$ and $\eta(e_i) = (p^\bullet e_i, e_i{}^\bullet p)$ determine a region (σ, η) of T. Two conditions should be fulfilled for this purpose. Let $U \subseteq T$ be a minimal subset of transitions such that every state $s \in S$ may be reached from s_0 in U, hence $U = (S, E, U, s_0)$ is a tree with one path (or branch) β_s from s_0 to s for each state $s \in S$, and for every transition $t \notin U$ (chord of the tree) there is exactly one cycle γ_t in $U \cup \{t\}$ (with an arbitrary orientation). In order that p represents a region, the first condition is that

$$\sum_{i=1}^{n} \gamma_t(e_i) \times (e_i{}^\bullet p - p^\bullet e_i) = 0$$

for every cycle γ_t, letting $\gamma_t(e_i)$ count the direct transitions labelled e_i minus the inverse transitions labelled e_i on the cycle γ_t. The second condition is that

$$M_0(p) + \sum_{i=1}^{n} \beta_s(e_i) \times (e_i{}^\bullet p - p^\bullet e_i) - p^\bullet e \geq 0$$

for every branch β_s and for every event e enabled in T at its extremity s, letting $\beta_s(e_i)$ count the transitions labelled e_i on the branch β_s.

The above conditions amount altogether to a finite system of linear homogeneous inequalities in the integer variables $M_0(p)$, $p^\bullet e_i$ and $e_i{}^\bullet p$ (constrained to be greater than or equal to 0). As a consequence, the vectors p that represent the regions of T are all the integer vectors within a polyhedral cone

$$\mathcal{C} = \{\sum_{j=1}^{m} q_j \cdot p_j \mid q_j \in \mathbb{Q}_+\}$$

generated from a finite family of integer vectors p_j (which are thus regions) using non negative rational coefficients q_j. The minimal family of generators, unique here up to integer multiples, may be computed using Chernikova's algorithm.

It follows from the convexity of $\mathcal{C}$ that whenever a state s may be separated from an event e (resp. from another state s') by some region in $\mathcal{C}$, it may also be separated from e (resp. from s') by some region p_j chosen among generators. One may therefore decide upon the satisfaction of the separation axioms by looking for an admissible subset of the finite set $\{p_1, \ldots, p_m\}$. Moreover, the canonical net with the set of places $\{p_1, \ldots, p_m\}$ is equivalent to the saturated net with all integer vectors in $\mathcal{C}$ as places: the reachable state graph of the canonical net is the best approximation (by state graphs) of the given transition system T.

P/T-net synthesis has potential applications to the distribution of reactive automata, and successful experiments have been conducted in this field on simple communication protocols, using the tool SYNET. The key is to set distribution constraints on the synthesized nets such that conflicts never occur between events located at different sites. This idea has been embodied by Caillaud in distributable P/T-nets where both places and events have locations and events cannot consume distant tokens. Locations are fixed by a placement map $\lambda : (P \cup E) \to \Lambda$ with a finite range $\Lambda = \{1, \ldots, l\}$ identifying sites on a network. The distribution constraints write:

$$p^\bullet e \neq 0 \Rightarrow \lambda(e) = \lambda(p)$$

for all $e \in E$ and $p \in P$. The dissymmetry between input and output reflects the intuition that producing tokens and sending them to distant places where they may arrive after some delay is not a problem, whereas consuming tokens not yet arrived is nonsense. As the sending of tokens to distant places may be achieved by asynchronous message passing, a distributable net may be split and compiled to a (branching bisimilar) collection of l automata communicating by asynchronous messages. An indirect method follows for distributing automata: *rst synthesize a distributable net then split it.* Now, it is pretty easy to specialize synthesis to distributable nets: in order to accomodate distribution constraints, it suffices to replace the cone $\mathcal{C}$ of all regions by smaller cones $\mathcal{C}_k$, one for each $k \in \{1, \ldots, l\}$, representing places at location k. Thus, $\mathcal{C}_k$ is just the subset of vectors p in $\mathcal{C}$ such that $p^\bullet e = 0$ for $\lambda(e) \neq k$.

This application of net synthesis would be less attractive in the framework of one-safe nets or elementary nets: less room would be left for asynchronous message passing since bounds on places mean bounds on synchronic distances. The methods sketched in this section may be adapted to work on infinite systems whose transition graphs are context-free (they may be generated from context-free graph grammars).

3 A Method for Synthesizing P/T-Nets from Languages, with Potential Applications to Distributed Control

The net synthesis problem for languages consists in approximating at best a nonempty prefix closed language $L \subseteq E^*$ (taken as input) by the language of some initialized net in a specified class of Petri nets. Labelings of nets are injective and λ-free, and all markings are accepting states. Approximations are from above w.r.t. the inclusion of languages. With these assumptions, the language of an initialized net is the intersection of the languages of all atomic subnets produced as induced restrictions of the net on individual places. The intersection of all net languages larger than L is the closure of L w.r.t. Petri net languages.

The closure $\overline{L}$ of L w.r.t. Petri net languages is generally not the language of a nite Petri net (with finite set of places). A main problem is to determine sufficient conditions on L ensuring that $\overline{L}$ is realized by a finite net. A second problem is to compute this finite net from a grammar or from another device generating L, and to decide when the approximation is exact.

Assuming now a fixed class of Petri nets, let $T = (\mathcal{S}, \mathcal{E}, \mathcal{T})$ be the state graph of the net $\mathcal{N}$ representative of this class. A nonempty prefix closed language $L \subseteq E^*$ may be identified with the transition system $T(L) = (S, E, T)$ such that $S = L$ and

$$T = \{u \xrightarrow{e} v \mid u, v \in L \wedge ue = v\}$$

Even better, L may be identified with the initialized transition system $T(L)$ with initial state ε (the empty word). Let us call regions of L the regions of $T(L)$. It comes almost immediately that L coincides with the language of some initialized net if and only if the event state separation axiom is satisfied in $T(L)$. This yields a uniform but not practical characterization of Petri net languages.

In order to make the characterization practical, let us focus on P/T-nets. A region of L or $T(L)$ is then a pair (σ, η) where $\sigma : L \to \mathbb{N}$ and $\eta : E \to \mathbb{N} \times \mathbb{N}$ are maps such that, letting $\eta(e) = ({}^\bullet\eta(e), \eta^\bullet(e))$, one has $\sigma(u) \geq {}^\bullet\eta(e)$ and $\sigma(v) = \sigma(u) - {}^\bullet\eta(e) + \eta^\bullet(e)$ whenever $u, v \in L \wedge ue = v$. A region (σ, η) of L or $T(L)$ may now be identified with a vector of non negative integers

$$p = (M_0(p), p^\bullet e_1, e_1{}^\bullet p, \ldots, p^\bullet e_n, e_n{}^\bullet p)$$

where $M_0(p) = \sigma(\varepsilon)$ and $p^\bullet e_i = {}^\bullet\eta(e_i)$ and $e_i{}^\bullet p = \eta^\bullet(e_i)$ for all i. Seeing that $T(L)$ is a tree whose branches are the words of L, a vector p as above represents a region of L if and only if, for all $u \in L$ and $e \in E$ such that $ue \in L$,

$$M_0(p) + \sum_{i=1}^{n} u(e_i) \times (e_i{}^\bullet p - p^\bullet e_i) - p^\bullet e \geq 0$$

where $u(e_i)$ counts occurrences e_i in u. This condition unfolds to an infinite system of linear inequalities that does not help to compute regions practically.

A special situation is met when L is regular or context-free and more generally when the Parikh images of the right residuals $L/e = \{u \mid ue \in L\}$ are semilinear, i.e. when for each event $e \in E$ the set of firing counts

$$\Psi(L/e) = \{(u(e_1), \ldots, u(e_n)) \mid u \in L/e\}$$

is a nite union of linear subsets $\Psi(v\,W^*)$ where v is a word in E^* and W is a nite set of words in E^*. In this case, a vector as above represents a region of L if and only if for each event e, for each linear subset $\Psi(v\,W^*)$ of L/e, and for each word w in the finite set W, the following conditions are satisfied:

$$M_0(p) + \sum_{i=1}^{n} v(e_i) \times (e_i{}^\bullet p - p^\bullet e_i) - p^\bullet e \geq 0$$

$$\sum_{i=1}^{n} w(e_i) \times (e_i{}^\bullet p - p^\bullet e_i) \geq 0$$

Altogether, one gets a finite system of linear homogeneous inequalities, hence the regions of L are all integer vectors in a polyhedral cone. The canonical net $\mathcal{N}(L)$ assembled from the minimal set of generating regions $\{p_1, \ldots, p_m\}$ realizes the closure $\overline{L}$ of L w.r.t. Petri net languages. This finite net yields the best approximation of L by a Petri net language, but in general one cannot decide whether the approximation is exact, that is to say, i.e. whether $\{p_1, \ldots, p_m\}$ is an admissible subset of regions w.r.t. event-state separation.

The decision is effective in the special case when L is regular or *deterministic* context-free and more generally when the Parikh-images of all the refusal sets $(L \ominus e) = \{u \mid u \in L \wedge ue \notin L\}$ are semilinear. In this case, checking event state separation amounts to check, for each event e and for each linear subset $\Psi(v\,W^*)$ of $(L \ominus e)$, that the following conditions are satisfied for some place $p \in \{p_1, \ldots, p_m\}$ and for all words w in the finite set W:

$$M_0(p) + \sum_{i=1}^{n} v(e_i) \times (e_i{}^\bullet p - p^\bullet e_i) - p^\bullet e < 0$$

$$\sum_{i=1}^{n} w(e_i) \times (e_i{}^\bullet p - p^\bullet e_i) = 0$$

Alternaltively, when L is regular, one may decide directly upon the inclusion $\overline{L} \subseteq L$ using Jancar and Moeller's procedure deciding whether the language of a finite P/T-net is included in a regular language taken as input. This procedure may also be used to answer the net synthesis problem with tolerance: given regular languages L and L' such that $L \subseteq L'$ decide whether exists and construct a finite P/T-net N with language $\mathcal{L}(N)$ in the range delimited by L and L'. Indeed, solutions exist if and only if the canonical net $\mathcal{N}(L)$ is a solution, i.e. if and only if $\overline{L} \subseteq L'$

A similar problem in discrete control is as follows: given a net N_P representing a plant and two regular languages L and L' specifying the minimal resp. maximal expected behaviours of the plant, construct a net N_K controlling this plant such that the specifications are met. The problem is thus to construct N_K such that $L \subseteq \mathcal{L}(N_P \times N_K) \subseteq L'$, where $N_P \times N_K$ is the net obtained by amalgamating N_P and N_K on events. It is easily seen that solutions exist if and only if the canonical net $\mathcal{N}(L)$ is a solution. In the setting of finite automata, Rudie and Wonham dealt with a refined form of the controller synthesis problem with tolerance, called decentralized controller synthesis: the controller K is asked to be the synchronous product of local controllers with specific subsets of observable and controllable events. One topic of ongoing research is to reformulate the decentralized controller synthesis problem in the setting of P/T-nets and to address the problem of asynchronous control in this alternative framework. The idea is to set distribution constraints on the synthesized controller N_K so that it may be split accordingly into a collection of local controllers communicating by asynchronous message passing.

References

1. Chernikova, N.V.: Algorithm for nding a general formula for the non-negative solutions of a system of linear inequalities. U.S.S.R. Computational Math. and Math. Physics **5** no.2 (1965) 228 233
2. Arnold, A., Nivat, M.: Comportements de processus. Les mathematiques de l'informatique, colloque AFCET (1982) 35 68
3. Ehrenfeucht, A., Rozenberg, G.: Partial (Set) 2-Structures; *part I:* Basic Notions and the Representation Problem. Acta Informatica **27** (1990) 315 342
4. Ehrenfeucht, A., Rozenberg, G.: Partial (Set) 2-Structures; *part II:* State Spaces of Concurrent Systems. Acta Informatica **27** (1990) 343 368
5. Mukund, M.: Petri Nets and Step Transition Systems. Int. Journal of Found. of Comp. Science **3** no.4 (1992) 443 478
6. Rudie, K., Wonham, W.M.: Think Globally, Act Locally: Decentralized Supervisory Control. IEEE Trans. on Automatic Control **37** no.11 (1992) 1692 1707
7. Droste, M., Shortt, R.M.: Petri Nets and Automata with Concurrency Relations - an Adjunction. Semantics of Programming Languages and Model Theory, M. Droste and Y. Gurevich eds (1993) 69 87
8. Badouel, E., Bernardinello, L., Darondeau, Ph.: Polynomial algorithms for the synthesis of bounded nets. Proc. CAAP, LNCS **915** (1995) 364 378
9. Jancar, P., Moeller, F.: Checking Regular Properties of Petri Nets, Proc. Concur, LNCS **962** (1995) 348 362
10. Badouel, E., Darondeau, Ph.: Dualities between nets and automata induced by schizophrenic objects. Proc. CTCS, LNCS **953** (1995) 24 43
11. Desel, J., Reisig, W.: The Synthesis Problem of Petri Nets. Acta Informatica **33** (1996) 297 315
12. Badouel, E., Darondeau, Ph.: Theory of regions. Lectures on Petri Nets I: Basic Models, LNCS **1491** (1998) 529 586
13. Bernardinello, L.: Proprietes algebriques et combinatoires des regions dans les graphes et leur application a la synthese de reseaux. Theses, Univ. Rennes (1998)
14. Darondeau, Ph.: Deriving Unbounded Petri Nets from Formal Languages. Proc. Concur, LNCS **1466** (1998) 533 548
15. Badouel, E.: Automates reversibles et reseaux de Petri, dualite et representation : le probleme de synthese. Document d'habilitation **31**, Irisa (1999) available from http://www.irisa.fr/EXTERNE/bibli/habilitations.html
16. Caillaud, B.: Bounded Petri-net Synthesis Techniques and their Applications to the Distribution of Reactive Automata. JESA **9 10** no.33 (to appear)
17. Darondeau, Ph.: On the Petri Net Realization of Context-Free Graphs. (to appear)
18. Synet: http://www.irisa.fr/pampa/LOGICIELS/synet/synet.html

UML - A Universal Modeling Language?

Gregor Engels, Reiko Heckel, and Stefan Sauer

University of Paderborn, Dept. of Computer Science, D 33095 Paderborn, Germany
engels|reiko|sauer@upb.de

Abstract. The Unified Modeling Language (UML) is the de facto industrial standard of an object-oriented modeling language. It consists of several sublanguages which are suited to model structural and behavioral aspects of a software system. The UML was developed as a general-purpose language together with intrinsic features to extend the UML towards problem domain-specific profiles. The paper illustrates the language features of the UML and its adaptation mechanisms. As a conclusion, we show that the UML or an appropriate, to be defined core UML, respectively, may serve as a universal base of an object-oriented modeling language. But this core has to be adapted according to problem domain-specific requirements to yield an expressive and intuitive modeling language for a certain problem domain.

Keywords: object-oriented model, UML, OCL, profile, class diagram, interaction diagram, statechart

1 Introduction

Main objectives of the software engineering discipline are to support the complex and hence error-prone software development task by offering sophisticated concepts, languages, techniques, and tools to all stakeholders involved.

An important and nowadays commonly accepted approach within software engineering is the usage of a software development process model where in particular the overall software development task is separated into a series of dedicated subtasks. A substantial constituent of such a stepwise approach is the development of a *system model*. Such a model describes the requirements for the software system to be realized and forms an abstraction in two ways (cf. Fig. 1). First, it abstracts from real world details which are not relevant for the intended software system. Second, it also abstracts from the implementation details and hence precedes the actual implementation in a programming language.

Thus, the system model plays the role of a *contract* between a client, ordering a software system, and a supplier, building and delivering a software system. Therefore, the contract has to be presented in a language which is understandable by both the client, generally not being a computer scientist, and the supplier, hopefully being a computer scientist. This requirement excludes cryptic, mathematical, or machine-oriented languages as modeling languages and favors diagrammatic, intuitively understandable, *visual languages*.

M. Nielsen, D. Simpson (Eds.): ICATPN 2000, LNCS 1825, pp. 24 38, 2000.

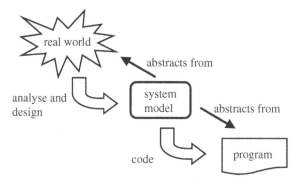

Figure 1. Role of System Model

Besides acting as a contract document between client and supplier of a software system, a system model may serve as a *documentation* document for the realized software system. The existence of such a documentation, which is consistent with the software system, substantially facilitates any required change of the system in the maintenance phase.

In case of missing documentation, respectively system model, a reverse transformation from the software system to the system model has to be performed in order to yield a model of the system on a more abstract and better understandable level where any required change can be discussed with the client and the supplier.

Thus, the system model does not only play an important role in the *forward engineering* process of developing software, but also in *reverse engineering*. Hence techniques are studied and developed to understand and document existing legacy systems, to update their functionality or to integrate them into larger systems. The rebuilding of a generally not existing system model for a legacy system eases the understanding of the consequences of any system update in contrast to dangerous ad-hoc updates of the existing system itself.

The usefulness of an abstract system model was already recognized in the 1970s, when *structured methods* were proposed as software development methods [23]. These methods offered Entity-Relationship diagrams [3] to model the data aspect of a system, and data flow diagrams or functional decomposition techniques to model the functional, behavioral aspect of a system. The main drawbacks of these structured approaches were the often missing horizontal consistency between the data and behavior part within the overall system model, and the vertical mismatch of concepts between the real world domain and the model as well as between the model and the implementation.

As a solution to these drawbacks, the concept of an *abstract data type*, where data and behavior of objects are closely coupled, became popular within the 1980s. This concept then formed the base for the object-oriented paradigm and for the development of a variety of new object-oriented programming languages, database systems, as well as modeling approaches. Nowadays, the *object-oriented paradigm* has become the standard approach throughout the whole software development process. In particular, object-oriented languages like C++ or Java have become the de facto standard for programming. The same holds for the analysis and design phases within a software development process where object-oriented modeling approaches are more and more becoming the standard ones.

The success of object-oriented modeling approaches was hindered in the beginning of the 90s by the fact that surely more than fifty object-oriented modeling approaches claimed to be the right one, the so-called object-oriented method war. This so-called method war came to a (temporary) end by an industrial initiative which pushed the development of the meanwhile standardized object-oriented modeling language UML (Unified Modeling Language) [18].

UML aims at being a general purpose language. Thus, the question arises whether UML can be termed a *universal language* which is usable to model all aspects of a software system in an appropriate way. In particular, it has to be discussed

which language features are offered to model a certain aspect like structure or behavior,

whether horizontal consistency problems are resolved in order to yield a complete and consistent model,

whether all vertical consistency problems are resolved, such that

– real world domain-specific aspects can be modeled in an appropriate, intuitive way, and that

– a transition from a UML model towards an implementation is supported.

It is the objective of this article to discuss these issues. Section 2 will provide an overview on UML and will explain the concepts offered by UML to model system aspects. Section 3 illustrates briefly the UML approach to define the syntax and (informal) semantics of UML. This shows how horizontal consistency between different model elements can be achieved. In addition, extensibility mechanisms of UML are explained which allow to adapt UML to a certain problem domain. Section 4 discusses current approaches to define domain-specific adaptations of UML, so-called *pro les*. The article closes with some conclusions in sect. 5 and a reference list as well as a list of related links to get further information.

2   Language Overview

Object-oriented modeling in all areas is nowadays dominated by the Unified Modeling Language (UML) [18]. This language has been accepted as industrial standard in November 1997 by the OMG (Object Management Group). UML was developed as a solution to the object-oriented method war mentioned above. Under the leadership of the three experienced object-oriented methodologists Grady Booch, Ivar Jacobson, and James Rumbaugh, and with extensive feedback of a large industrial consortium, an agreement on one object-oriented modeling language and, in particular, on one concrete notation for language constructs was reached in an incremental and iterative decision process. For today, UML version 1.3 represents the currently accepted industrial standard [2,16].

Main objectives for designing the Unified Modeling Language (UML) were the following:

UML was intended as a general purpose object-oriented modeling language instead of a domain-specific modeling language.

It was intended to be complete in the sense that all aspects of a system can be described and modeled in an appropriate way.

It was intended to be a visual, diagrammatic language, as such a language is generally better intuitively understandable than a textual one.

UML was not intended to be a new language, but an appropriate reuse of best practices of already existing modeling languages which are suited to model certain aspects of a system.

An important objective was to agree on a formalized syntax and standard notation for modeling constructs, while an informally given semantics definition was found to be acceptable (at least in the beginning of the standardization process).

As the name says, UML was only intended to be a language. A discussion of an appropriate method or process for deploying UML was intended to be separated from defining the language.

Despite the fact that UML was intended to be a general purpose language, concepts should already be included in the language which allow to adapt the language towards particular problem domains.

As it is common in structured as well as in object-oriented modeling approaches and in order to meet all objectives stated above, the Unified Modeling Language (UML) was defined as a family or even better <u>U</u>nion of <u>M</u>odeling <u>L</u>anguages. In particular in order to cover all aspects of a system, several modeling languages are combined where each of them is suited for modeling a specific system aspect. This means that a system model is given by a set of submodels or views where each submodel concentrates on a specific system aspect. On the other hand, the same aspect may be modeled from different perspectives. Thus, the different submodels may overlap and even provide a redundant or conflicting specification of certain system aspects. This approach of providing overlapping, non-orthogonal sublanguages eases the specification process, as the designers may describe the same issue incrementally by interrelating it to other issues. In contrast, the usage of different, even non-orthogonal sublanguages for developing a system model increases the danger of inconsistencies between the different submodels, and thus enforces additional means to handle and to prevent from inconsistencies.

Originating from the concept of an abstract data type, the traditionally distinguished system aspects are the *structural* aspect and the *behavioral* aspect of a system. UML follows this distinction and offers the following sublanguages to specify structural aspects on one side and behavioral aspects on the other side.

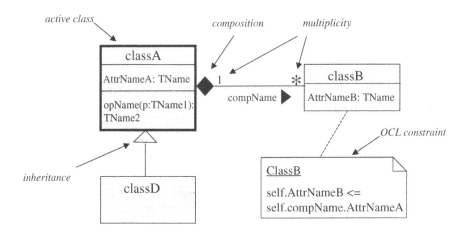

Figure 2. Class Diagram

Modeling the Structural Aspect. UML provides *class* and *object diagrams*, respectively, to model all structural aspects of a system on a type and instance level, respectively. These diagrams originate from Entity-Relationship diagrams [3] and offer means to specify the structure of objects and the possible structural relationships between objects. Objects are described by their attributes as well as by the signatures of operations which may change the state of an object. Structural relationships can be described as general associations or as a weak or strong aggregation relationship between objects, the latter kind of aggregations being so-called compositions. In addition, objects may be specified as passive or active objects, the latter ones having their own, permanently active thread of control. Figure 2 shows the standard notation of UML for these language features in an abstract example.

Allowed object societies, i.e. objects with their interrelations, may be further restricted by additional integrity constraints. These constraints may be formulated in a graphical way and attached to e.g. class diagrams (as for instance, multiplicity constraints of relationships) or may be formulated in the more expressive textual language OCL (*Object Constraint Language*) [22]. OCL is based on predicate logic and may be used in a UML model to specify integrity constraints or invariants for object societies, but also e.g. pre-/post-conditions for operations. For instance, the OCL constraint in fig. 2 states that the value of attribute AttrNameB of an object of ClassB has to be less than or equal to the value of attribute AttrNameA of the object of ClassA reachable via the compName link between these two objects.

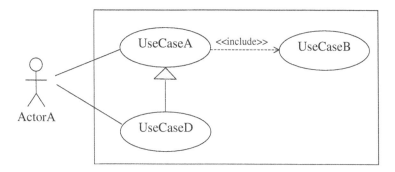

Figure 3. Use Case Diagram

Modeling the Behavioral Aspect. UML offers several diagram types to model the behavioral aspect of a system. Each of them focuses on a certain view on a system and offers appropriate language features.

A global, coarse-grained, sometimes also called external view on a system can be modeled by *use case diagrams*. This view is restricted to the identification of the main functionality or processes of a system, called use cases, and to (external) actors participating in these use cases. Use cases are only described by their name and an optional textual explanation. Use case diagrams may be structured by include, extend, or inheritance relationships between use cases (cf. fig. 3).
Use case diagrams have to be refined by the usage of other behavioral diagrams (see below) in order to describe what happens during the execution of a use case (i.e. process).

The behavior of single objects over time is described by *state machines*. Objects have a control state (in contrast to a data state) which may change in reaction to received events (triggering state transitions). Such an event may be a signal or call event from another object or a time signal which causes the object to change its state. Thus, state machines are used to model the lifecycle of an object and provide a so-called intra-object view. State machines in UML are based on Harel's statecharts [11] and offer means like concurrent and sequential composite states, history states, or junction states to model complex behavior of an object. Figure 4 gives an abstract example of such a state machine.
Summarizing, state machines are mainly used to model a state- and event-based view on a system. What is missing in such a description is a model of the cooperation and interaction between different objects in a system. This is provided by the following three behavior diagrams where each of them focuses on a certain aspect.

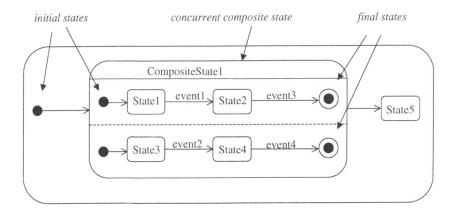

Figure 4. State Machine

The inter-object view on a system, i.e. the communication between and collaboration of different objects, can be described by a UML activity, sequence or collaboration diagram:

A control-flow oriented description can be given by a UML *activity diagram*. Syntactically, an activity diagram is a special form of a state machine where states are interpreted and labelled by activities. In contrast to usual state machines, a state change is automatically triggered when the execution of an activity has been finished. Activity diagrams do not relate activities to certain objects and represent mainly a procedural, possibly concurrent flow within a system. Objects may be exchanged as in-/out-parameters between different activities, which may be indicated by additional object flow links between states. Figure 5 illustrates the used notations in activity diagrams.

A scenario-oriented description of the interaction between objects can be given by *sequence diagrams*. They originate from message sequence charts (MSC) [12] and focus mainly on the sequence of message exchanges over time between objects involved in a certain activity. Each object is represented by a vertical life line on which the active and passive periods of an object are shown. Different forms of message exchange like synchronous or asynchronous ones can be indicated by different shapes of arrows between object life lines. Figure 6 illustrates the used notations in sequence diagrams.

An object structure-oriented description of the interaction between objects can be given by *collaboration diagrams*. They represent mainly the same information as sequence diagrams, but focus on the objects and their structural

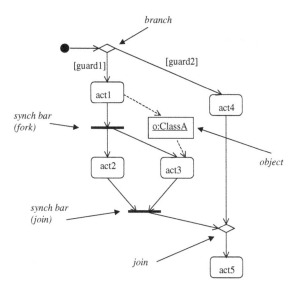

Figure 5. Activity Diagram

interrelations. Thus, the base of a collaboration diagram is an object diagram where the links between objects are additionally labelled by messages which are sent between a sending object and a receiving object. Links can be distinguished into those based on structural relationships, parameters, local variables, etc. The sequencing of messages is described by sequence numbers which are attached to messages and describe a sequential, nested, or concurrent sending of messages. Figure 7 illustrates the used notations in collaboration diagrams.

In addition to these diagrams for modeling the structural as well as behavioral aspects of a system, UML provides two diagram types to describe the transition from a model to the corresonding implementation. These so-called implementation diagrams are the component diagram and the deployment diagram.

The *component diagram* describes the software architecture of a system which consists of components, their interfaces and their interrelations. A component itself encapsulates the implementation of elements as e.g. classes from the system model.

The *deployment diagram* goes even one step further and describes the hardware architecture of a system, consisting of nodes as physical objects and their interrelations. The deployment diagram describes the distribution of objects and components to nodes, and thus links the software architecture to the hardware architecture.

Compared to the UML diagrams explained above to model the structure and behavior of a system, these two implementation diagrams are still in a very rudimentary form in the current UML version. Ongoing discussions to combine UML with language features from the ROOM (Real-time Object-Oriented Modeling)

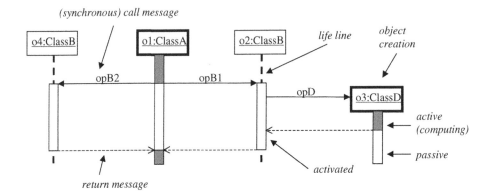

Figure 6. Sequence Diagram

approach [19] to model real-time, embedded systems will result in an improved, more expressive form of these implementation diagrams.

This concludes the overview on the UML language. UML offers a lot of additional features which could not be explained in this brief overview. An example is the package concept which allows to divide a model into smaller parts with clearly defined dependencies and thus supports to manage huge models also. The interested reader is referred to [2,18] and to the links to related web-sites (at the end of this article).

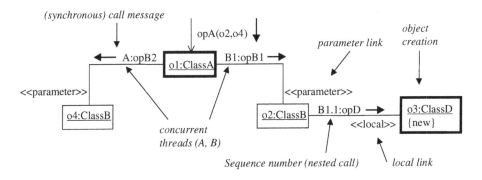

Figure 7. Collaboration Diagram

3 Language De nition

The main focus of the OMG standardization effort so far was an agreement on a commonly accepted concrete notation as well as abstract syntax for all these diagram types of the UML. The semantics of the UML is currently only informally, textually defined, and its further development towards a precise semantics is postponed to the next standardization phase. In this section, we will briefly sketch the syntax definition approach followed by the OMG and explain UML-intrinsic features to adapt UML to a problem-domain specific modeling language.

For the definition of the abstract syntax of the UML, the OMG follows a four-layered meta modeling approach. These four layers are the following:

The *MetaMetaModel (M3)* layer provides a so-called Meta Object Facility (MOF) to define meta models on the next lower layer. The MOF consists of language features for defining an Entity-Relationship diagram or class diagram as well as a constraint language to define additional integrity constraints.

The MOF is used on the *MetaModel (M2)* layer to define a concrete meta model for a modeling language. This meta model consists of a concrete class diagram with additional integrity constraints which defines the allowed abstract syntax features of a modeling language and their interrelations. Thus, the meta model defines the abstract syntax of a modeling language.

A concrete UML model is an element of the *Model (M1)* layer and is an instance of the meta model layer M2.

Finally, a conrete runtime extension of a UML model is an element of the *Objects (M0)* layer and is an instance of the model layer M1.

Thus, the UML meta model on layer M2 is specified as an instance of the meta-meta model of layer M3 by a UML class diagram together with OCL constraints, i.e., partly deploying the UML itself. While the class diagram part defines the abstract context-free syntax, the OCL constraints define the context-sensitive syntax of UML. By providing one overall meta model for all sublanguages of the UML, the horizontal consistency problem between different submodels is resolved. Particularly, the OCL constraints, also called well-formedness rules, take care that the different submodels written in different sublanguages of UML are syntactically well integrated.

This agreement on a well-defined abstract as well as concrete syntax together with a (yet informally) defined semantics has the advantage that all users of UML have the same understanding of a system model described by UML. The disadvantage of such an approach is that one has to agree on a general-purpose language with high-level language features which might not be expressive enough to model problem-domain specific details in an appropriate, intuitive way. Therefore, two types of language extensions have been discussed within the OMG to adapt the UML to problem-domain specific needs [17]. These are the *heavyweight* and the *lightweight extension* mechanisms.

The heavyweight extension mechanism is provided by the MOF which means that it is possible in principle to change and adapt the UML by modifying the UML metamodel on layer M2. As the name says, this kind of extension has great impact on the UML language and, therefore, is not possible for an individual user of UML.

In contrast to this, the lightweight extension mechanisms are built-in mechanisms of UML and allow any individual modeler to tailor the UML to her needs. This provides the opportunity to adapt the UML to the requirements of a certain problem domain by tailoring the general-purpose, universal UML to a problem domain-specific modeling language. As this tailoring means that the syntax as well as the semantics of UML constructs might be changed, it is obvious that it has to be done with great care. Thus, it is unlikely and not intended that an individual user starts to adapt the UML. It will be mainly the task of, e.g., a user group or a tool vendor, to propose and to do such an adaptation of the UML for a specific problem domain. The result of such an adaptation is a UML dialect which is termed a *UML pro le*.

Three lightweight extension mechanisms are distinguished for the UML. These are constraints, tagged values, and stereotypes.

> *Constraints* are expressed in the OCL (Object Constraint Language) and specify additional restrictions on a UML system model. They are comparable to integrity constraints in the database field and can be added to any model part in an UML system model.
>
> *Tagged values* are pairs of strings - a tag string and a value string - which can be added to any model element in a UML system model. This feature allows to attach additional information to a UML system model which can not directly be expressed by UML language features.
>
> The most powerful and thus most heavily discussed extension mechanism are *stereotypes*. Stereotypes allow to give existing UML model elements an additional classification, and thus to tailor them for a specific purpose. Stereotyped model elements of UML are indicated by an additional annotation or they may even have a different concrete notation. Stereotypes can range from modification of concrete syntax to redefinition of original semantics of model elements (cf. [1]).

Summarizing, it can be stated that the abstract and concrete syntax of UML has been formally defined by following a four-layered meta model approach. In addition, lightweight extension mechanisms are provided to adapt the general-purpose UML to a specific problem domain.

4   Language Adaptations

The main advantage and at the same time the main drawback of UML is that it is a general-purpose language. This objective of the UML designers automatically yielded a language which is not capable of providing features which are appropriate to express problem domain-specific situations. This drawback of being a general-purpose language and, in addition, the lacking of a precise semantics has been identified by the UML standardization groups and has led to establishing corresponding task groups and related RFPs (Request For Proposals) by the OMG (cf. [14]) to overcome these shortcomings. It has to be expected that

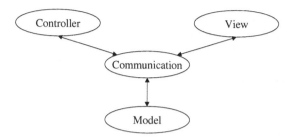

Figure 8. MVCC Architecture

further versions of UML as well as proposals for domain-specific profiles will be developed and published within the next years.

As an example of such a problem domain, we discuss the requirements for applications in the field of embedded, interactive software systems. Such software systems are typical for today's software systems and can be found, for instance, in banking terminals, as infotainment software in advanced automotive systems, or in production control systems. These systems support on one hand a window-based user interface and at the same time the connection to hardware components. Often, these software systems have to obey real-time constraints, too, in order to react to events caused by the user or an embedded system component appropriately.

A closer look at the architecture of those systems shows that the classical Model-View-Controller (MVC) architecture [15] has to be extended to a Model-View-Communication-Controller (MVCC) architecture [20] where the communication between the different components is treated as a first-class object, too (cf. fig. 8).

Summarizing, a modeling language which offers appropriate support for all aspects of such an interactive, embedded system has to provide language features to model

the model part, i.e., the problem domain specific objects,
the view part, i.e., the user interface of the system,
the communication part, i.e., the interaction between the different constituents of the system, and
the controller part which might include a real-time behavior.

In addition, appropriate means should be provided to model the component-based style of the software architecture which reflects the hardware architecture of those embedded systems.

A comparison with the language features of UML shows that the language elements are mainly suited to describe the model part. Specific language features are missing in the UML to specify the user interface and particularly the layout of a user interface, to specify the interaction between different components in a fine-grained way, or to specify real-time aspects.

This observation caused several groups to work on the development of specific profiles. A prominent example is the work on a profile for component-based, real-time systems based on ROOM [19]. Ongoing own research work comprises the investigation of fine-grained modeling of interaction [7] based on SOCCA [5,6], and the development of a profile for multimedia applications which, in particular, supports the modeling of the layout of the user interface [21].

5   Conclusion and Perspectives

In this paper, we have discussed the role of a system model within the software development process and the appropriateness of the Unified Modeling Language (UML) as a language to specify these system models. In particular, we have illustrated the main features of the sublanguages of the UML to model structural as well as behavioral aspects of a system.

UML has been designed as a general-purpose language, but also as a language which can easily be extended to a problem domain-specific language. The development of so-called problem domain-specific profiles is supported by UML-intrinsic extension mechanisms as constraints, tagged values, and particularly stereotypes.

Thus, returning to the question posed in the title of this paper, whether the U in UML can also be interpreted as *universal*, the following can be concluded:

At first glance, UML is a *union* of modeling languages as several already existing modeling languages have been gathered under one umbrella. But it is more than a disjoint union of these languages, as due to the definition of one common meta model for all sublanguages, an integration, at least on the syntactical level, has taken place. Thus, the UML can be termed a *uni ed* language on the abstract syntax level, but also on the concrete syntax level, since an agreement on concrete notations has taken place, too.

Discussing the question, whether the UML is a *universal* language which can be deployed for modeling all of today's software systems, it has to be concluded that the UML forms an ideal base for any problem domain-specific language. But the UML can not fulfill the role of a universal language which should or could be deployed in any problem domain. In order to provide intuitive and expressive modeling features for a certain domain, UML has to be extended and adapted by appropriate profiles.

This conclusion is also reflected by ongoing research and development work in the UML field. In particular, it is currently discussed which constituents of the UML belong to a *core UML* which then might serve as the base for developing problem domain-specific profiles [9].

Besides that, the embedding of the UML into the software development process is an important topic. This includes the definition of a software process model [13,4] as well as techniques to transform a UML model into a corresponding implementation in a programming language [8].

References

1. St. Berner, M. Glinz, St. Joos: A Classification of Stereotypes for Object-Oriented Modeling Languages. In [10], 249-264.
2. G. Booch, J. Rumbaugh, I. Jacobson: The Unified Modeling Language User Guide. Addison-Wesley, Reading, MA, 1999.
3. P. Chen: The Entity-Relationship Model - Toward a Unified View of Data. ACM Transactions on Database Systems, 1(1), 1976, 9-36.
4. D. D'Souza, A. Wills: Objects, Components, and Frameworks with UML - the Catalysis Approach. Addison-Wesley, 1998.
5. G. Engels, L.P.J. Groenewegen: SOCCA: Specifications of Coordinated and Co-operative Activities. In A. Finkelstein, J. Kramer, B.A. Nuseibeh (eds.): Software Process Modelling and Technology. Research Studies Press, Taunton, 1994, 71-102.
6. G. Engels, L.P.J. Groenewegen, G. Kappel: Object-Oriented Specification of Coordinated Collaboration. In N. Terashima, Ed. Altman: Proc. IFIP World Conference on IT Tools, 2-6 September 1996, Canberra, Australia. Chapman & Hall, London, 1996, 437-449.
7. G. Engels, L.P.J. Groenewegen, G. Kappel: Coordinated Collaboration of Objects. In M. Papazoglou, St. Spaccapietra, Z. Tari (eds.): Object-Oriented Data Modeling Themes. MIT Press, Cambridge, MA, 2000.
8. G. Engels, R. Hücking, St. Sauer, A. Wagner: UML Collaboration Diagrams and Their Transformation to Java. In [10], 473-488.
9. A. Evans, St. Kent: Core Meta-Modelling Semantics of UML: The pUML Approach. In [10], 140-155.
10. R. France, B. Rumpe (eds.): UML '99 - The Unified Modeling Language - Beyond the Standard. Second Intern. Conference. Fort Collins, CO, October 28-30, 1999. LNCS 1723. Springer, Berlin, 1999.
11. D. Harel: Statecharts: A Visual Formalism for Complex Systems. Science of Comp. Prog., 8 (July 1987), 231-274.
12. ITU-TS Recommendation Z.120: Message Sequence Chart (MSC). ITU-TS, Geneva, 1996.
13. I. Jacobson, G. Booch, J. Rumbaugh: The Unified Software Development Process, Addison-Wesley, Reading, MA, 1999.
14. C. Kobryn: UML 2001: A Standardization Odyssey. CACM, 42(10), October 1999, 29-37.
15. G.E. Krasner, S.T. Pope: A cookbook for using the model-view-controller user interface paradigm in Smalltalk-80. Journal of Object-Oriented Programming, 1(3), August/September 1988, 26-49.
16. Object Management Group. OMG Unified Modeling Language Specification, Version 1.3. June 1999.
17. Object Management Group, Analysis and Design Platform Task Force. White Paper on the Profile Mechanism, Version 1.0. OMG Document ad/99-04-07, April 1999.
18. J. Rumbaugh, I. Jacobson, G. Booch: The Unified Modeling Language Reference Manual. Addison-Wesley, Reading, MA, 1999.
19. B. Selic, G. Gullekson, P. Ward: Real-Time Object-Oriented Modeling. Wiley, New York, 1994.
20. St. Sauer, G. Engels: MVC-Based Modeling Support for Embedded Real-Time Systems. In P. Hofmann, A. Schürr (eds.): OMER Workshop Proceedings, 28-29 May, 1999, Herrsching (Germany). University of the German Federal Armed Forces, Munich, Technical Report 1999-01, May 1999, 11-14.

21. St. Sauer, G. Engels: Extending UML for Modeling of Multimedia Applications. In M. Hirakawa, P. Mussio (eds.): Proc. 1999 IEEE Symposium on Visual Languages, September 13-16, 1999, Tokyo, Japan. IEEE Computer Society 1999, 80-87.
22. J. Warmer, A. Kleppe: The Object Constraint Language: Precise Modeling with UML. Addison-Wesley, Reading, MA, 1998.
23. E. Yourdon, L.L. Constantine: Structured Design: Fundamentals of a a Discipline of Computer Program and Systems Design. Prentice-Hall, Englewood Cliffs, NJ, 1979.

LINKS

www.omg.org - OMG home page

www.cs.york.ac.uk/puml - precise UML group

www.rational.com/uml/index.jtmpl - UML literature

uml.shl.com - UML RTF home page

Veri cation of Timed and Hybrid Systems

Kim Guldstrand Larsen

BRICS , Dep. of Computer Science, Aalborg University, Denmark
kgl@cs.auc.dk

Abstract. UPPAAL [UPP, BLL^{+}98] is an integrated tool environment
for modelling, simulating and veri cation of real-time and hybrid systems, developed jointly by BRICS at Aalborg University in Denmark and
by DoCS at Uppsala University in Sweden. In this talk we will review
the status of the currently distributed version of UPPAAL and describe
in more detail the ongoing developments which are to be incorporated
in future releases of the tool.

Extended Modelling Language

The modelling language of UPPAAL supports model-checking of safety and bounded liveness properties of systems that can be modeled as a collection of timed
automata communicating through channels or shared variables. Typical application areas include real-time controllers and communication protocols in particular those where timing aspects are criticial.

In the currently distributed version the modelling language is somewhat
richer compared to that of its predecessors. The new language supports process
templates and more complex (bounded) data structures, such as data variables,
constants, arrays etc. A process template in the new language is a timed automaton extended with a list of formal parameters and a set of locally declared
clocks, variables and constants. Typically, a system description will consist of
a set of instances of timed automata declared from the process templates, and
some global data, such as global clocks, variables, synchronisation channels etc.

The above extensions do not increase the expressive power of the modelling
formalism but "merely" permit descriptions to be more concise and flexible. In
contrast to this, we are currently experimenting with an extension which allows
clocks to be stopped in certain situations — so-called *stop-watches*. Though a
seemingly minor upgrade from the model of timed automata, we have shown
[CL00] that the introduction of stop-watches yields the full expressive power of
linear hybrid automata [Hen96]. In addition, the existing efficient analysis for
timed automata may be extended to an (approximate) analysis in the presence
of stop-watches. Thus, linear hybrid automata may be analysed without the need
for representing and manipulating general polyhedra.

BRICS: Center for Basic Research in Computer Science at Aarhus and Aalborg
Univeristy

M. Nielsen, D. Simpson (Eds.): ICATPN 2000, LNCS 1825, pp. 39 42, 2000.

Beyond Model-Checking

A new application area of UPPAAL is that of *scheduling*, and a number of such problems have been encountered as case-studies in the ESPRIT project VHS [VHS] (e.g. [BS99, KLPW99]). Modelling the tasks to be scheduled as well as the constraining, shared resources involved as interacting timed automata allows the scheduling problem to be stated as a (time-bounded) reachability question. The diagnostic trace potentially provided by UPPAAL offers a valid schedule to the problem. However, often one wants not just an arbitrary valid schedule but a schedule which is *optimal* with respect to some suitable cost measure (e.g. in terms of total elapsed time). An experimental implementation of UPPAAL that allows for optimal schedules to be generated with respect to user-defined optimization criteria is currently under development and investigation in collaboration with researchers at BRICS@Aarhus and Nijmegen University.

Improvement of Veri er

A main focus of the UPPAAL project is to develop efficient algorithms and data structures for the verification of timed systems.

We have recently developed a new data structure called *Clock Di erence Diagrams*, CDDs [LWYP99, BLP+99]. The new structure is BDD-like (i.e. it allows for sharing of isomorphic sub trees) but intended for representing and efficiently manipulating the non-convex subsets of the Euclidian space encountered during verification of timed automata. In the currently distributed version of UPPAAL the symbolic state-space is represented using so-called Difference Bounded Matrices, DBMs. In an experiment using eight industrial examples, we have found that use of CDDs instead of DBMs led to space savings between 46% and 99% with a moderate increase in run time. For a related data structure we refer the reader to [MLAH99].

A distributed implementation of the reachability algorithm running on a number of platforms has recently been completed [BHV00] and will soon be available. The experiments with the implementation have been succesfull not only in providing essentially linear speed-up in the number of processors available but also to point to alternative search-orders compared with the standard breadth- and depth-first search orders.

Despite the increasing number of succesfull applications of UPPAAL the state-explosion problem is not just a theoretical limitation of the technology[1] but also a phenomena encountered in practice [HLS99]. Thus, to truely scale up, it is necessary to complement the tool with methods for *abstraction* and *compositionality*. In [JLS00] we have developed and applied such a methodology and demonstrated how all steps of the methods may be supported by UPPAAL.

---

[1] in the sence that model-checking a single timed automaton is either EXPTIME- or PSPACE-complet depending on the expressiveness of the logic considered

New Case Studies

In this section we briefly describe some recent case studies performed using UPPAAL.

In [HLP00] we address the problem of synthesising production schedules and control programs for a batch production plant model built in LEGO   MIND-STORM   (in fact this plant is model of the real Steel Production Plant SID-MAR in Ghent, Belgium). To deal with the complexity of the model a general methodology of *adding guidance* to a model by augmenting it with additional guidance variables and transition guards is presented. Applying this technique made synthesis of control program feasible for a plant producing as many as 60 batches. In comparison, only two batches could be scheduled without guides.

The paper [Hun99] and more recently [IKL$^+$] also considers systems controlled by LEGO   RCX   bricks. Here the studied problem is that of checking properties of the actual programs, rather than abstract models of programs. It is shown how UPPAAL models can be automatically synthesised from RCX programs, written in the programming language *Not Quite C* (NQC). Moreover, a protocol to facilitate the distribution of NQC programs over several RCX bricks is developed and proved to be correct. See also [LPL99] for a course exploiting the relationships between LEGO   MINDSTORM   constructions, NQC programs and UPPAAL-models.

In [KLPW99] an analysis of an experimental batch plant using UPPAAL is presented. The plant consists of a number of physical components (valves, pumps, tanks, etc.) each modelled as timed automata. To model the actual levels of liquid in the tanks, integer variables are used in combination with real/valued clocks which control the change between (discrete) levels at instances of time which may be predicted from a more accurate hybrid automata model. A crucial assumption of this discretisation is that the interaction between the tanks and the rest of the plant must be such that any plant event affecting the tanks only occurs at these time instances. If this assumption can be guaranteed (which is one of the verification efforts in this framework), the verification results are exact and not only conservative with respect to a more accurate model, where the continuous change of the levels may have been given by some suitable differential equation.

References

[ABBL98]   L. Aceto, P. Bouyer, A. Burgueno, and K. Larsen. The power of reachability testing for timed automata. In *Appears in Arvind and Ramanujam, editors, Foundations of Software Technology and Theoretical Computer Science: 18th Conference, FST&TCS '98 Proceedings, LNCS 1530*, pages 245 256. Springer Verlag, December 1998.

[BHV00]   Gerd Behrmann, Thomas Hune, and Frits Vaandrager. Distributed timed model checking   how the search order matters. *Lecture Notes of Computer Science*, 2000. To appear in Proceedings of Computer Aided Veri - cation.

[BJLY98]   Johan Bentsson, Bengt Jonsson, Johan Lilius, and Wang Yi. Partial Order Reductions for Timed systems. *Lecture Notes in Computer Science*, 1998.

[BLL+98]   Johan Bengtsson, Kim G. Larsen, Fredrik Larsson, Paul Pettersson, Yi Wang, and Carsten Weise. New Generation of UPPAAL. In *Int. Workshop on Software Tools for Technology Transfer*, June 1998.

[BLP+99]   G. Behrmann, K.G. Larsen, J. Pearson, C. Weise, and W. Yi. E cient Timed Reachability Analysis Using Clock Di erence Diagrams. *Lecture Notes in Computer Science*, 1633, 1999. In Proceedings of Computer Aided Veri cation 1999.

[BS99]   Rene Boel and Geert Stremersch. VHS case study 5: modelling and veri - cation of scheduling for steel plant at SIDMAR. draft, 1999.

[CL00]   Franck Cassez and Kim G. Larsen. The Impressive Power of Stopwatch Automata. Submitted for publication., 2000.

[Feh99]   Ansgar Fehnker. Scheduling a steel plant with timed automata. Technical Report CSI-R9910, Computing Science Institute Nijmegen, 1999.

[Hen96]   T. Henzinger. The theory of hybrid automata. In *Proceedings of 11th Annual IEEE Symposium on Logic in Computer Science*. IEEE Computer Society Press, 1996.

[HLP00]   Thomas Hune, Kim G. Larsen, and Paul Pettersson. Guided Synthesis of Control Programs Using UPPAAL. *To appear in Processsedings of Workshop on Distributed Systems Veri cation and Validation*, 2000.

[HLS99]   Klaus Havelund, Kim G. Larsen, and Arne Skou. Formal Veri cation of a Power Controller Using the Real-Time Model Checker UPPAAL. *Lecture Notes in Computer Science*, 1601, 1999. In proceedings of 5th International AMAST Workshop, ARTS'99.

[Hun99]   Thomas Hune. Modelling a Real-Time Language. *In proceedings of 4th Workshop on Formal Methods for Industrial Critical Systems, FMICS'99.*, 1999.

[IKL+]   Torsten K. Iversen, Kre J. Kristo ersen, Kim G. Larsen, Morten Laursen, Rune G. Madsen, Ste en K. Mortgensen, Paul Pettersson, and Chris B. Thomasen. Model-Checking Real-Time Control Programs. To be published in Proceedings of Euromicro 2000.

[JLS00]   Henrik E. Jensen, Kim G. Larsen, and Arne Skou. Scaling Up UPPAAL - Automatic Veri cation of Real-Timed Systems Using Compositionality and Abstraction. Submitted for publication, 2000.

[KLPW99]   K. Kristo ersen, K. Larsen, P. Pettersson, and C: Weise. Experimental batch plant - VHS case study 1 using timed automata and UPPAAL. BRICS, University of Aalborg, Denmark, May 1999.

[LPL99]   Kim G. Larsen, Paul Pettersson, and Hans Henrik L vengreen. Real-time systems   course, 1999. http://www.cs.auc.dk/ kgl/DTU00/Plan.html.

[LWYP99]   Kim G. Larsen, Carsten Weise, Wang Yi, and Justing Pearson. Clock Di erence Diagrams (extended version). *Nording Journal of Computing*, 6, 1999.

[MLAH99]   J. M ller, J. Lichtenberg, H. R. Andersen, and H. Hulgaard. Fully symbolic model checking of timed systems using di erence decision diagrams. In *Workshop on Symbolic Model Checking*, The IT University of Copenhagen, Denmark, June 1999.

[UPP]   The UPPAAL home page. http://www.uppaal.com.

[VHS]   The VHS project home page. http://www-verimag.imag.fr/VHS.

Parametric Stochastic Well-Formed Nets and Compositional Modelling

Paolo Ballarini[1], Susanna Donatelli[1], and Giuliana Franceschinis[2]

[1] Dip.to di Informatica, Universita di Torino, Italy
susi@di.unito.it
[2] DSTA, Universita del Piemonte Orientale, Alessandria, Italy
giuliana@di.unito.it

Abstract. Colored nets have been recognized as a powerful modelling paradigm for the validation and evaluation of systems, both in terms of compact representation and aggregate state space generation. In this paper we discuss the issue of adding compositionality to a class of stochastic colored nets named Stochastic Well-formed Nets, in order to increase modularity and reuse of the modelling e orts. This requires the notion of Parametric Stochastic Well-formed net: nets in which a certain amount of information is left unspeci ed, and is instantiated only upon model composition. The choice of the compositional rule has been based on previous work on layered models for integrated hardware and software systems (the processes, services and resources methodology), and an example of layered modelling with Parametric Stochastic Well-formed net is presented to show the e cacy of the proposed formalism.

1 Introduction and Motivations

Petri nets have been accepted in the industrial world as a formalism for studying complex systems both for correctness assessment and for quantitative evaluation. To answer to the industrial needs in an efficacious manner it is important that tools are available that support modular modelling, as well as model reuse.

Our recent experience in the European project TIRAN, where a library of "fault tolerant mechanisms" based on a layered software architecture, has been designed, validated and implemented, has required fairly detailed models. As a consequence colored nets [12, 13], with efficient solution mechanism [6, 7] have been a must, as well as a methodology that supports modularity, reuse and easy modification of models. A candidate for such a methodology is $\mathcal{PSR}$ [10], a proposal that was motivated by the need of modelling both hardware and software aspects of systems that, again, was motivated by an industrial project [2] in the field of embedded system. The $\mathcal{PSR}$ being based on GSPN [1], it was a natural choice to use Stochastic Well-formed nets [6] (SWN) for the colored extension of the $\mathcal{PSR}$. This implies that SWN need to be enhanced by a notion of compositionality that allows: **1)** to exchange information between interacting

We acknowledge contribution of the EEC project 28620 TIRAN.

M. Nielsen, D. Simpson (Eds.): ICATPN 2000, LNCS 1825, pp. 43 62, 2000.

models; **2)** to build each model without knowing the details of how the colors have been defined in the models to be composed with it; **3)** to properly treat problems arising from the introduction of colors (wrong-match problem); **4)** to model alternative ways of performing an abstract operation (in $\mathcal{PSR}$this corresponds, for example, to different implementation of the same service) To achieve this goal it was necessary to extend the SWN class to that of Parametric SWN (PSWN), where colors classes can be undefined and variables belong to undefined classes, and to define a compositional rule that allows to exchange color classes and color values.

Several techniques have been proposed in the literature for the composition or compositional analysis of high level models (see for example [11, 9, 3, 5, 4], while a very thorough survey of these methods can be found in [14]), however none of them provided us the flexibility of parametric color classes.

This paper contributes to the definition of a colored $\mathcal{PSR}$ by defining PSWN (in Section 3), and a suitable compositional rule (in Section 4). An example of application of PSWN and of the compositional rule is presented in Section 5, while Section 6 concludes the paper. The definition of $\mathcal{PSR}$ for SWN being the motivation for this work, we begin the paper by an informal summary of what $\mathcal{PSR}$ is.

2 $\mathcal{PSR}$ Methodology

$\mathcal{PSR}$ is a methodology that has been defined in [10] to guide the user in building integrated GSPN models of hardware and software. Hardware and software are modelled in two separate *levels*, called $\mathcal{P}$ (for processes) and $\mathcal{R}$ (for resources), while complex services based on resources to be used by the processes are modelled by a third level, called $\mathcal{S}$ (for services), and the global model is built using a set of operators. Level $\mathcal{R}$ and level $\mathcal{S}$ are composed to produce and intermediate level called $\mathcal{SR}$, that is finally composed with level $\mathcal{P}$. Timed activities are only present in the $\mathcal{R}$ level.

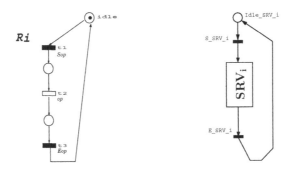

Fig. 1. A resource operation *op*, and a service *srv*

R-level. Since the hardware can be seen as a set of interacting components, then level $\mathcal{R}$ is built as the parallel composition of a set $\{R_i\}$ of GSPN models, one per component. For each operation op of a hardware component i, there is in its model R_i a sequence $t_1 t_2 t_3$, as shown in Fig. 1, with transition labeling   defined as: $(t_1) = S_{op}$, $(t_2) = op$, and $(t_3) = E_{op}$, t_1 and t_3 are immediate transitions that model the start and end of the operation, while t_2 is timed and models the operation itself. Transition labels are partitioned in two subsets: L_{loc} (the set of labels associated to timed transitions representing operations) and L_{off} (the set of labels representing start and end of operation). Level $\mathcal{R}$ is obtained as the composition of the various R_i models by transition superposition on the set of labels L_{loc}.

S-level. Level $\mathcal{S}$ is a GSPN model that describes how simple operations of different resources offered by the $\mathcal{R}$ level are combined into more complex ones, called services, to be offered to the $\mathcal{P}$ level. This allows level $\mathcal{P}$ to be fully independent of resources. Each subnet describing a service srv starts and ends with an immediate transition labeled S_srv and E_srv, that models the start and end of the service. Transitions in a service subnet that model the start or the end of operation op on a resource are labeled S_{op} (E_{op} respectively). Labels of level $\mathcal{S}$ are also partitioned in two subsets: $L_{req}(\mathcal{S})$, the set of S_{op} and E_{op} for operations of $\mathcal{R}$ requested by $\mathcal{S}$, and $L_{off}(\mathcal{S})$, the set of labels S_srv and E_srv offered by $\mathcal{S}$ to $\mathcal{P}$. A model of service srv_i is shown in Fig. 1.

The composition of $\mathcal{S}$ with $\mathcal{R}$ is realized by an operator for the synchronization of labeled GSPN: a non-injective, multi-labeled, safe GSPN ($\mathcal{S}$-level) is composed via transition's superposition with a safe, injective labeled GSPN ($\mathcal{R}$-level) over a subset E of labels (the operator is denoted by $\|\|\|_E$). $\mathcal{SR} = (\mathcal{R} \|\|\|_{L_{\text{off}}(\mathcal{R})} \mathcal{S})$

P-level. The GSPN at level $\mathcal{P}$ models the software running on the hardware in terms of the services requested to level $\mathcal{S}$, therefore transitions of $\mathcal{P}$ may be labeled with the name of a service srv, and an appropriate operator is defined that composes level $\mathcal{P}$ with the $\mathcal{SR}$ level, to obtain $\mathcal{PSR} = \mathcal{SR} \diamond^{L_{\text{off}}(\mathcal{S})} \mathcal{P}$ where the $\diamond$ operator should also take into account that each request for service srv in $\mathcal{P}$ should be translated into a pairs of S_srv and E_srv requests.

A number of constraints applies to the GSPN models of the single levels: all levels are safe; labeling of level $\mathcal{R}$ is injective (all primitive operations are different), while levels $\mathcal{S}$ and $\mathcal{P}$ may be non injective (more than one implementation for a given service and the same service requested more than once), GSPN for levels $\mathcal{P}$ and $\mathcal{R}$ are LGSPN (at most one label per transition) while GSPN for level $\mathcal{S}$ is a MGSPN (a set of labels per transition, since a service may require the simultaneous acquisition of more than one resource)

3   PSWN De nition

The constraint of safeness, introduced in [10] to solve the so-called "min-match problem" implies that, if there is more than one copy of a resource, or of a service, then the corresponding net should be duplicated, which quickly makes the global $\mathcal{PSR}$ model, and even the models of the separate levels, difficult to build and debug, and of course to solve, since the state space grows much more quickly. In [10] colors where suggested as a viable solution.

A straightforward extension of $\mathcal{PSR}$ to colored nets is to define a single colored model for n replicas of a given resource or service, and to use colors to distinguish replicas. The requirements of safeness on GSPN is translated into the requirement of color-safeness for SWN. Unfortunately this simple approach is not correct, since a problem, named WRONG-MATCH, can arise, that suggests that compositional modelling for SWN should allow color classes to be of parametric types.

In this section the SWN formalism is first briefly recalled, the WRONG-MATCH problem is discussed, and an extension of SWN to parametric nets called PSWN is finally presented.

3.1   A Quick Reminder of SWN

The starting point in the structured definition of the SWN color syntax is the set of *basic color classes* $\mathcal{BCC} = \{C_1 \quad C_n\}$. A basic color class C_i is a nonempty, finite set of colors. Examples are the class of processors, the class of memories, the class of busses, etc. Basic color classes are disjoint, moreover, a class may be partitioned into several *static subclasses* ($C_i = C_{i\,1} \cup \quad \cup C_{i\,n_i} \forall j\,k : j \neq k\ C_{i\,j} \cap C_{i\,k} = \emptyset$): colors belonging to different static subclasses represent objects of the same type but with possibly different behaviour, for example the basic color class of processors could be partitioned into two (disjoint) static subclasses, one containing the *fast* processors and the other containing the *slow* ones. Fig. 2 shows two SWN nets with $\mathcal{BCC} = \{A\ B\ C\}$; in the next subsection we shall take $A = \{a_1\ a_2\}$, $B = \{b_1\ b_2\}$ and $C = \{c_1\ c_2\}$.

Place *color domains* are defined by composition through the Cartesian product operator of basic color classes. For example, in net $\mathcal{N}_2$ of Fig. 2, place p_1 has color domain A, a basic color class, while place PP has color domain $A \times B \times C$.

Arcs are labeled with functions that specify the input/output behaviour of a transition: they are *variables* or simple arithmetic expressions built out of tuples of variables; variables are typed, and their types must match the classes in the color domain of the place to which the arc is connected. In net $\mathcal{N}_2$ of Fig. 2 the function on the arc from p_1 to t_1 is variable $\langle x \rangle$, of color A, while function $\langle h\ i\ j \rangle$, on the arc out of place PP is a triple of variables of color A, B, and C respectively. Arc expressions use the operators $+$ and $-$ (for set union and difference), a successor function $!$ for ordered classes, and the diffusion function S_{sc} (the set with one token per color of the static subclass, or class, sc).

A transition fires for a given assignment of colors to variables of suitable type. For example, in net $\mathcal{N}_1$ of Fig. 2, transition t_1 can fire for any triple of colors of $A \times B \times C$, if there is at least one token per color in the input places of t_1.

The scope of a variable is the set of input and output arcs of a transition, so that, for example, the two x variables on the arcs into and out of p_4 may represent unrelated colors.

Guards can be associated to transitions to create dependencies among variables: for example in the case of transition t_2 of net $\mathcal{N}_2$ of Fig. 2 a firing instance is the assignment of a value to the 6 variables appearing in the input arcs, but, assuming again that each input place has at least one token per color, not all combinations are possible, since the guard associated to t_2 limits the possible assignments to those in which the value assigned to h, i, and j is equal to that assigned to x, y, and z respectively.

The transition color domain is therefore defined in terms of the basic color classes of the variables appearing on its input and output arcs, and on its guard.

A peculiarity of SWN is that color domains can be structured as Cartesian product, but variables are "typed" according to the basic color classes.

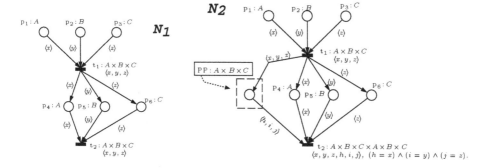

Fig. 2. SWNs and the WRONG-MATCH problem

3.2   A Motivation for PSWN: The WRONG-MATCH Problem

Let us assume that in the SWN $\mathcal{N}_1$ of Fig. 2 the initial marking M_0 assigns a token per color to places p_1 p_2 and p_3. We can then consider the following firing sequence $M_0 \xrightarrow{\langle t_1 \ \langle a_1 \ b_1 \ c_1 \rangle \rangle} M_1 \xrightarrow{\langle t_1 \ \langle a_2 \ b_2 \ c_2 \rangle \rangle} M_2$ that removes all tokens from the input places of t_1 and assigns a token per color to places p_4 p_5 and p_6. The transition instances of t_2 enabled in M_2 are then all $\langle t_2 \ \langle x \ y \ z \rangle \rangle$ such that $\langle x \ y \ z \rangle \in M_2[p_4] \times M_2[p_5] \times M_2[p_6]$:

$$
\begin{aligned}
E(M_2 \ t_2) = \{ & \langle t_2 \ \langle a_1 \ b_1 \ c_1 \rangle \rangle \ \ \langle t_2 \ \langle a_1 \ b_1 \ c_2 \rangle \rangle \ \ \langle t_2 \ \langle a_1 \ b_2 \ c_1 \rangle \rangle \\
& \langle t_2 \ \langle a_1 \ b_2 \ c_2 \rangle \rangle \ \ \langle t_2 \ \langle a_2 \ b_1 \ c_1 \rangle \rangle \ \ \langle t_2 \ \langle a_2 \ b_1 \ c_2 \rangle \rangle \\
& \langle t_2 \ \langle a_2 \ b_2 \ c_1 \rangle \rangle \ \ \langle t_2 \ \langle a_2 \ b_2 \ c_2 \rangle \rangle \}
\end{aligned}
$$

If we now interpret A as service identities, B as cpu's, C as memories, t_1 as a start operation over cpu and memory, and t_2 as the corresponding end, then we have modelled a situation in which service $a1$ acquires cpu $b1$ and memory $c1$, but can give back cpu $b1$ and memory $c2$, if, in the meantime, a service $a2$ has acquired cpu $b2$ and memory $c2$: we have a so-called WRONG-MATCH problem. A possible solution is shown in net $\mathcal{N}_2$ of Fig. 2: a new place PP is used to keep track of the triples (service, cpu, memory); we shall term it "propagation place", since it is used to propagate the triple's identities to subsequent transitions. The guard on transition t_2 ensures that service, memory and cpu are matched in t_2 as they were created by t_1, indeed

$$E(M_2\ t_2) = \{\langle t_2\ \langle a_1\ b_1\ c_1\ a_1\ b_1\ c_1\rangle\rangle\ \langle t_2\ \langle a_2\ b_2\ c_2\ a_2\ b_2\ c_2\rangle\rangle\}$$

The solution proposed works fine, but *it lacks compositionality.* If the model of service/cpu's/memories is built as the composition of two models, one for the service level, and one for the resource level, then where should PP belong to? There are three possibilities: PP is added after the composition, on the complete $\mathcal{PSR}$ model, PP belongs to $\mathcal{R}$, or PP belongs to $\mathcal{S}$. The first choice requires a global knowledge of the system behaviour, and a manipulation of the final model, that may be huge and even not available in graphical form (depending on its size). The second and third choices destroy modularity: since PP has color domain $A \times B \times C$, either the resource level knows about the service level colors, or the service level knows about the resource level colors. Indeed a service of level $\mathcal{S}$ *should not be aware of the color class and on the color class structure* of the resources at level $\mathcal{R}$, and the same is true for resources with respect to services. If reuse, modularity and modifiability are an issue then it is necessary to allow color classes of undefined type, that will be instantiated when the levels are composed. Undefined colors will be used to represent unknown resources at level $\mathcal{S}$ and unknown services at level $\mathcal{R}$ or $\mathcal{P}$, and the composition operator should take care of instantiating undefined colors to the basic color classes. SWN where undefined color classes are allowed are called Parametric SWN.

3.3   Parametric SWN De nition

Let $cd(t)$ be the color domain of transition t expressed as set of pairs $(C_i\ x)$, where x is a variable of t, and C_i is its basic color class, then we can define:

De nition 1 (Parametric Stochastic Well-formed Nets). *A Parametric Stochastic Well-formed Net (PSWN) is an nine-tuple:*

$$\mathcal{N} = \langle P\ T\ \textbf{Pre Post Inh pri}\ \mathcal{C}\ cd\ \mathbf{w}\rangle$$

*$P\ T$ **Pre Post Inh pri**, and w, (that stay for places, transitions, input, output, inhibitor, and priority functions, and weight assignment respectively) are de ned as for classical SWN,*
$\mathcal{C} = \mathcal{BCC} \cup \mathcal{PC}$, where $\mathcal{BCC} = \{C_1\quad C_n\}$ is the nite set of nite basic color classes, ($\forall i\ C_i = \{c_{i_1}\quad c_{i_{n_i}}\}$) and $\mathcal{PC} = \{XC_1\quad XC_n\}$ is the nite set of parametric color classes
cd de nes the color domain of places and transitions, with $cd : P \to (\mathcal{BCC} \cup \mathcal{PC})^n$ $n \geq 0$, and $cd : T \cup \left((\mathcal{BCC} \cup \mathcal{PC}) \times VAR\right)^m$ $m \geq 0$, where VAR is the set of variables.

PSWN are therefore SWN in which the color classes can be either fully defined ($\mathcal{BCC}$) or undefined ($\mathcal{PC}$): since we are working in a compositional environment, the undefined classes will be instantiated through net composition, based on superposition of transitions of equal label. To this aim we need to define labeled PSWN, in which a function assigns a set of labels to each transition, and import and export functions are associated to labeled transitions, to import/exports basic color classes and color values.

Definition 2 (Multilabeled PSWN). *A* Multilabeled PSWN *(MPSWN) is a twelve-tuple:*

$$\mathcal{N} = \langle P\ T\ \textbf{Pre Post Inh pri}\ \mathcal{C}\ cd\ \textbf{w}\ imp\ exp\ \rangle$$

where:

> $P\ T\ \textbf{Pre Post Inh pri}$, $\mathcal{C}\ cd\ and\ \textbf{w}$, *are defined as for PSWN,*
> *is a labeling function. If L is a set of labels, then the labeling is defined as*
> $: T \to \{\mathcal{P}(L) \cup\ \}$, *where $\mathcal{P}(L)$ is the powerset of L,*
> $imp(t\ l)$, *with $l \in (t)$ is a pair $(C\ x)$ of $cd(t)$, or the empty pair $(- -)$, and it is named import function of* t *over label l, and*
> $exp(t\ l)$, *with $l \in (t)$ is a list of pairs $(C_i\ x_i)$, with $(C_i\ x_i) \in cd(t)$, or the empty pair $(- -)$, and is named export function of* t *over label l,*

Observe that more than one label can be assigned to a single transition; MPSWN for which at most one label is associated to each transition are termed "simple", and are indicated as LPSWN.

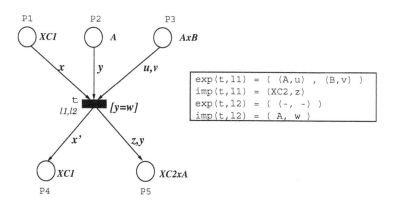

Fig. 3. A Multilabeled PSWN.

PSWN are meant to import/export values and colors: to instantiate parametric colors into colors built out of $\mathcal{BCC}$, and to unify variables of parametric color type with imported variables of color built out of $\mathcal{BCC}$. Clearly a certain consistency in the use of parametric colors and import and export functions should be respected in the model. To this aim we define an "instantiatable" MPSWN, as an MPSWN in which the following constraints are satisfied:

C1 Parametric color classes cannot appear in $\exp(\mathsf{t}\ l)$, for any $\mathsf{t} \in T$ and $l \in L$

C2 For any transitions t, if $XC \in \mathcal{PC}$ is a parametric color appearing in the color domain of t ($XC \in \mathcal{PC}$ and $XC \in cd(\mathsf{t})$), then:

    import if XC is in the color domain of an output place of t, and not in the color domain of any of its input places, then color XC must be in exactly one of the import lists of t,

    propagation if XC is in the color domain of an input place p of t, then it does not appear in the import list of t,

        and there must exists exactly one directed path from a transition t', that imports XC, to p, passing through places and transitions that have XC in their color domain.

From now on we write MPSWN for instantiatable MPSWN.

Fig. 3 depicts a MPSWN net. Let A B be basic color classes (A $B \in \mathcal{BCC}$) and $XC1$ $XC2$ be parametric ones ($XC1$ $XC2 \in \mathcal{PC}$). Transition t is labeled $\{l1\ l2\}$, and has color domain:
$$cd(t) = \{(XC1\ x)\ (A\ y)\ (A\ u)\ (A\ w)\ (B\ v)\ (XC2\ z)\}$$
$XC1$ is a parametric color class that does not appear in the import lists of t, therefore, for the net to be instantiatable, there must be a propagation path to place P1 from a transition t', not shown in the figure. $XC2$ is instantiated through label $l1$, and variables z and w will be unified through label $l1$ and $l2$ respectively.

Observe that the same parametric color cannot be used in two unconnected part of a MPSWN, due to the requirement that there should exists one directed path from an import transition for that color, and therefore each parametric color of $\mathcal{PC}$ can be associated with the unique transition that imports it. Another important observation is that an import list contains a single pair (color, variable), while the export list may contain many: this implies that, when the import lists of a MPSWN are going to be matched with the export lists of another MPSWN, then a parametric color can be substituted by a structured color class.

Finally, let us observe that it is possible to have an import function of the type $\mathrm{imp}(\mathsf{t}\ l) = (C\ x)$, where C is a basic color class: in this case only a variable will be imported, and not its color class, since C is not parametric, as it is the case for variable w in the net of Fig. 3.

4 Compositional Rule

Colors are instantiated, and values are exchanged, through superposition of transitions of equal label. To this aim, we need to describe how import and export lists are *uni ed* by superposition of the transitions of equal label. There is a double effect of the unification, namely:

renaming of variables (that will produce as an effect the exchange of values)
instantiation of parametric colors (that may require a propagation along the
net)

Let us introduce the compositional rule by means of a few examples.

Fig. 4 shows two MPSWN $\mathcal{N}_1$ and $\mathcal{N}_2$ that are superposed on label l, on
which $N1$ exports color B and its associated variable y, and $N2$ imports color
XC and its associated variable u, to produce net $\mathcal{N}$, in which the output place
P6 has now color B, and the variable on the arc from t to P6 is now y.

Observe that it may happen that names of the variables in $N1$ and $N2$
are not disjoint. Since variables in different nets have no relationships between
them, then it is necessary to introduce a renaming to make sure that names are
disjoint. For simplicity, we assume that the renaming has been done before the
composition, and therefore, from now on, we shall assume that *the sets of the
names of the variables of the nets that are composed are disjoint.*

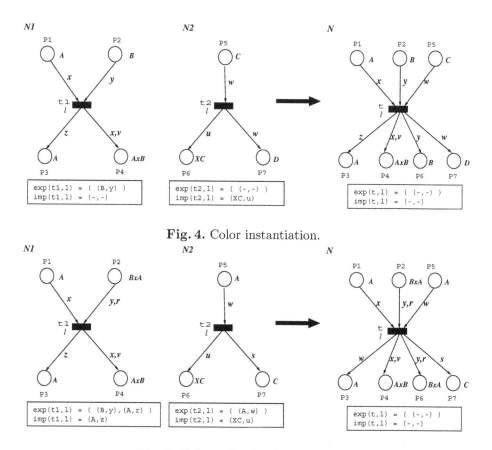

Fig. 4. Color instantiation.

Fig. 5. Color and value instantiation.

Fig. 5 shows instead a case in which the same label is used to import a color class and export a value at the same time. Label l is used to instantiate XC to $B \times A$, and its variable u to $(y\ r)$, but also to substitute variable z with variable w, which results, in the composed net, in transition t adding to place P3 a token of the same color as that taken from P5.

Let us consider the case of a non-injective labeling: net $\mathcal{N}_2$ of Fig. 6 has two transitions labeled l, that, when superposed with t1 of net $\mathcal{N}_1$ give rise to transitions t1-2 and t1-3. The export of color B by transition t1 modifies color $XC2$ of P10 to B, and variable u to y. More interesting is the case of the import, by transition t1 of color $XC1$: since two different transitions with label l exists in $N2$, each of them with an export color, then place P3 is replicated, replica P3-C is for the assignment of color C to $XC1$ (import through t2), and replica P3-E is for the assignment of color E to $XC1$ (import through t3).

Observe that, in general, place P3 in $\mathcal{N}_1$ can be connected to a subnet in which color $XC1$ appears on places and transitions , and variables of color $XC1$ on input and output arcs (propagation), which implies that the all subnet should be replicated for the two different color assignment. Thanks to constraint **C2-propagation** of the definition of instantiatable MPSWN, this operation is well defined and easy to implement, since it is possible to determine, by a syntactic checking of the two nets, what are the different assignments of color classes (or Cartesian product of color classes) to $XC1$. Once the set of all possible assignment is determined than the subnet replication is similar to the unfold operation (the one that allows to obtain P/T nets from colored ones).

As a final example we present in Fig. 7 a case of composition of MPSWN nets where two labels are assigned to the same transition: this model arises in all situations in which a service (net $\mathcal{N}_1$) needs to acquire two resources ($l1,l2$) simultaneously. Net $\mathcal{N}_1$ keeps in place P3 the identifier of the resource acquired through $l1$, and in place P4 the pair (resource acquired through $l2$, service id). The two resources of $\mathcal{N}_2$ export colors C and E, and their identity through the variables w and r. The single resulting transition is shown in the bottom part of Fig. 7.

It is obvious that the composition of $N1$ and $N2$ could be done in two steps, one to superpose t1 with t2 on label $l1$, and the other to superpose the resulting one with t3 over $l2$, thanks to constraint **C2-import**.

4.1   Composition of Two MPSWN

In this part we shall define the compositional operator, for the simplest case, that of LPSWN that have an injective labeling. This makes the definition much simpler to write and understand, since for each transition of $\mathcal{N}_1$ there is at most one transition of $\mathcal{N}_2$ with which to synchronize. As said above the extension to multilabeling is trivial, while the non-injective one is more complicated, due to the need to replicate subnets.

De nition 3 (Composition $\|\|_E$). *Let's take two LPSWN with injective labeling* $\mathcal{N}_i = \langle P_i\ T_i\ \mathbf{Pre}_i\ \mathbf{Post}_i\ \mathbf{Inh}_i\ \mathbf{pri}_i\ \mathcal{C}_i\ cd_i\ \mathbf{w}_i\ \mathrm{imp}_i\ \exp_i\ _i\rangle$ *with* $i \in$

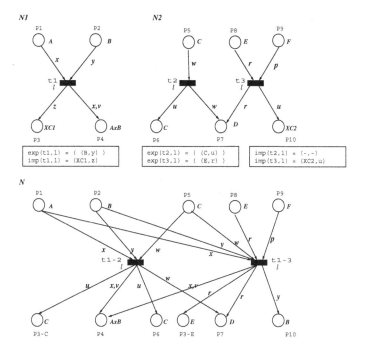

Fig. 6. Import-export for a non injective labeling.

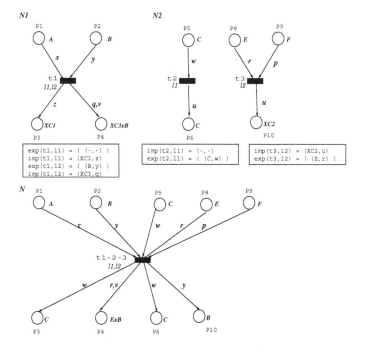

Fig. 7. Import-export for multilabeling.

$\{1\ 2\}$. *The LPSWN* $\mathcal{N} = \mathcal{N}_1\ \|\|_E\ \mathcal{N}_2$ *resulting from the composition of* $\mathcal{N}_1$ *and* $\mathcal{N}_2$ *on the set of labels* $E \subseteq\ _1(T_1) \cap\ _2(T_2)$, *is the following LPSWN:*

$$\mathcal{N} = \langle P\ T\ \textbf{Pre Post Inh pri}\ \mathcal{C}\ cd\ w\ \text{imp exp}\ \rangle$$

with: $P = P_1 \bigcup P_2$ $T = T_1\backslash T_1^E \bigcup T_2$ *(all transitions of* $T1$ *that do not synchronize, plus all that of* T_2 *where* T_i^E *is the subset of transitions* $\mathsf{t} \in T_i$ *such that* $(\mathsf{t}) \in E$*);*
$\mathcal{PC} = \mathcal{PC}_1\backslash \mathcal{PC}_1^E \bigcup \mathcal{PC}_2\backslash \mathcal{PC}_2^E$ *where* $\mathcal{PC}_i^E$ *is the subset of* $\mathcal{PC}$ *that is instantiated to basic color classes thanks to import operations on the labels of* E*;*
$\text{imp} = \text{imp}_1\backslash \text{imp}_1^E \bigcup \text{imp}_2\backslash \text{imp}_2^E$*, where* imp_i^E *represents the restriction of* imp_i *to the subset of labels* E *and* $\text{exp} = \text{exp}_1 \backslash \text{exp}_1^E \bigcup \text{exp}_2 \backslash \text{exp}_2^E$*, where* exp_i^E *represents the restriction of* exp_i *to the subset of labels* E*, moreover* $\mathcal{BCC} = \mathcal{BCC}_1 \bigcup \mathcal{BCC}_2$

$$cd(t) = \begin{cases} cd_1(t) & \underline{\textbf{if}}\ \ t \in T_1\backslash T_1^E \\ cd_2(t) & \underline{\textbf{if}}\ \ t \in T_2\backslash T_2^E \\ cd_1(t^*)[\text{imp}_1(t^*\ l) \leftrightarrow \text{exp}_2(t\ l)]\cup \\ \quad cd_2(t)[\text{imp}_2(t\ l) \leftrightarrow \text{exp}_1(t^*\ l)]\ \underline{\textbf{if}}\ \ t \in T_2^E & _2(t) = \ _1(t^*) = l \end{cases}$$

where $[\text{imp}_i(t^*\ l) \leftrightarrow \text{exp}_j(t\ l)]$ *represents the substitutions due to the uni cation of color classes and variables caused by the import of* i *and the export of* j*, with* $i \neq j$*;*

$$cd(p) = \begin{cases} cd_1'(p)\ \underline{\textbf{if}}\ \ p \in P_1 \\ cd_2'(p)\ \underline{\textbf{if}}\ \ p \in P_2 \end{cases}$$

where $cd_i'(p)$ *is* $cd_i(p)$ *where all parametric classes in* $\mathcal{PC}_i^E$ *have been substituted by the corresponding basic color classes of* $\mathcal{BCC}_j$ $i \neq j$ *imported through* L^1*; for* $F_i \in \{Pre_i\ Post_i\ Inh_i\}$*, we de ne:*

$$F(t) = \begin{cases} F_1'(t) & \underline{\textbf{if}}\ \ t \in T_1\backslash T_1^E \\ F_2'(t) & \underline{\textbf{if}}\ \ t \in T_2\backslash T_2^E \\ F_1'(t) \bigcup F_2'(t)\ \underline{\textbf{if}}\ \ t \in T_2^E \end{cases}$$

and, again, F_i' *stands for* F_i *where variables of colors belonging to* $\mathcal{PC}_i^E$ *have been substituted by the imported variables.*

The extension of the definition to the case of injective multilabeling is quite straightforward, since it requires only to take into account that one transition in the composed net may result from the composition of more than two transitions, while the extension to non injective labeling, although conceptually similar, is more complicated to define and implement, since places and transitions can get duplicated to account for different colors imported through the same label.

---

[1] Observe that this operation is well defined since, due to injectivity of the labeling function and to the constraints imposed on PSWN, each parametric color is instantiated to exactly one basic color class

5   Example of PSWN Use

The model of the system under study is obtained by composing three levels, which correspond to levels $\mathcal{P}$, $\mathcal{S}$ and $\mathcal{R}$ of the $\mathcal{PSR}$ methodology. The complete model shall represent (at a high abstraction level) the behaviour of a e-mail service offered by a provider to its customers. The $\mathcal{P}$ level represents the behaviour of the customers, the $\mathcal{S}$ level represents the implementation of the mailbox service, while the $\mathcal{R}$ level represents the resources (modems and disks) needed to implement the service.

For the sake of space, $\mathcal{P}$ is not described in detail, instead only isolated subnets to be composed with the $\mathcal{SR}$ net are illustrated in Fig. 8. The color classes of level $\mathcal{P}$ are $CUST$, the set of customers, $RES= NOMSG \cup MSG = \{nomsg\} \cup \{somemsg\}$, used to distinguish the case of a receive message service request completed without receiving any message (no messages where present in the mailbox of customer x) from the case in which messages are actually received. $XCs1$ is a parametric color class representing a mailbox address, imported from the $\mathcal{S}$ level. There is only one propagation place, namely AddressBook, used to keep the association between a customer and her mailbox.

The transitions representing a service request are: GetAddress, ReleaseAddress, SendMsg and ReceiveMsg.

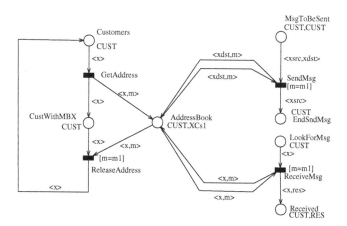

Fig. 8. Some portions of the processes level

5.1   Level $\mathcal{S}$: The Services

The level $\mathcal{S}$ subnet, depicted in Fig. 9, models the services offered to the $\mathcal{P}$ level: it represents an electronic mailbox service offered by a provider. Customers can connect to the provider through a modem, to obtain four type of services: (1) a

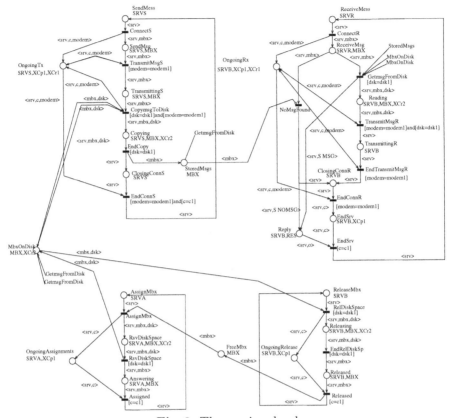

Fig. 9. The services level

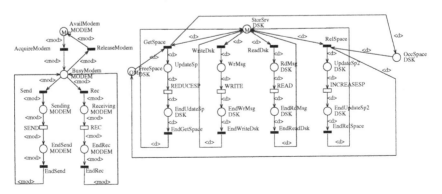

Fig. 10. The resources level

mailbox assignment service (required by a customer the first time she connects to the provider), (2) a mailbox release service (perhaps because the customer has found a more convenient provider), (3) a send message and (4) a receive messages service.

The $\mathcal{S}$ level models how mailboxes (which can be seen as *logical resources*) are administrated. The submodels corresponding to each service can be easily identified on the model: *Assign Mailbox* corresponds to the leftmost, lower subnet, *Release Mailbox* corresponds to the rightmost, lower subnet, *Send Message* corresponds to the leftmost, upper subnet, finally *Receive Messages* corresponds to the rightmost, upper subnet. When a new mailbox is assigned, some disk space is reserved for the new mailbox on one of the provider's storage servers: a disk identifier is hence imported from level $\mathcal{R}$ to keep track of the mailbox-disk association. When a mailbox is released, a notification is sent to the provider who marks the space reserved to that mailbox as free. A *Send Message* service comprises four steps: first a modem connection to the provider is established, then the message is sent to the provider through the modem, next the message is stored on the disk space reserved for the destination mailbox, finally the modem connection is released. Place **StoredMsg** is used to keep track of the messages in each mailbox which have not yet been received. Similarly a *Receive Messages* service starts with a modem connection with the provider, followed by a read messages from disk operation (provided that at least one message was sitting in the mailbox) and a transmission of the messages from the provider to the client through the modem. It might be the case that no messages are present in the mailbox: to take into account this aspect, the same color class *RES* already defined for the $\mathcal{P}$ is included, and *the value of variable o* is exported from level $\mathcal{S}$ (transition **EndSrv**) to level $\mathcal{P}$ to allow the information transfer between the two levels.

The colour classes defined in the $\mathcal{S}$ level are: *MBX*, the set of mailbox identifiers, *SRVS*, the set of servers that can perform a "send message" service (we may think at them as several threads of a send message server process), *SRVR*, the set of servers that can perform a "receive message" service (again, we may think at them as several threads of a receive message server process), *SRVA*, the set of servers that can assign a new mailbox identifier, *SRVE*, the set of servers that can accept a "release mailbox" request, and *RES*, present also in the $\mathcal{P}$ level, introduced above. Moreover there are a number of parametric classes:

XCp1: used to keep track of the identity of the customer that a given server is working for (to avoid WRONG-MATCH problems); there are five propagation places (one for each service subnet, plus an additional place in the Receive message subnet) including this parametric class in their colour domain: OngoingTx,OngoingRx, EndSrv,OngoingAssignment, and OngoingRelease.

XCr1: used to keep track of the identity of the provider modem that is currently supporting a given data transfer: there are two propagation places associated with this parametric class, namely OngoingTx, OngoingRx.

XCr2: used to keep track of the identity of the disk (or storage server) where the messages of a given mailbox are stored. There are several propagation places including this class in their colour domain: MbxOnDisk used to keep track of the association mailbox-disk during the whole life of a mailbox (since its assignment until it is released), RsvDiskSpace, Releasing, Copying and Reading, all representing a phase of one service in which a disk operation is required.

The *o ered labels* that represent the interface between level $\mathcal{S}$ and level $\mathcal{P}$, and the corresponding import/export functions are summarized in Tab. 1. The (1) and (2) annotations close to the level $\mathcal{P}$ transitions names (see column two) indicate the first and second transition in the subnet resulting from the vertical expansion of the service request transition in the $\mathcal{P}$ level (see [10]).

Label	Trans. in $\mathcal{P}$ - Trans. in $\mathcal{S}$	$\mathcal{P}$ Imp./Exp.	$\mathcal{S}$ Imp./Exp.
S_GetAddress	GetAddress(1) - AssignMbx	exp $(CUST,x)$	imp $(XCp1,c)$
E_GetAddress	GetAddress(2) - Assigned	imp $(XCs1,m)$	exp (MBX,mbx)
		exp $(CUST,x)$	imp $(XCp1,c1)$
S_ReleaseAddress	ReleaseAddress(1) - RelDiskSpace	exp $(CUST,x)$	imp $(XCp1,c)$
		imp $(XCs1,m1)$	exp (MBX,mbx)
E_ReleaseAddress	ReleaseAddress(2) - Released	exp $(CUST,x)$	imp $(XCp1,c1)$
S_SendMessage	SendMessage(1) - ConnectS	exp $(CUST,xdst)$	imp $(XCp1,c)$
		imp $(XCs1,m1)$	exp (MBX,mbx)
E_SendMessage	SendMessage(2) - EndConnS	exp $(CUST,xdst)$	imp $(XCp1,c1)$
S_RecMessage	ReceiveMessage(1) - ConnectR	imp $(XCs1,m1)$	exp (MBX,mbx)
		exp $(CUST,x)$	imp $(XCp1,c)$
E_RecMessage	ReceiveMessage(2) - EndSrv	exp $(CUST,x)$	imp $(XCp1,c1)$
		imp(RES,res)	exp (RES,o)

Table 1. Interface between levels $\mathcal{P}$ and $\mathcal{S}$

Label	Trans. in $\mathcal{S}$ - Trans. in $\mathcal{R}$	$\mathcal{S}$ Imp./Exp.	$\mathcal{R}$ Imp./Exp.
S_Conn_to_Pr	ConnectS - AcquireModem	imp $(XCr1,modem)$	exp $(MODEM,mod)$
	ConnectR - AcquireModem	imp $(XCr1,modem)$	exp $(MODEM,mod)$
E_Conn_to_Pr	EndConnS - ReleaseModem	imp $(XCr1,modem1)$	exp $(MODEM,mod)$
	EndConnR - ReleaseModem	imp $(XCr1,modem1)$	exp $(MODEM,mod)$
S_ModemRec	TransmitMsgR - Rec	imp$(XCr1,modem1)$	exp $(MODEM,mod)$
E_ModemRec	EndTransmitMsgR - EndRec	imp $(XCr1,modem1)$	exp $(MODEM,mod)$
S_ModemSend	TransmitMsgS - Send	imp $(XCr1,modem1)$	exp $(MODEM,mod)$
E_ModemSend	CopymsgToDisk - EndSend	imp $(XCr1,modem1)$	exp $(MODEM,mod)$
S_GetDiskSpace	AssignMbx - GetSpace	imp $(XCr2,dsk)$	exp (DSK,d)
E_GetDiskSpace	RsvDiskSpace - EndGetSpace	imp $(XCr2,dsk1)$	exp (DSK,d)
S_RelDiskSpace	RelDiskSpace - RelSpace	imp $(XCr2,dsk1)$	exp (DSK,d)
E_RelDiskSpace	EndRelDiskSp - EndRelSpace	imp $(XCr2,dsk1)$	exp (DSK,d)
S_WriteDisk	CopymsgToDisk - WriteDsk	imp $(XCr2,dsk1)$	exp (DSK,d)
E_WriteDisk	EndCopy - EndWriteDsk	imp $(XCr2,dsk1)$	exp (DSK,d)
S_ReadDisk	GetmsgFromDisk - ReadDsk	imp $(XCr2,dsk1)$	exp (DSK,d)
E_ReadDisk	TransmitMsgR - EndReadDsk	imp $(XCr2,dsk1)$	exp (DSK,d)

Table 2. Interface between levels $\mathcal{S}$ and $\mathcal{R}$

5.2   Level $\mathcal{R}$: The Resources Level

The available resources at this level are the modems and the storage-servers, offering the following operations (see Fig. 10): **modem**: connect to provider, transmit from client to provider, transmit from provider to client, disconnect; **storage server**: allocate space, deallocate space,write disk, read disk.

In this example there is no explicit representation of cooperation between resources (see [10]). Actually we have kept this level very simple for the sake of space, and the resources represented at this level are very abstract, and are only those at the provider site. It might be reasonable to explicitly represent also the resources at the client site (both the computer and the modem). In this case we should have merged the client modem and provider modem timed transitions representing the fact that they must cooperate for the data transfer to take place: of course the resulting transition would be assigned the speed of the slowest modem.

The colour classes defined at this level are $MODEM$, the set of modems available at the provider's site, DSK, the set of storage devices available for keeping the messages contained in the mailboxes (place FreeSpace models the available space, in terms of new mailboxes that can be allocated, for each disk);

The o ered labels that represent the interface between level $\mathcal{R}$ and level $\mathcal{S}$, and the corresponding import/export functions are summarized in Tab. 2.

5.3   Composition

Some composition steps are now shown to give the idea of how they work: only parts of the resulting net are shown for the sake of space and readability. Of course in practice the composition should be automatized and the modeller should probably avoid to ever look at the composed model (at least in its whole, to save her eyes).

The $\mathcal{PSR}$ methodology prescribes to first compose the $\mathcal{S}$ and $\mathcal{R}$ levels obtaining the $\mathcal{SR}$ subnet to be composed next with $\mathcal{P}$. Here we only show the composition on label S_Connect_to_Provider and on labels S_GetDiskSpace, and E_GetDiskSpace (see Fig. 11). Observe that transition AcquireModem of level $\mathcal{R}$ composes with the two transitions ConnectS and ConnectR of level $\mathcal{S}$ labeled with S_Connect_to_Provider. In both cases, the imported variable $modem$ and the exported variable mod are unified (see Tab. 2), so that the parametric class $XCr1$ is instantiated to $MODEM$, moreover in the composed model the variable names appearing on the arcs are changed to reflect the fact that mod and $modem$ represent the same object (the arc from ConnectS to place OngoingAssignments in the composed model is now inscribed with function $< srv\ c\ mod >$, because import variable $modem$ has been renamed into mod, the corresponding export variable name). Similarly, when composing transition GetSpace in $\mathcal{R}$ with transition AssignMbx through label S_GetDiskSpace, the export variable d matches the import variable dsk, so that the parametric class $XCr2$ is instantiated to DSK, moreover the variable names appearing on the arcs incident to the new transition AssignMbx in the $\mathcal{SR}$ level are changed accordingly.

Observe now what happens when transitions EndGetSpace (level $\mathcal{R}$) and RsvDiskSpace (level $\mathcal{S}$) are composed. In this case there are two variables of class $XCr2$ in the colour domain of transition RsvDiskSpace: dsk is propagated (the corresponding object comes from propagation place RsvDiskSpace) while $dsk1$ is imported from level $\mathcal{R}$ for comparison purposes (i.e., it is not propagated to any output place after being imported): the transition guard $[dsk = dsk1]$ ensures that the end of operation on the $\mathcal{S}$ level part and on the $\mathcal{R}$ part match correctly (so that WRONG-MATCH problems are avoided). Again, after the composition, the variables involved in the import operation are renamed in the composed model to reflect the fact that d in $\mathcal{R}$ and $dsk1$ in $\mathcal{S}$ represent the same object.

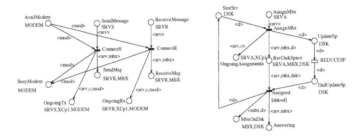

Fig. 11. $\mathcal{SR}$ composition (part of the composed model)

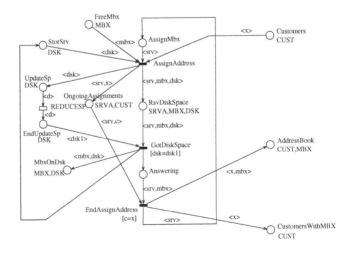

Fig. 12. PSR model (part of the composed model)

Finally to give a flavor of how the complete $\mathcal{PSR}$ model shall look like, let us consider the composition of the GetAddress transition of level $\mathcal{P}$ with the $\mathcal{SR}$ composed subnet (see Fig. 12). First of all GetAddress undergoes a vertical expansion transformation so that the level $\mathcal{P}$ transitions to be composed are two, say GetAddress1 with associated label S_GetAddress and GetAddress2 with associated label E_GetAddress. The first transitions exports the customer identifier x (and the corresponding type $CUST$) to the $\mathcal{SR}$ level, while the second, besides exporting again the customer identifier (for comparison purposes, to avoid WRONG-MATCH problems) imports the identity of the mailbox that has been assigned to the customer. Fig. 12 shows the result of the composition.

6   Conclusions

In this paper we have presented an extension of the SWN formalism to allow for color classes and variables to be of undefined (parametric) type. A compositional operator based on transition superposition has been defined that allows parametric color classes to be instantiated into basic color classes or structured color classes, using import and export lists.

An example has been discussed to show the adequacy of the proposed formalism and compositional rule to the case of layered modelling in the context of the $\mathcal{PSR}$ methodology. Indeed the proposed compositional rule is not as general as it could be, since it is intended to be used in layered models and it makes use, and it takes advantage, of the limitations imposed by the $\mathcal{PSR}$ definition; in particular, the composition is intended to act "in one shot", since in the $\mathcal{PSR}$ definition for GSPN, once two layers L1 and L2 have been composed, there is no way of composing an additional piece of L2. If L2 grows, we must redo the composition L1-L2 from scratch.

Future works will follow three lines: extension of PSWN, definition of the $\mathcal{PSR}$ methodology using PSWN, and implementation.

A straightforward extension of the PSWN presented in this paper is to allow the import function to specify a set of pairs, and not only a single one as it is now, so that more than one parametric class can be instantiated through a single label. The associated semantics is that all color classes and variables specified in the set of pairs are assigned the same imported class and/or variable. Observe that for import a set should be defined instead of a list, as for the export case, since the list of pairs in the export function identifies a *single* structured color class.

An important part of the definition of the $\mathcal{PSR}$ using MPSWN will consist in taking care of the compositional definition of the rate function **w** and of the priority function **pri**, for example along the lines of the work in [15].

A prototype implementation of MPSWN and its compositional rule into the GreatSPN package [8] is under development as part of the effort for the European project TIRAN mentioned in the introduction.

References

[1] M. Ajmone Marsan, G. Balbo, G. Conte, S. Donatelli, and G. Franceschinis. *Modelling with Generalized Stochastic Petri Nets*. J. Wiley, 1995.

[2] C. Anglano, S. Donatelli, and R. Gaeta. Parallel architectures with regular structure: a case study in modelling using SWN. In *Proc. 5th Intern. Workshop on Petri Nets and Performance Models*, Toulouse, France, October 1993. IEEE-CS Press.

[3] E. Battiston, O. Botti, E. Crivelli, and F. De Cindio. An incremental specification of a hydroelectric power plant control system using a class of modular algebraic nets. In *Proc. of the 16th international conference on Application and Theory of Petri nets 1995*, Torino, Italy, 1995. Springer Verlag. Volume LNCS935.

[4] E. Best, H. Flrishhacl ANF W. Fraczak, R. Hopkins, H. Klaudel, and E. Pelz. A class of composable high level Petri nets with an application to the semantics of B(PN)². In *Proc. of the 16th international conference on Application and Theory of Petri nets 1995*, Torino, Italy, 1995. Springer Verlag. Volume LNCS935.

[5] P. Buchholz. A hyerarchical view of GCSPN and its impact on qualitative and quantitative analysis. *Journal of Parallel and Distr. Computing*, (15), July 1992.

[6] G. Chiola, C. Dutheillet, G. Franceschinis, and S. Haddad. On Well-Formed coloured nets and their symbolic reachability graph. In *Proc. 11th Intern. Conference on Application and Theory of Petri Nets*, Paris, France, June 1990. Reprinted in *High-Level Petri Nets. Theory and Application*, K. Jensen and G. Rozenberg (editors), Springer Verlag, 1991.

[7] G. Chiola, C. Dutheillet, G. Franceschinis, and S. Haddad. A Symbolic Reachability Graph for Coloured Petri Nets. *Theoretical Computer Science B (Logic, semantics and theory of programming)*, 176(1&2):39–65, April 1997.

[8] G. Chiola, G. Franceschinis, R. Gaeta, and M. Ribaudo. GreatSPN 1.7: Graphical Editor and Analyzer for Timed and Stochastic Petri Nets. *Performance Evaluation, special issue on Performance Modeling Tools*, 24(1&2):47–68, November 1995.

[9] S. Christensen and L. Petrucci. Modular state space analysis of coloured Petri nets. In *Proc. of the 16th international conference on Application and Theory of Petri nets 1995*, Torino, Italy, 1995. Springer Verlag. Volume LNCS935.

[10] S. Donatelli and G. Franceschinis. The PSR methodology: integrating hardware and software models. In *Proc. of the 17th International Conference in Application and Theory of Petri Nets, ICATPN '96*, Osaka, Japan, june 1996. Springer Verlag. LNCS, Vol 1091.

[11] S. Haddad and P. Moreaux. Evaluation of high level Petri nets by means of aggregation and decomposition. In *Proc. of the 6th international workshop on Petri nets and performance models*, Durham, North Carolina, U.S.A, October 1995.

[12] K. Jensen. *Coloured Petri Nets, Basic Concepts, Analysis Methods and Practical Use. Volume 1*. Springer Verlag, 1992.

[13] K. Jensen. *Coloured Petri Nets, Basic Concepts, Analysis Methods and Practical Use. Volume 2*. Springer Verlag, 1995.

[14] Isabel C. Rojas M. *Compositional construction and Analysis of Petri net Systems*. PhD thesis, University of Edinburgh, 1997.

[15] E. Teruel, G. Franceschinis, and M. De Pierro. Clarifying the priority specification of gspn: Detached priorities. In *Proc. 8th Intern. Workshop on Petri Nets and Performance Models*, Zaragoza, Spain, September 1999. IEEE-CS Press.

Reducing k-Safe Petri Nets
to Pomset-Equivalent 1-Safe Petri Nets

Eike Best[1] and Harro Wimmel[2]

[1] Carl von Ossietzky Universität Oldenburg, D-26111 Oldenburg
[2] Universität Koblenz-Landau, D-56075 Koblenz

Abstract. It is a well-known fact that for every k-safe Petri net, i.e. a Petri net in which no place contains more than $k \in \mathbb{N}$ tokens under any reachable marking, there is a 1-safe Petri net with the same interleaving behaviour. Indeed these types of Petri nets generate regular languages. In this paper, we show that this equivalence of k-safe and 1-safe Petri nets holds also for their pomset languages, a true-concurrency semantics.

Keywords: Causality / partial order theory of concurrency, Petri nets, Pomsets.

1 Introduction

The well-known straightforward constructions of turning a k-safe Petri net into an 'equivalent' 1-safe net preserve its interleaving behaviour, but not its partial order behaviour [Bes84]. Indeed, it has been an open question whether or not the pomsets [Pra84] of a k-safe Petri net are also the pomsets of a 1-safe Petri net. The present paper answers the question affirmatively. It does so via coloured Petri nets [Jen92, Rei98, Smi98]. A k-safe Petri net is first translated into a coloured Petri net whose unfolding is 1-safe and back into an uncoloured net, and it is shown that all nets involved in this construction have the same pomsets.

This result is presented here in a self-contained fashion, but is related strongly to [PW97], [PW98], and [Wim00], where the pomset languages of general Petri nets are investigated.

2 Nets and Their Pomset Languages

Let Σ denote some fixed alphabet of actions. We frequently use a, b, ... as names for actions. We may or may not use a silent action $\tau \notin \Sigma$; it will not have any influence on the result of this paper.

For a relation R on some set A, $R \subseteq A \times A$, let $R^+ = R \cup R \circ R \cup R \circ R \circ R \ldots$ denote the transitive closure of R. We call R a partial order on A if it is irreflexive and

M. Nielsen, D. Simpson (Eds.): ICATPN 2000, LNCS 1825, pp. 63 82, 2000.

transitive. Given a partial order R on A, we call $b, c \in A$ *concurrent*, $b \, co_R \, c$, if neither bRc nor cRb. A set $C \subseteq A$ is called a *chain* if bRc or cRb for all $b, c \in C$, and it is a *line* if additionally all $a \in A \backslash C$ are concurrent with some $c \in C$; i.e., a line is a maximal chain. Further, we call $C \subseteq A$ an *antichain* if all $b, c \in C$ are concurrent and a *cut* if additionally for all $a \in A \backslash C$ there is some $c \in C$ with aRc or cRa; i.e., a cut is a maximal antichain. $C \subseteq A$ is a B-cut for some $B \subseteq A$ if C is a cut in A and $C \subseteq B$.

Any mapping $h \colon A \to B$ with finite sets A and B may be canonically extended to a mapping on multisets $h' \colon \mathbb{N}^A \to \mathbb{N}^B$ by defining

$$h'(f)(b) = \sum_{a \in h^{-1}(b)} f(a)$$

for any $f \colon A \to \mathbb{N}$ and $b \in B$. We will not distinguish between h and h'. If in doubt, a mapping should be interpreted as multiset mapping. Moreover, besides union, intersection and difference, the following operations are defined on multisets $X, Y \in \mathbb{N}^U$:

- add them: $(X+Y)(a) = X(a)+Y(a)$ for all $a \in U$,
- subtract them: $(X-Y)(a) = \max\{0, X(a)-Y(a)\}$,
- multiply them by a scalar $k \in \mathbb{N}$: $(k \cdot X)(a) = k \cdot X(a)$,
- or compare them: $(X \geq Y) \iff X(a) \geq Y(a)$ for all $a \in U$.

Definition 1. *Posets, causality structures, pomsets*

A *partially ordered set (poset)* is a pair (E, R) of a finite set E and a partial order $R \subseteq E \times E$. We call e_1 a *predecessor* of e_2 (and e_2 a *successor* of e_1) if $e_1 R e_2$ holds. Further, e_1 is a *direct predecessor* of e_2 (and e_2 is the *direct successor* of e_1) if additionally there is no e with $e_1 R e$ and $e R e_2$.

A *causality structure* κ *over* Σ is a triple $\kappa = (E_\kappa, R_\kappa, \ell_\kappa)$ consisting of a poset (E_κ, R_κ) and a labelling function $\ell_\kappa \colon E_\kappa \to \Sigma$. Two causality structures κ_1, κ_2 are *isomorphic* if there is a bijection $\beta \colon E_{\kappa_1} \to E_{\kappa_2}$ with $\beta \circ R_{\kappa_2} \circ \beta^{-1} = R_{\kappa_1}$ and $\beta \circ \ell_{\kappa_2} = \ell_{\kappa_1}$.[1] A *pomset (partially ordered multiset)* φ is an isomorphism class of causality structures. By $[\kappa]$ we denote the pomset containing the causality structure κ as a representative.   □ 1

Graphically, a pomset $[(E, R, \ell)]$ will be presented as a directed graph whose nodes are labelled by elements of the multiset $\ell(E)$ and whose directed arcs, $\longrightarrow$, represent the direct predecessor relation. For an example see figure 1.

We use pomsets as a generalisation of finite words, and thus we do not consider pomsets with an infinite number of events.

---

[1] We interpret $\circ$ 'relationally', i.e. $(x, y) \in R \circ Q \iff \exists z \colon (x, z) \in R \land (z, y) \in Q$.

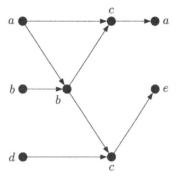

$\varphi = [(E, R, \ell)]$ with

$E = 1, 2, 3, 4, 5, 6, 7, 8$

$R = (1, 4), (2, 4), (1, 5), (4, 5)$
$(5, 7), (4, 6), (3, 6), (6, 8)^+,$

$\ell : 1, 7 \quad a; \ 2, 4 \quad b; \ 5, 6 \quad c;$
$\quad \ 3 \quad d; \ 8 \quad e.$

Fig. 1. Formal and graphical view of a pomset φ.

In this paper we will use (finite, labelled) coloured Petri nets [Jen92, Rei98, Smi98], which will simply be called nets below. Our definitions are parametrised with respect to arbitrary but fixed, mutually disjoint, sets Σ (of actions), VAL (of values) and MOD (of modes). The first set is the target of transition labellings, and values and modes play the role of place colours and transition colours, respectively.

Definition 2. *Petri net*

An (unmarked, unlabelled, finite) *net* is a triple (S, T, ι) where S and T are finite sets with $S \cap T = \emptyset$ and ι is a function

$$\iota : S \cup (S \times T) \cup (T \times S) \cup T \to 2^{VAL} \cup \mathbb{N}^{VAL \times MOD} \cup \mathbb{N}^{MOD \times VAL} \cup 2^{MOD}$$

such that:

$\forall s \in S : \iota(s) \subseteq VAL$ and $\forall t \in T : \iota(t) \subseteq MOD$.
$\forall (x, y) \in (S \times T) \cup (T \times S) : \iota((x, y)) \in \mathbb{N}^{\iota(x) \times \iota(y)}$.

A net is called *simple* if all $\iota((x, y))$ are subsets of $\iota(x) \times \iota(y)$ (rather than true multisets). A (transition) *labelling* is a function $\lambda : T \to \Sigma \cup \{\tau\}$. A *marking* is an element $M \in \mathbb{N}^{S \times VAL}$ such that $\forall s \in S, v \in VAL : M(s, v) > 0 \Rightarrow v \in \iota(s)$. Sometimes a net is *marked* with an initial marking M_0. Sometimes, also, a finite set of final markings $\mathcal{M}_f \subseteq \mathbb{N}^{S \times VAL}$ is provided. As the need may be, we enlarge the definition by such parameters; thus, (S, T, ι, λ) is a labelled unmarked net, while $(S, T, \iota, \lambda, M_0, \mathcal{M}_f)$ is a marked labelled net with an additional set of final markings (and for the purposes of this paper, we call such an object a *coloured net*). □ 2

Definition 2 contains the class of (arc-weighted) *place/transition nets* (*PT-nets*, for short) as a special case: consider $\bullet \in VAL$ and $\blacksquare \in MOD$ and $\iota(s) = \{\bullet\}$ for every

$s \in S$ and $\iota(t) = \{\blacksquare\}$ for every $t \in T$. By the fact that ι on arcs may be a multiset (rather than just a set), arbitrary non-negative integer arc weights (including 0, for no arc) can be modelled. We denote the class of finite, labelled, not necessarily simple, PT-nets with an initial and a finite set of final markings by $\mathcal{N}$.

The transition rule for nets is defined as follows. A marking M *activates* a transition t in mode $m \in \iota(t)$ (in symbols: $M[(t, m)\rangle$) iff $M \geq \iota(., t)(., m)$.[2]Moreover, M leads to a marking M' through the *occurrence* of t in mode m (in symbols: $M[(t, m)\rangle M')$ iff $M[(t, m)\rangle$ and $M' = M - \iota(., t)(., m) + \iota(t, .)(m, .)$. A marking M is *reachable* from M_0 iff there exists a series of occurrences from M_0 to M. Such sequences are often called the *interleaving behaviour*. A marked net (S, T, ι, M_0) is k-*safe* (for $k > 0$) iff in every marking M reachable from M_0 all places $s \in S$ contain at most k same tokens (i.e., M is a multiset with multiplicities at most k). Let $\mathcal{N}_k \subseteq \mathcal{N}$ denote the class of all k-safe PT-nets in $\mathcal{N}$.

We assume all our nets to be *T-restricted*, that is, every transition has at least one preplace and at least one postplace, formally: $\forall t \in T : \iota(., t) \neq \emptyset \neq \iota(t, .)$.[3]

Our formalisation of the pomset behaviour of a net is very similar to that of Pomello et al. [PRS92]. Pomsets are seen as an abstraction of processes, which in turn are based on occurrence nets. A process reflects an actual simulation of the token game of a net in a 'truly concurrent' way.

Definition 3. *Occurrence net*

A (finite) *occurrence net*, *O-net* for short, is a simple net $O = (B, E, F)$ satisfying the following conditions:

(1) For all $x \in B \cup E$, the set $F(x)$ is a singleton.
(2) The relation F (which, by simpleness and property (1), may as well be viewed as a relation on $(B \times E) \cup (E \times B)$) is acyclic (i.e. F^+ is irreflexive).
(3) $\forall b \in B : |F(., b)| \leq 1 \geq |F(b, .)|$.

Elements of B and E are called *conditions* and *events*, respectively. Further, let $\min O = \{ b \in B \mid F(., b) = \emptyset \}$ and $\max O = \{ b \in B \mid F(b, .) = \emptyset \}$.   □ 3

The set of B-cuts of an O-net (that is, the B-cuts of the partial order $(B \cup E, F^+)$) can itself be partially ordered (yielding, in fact, a finite lattice) by stipulating that $C_1 \sqsubseteq C_2$ iff $\forall b_1 \in C_1, b_2 \in C_2 : (b_1, b_2) \in F^+ \vee (b_1, b_2) \in co_{F^+}$.

---

[2] We use place-holder (dot) notation (.) to denote derived multisets. An expression with k place-holders denotes a multiset of k-tuples of the type determined by the respective parts of the expression. Thus, $X(.)$ denotes the same multiset as X, whereas $\iota(., t)(., m)$ denotes a multiset of pairs (s, v), with $s \in S$ and $v \in VAL$, where each pair (s, v) has multiplicity $(\iota(s, t))(v, m)$.

[3] This requirement greatly simplifies the definition of concurrent behaviour, and it is moreover not a severe restriction in the case of k-safe nets: transitions without any preplaces and postplaces are uninteresting; transitions without preplaces but with postplaces will not occur since the net is k-safe; and transitions with preplaces but without postplaces can be made T-restricted without changing concurrent behaviour, by adding a postplace artificially.

Definition 4. *Process*

A *process* π of a net $N = (S, T, \iota, M_0, \mathcal{M}_f)$ is a tuple (B, E, F, r), where $r : (B \cup E) \rightarrow (S \cup T)$ and:

- $O = (B, E, F)$ is an O-net.
- $r(B) \subseteq S$ and $r(E) \subseteq T$, i.e. r 'folds' conditions to places and events to transitions.
- r satisfies an initiality condition: $r(\min O) = M_0$; a 'progress' condition:

$$\forall e \in E : r(F(., e)) = \iota(., r(e)) \wedge r(F(e, .)) = \iota(r(e), .),$$

and a finality condition: $r(\max O) \in \mathcal{M}_f$. □4

The *length* of a process π is given as $|\pi| = |E|$.[4] Let $Proc(N)$ denote the class of all processes of N. We call two processes *isomorphic* if one of them results from the other by renaming conditions and events.

The upper part of figure 2 shows an example. We have given a PT-net (shown here in full as a coloured net) as an example, since the major result of this paper holds for such nets. The function r is shown as inscription inside conditions and events of π. The sets below the places/conditions and transitions/events x are the $\iota(x)$, and the sets next to arcs (x, y) are the $\iota((x, y))$.

It is a well-known fact (see e.g. [GV83, Jen92]) that a net $N = (S, T, \iota, M_0)$ is k-safe iff for all B-cuts C of processes $\pi = (B, E, F, r)$ of N and all places $s \in S$ and values $v \in \iota(s)$ the following holds: $|\{b \in B \mid r(b) = s \wedge F(b) = \{v\}\} \cap C| \leq k$. This means that any value may 'occur' in any B-cut at most k times. For PT-nets, this specialises to the fact that N is k-safe iff for all π and C and for all places s: $|r^{-1}(s) \cap C| \leq k$.

Possible observable behaviour is specified by an additional labelling λ. As a pomset is supposed to describe such a behaviour, we consider only the observable events and their relations in a process.

Definition 5. *Pomset of a process*

Let $N = (S, T, \iota, \lambda, M_0, \mathcal{M}_f)$ and let $\pi = (B, E, F, r)$ be a process of N. The pomset $\Phi(\pi)$ is defined as $[(E_\varphi, R_\varphi, \ell_\varphi)]$ with $E_\varphi = \{e \in E \mid \lambda(r(e)) \neq \tau\}$, $R_\varphi = F^+|_{E_\varphi \times E_\varphi}$, and $\ell_\varphi = r|_{E_\varphi} \circ \lambda$. □5

Thus, the pomset $\Phi(\pi)$ is an abstraction from the process π. We allow for autoconcurrency here, i.e., multiple instances of the same action may occur in a pomset $\Phi(\pi)$ without any causal ordering between them.

---

[4] Recall that we consider only finite O-nets.

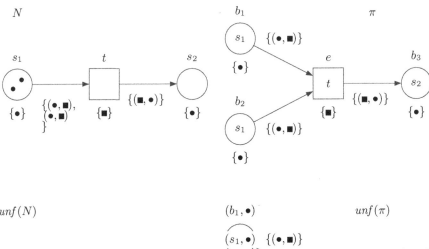

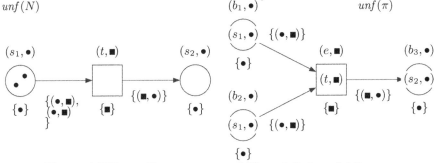

Fig. 2. A PT-net N, a process π of N, and their unfoldings.

Definition 6. *Pomset language*

We define the set of all pomsets, or *pomset language*, of a given net N by $\mathcal{P}(N) = \{\Phi(\pi) \mid \pi$ is a process of $N\}$. Moreover, $\mathcal{P}_k = \{\mathcal{P}(N) \mid N \in \mathcal{N}_k\}$ denotes the class of pomset languages of k-safe PT-nets for $k > 0$. □ 6

Some examples of PT-nets with typical pomsets from their respective languages can be seen in figure 3. Note that while N_1 is 2-safe and N_3 and N_4 are 8-safe, the net N_2 is not k-safe for any $k > 0$. The net N_1 is the same as the net N of figure 2, written in more traditional PT-net notation.

3   Balanced Coloured Nets

In this section, we will consider specific values VAL and, related to them, specific modes MOD. Let $\mathcal{F}$ be a finite set. At the present time, $\mathcal{F}$ is arbitrary, but we will later specialise it in order to prove our main result. Put $VAL = 2^{\mathcal{F}}$, i.e., every value is a subset of $\mathcal{F}$. Let $N = (S, T, \iota)$ be a simple net such that $\iota(s) = VAL = 2^{\mathcal{F}}$ for every place s in S. Let t be a transition in T and let m be a mode in $\iota(t)$. We will call m *balanced* w.r.t. t iff the following hold:

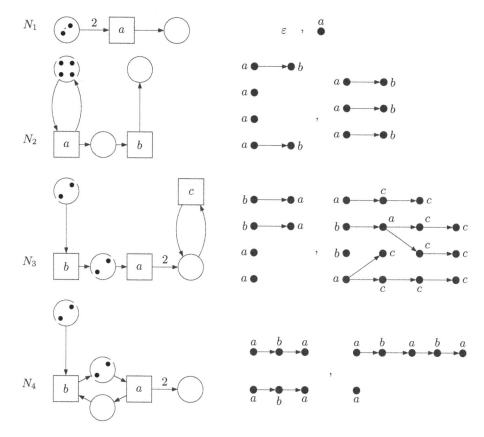

Fig. 3. Some PT-nets with some pomsets (without considering final markings).

(A) $\forall s\in\iota(.,t)\,\exists v\in\iota(s)\colon \iota(s,t)(v,m)>0$ and $\forall s\in\iota(t,.)\,\exists v\in\iota(s)\colon \iota(t,s)(m,v)>0$.

(B) $\bigcup_{s\in\iota(.,t),(s,v)\in\iota(.,t)(.,m)} v$ and $\bigcup_{s\in\iota(t,.),(s,v)\in\iota(t,.)(m,.)} v$ are sets.

(C) $\bigcup_{s\in\iota(.,t),(s,v)\in\iota(.,t)(.,m)} v = \bigcup_{s\in\iota(t,.),(s,v)\in\iota(t,.)(m,.)} v.$

Condition (A) means that m 'covers' all input and output places of t in the sense that it removes a value from every input place and puts a value on every output place. Condition (B) means that any two values m takes from input places of t must be disjoint,[5] and similarly for two values m puts on output places of t must be disjoint. Condition (C) means that m 'redistributes' the values it gets from input places of t onto the output places of t, in the sense that the union of the former is split into the latter; hence the term 'balanced'. Intuitively, elements of values cannot be created or lost, although the actual values may actually change.

---

[5] Remember that values are sets.

The balancedness conditions tell us how the elements of values are distributed through the events of a process. Such elements cannot get lost or created on the way; each element, as it turns out, must follow some line through the process.

A marking $M \in \mathbb{N}^{S \times 2^{\mathcal{F}}}$ of N is called *spread* if all its tokens are mutually disjoint, that is, if $\sum_{s \in S} \sum_{(s\,v) \in M} v$ is a set. Clearly, a spread marking is 1-safe. What is more:

Lemma 1.

Let $N = (S, T, \iota, M_0)$ be a simple net in which all modes are balanced and let M_0 be spread. Then N is 1-safe.

Proof: By induction. M_0 is clearly spread and 1-safe. Suppose $M[(t, m)\rangle M'$ such that M is spread. Since m is balanced, it can redistribute only already existing values by (B) and (C). Thus it cannot create any set that has an element in common with other (created or untouched) sets, whence M' is also spread and thus 1-safe.   ∎ 1

Definition 7. *Net unfolding*

Let $N = (S, T, \iota, \lambda, M_0, \mathcal{M}_f)$ be a coloured net. Its *unfolding*, $unf(N) = (S', T', \iota', \lambda', M_0', \mathcal{M}_f')$, is defined as follows: its places S' are all the tuples (s, v) with $s \in S$ and $v \in \iota(s)$, and $\iota'((s, v)) = \{v\}$; its transitions T' are all the tuples (t, m) with $t \in T$ and $m \in \iota(t)$, and $\iota'((t, m)) = \{m\}$; its arcs are $\iota'(((s, v), (t, m))) = \iota(s, t)(v, m) \cdot \{(v, m)\}$ and $\iota'(((t, m), (s, v))) = \iota(t, s)(m, v) \cdot \{(m, v)\}$ for any $(s, v) \in S'$ and $(t, m) \in T'$; its labelling is $\lambda'((t, m)) = \lambda(t)$; and the marking M_0' are all the tokens $((s, v), v)$ with (s, v) being a token of M_0. The final markings of $\mathcal{M}_f'$ are constructed analogously from the final markings of $\mathcal{M}_f$.   □ 7

The lower part of figure 2 shows an example. The operation of unfolding just serves to 'flatten out' the net (introducing new names for places and transitions), but it leaves its structure, i.e. the connectivity of pairs (s, v) and (t, m), intact. More precisely, we have the following lemma which states that N and $unf(N)$ have exactly the same (concurrent) behaviour. First, we define the unfolding of a process $\pi = (B, E, F, r)$ of N as $unf(\pi) = (B', E', F', r)$, where $(B', E', F') = unf((B, E, F))$ and $r'((b, \bullet)) = (r(b), \bullet)$ and $r'((e, \blacksquare)) = (r(e), \blacksquare)$.

Lemma 2.

π is a process of the coloured net N if and only if $unf(\pi)$ is a process of $unf(N)$, and what is more, π and $unf(\pi)$ are isomorphic.

Proof: The proof follows standard theory [GV83, Jen92, Rei98, Smi98]. First, we note that if all value sets $\iota(s)$ and all mode sets $\iota(t)$ of N are singletons, then

$unf(N)$ is isomorphic to N (i.e., the same as N, up to renaming of places and transitions). This follows directly from the definition of unfolding, and it has as a consequence that the unfolding of any PT-net is isomorphic to that PT-net, as well as that the unfolding of any O-net or process is isomorphic to the original O-net, resp. process. (It also has the consequence that $unf(N)$ is isomorphic to $unf(unf(N))$.)

Now let π be a process of N. Since the definition of r depends only on (the initial and final markings and) the connectivity of tuples (s, v) with tuples (t, m), and because this connectivity is not changed by unfolding, it is not hard to check that $unf(\pi)$ is indeed a process of $unf(N)$, and vice versa.   ■ 2

We will need the following two corollaries:

Corollary 1.

For any coloured net N, $\mathcal{P}(N) = \mathcal{P}(unf(N))$.   ■ 1

Corollary 2.

For any coloured net N, N is k-safe iff $unf(N)$ is k-safe. In particular, for a simple N with only balanced modes and a spread initial marking M_0, $unf(N)$ is 1-safe.

Proof: The second claim follows with lemma 1.   ■ 2

There is also a reverse relationship between coloured nets N whose value and mode sets are singletons (such as unfoldings) and PT-nets; let us call nets whose value and mode sets are singletons *basic*. Since all types and mode sets of basic nets are singletons, their exact identity does not matter as long as we keep the place and transition sets intact. Thus, we may translate N into a PT-net, $PT(N)$, with the same behaviour, by replacing all value sets by the standard singleton set $\{\bullet\}$, all mode sets by $\{\blacksquare\}$, and modifying arc multisets accordingly (but keeping places, transitions and arcs unchanged otherwise). We may also apply this operation PT to non-basic nets, but there will generally be a change in the behaviour: the resulting net behaves as if every mode was allowed for every transition.

Lemma 3.

Let N be basic. Then π is a process of N if and only if $PT(\pi)$ is a process of $PT(N)$, and they are isomorphic.

Proof: Similar to the one above.   ■ 3

Again there are two corollaries:

Corollary 3.

Let N be basic. Then $\mathcal{P}(N) = \mathcal{P}(PT(N))$.   ■ 3

Corollary 4.

Let N be basic. If N is 1-safe, then so is $PT(N)$.   ■ 4

These considerations will allow us to convert a k-safe PT-net into a simple, balanced coloured net with a spread marking, and apply the *unf* and *PT* operations to the latter to get a 1-safe net with the same pomsets.

Let S be any set of places and let M be any function in $\mathbb{N}^{S \times 2^{\mathcal{F}}}$. We define from M the function *discolour*(M) in $\mathbb{N}^{S \times \{\bullet\}}$ by

$$\forall s \in S: \; discolour(M)(s, \bullet) = \sum_{v \in 2^{\mathcal{F}}} M(s, v).$$

That is, the marking *discolour*(M) has as many tokens on each place as there have been values in M on that place. In the next definition, c is to be understood informally as colouring the initial marking M_0 of an uncoloured net and discolouring (through c^{-1}) the set of final markings of the resulting coloured net.

Definition 8. *Colouring a net*

Let $N = (S, T, \iota, \lambda, M_0, \mathcal{M}_f)$ be a PT-net. Any relation $c \subseteq (\{M_0\} \cup \mathcal{M}_f) \times (\mathbb{N}^{S \times 2^{\mathcal{F}}})$ is valid *colouring* of N if

(a) $c(M_0) = \{M_0'\}$ such that M_0' is spread and $M_0 = discolour(M_0')$.
(b) $\forall M \in \mathcal{M}_f : c(M) = \{M' \in \mathbb{N}^{S \times 2^{\mathcal{F}}} \mid M' \text{ is spread and } discolour(M') = M\}$.

For a valid colouring c, the coloured net $colour(N, c) = (S, T, \iota', \lambda, M_0', c(\mathcal{M}_f))$, where $c(M_0) = \{M_0'\}$, is defined as a coloured version of N, if for every $s \in S$, $\iota'(s) = 2^{\mathcal{F}}$; for every $t \in T$, the set of $\iota'(t)$ consists of *all* modes that are balanced w.r.t. t; and the arcs satisfy $\iota'((x, y)) \neq \emptyset \iff \iota((x, y)) \neq \emptyset$ (and are otherwise defined uniquely by the balanced modes).   □ 8

Thus, the coloured version of a net contains the maximal set of balanced modes for every transition. It is easy to see that these mode sets are still finite, since every mode can be determined by the transition it belongs to and by the way it combines and distributes the pre-values and the post-values (and all sets involved are finite, due to the finiteness of $\mathcal{F}$). Note also that condition (b) ensures that c is uniquely determined on $\mathcal{M}_f$, and thus in order to define a valid colouring, it is only necessary to specify M_0' with $c(M_0) = \{M_0'\}$ and verify (a).

Lemma 4.

For all PT-nets $N \in \mathcal{N}$ and valid colourings c, $\mathcal{P}(colour(N, c)) \subseteq \mathcal{P}(N)$.

Proof: Take any pomset $\varphi \in \mathcal{P}(colour(N, c))$. There is a process $\pi = (O, r)$ of $colour(N, c)$ with $\Phi(\pi) = \varphi$. Now consider $PT(colour(N, c))$. Clearly, this is the original PT-net N and $PT(\pi)$ is a process of N. π is easily seen to be a coloured version of $PT(\pi)$, and as a colouring does not influence the way pomsets are constructed from processes, both processes give rise to the same pomset. Overall, $\mathcal{P}(colour(N, c)) \subseteq \mathcal{P}(N)$.   ■ 4

The reverse inclusion also holds with a carefully chosen c, provided that N is k-safe, but it is not done as easily. Take a situation in the coloured net $colour(N, c)$ where a transition t with, say, two places s_1 and s_2 in its preset and three places in its postset is supposed to occur, and assume there are tokens $\{1\}$ in s_1 and $\{2\}$ in s_2. Clearly, t cannot occur, as it cannot redistribute two singleton sets into three mutually disjoint nonempty sets (remember that $colour(N, c)$ is balanced). In the 'uncoloured' net N though, there are just two 'black tokens' • in the preset, and the transition t can occur.

At this point, our undetermined set $\mathcal{F}$ from the beginning of this section comes to the rescue. We will choose $\mathcal{F}$ large enough (but still, importantly, finite) so that it is always possible to redistribute the tokens in a way such that the transition t we have in mind can eventually occur (in the coloured net, of course).

4   A Pumping Lemma for Processes

For a single process π we can easily determine the set $\mathcal{F}$ needed to enable all the necessary transitions at the correct moments: Choose some possible linear order in which the transitions can occur to generate this process π and proceed through the transitions. If a transition can occur, just let it occur and redistribute the token sets arbitrarily. If there are not sufficiently many elements of $\mathcal{F}$ in order for the transition to occur, just add new elements to tokens of the initial marking so that they can be sent through the process so far and get to lie in the preset of our hitherto unfireable transition. Repeat if necessary, until there are enough elements in all presets. When this is done all through the sequence, we get a colouring for the initial marking under which the process π can be simulated. It is clearly visible in this algorithm that the (new) elements follow some lines through our process. These lines become important now.

Definition 9. *Line system*

Let $\pi = (B, E, F, r)$ be a process of a N. A *line system* L for π is a minimal set of lines of F^+ that covers all conditions $b \in B$ of π. Let $\mathcal{L}(\pi)$ denote the set of all line systems for π.   □ 9

It is known [GR83] that every line and every cut of π have exactly one element in common. The basic idea is to get a line system L for π, generate one element of $\mathcal{F}$ for each line in L, and put it onto the condition on the beginning of its line. Since more than one line may lie on a condition, this may yield non-singleton sets, but since every condition lies on some line, we always get nonempty sets. The minimal conditions exactly represent the initial marking, and thus each token of the initial marking gets the elements we have put onto the corresponding condition, and since each line can lie on exactly one condition, this marking will be spread. When a transition occurs, each element that is processed through this transition just follows 'its' line, and thus the balancedness conditions are satisfied as well.

The algorithm remains easy still if we have a finite number of processes. We just apply it to each process and ensure that no element appears for two different processes. A simple union then guarantees that all processes can be simulated.[6]

A problem arises clearly when we consider an infinite number of processes. In fact, we have two problems: first, will the number of elements that we must assign to a single (finite, but arbitrarily large) process be bounded by some number, and second, is there a way we can use an assignment of elements for a whole (arbitrarily large) class of processes, and will the number of classes be finite? If both questions can be answered with yes, we can indeed find the set $\mathcal{F}$ and a colouring for the initial marking that we are looking for.

The answer to the first question is simply that a process, no matter how long, of a k-safe PT-net $N = (S, T, \iota, \lambda, M_0, \mathcal{M}_f)$ has a line system with at most $k{\cdot}|S|$ lines. To prove this, we use a theorem from graph theory, compare [EHK96, Tve67].

Theorem 1. DILWORTH 1950

In a partial order, the maximal cardinality of a cut is equal to the minimal number of chains into which the partial order can be partitioned.   ∎ 1

Lemma 5. *Maximal Number of Lines*

Let $\pi = (B, E, F, r)$ be a process of a k-safe PT-net $N = (S, T, \iota, \lambda, M_0, \mathcal{M}_f) \in \mathcal{N}$. Then, there is a line system L for π with $|L| \leq k{\cdot}|S|$.

Proof: N is k-safe, so we know that no marking can have more than $k{\cdot}|S|$ tokens and, accordingly, no B-cut can contain more than $k{\cdot}|S|$ elements. By Dilworth's theorem then, we know that the minimal number of chains into which the partial order $(B, F^+|_{B \times B})$ can be partitioned into is $k{\cdot}|S|$ or less. We enlarge each of these chains to a line arbitrarily and get a set of lines covering all conditions

---

[6] We do not need to consider isomorphic processes here, as it is not necessary to handle more than one representative from each isomorphism class in this way.

in B. Clearly, this set of lines contains a line system for π with at most $k \cdot |S|$ lines.   ∎ 5

In fact, the set of lines constructed in this proof is already a line system, i.e. no line may be removed. For suppose we could remove one line. This means that all conditions b in this line are also element in some other line. We remove this redundant line. We convert our set of lines back to a set of chains by removing conditions appearing in more than one line in all but one of them. Clearly, we get a set of chains partitioning $(B, F^+|_{B \times B})$ with one chain less than the minimal number, which is a contradiction.

Dilworth's theorem thus provides us with an upper bound on the number of lines covering an arbitrary process, such that this bound is independent of the size of that process, and thus with the answer 'yes' to the first question above.

For the next step we introduce the notion of equivalence of cuts in a process, relative to a given line system of that process. The rationale behind this definition is to provide a means by which a process can be 'cut' into pieces, such that the relevant pieces are bounded in their size.

Definition 10. *Equivalence of cuts*

Let $N = (S, T, \iota, \lambda, M_0, \mathcal{M}_f)$ be a PT-net, $\pi = (O, r)$ a process of N and $L \in \mathcal{L}(\pi)$ a line system for π. Let C_1 and C_2 be B-cuts of π. We call the cuts C_1 and C_2 equivalent with respect to L, $C_1 \equiv_L C_2$ iff the relation $\rho \subseteq C_1 \times C_2$, defined by

$$(b_1, b_2) \in \rho \iff \exists l \in L : l \cap C_1 = \{b_1\} \wedge l \cap C_2 = \{b_2\}$$

is an r-preserving bijection.   □ 10

It is easy to check that $\equiv_L$ is indeed an equivalence relation. Clearly, if $\min \pi \equiv_L \max \pi$ for some process π with line system L, then π can be concatenated (denoted by $\cdot$) with itself (by identifying conditions via $\equiv_L$) to yield processes $\pi \cdot \pi$, $\pi \cdot \pi \cdot \pi$ etc., which are larger than π (unless π has no events, in which case $\pi = \pi \cdot \pi$). Such a π is called *repeatable*. Also, if some process π with line system L contains two B-cuts C_1 and C_2 with $C_1 \sqsubseteq C_2$ and $C_1 \equiv_L C_2$, then π can be split into an initial part π_1 (up to and including C_1), a middle part π_2 (from C_1 up to C_2, inclusively) and a final part π_3 (from C_2, inclusively, onwards), such that π_1 and π_3 can be concatenated to yield a process $\pi_1 \cdot \pi_3$ which is smaller than π (unless C_1 and C_2 are the same cut, in which case $\pi = \pi_1 \cdot \pi_3$).

The following lemma shows that there is a bound on the number of non-equivalent cuts in a process, which does not depend on the size of the process.

Lemma 6. *Number of non-equivalent cuts*

Let $N = (S, T, \iota, \lambda, M_0, \mathcal{M}_f)$ be a k-safe PT-net, $\pi = (B, E, F, r)$ a process of N, and $L \in \mathcal{L}(\pi)$ a line system for π with $|L| \le k\cdot|S|$. Any set of cuts of π that are pairwise non-equivalent with respect to L contains no more than $(k\cdot|S|^2)^{k\cdot|S|} \cdot k \cdot |S|$ elements.

Proof: Let $\mu(|C|, |L|)$ denote the maximal number of different ways in which the lines of the line system L can cross the fixed cut C.[7] We have the following recurrence for $\mu(|C|, |L|)$:

$$|C| = 1 : \mu(|C|, |L|) \qquad = 1$$
$$|C| > 1 : \mu(|C|, |L|+1) = |C| \cdot \mu(|C|, |L|).$$

The first line comes from the fact that all lines must touch the single element of the cut. The second line comes from the fact that the $|L|+1$'st line has $|C|$ possibilities, and that each time all other lines have all possibilities. Evidently, $\mu(|C|, |L|) = |C|^{|L|}$ solves the recurrence.

This implies that any set of non-equivalent B-cuts, each of cardinality m (where m varies between 1 and $k\cdot|S|$), can contain at most $|S|^m\cdot m^{k\cdot|S|}$ elements, where the first factor bounds the number of possible labellings of conditions in such a cut, and the second number bounds the number of different ways the line system can cross the cut (since $|L|$ is bounded by $k\cdot|S|$). Since two cuts of different cardinality can never be equivalent, it follows that the number of possible different cuts is bounded by the sum

$$|S|^1\cdot 1^{k\cdot|S|} + |S|^2\cdot 2^{k\cdot|S|} + \ldots + |S|^{k\cdot|S|}\cdot(k\cdot|S|)^{k\cdot|S|}.$$

Bounding each summand by the last one yields

$$(|S|^{k\cdot|S|}\cdot(k\cdot|S|)^{k\cdot|S|}) \cdot k\cdot|S| = (k\cdot|S|^2)^{k\cdot|S|} \cdot k \cdot |S|$$

as an upper bound for the overall number of non-equivalent cuts, as claimed.

∎ 6

We may now formulate our pumping lemma.

Lemma 7. *Pumping Lemma*

Let $k>0$ and let $N \in \mathcal{N}_k$. Then there exists a number $n_0 \in \mathbb{N}$ such that for all processes $\pi \in Proc(N)$ with $|\pi|>n_0$ there are a line system $L \in \mathcal{L}(\pi)$ and processes x, y, z such that:

---

[7] We may assume $|C| \ge 1$ (and thus $|S| \ge 1$), which is a consequence of T-restrictedness.

- $\pi = x \cdot y \cdot z$.
- $x \cdot z$ is defined and $|x \cdot z| \leq n_0$.
- y is repeatable and $\forall n \in \mathbb{N}: x \cdot y^n \cdot z \in Proc(N)$.

Proof: For $N = (S, T, \iota, \lambda, M_0, \mathcal{M}_f)$ define $n_0 = (k \cdot |S|^2)^{k \cdot |S|} \cdot k \cdot |S|$. Let now $\pi = (B, E, F, r)$ be a process of N with $|\pi| > n_0$ and let R be any total ordering of E with $F^+|_{E \times E} \subseteq R$, i.e. a linearisation of the events of π. We denote this ordering by calling the events $e_1, \ldots, e_m$. We construct then a sequence of B-cuts $C_0, C_1, \ldots, C_m$ with $C_0 = \min \pi$ and $C_{i+1} = (C_i \backslash F(., e_{i+1})) \cup F(e_{i+1}, .)$ for $1 \leq i < m$. Clearly, $C_i \sqsubseteq C_{i+1}$ for $0 \leq i < m$. By lemma 5 there is a line system $L \in \mathcal{L}(\pi)$ with at most $k \cdot |S|$ lines, and by lemma 6 there are at most n_0 different equivalent cuts in π. Therefore, there must be two cuts C_i and C_j with $i < j$, $C_i \sqsubseteq C_j$ and $C_i \equiv_L C_j$. Choose i and j such that $j - i$ becomes maximal with this property, and then split π at the cuts C_i and C_j obtaining x, y, and z with $\pi = x \cdot y \cdot z$. The equivalence of C_i and C_j guarantees that $x \cdot y^n \cdot z$ is well defined (also for $n=0$). Due to the maximality of $j - i$ there are no equivalent cuts in the sequence $C_0, \ldots, C_i, C_{j+1}, \ldots, C_m$ of cuts remaining from the original sequence in $x \cdot z$, thus $|x \cdot z| \leq n_0$ by lemma 6 (and the fact that a process contains at least as many cuts as events). ■ 7

Our pumping lemma guarantees that each process can be 'cut down' to a process (of the form $x \cdot z$) below a certain length in just one step, i.e. the pumping lemma will need to be applied only once. Note that it is a true generalisation of the ordinary pumping lemma for regular languages, as it applies, in particular, to a 1-safe net each of whose transitions has exactly one pre-place and one post-place.

5 The Equivalence of k-Safe and 1-Safe Nets

Using the pumping lemma for processes we may now show that each k-safe PT-net can be transformed into a simple coloured net with a spread initial marking and balanced modes (and hence, a 1-safe PT-net) with the same pomset language.

Lemma 8.

For all k-safe PT-nets N, there exists a colouring c such that the net $colour(N, c)$ is well defined and, moreover, $\mathcal{P}(N) \subseteq \mathcal{P}(colour(N, c))$.

Proof: In the first part of the proof, we show how such a coloured net can be constructed. We need only to specify the colouring function c.

Let $N = (S, T, \iota, \lambda, M_0, \mathcal{M}_f)$ be a k-safe PT-net, $n_0 \in \mathbb{N}$ the constant of the pumping lemma for N and $\Pi = \{\pi_1, \ldots, \pi_m\} \subseteq \{\pi \in Proc(N) \mid |\pi| \leq n_0\}$ a finite set of non-pumpable processes of N containing one representative from each isomorphism class of processes of $\{\pi\in Proc(N) \mid |\pi|\leq n_0\}$. We assume that the conditions of these processes are renamed such that $\min\pi_1 = \min\pi_2 = \ldots = \min\pi_m = B_{\min}$ holds; let r be their common labelling function on the set $B_{\min}$.

For $1\leq i\leq m$ and $1\leq j\leq k\cdot|S|$ let $\Psi_{i\,j}$ be pairwise disjoint sets with $|\Psi_{i\,j}| = k\cdot|S|$. Let

$$\mathcal{F} = \bigcup_{1\leq i\leq m\ 1\leq j\leq k\cdot|S|} \Psi_{i\,j}.$$

For each $\pi_i\in\Pi$ determine a line system $L_{\,i}\in\mathcal{L}(\pi_i)$ with $|L_{\,i}| \leq k\cdot|S|$ (as in lemma 5), and enumerate the lines by a bijection $\alpha_{\,i}\colon L_{\,i} \to \{1,\ldots,|L_{\,i}|\}$.

Now define

$$\beta\colon B_{\min} \to 2^{\mathcal{F}} \qquad \text{by}\quad \beta(b) = \bigcup_{1\leq i\leq m}\bigcup_{\ell\in L_{\,i},\, b\in\ell} \Psi_{i\ \alpha_{\,i}(\ell)}$$
$$c \subseteq (\{M_0\}\cup\mathcal{M}_f) \times (\mathbb{N}^{S\times 2^{\mathcal{F}}}) \text{ by } c(M_0) = \{M_0'\}$$
$$\text{and } M_0'(s,v) = |\{b \in B_{\min} \mid r(b)=s \wedge \beta(b)=v\}|.$$

The marking M_0' is spread as a consequence of the requirement that the sets $\Psi_{i\,j}$ are mutually disjoint, and $M_0 = discolour(M_0')$ by the initiality condition of definition 4. Recall (from the remark after definition 8) that is only necessary to specify $c(M_0)$ and verify 8(a). Hence $colour(N,c)$ is well defined.

The mapping β defines a front end for mode distributions, by which we can obtain whichever process we like. In this way, we show in the second part of the proof that $\mathcal{P}(N) \subseteq \mathcal{P}(colour(N,c))$ with the above c.

Let $\varphi\in\mathcal{P}(N)$ a pomset, and $\pi=(O_{\,},r_{\,})\in Proc(N)$ a process of N with $\Phi(\pi)=\varphi$.

Case 1: $|\pi| \leq n_0$. Clearly, there is a process $\pi_i = (B,E,F,r) \in \Pi$ with $\Phi(\pi_i) = \Phi(\pi)$. We will construct a labelling of π_i such that the resulting object is a process of $colour(N,c)$. To this end, we extend the mapping β, which associates a value (i.e., a subset of $\mathcal{F}$) with every condition in $B_{\min}$ to the set of all conditions of π_i. Initially, every condition in $B_{\min}$ is provided with a value that is a union of certain sets $\Psi_{k\,j}$ (including $k{=}i$). We will not split the sets $\Psi_{k\,j}$ themselves, but only their distribution in these unions. In doing so, we make sure that for each line $\ell \in L_{\,i}$ and each condition $b \in B$ we keep the property $b \in \ell \iff \Psi_{i\ \alpha_{\,i}(\ell)} \subseteq \beta(b)$ (which is trivially true for $\min\pi_i$ by definition of β). More precisely, if e is an event, ℓ is a line of $L_{\,i}$ and b is an input condition of e whose value β is already known, then by construction $\Psi_{i\ \alpha_{\,i}(\ell)} \subseteq \beta(b)$. Now let b' be the (unique, if any) output condition of e which also lies on l; then the set $\Psi_{i\ \alpha_{\,i}(\ell)}$ becomes a subset of $\beta(b')$. We also get that $\beta(b) \neq \emptyset$ for all $b \in B$, since the line system $L_{\,i}$ covers all conditions in B. For each $b \in \min\pi_i$ and each remaining set $\Psi_{k\,j}$ of $\beta(b)$ with $k\neq i$ we select some line ℓ' in π_i and make sure that for all $b' \in B$ we get $b' \in \ell' \iff \Psi_{k\,j} \subseteq \beta(b')$, thus rendering $\bigcup_{b\in F(\,e)}\beta(b) = \bigcup_{b\in F(e\,)}\beta(b)$ valid. We then associate the (singleton) value set $\{\beta(b)\}$ with every b, and define the (singleton) set of modes for every event e in the way it was just constructed. By construction, all modes are balanced, and therefore, each event corresponds to a mode in the net $colour(N, c)$ and the so labelled π_i satisfies initiality and progress properties of a process of $colour(N, c)$. It is easily seen that the finality property is also satisfied. Overall, $\varphi = \Phi(\pi_i) \in \mathcal{P}(colour(N, c))$.

We see additionally that for each condition $b \in B$, $\beta(b)$ has at least $k \cdot |S|$ elements – namely those of at least one of the $\Psi_{i,j}$.[8]

Case 2: $|\pi| > n_0$. We can split $\pi = x \cdot y \cdot z$ by the pumping lemma and apply the proof above to $\pi' = x \cdot z$, again with the result that $\beta(b)$ contains at least $k \cdot |S|$ elements for each $b \in B$. This holds especially for those b of the splitting cut between x and z. We fit y back into the process and define β on y in the obvious way, choosing the line system from the pumping lemma for y. It is clear that we can attach at least one element to each line, even if all lines lead through the same condition: there are at most $k \cdot |S|$ lines and each condition of $\min y$ has $k \cdot |S|$ elements attached to it by β. The pumping lemma guarantees now that every line ends in $\max y$ at exactly the same condition it starts in at $\min y$, therefore the distribution of the elements in $\min y$ is exactly as in $\max y$, so β is well-defined at the connecting cut between y and z. Overall, we get $\varphi \in \mathcal{P}(colour(N, c))$.

$\blacksquare$ 8

The construction defined in the proof of lemma 8 may be illustrated on figure 4. There are four (non-isomorphic) processes, and the number $k \cdot |S|$ is 4 (since the net is 2-safe and has two places). Hence we define 16 sets $\Psi_{i,j}$ of four elements each. In all, $\mathcal{F}$ has a total of 64 elements. The colouring c is in this case $c(M_0) = \{M_0'\}$ with:

$$
\begin{aligned}
M_0'(s, \Psi_{1,1} \cup \Psi_{2,1} \cup \Psi_{3,1} \cup \Psi_{4,1}) &= 1 \quad \text{(because of } b_1) \\
M_0'(s, \Psi_{1,2} \cup \Psi_{2,2} \cup \Psi_{3,2} \cup \Psi_{4,2}) &= 1 \quad \text{(because of } b_2) \\
M_0'(s, v) &= 0 \quad \text{for any other } v.
\end{aligned}
$$

For instance, in order to execute π_4 in the so coloured net, the eight sets $\Psi_{1,1}$–$\Psi_{4,2}$ will be combined (by forming their union) at condition b_3.

Finally, combining the previous considerations, we obtain:

Theorem 2. MAIN RESULT

For every k-safe PT-net N there is a 1-safe PT-net N' with $\mathcal{P}(N) = \mathcal{P}(N')$.

---

[8] This property will be used in 'Case 2' of the proof; if only 'Case 1' would have had to be considered, we could have restricted all $\Psi_{i,j}$ to be singleton sets.

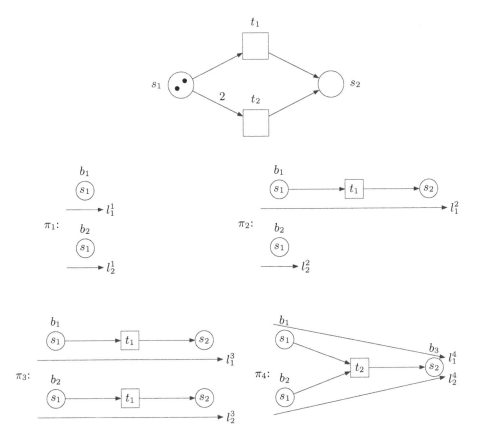

Fig. 4. A PT-net, its processes, and a line system for each of them.

Proof: Let c be a colouring for the initial marking of N as in the previous lemma. Define $N' = PT(unf(colour(N, c)))$. Then,

$$\mathcal{P}(N) = \text{(lemma 4 and lemma 8)}$$
$$\mathcal{P}(colour(N, c))$$
$$= \text{(corollary 1)}$$
$$\mathcal{P}(unf(colour(N, c)))$$
$$= \text{(corollary 3)}$$
$$\mathcal{P}(PT(unf(colour(N, c)))).$$

Thus, $\mathcal{P}(N) = \mathcal{P}(N')$, and moreover, N' is 1-safe by corollaries 2 and 4.   ■ 2

Corollary 2 applies because $colour(N, c)$ is by definition simple and has only balanced modes and a spread initial marking. Corollaries 3 and 4 apply because $unf(colour(N, c))$ is basic by definition.

6   Concluding Remarks

We have shown that 1-safe Petri nets have the same expressive power with respect to a true-concurrency semantics as k-safe Petri nets for arbitrary $k>1$. Proofs for 1-safe Petri nets not dealing with complexity thus also hold for k-safe Petri nets. As an example one may consider an algebraic characterisation for 1-safe Petri nets with unobservable τ-transitions by Grabowski [Gra81] which is now known to be valid for the class of bounded Petri nets, too.

An earlier result of [Pe92] (stating that every k-safe net has a pomset-equivalent net whose initial and final markings are 1-safe) can be recovered by restricting attention to initial and final markings. Conversely, the transformations described in [Pe92] cannot apparently be generalised to our present problem, since they work inductively.

Acknowledgments

We are grateful to the reviewers for their comments. Part of this work has been done within the DAAD-sponsored projects BAT (Arc programme) and PORTA (Procope programme).

References

[Bes84] E. Best: In Quest of a Morphism. Petri Net Newsletters Vol. 18, pp.14–18, 1984.

[EHK96] T. Emden-Weinert, S. Hougardy, B. Kreuter, H.J. Prömel, A. Steger: *Einführung in Graphen und Algorithmen*. Skriptum der Humboldt-Universität Berlin, 1996.

[GV83] U. Goltz, U. Vogt: Processes of Relation Nets. Petri Net Newsletters Vol. 14, pp.10–19, 1983.

[GR83] U. Goltz, W. Reisig: The Non-sequential Behaviour of Petri Nets. Information and Control Vol. 57, pp.125–147, 1983.

[Gra81] J. Grabowski: On Partial Languages. Annales Societatis Mathematicae Polonae, Fundamenta Informaticae Vol. IV.2, pp.428–498, 1981.

[Jen92] K. Jensen: *Coloured Petri Nets. Basic Concepts, Analysis Methods and Practical Use. Volume 1*. EATCS Monographs on Theoretical Computer Science, Springer-Verlag, 1992.

[Pe92] E. Pelz: Normalization of Place/Transition Systems Preserves Net Behaviour. Informatique Théorique et Applications Vol. 26/1, pp.19–44, 1992.

[PRS92] L. Pomello, G. Rozenberg, C. Simone: A Survey of Equivalence Notions for Net Based Systems. Lecture Notes in Computer Science Vol. 609, pp.410–472, 1992.

[Pra84] V. Pratt: Modeling Concurrency with Partial Orders. International Journal of Parallel Processing no. 15, pp.33–71, 1986.

[PW97] L. Priese, H. Wimmel: Algebraic Characterization of Petri Net Pomset Languages. Proceedings of CONCUR'97, Lecture Notes in Computer Science Vol. 1243, pp.406–420, 1997.

[PW98] L. Priese, H. Wimmel: A Uniform Approach to True-Concurrency and Inter-leaving Semantics for Petri Nets. Theoretical Computer Science Vol. 206, pp.219–256, 1998.

[Rei98] W. Reisig: Elements of Distributed Algorithms. Springer-Verlag, 1998.

[Smi98] E. Smith: Principles of High-level Net Theory. in: Lectures on Petri Nets I, Basic Models, Lecture Notes in Computer Science Vol. 1491, pp.174–210, 1998.

[Tve67] H. Tverberg: On Dilworth's Decomposition Theorem for Partially Ordered Sets. J. of Combin. Theory Vol. 3, pp.305–306, 1967.

[Wim00] H. Wimmel: Algebraische Semantiken für Petri-Netze. Ph.D. Thesis, Univer-sität Koblenz-Landau, 2000.

Executing Transactions in Zero-Safe Nets

Roberto Bruni and Ugo Montanari

Dipartimento di Informatica, Università di Pisa, Italia
bruni,ugo @di.unipi.it
http://www.di.unipi.it/ bruni
and http://www.di.unipi.it/ ugo/ugo.html

Abstract Distributed systems are often composed by many heterogeneous agents that can work concurrently and exchange information.
Therefore, in their modeling via PT nets we must be aware that the
basic activities of each system can vary in duration and can be constituted by smaller internal activities, i.e., transitions are conceptually
refined into *transactions*. We address the issue of modeling transactions
in distributed systems by using *zero-safe nets*, which extend PT nets
with a simple mechanism for transition synchronization. In particular,
starting from the zero-safe net that represents a certain system, we give
a distributed algorithm for executing the transactions of the system as
transitions of a more abstract PT net. Among the advantages of our approach, we emphasize that the zero-safe net can be much smaller than
its abstract counterpart, due to the synchronization mechanism.

Keywords: PT nets, zero-safe nets, distributed transactions, net unfolding, reachability.

Introduction

A distributed system can be viewed as a collection of several components that
can evolve concurrently by performing local actions, but can also exchange information according to some predefined protocols that specify how two or more
components can interact.

Computational models for distributed systems are often defined using transition systems labeled over certain elementary actions. *Place/transition Petri
nets* [21,23] (abbreviated as PT *nets*) can be viewed as particular structured
transition systems, where the additional algebraic structure (i.e., monoidal composition of states and runs) offers a suitable basis for expressing the concurrency
of such elementary actions. In fact they have been extensively used both as a
foundational model for concurrent computations and as a specification language,
due to their well assessed theory, an intuitive graphical presentation and several

Research supported by CNR Integrated Project *Progettazione e Verifica di Sistemi
Eterogenei Connessi mediante Reti*; by Esprit Working Groups *CONFER2* and *CO-
ORDINA*; and by MURST projects *Tecniche Formali per Sistemi Software* and
TOSCA: Tipi, Ordine Superiore e Concorrenza.

supporting tools. Notice that the PT net model assumes that each transition synchronizes the consumption and production of its pre- and post-set. This assumption has several serious consequences, since, in general, it requires that a local activity can lock several resources, thus inhibiting the behavior of other transitions.

When designing large and complex systems via PT nets, the more convenient approach is of course to start by outlining a very abstract model and then to refine each transition (that might model a complex activity of the system) into a net that offers a better understanding of the modeled activity. For example, communication protocols for passing and retrieving information cannot leave aside, during the refinement, the fact that agent synchronization is built on finer actions (e.g., for sending message requests, acknowledgments). Moreover, such actions must be executed according to certain local/global strategies that must be completed before the interaction is closed. This means that the abstract transition is seen, at the refined level, as a possibly complex distributed computation that succeeds if and only if all the involved component accomplish their tasks, and that we call *transaction*. In particular the *commit* of the transaction synchronizes all the termination operations of local tasks. Of course, for the refinement to be correct, we must assume that the transaction is executed *atomically*, as if it were a transition. Thus the execution strategy can be only partially distributed, since certain local choices must be synchronized. However, this is similar to what happens for ordinary PT nets.

In fact, let us consider a generic interpreter for PT nets. Then, before executing any transition t, the interpreter must lock all the distributed resources that t will consume and this must be done atomically, otherwise a different transition t' could lock some of the resources needed by t. Therefore the interpreter can afford only a certain degree of distribution. There are at least two cases where atomic locking and distribution of the interpreter do not interfere: 1) nets under consideration are deterministic; 2) nets under consideration are *free choice*.[1] In the first case, at a given instant there is only one transition that can consume a certain resource and therefore the locking can be performed non atomically. In the second case all the nondeterministic choices can be performed by looking at just one place, and thus locking is not needed.

Several approaches have appeared in the literature that present different refinement techniques for top-down design of a concurrent system (e.g., the work on *Petri Box calculus* [5,4]). An extensive comparison of the different approaches and an impressive list of references can be found in [6,16]. Typically at each step a single transition (say t) of the actual net N is refined into a suitable subnet M, yielding the net $N[t \rightarrow M]$ (the net M is usually equipped with initial and final transitions that, in $N[t \rightarrow M]$, become connected to the pre- and post-set of t). This is also related to the notion of *general net morphism* proposed by

---

[1] We recall that a net is *free choice* if for any transitions t_1 and t_2 whose pre-sets are not disjoint, then the pre-sets of t_1 and t_2 consist of exactly one place, or equivalently, a net is free choice if for any place s in the pre-set of two or more transitions, then the pre-set of any such transition is given by $\{s\}$.

Petri, that can be used to map the refined net into its abstract representative by collapsing the structure of M into the transition t. In general some constraints must be assumed on the "daughter" net M for its behavior to be consistent with that of t, as e.g. in [26,25,24].

The approach we pursue in this paper is slightly different, in the sense that all the transitions of the abstract net are refined by runs of *the same* zero-safe net. In particular, the refinements M_1 and M_2 of two transitions t_1 and t_2 of N might use the same kind of resources, whereas in the mentioned approaches, the nets M_1 and M_2 must be disjoint (except for the interfaces of t_1 and t_2). *Zero-safe nets* (ZS nets) have been introduced in [10] to provide a basic synchronization mechanism for transitions — PT net transitions can only synchronize tokens. Besides transitions and ordinary places (here called *stable* places), ZS nets include a set of *zero* places. These are a sort of idealized resources that remain invisible to external observers, whilst *stable markings*, which just consist of tokens in stable places, define the observable states of the system. The idea is that any evolution step of a ZS net starts at some stable marking, evolves through hidden states (i.e., markings with some tokens in zero places) and eventually terminates into a stable marking. Moreover, all the stable tokens produced during a certain step are released together at the end of the step, when the commit is executed.

The synchronization of transitions can thus be performed via zero tokens. Indeed, consider a transition t_1 that produces a token in a certain zero place z, then to complete the step there must be some transition, say t_2, which is able to consume it. Since the computation is atomic for an external observer, t_1 and t_2 look as synchronized at the abstract level. Pursuing this view, a "refined" ZS net and an "abstract" PT net are supposed to model the same given system. The latter, where only stable places are considered, offers the synchronized view, which corresponds to the abstraction from the mechanism for supervising the atomic production and consumption of zero tokens. This mechanism provides a unique computational strategy, which is expressive enough to model, e.g., a multicasting system and simple process algebras (cf. [7,12]). We remark that the notions of "atomicity" and "evaluation strategy" required by the methodology for designing systems via refinement, are built-in inside the ZS net model itself.

The operational and abstract semantics of ZS nets according to the two more diffused net philosophies (called *collective token* and *individual token*) have been presented in [10,11], where it is shown that they can be formulated as universal constructions in the language of category theory. A comparison between the two approaches together with several applications has been discussed in [12] and in the Ph.D. Thesis of first author [7]. A tutorial presentation of the material can be found in [13]. We focus here on a distributed interpreter for ZS nets.

Dining Philosophers and Free Choice Nets. A simple example that fits well the situation discussed so far can be illustrated by modeling the well-known problems of "dining philosophers:" there are n philosophers (with $n \geq 2$) sat on a round table; each having a plate in front with some food on it; between each couple of plates is put a fork, for a total of n forks on the table; each philosopher cyclically thinks and eats, but to eat he needs both the fork on the left and the one on the right of his plate; after eating a few mouthfuls, the philosopher returns the forks on the table and starts thinking again.

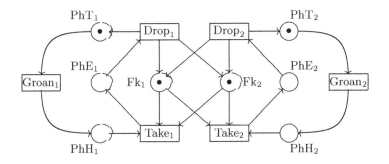

Fig.1. An abstract view for (two) dining philosophers.

The PT net for the case $n = 2$ is illustrated in Figure 1, where, as usual, places are represented by circles, transitions by rectangular boxes, tokens by bullets, and the flow relation by weighted directed arcs (unary weights are omitted). A token in one of the places PhH_i, PhE_i, and PhT_i, for $1 \leq i \leq 2$, means that the ith philosopher is hungry, is eating, and is thinking (soon after eating), respectively. A token in the place Fk_i means that the ith fork is on the table. The transitions $Take_i$, $Drop_i$, and $Groan_i$ represent that the ith philosopher takes the forks and starts eating, finishes eating and drops the forks, feels his stomach groaning and prepares to eat, respectively. In fact, note that $Take_i$ requires both forks and thus cannot be accomplished if the other philosopher is eating. The initial marking of the net consists of one token in each of the places PhT_1, PhT_2, Fk_1, and Fk_2 (i.e., both philosophers are thinking and both forks are on the table).

Of course, this model does not tell how the philosophers access the "resources" needed to eat, whereas the action $Take_i$ is not trivial and requires some atomic mechanism for getting the forks. At a more refined level, for example, the strategy for executing the action $Take_i$ could be specified as "take the ith fork (if possible), then the $((i \bmod 2) + 1)$th fork (if possible) and eat," hence it is not difficult to imagine a deadlock where each philosopher takes one fork and cannot continue, since conflict arises. The fact is that the coordination mechanism is hidden inside transitions whose granularity is too coarse.

The situation is completely different if one wants to model the system using *free choice* nets, where all decisions are made locally (i.e., by looking at just one place). To see this, let us concentrate our attention to a subpart of the net in Figure 1, depicted in Figure 2, which will suffice to illustrate the point.

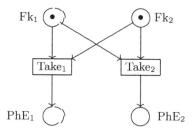

Fig.2. Centralized nondeterminism.

We can translate any net into a free choice net by adding special transitions that perform the local decisions required. For example, the free choice net corresponding to the net in Figure 2 is represented in Figure 3.

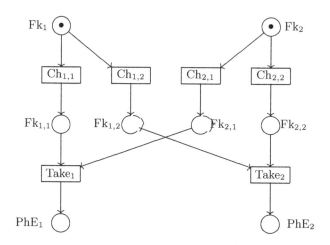

Fig.3. Local nondeterminism (free choice).

The free choice net in Figure 3 models a system where two decisions can take place independently: one decision concern the assignment of the first fork either to the first or the second philosopher, the other decision concerns the assignment of the second fork. Of course it might as well happen that the first fork is assigned to the first philosopher ($Ch_{1,1}$) and the second fork is assigned to the second philosopher ($Ch_{2,2}$), and in such case none of the $Take_i$ actions can occur. Thus, the translated nets admits computations not allowed in the abstract system of Figure 2, and is not a correct implementation.

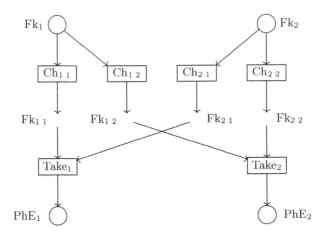

Fig.4. Atomic free choice.

Zero-safe nets allow for overcoming this problem by executing only certain atomic transactions, where tokens produced in low-level resources are also consumed. In our example, the invisible resources consist of places $Fk_{i,j}$ for $1 \leq i, j \leq 2$, that can be interpreted as zero-places. In this way the computation performing $Ch_{1,1}$ and $Ch_{2,2}$ is forbidden because it stops in an invisible state, i.e., a state that contains zero tokens. Figure 4 represents the low-level model as a ZS net. (Smaller circles stands for zero-places.)

This can be seen as a general procedure for moving from free choice nets to non deadlocking nets.

Concurrent Strategies, Operationally. Though the methodological framework of ZS nets is to some extent satisfactory from the theoretical point of view, from the operational point of view still many aspects must be faced. Indeed, we must specify the computational machinery for performing only correct transactions, recovering deadlocks and treating infinite low-level computations. Moreover, in many cases for such machinery performance could be more relevant than completeness, as it happens in most implementations of Prolog where depth-first goal resolution is preferred to the safer but more resource expensive breadth-first strategy.

In this paper we illustrate our proposal for equipping zero-safe nets with such an operational tool. In particular, we will answer positively to the question: Given a ZS net B and a stable marking u is it possible to compute in a distributed and efficient fashion the set of markings that can be reached from u via the execution of an atomic transaction? The solution relies on a modification of the interpreter for unfolding PT nets [20,27,19], which is extended with a commit rule that enforces the synchronous terminations of transactions. Since there can exist an infinite number of transactions enabled by a marking of a finite zs net, a termination problem is stated and proposed for further study.

Structure of the Paper. In Section 1.1 we recall some basic notion about PT nets, whereas in Section 1.2 we shift the attention to zs nets. In Section 2 we rephrase the ordinary net unfolding in terms of two simple inference rules and, as an original contribution, extend it to deal with transaction steps on zs nets. In particular, we show that a fair distributed algorithm based on the modified unfolding rules can be used to enumerate the reachable markings. We conclude by formulating in Section 3 some interesting open problems for which, at the moment, we have only some partial solutions.

1   Preliminaries

In this section we fix the notation and summarize the main concepts needed in the rest of the paper.

1.1   Place/Transition Petri Nets

Place/transition Petri nets (PT nets) can be regarded as graphs with distributed states described by finite distributions of typed resources (called *tokens*), where resources are typed over the set of *places* S. States are usually called *markings* and represented as multisets $u : S \to \mathbb{N}$, where $u(a)$ indicates the number of tokens that place a carries in u. A marking can be written either as $u = \{n_1 a_1, ..., n_k a_k\}$ with $n_i = u(a_i)$, or as the formal sum $u = n_1 a_1 \oplus \cdots \oplus n_k a_k$ denoting an element of the free commutative monoid $S^\oplus$ on the set of places S (the order of summands is immaterial and the monoidal composition is defined by taking $(\bigoplus_i n_i a_i) \oplus (\bigoplus_i m_i a_i) = (\bigoplus_i (n_i + m_i) a_i)$ and 0 as the neutral element). Of course, the monoid $(\mu(S), \cup, \emptyset)$ of finite multisets on S (with multiset union as monoidal operation and the empty multiset as unit) is isomorphic to $S^\oplus$. We denote multiset difference by the operator $\_ \ominus \_$.

Definition 1.1 (PT net). *A marked PT net is a 4-tuple $N = (S, T, F, u_{\text{in}})$ where S is the set of places, T is the set of transitions $(S \cap T = \emptyset)$, the multirelation $F : (S \times T) \cup (T \times S) \to \mathbb{N}$ is the flow relation and u_{in} is the initial marking of N, making precise the initial state of the system.*

When no confusion can arise we will omit to specify that our nets are "marked." For any transition $t \in T$, the *pre-set* and the *post-set* of t, denoted by $\text{pre}(t)$ and $\text{post}(t)$ respectively, are the multisets over S such that $\text{pre}(t)(a) = F(a, t)$ and $\text{post}(t)(a) = F(t, a)$, for all $a \in S$. The resources in $\text{pre}(t)$ are those required by the action t to be performed (*fired*), and corresponding tokens are consumed by any firing of t. The resources in $\text{post}(t)$ are fresh tokens produced by the action t after a firing. Therefore a PT net $N = (S, T, F, u_{\text{in}})$ can be equivalently defined as the (marked) graph

$$(S^{\oplus}, T, \text{pre}, \text{post}, u_{\text{in}})$$

where $\text{pre}(\_), \text{post}(\_): T \to S^{\oplus}$ define the source and target of transitions, respectively. As usual we will write $t: u \to v$ for a transition t with $\text{pre}(t) = u$ and $\text{post}(t) = v$.

The dynamics of a net can be easily expressed by the step relation $\_ \Rightarrow_N \_$ that is defined by the inference rules in Table 1: identities represent idle resources, generators represent the firing of a transition within the minimal marking that can enable it, and parallel composition provides concurrent execution of generators and idle steps.

$$\frac{u \in S^{\oplus}}{u \Rightarrow_N u} \qquad \text{(identities)}$$

$$\frac{t: u \to v \in T}{u \Rightarrow_N v} \qquad \text{(generators)}$$

$$\frac{u \Rightarrow_N v, \ u' \Rightarrow_N v'}{u \oplus u' \Rightarrow_N v \oplus v'} \qquad \text{(parallel composition)}$$

Table 1: The inference rules for $\_ \Rightarrow_N \_$.

The extension of this approach to computations $u_0 \Rightarrow u_1 \Rightarrow \cdots \Rightarrow u_n$ is not as straightforward as one would expect. Indeed, concurrent semantics must consider as equivalent all the computations where the same *concurrent* events are executed in different orders, and we cannot leave out of consideration the distinction between *collective* and *individual token philosophies* (noticed e.g., in [17], but see also [8,9] where it is explained how the two approaches can influence the *categorical, logical* and *behavioral* semantics of nets).

The simplest approach relies on the collective token philosophy (*CTph*), where semantics should not distinguish among different instances of the idealized resources. Of course, this is true only if any such instance is *operationally* equivalent to all the others. As a major drawback, this approach disregards that operationally equivalent resources may have different origins and histories, and may, therefore, carry different *causality* information. An alternative approach takes the name of individual token philosophy (*ITph*). According to the *ITph*, causal dependencies are a central aspect in the dynamic evolution of a net. As a consequence, only the computations that refer to isomorphic *Goltz-Reisig processes* [18] can be identified, and causality information is fully maintained, whereas for *CTph* the equivalence classes correspond to the *commutative processes* of Best and Devillers [3].

Whatever the preferred approach is, a more informative semantics is required, which takes into account suitable *proof terms* for computations. Such proof terms can be axiomatized to faithfully recover process semantics. If one is simply interested in "reachability" matters then proof terms can be ignored and consequently also the distinction between the *CTph* and *ITph* is not needed. In this case the obvious rules

$$\frac{u \Rightarrow_N v}{u \Rightarrow_N^* v} \qquad \textbf{(one-step computations)}$$

$$\frac{u \Rightarrow_N^* v, \ v \Rightarrow_N w}{u \Rightarrow_N^* w} \qquad \textbf{(sequential composition)}$$

can be introduced (transitive closure). In this sense, processes can be seen as (concurrent) *computation strategies*.

1.2   Zero-Safe Nets

Zero-safe nets augment PT nets with special places called *zero places*. Their role is to coordinate, via the token flow, the atomic execution of several transitions. From an abstract viewpoint, all the coordinated transitions will appear as being synchronized. However no new interaction mechanism is needed, and the coordination of the transitions participating in a step is handled by the ordinary behavioral rules.

De nition 1.2 (Zero-safe net). *A* zero-safe net *(also* ZS *net for short) is a tuple* $B = (S^\oplus, T, \text{pre}, \text{post}, u_{\text{in}}, Z)$ *where* $N_B = (S^\oplus, T, \text{pre}, \text{post}, u_{\text{in}})$ *is the underlying* PT *net and the set* $Z \subseteq S$ *is the set of* zero *places. The places in* $S \setminus Z$ *are called* stable *places. A* stable marking *is a multiset of stable places, and the initial marking* u_{in} *must be stable.*

Stable markings describe "observable" states of the system, whereas the presence of one or more zero tokens in a given marking denotes it as internal and hence unobservable. We call *stable tokens* and *zero tokens* the tokens that respectively belong to stable places and to zero places.

Remark 1.1. Since $S^\oplus$ is a free commutative monoid, it is isomorphic to the cartesian product $(S \setminus Z)^\oplus \times Z^\oplus$ and we can write $t: (u, x) \to (v, y)$ for a transition t with $\text{pre}(t) = u \oplus x$ and $\text{post}(t) = v \oplus y$, where u and v are stable multisets and x and y are multisets over Z.

As for PT nets, we can define the behavior of ZS nets by means of a step relation $\_ \Rightarrow_B \_$, defined by the inference rules in Table 2. An auxiliary relation $\_ \rightrightarrows_B \_$ is introduced for modeling *transaction segments*.

Notice that we can take advantage of the step relation $\_ \Rightarrow_{N_B} \_$ of the underlying net for concurrently executing several transitions (rule **underlying**). The rule **horizontal composition** acts as parallel composition for stable resources and as sequential composition for zero places. We call it "horizontal" because

$$\frac{u \oplus x \Rightarrow_{N_B} v \oplus y, \quad u,v \in (S \setminus Z)^{\oplus}, \quad x,y \in Z^{\oplus}}{(u,x) \rightrightarrows_B (v,y)} \qquad \text{(underlying)}$$

$$\frac{(u,x) \rightrightarrows_B (v,y), \ (u',y) \rightrightarrows_B (v',y')}{(u \oplus u',x) \rightrightarrows_B (v \oplus v',y')} \qquad \text{(horizontal composition)}$$

$$\frac{(u,0) \rightrightarrows_B (v,0)}{u \Rrightarrow_B v} \qquad \text{(commit)}$$

Table2: The inference rules for $\_ \Rrightarrow_B \_$.

we prefer to view it as a synchronization mechanism rather than as the ordinary sequential composition of computations, which flows vertically from top to down. The rule **commit** select the transaction segments that correspond to acceptable steps: they must start from a stable marking and end up in a stable marking. Note that, as a particular instance of the horizontal composition of two transaction segments $(u,0) \rightrightarrows_B (v,0)$ and $(u',0) \rightrightarrows_B (v',0)$, we can derive their parallel composition $(u \oplus u',0) \rightrightarrows_B (v \oplus v',0)$.

Abstract Net. At an abstract level, the system modeled via the ZS net B can be equivalently described via a PT net $\mathcal{N}(B)$ such that $S_{\mathcal{N}(B)} = S_B \setminus Z_B$ and $(\_ \Rightarrow_{\mathcal{N}(B)} \_) = (\_ \Rrightarrow_B \_)$. Among the several PT nets that verify these conditions we can choose the optimal one, which is freely generated [10,11]. Informally the transitions of such net must represent the proofs of transaction steps $u \Rrightarrow_B v$ taken up to equivalence (permutation of concurrent events) and that cannot be decomposed into shorter proofs. When these two conditions are verified, the concurrent kernel of the behavior has been identified, and all the steps can be generated by it.

It is worth remarking that the construction of (proved) transaction steps can be defined, in the language of category theory, as an adjunction, i.e., it is a free construction and thus preserves several net compositions (defined as colimits in the category of ZS nets). Moreover the construction of the abstract net defines a coreflection, whose universal properties confirm that it is the optimal such construction and that it is unique (up to isomorphic nets).

Remark 1.2. The constructions can be pursued according to either the *CTph* or the *ITph*. The two approaches yield the same step relation but different abstract nets. Whatever the chosen approach is, the categorical constructions given above are both possible.

2   An Operational De nition for Transactions

The question that we want to address in this paper is essentially how one can implement ZS nets. Since we have already defined their operational semantics, the answer should not be very complicated. But the problem is that the operational semantics relies on some sort of meta-definition, where one computes on the underlying net, builds transaction segments, and then can discard undesired behaviors and accept the good ones, acting as a filter. This means that there are important questions which can be asked for any actual interpreter — Is backtracking necessary? Is the implementation correct? And complete? Does any more efficient implementation exist?

We try to answer these questions (see the Conclusions) by defining a machinery for computing on ZS nets and studying its properties. The idea is to have a level where suitable computations of the underlying net are simulated and examined, in such a way that as soon as a connected transaction is computed, then the corresponding abstract behavior is generated (e.g., it can be used to compute on the system).

The idea is to adapt the classical net unfolding, which is used to extract the event structure that precisely represents the abstract behavior of the net, to pursue concurrently all the nondeterministic runs of the ZS net under inspection, in such a way that "commit" states are recognized and generated.

Whether one is interested in distinguishing between different concurrent proofs or is just interested in the step relation $\_ \Rightarrow_B \_$ is an important issue. For simplicity, we will consider the problem of computing the set of markings that can be reached in one transaction step, i.e., the set $\mathscr{R}(B, u) = \{v \mid u \Rightarrow_B v\}$. However, the solution we propose can be easily adapted to compute the abstract net of B.

2.1   PT Net Unfolding

The unfolding of a net gives a constructive way to generate all the possible computations, offering a satisfactory mathematical description of the interplay between nondeterminism and causality (and concurrency). In fact the unfolding construction allows for bridging the gap between PT nets and *prime algebraic domains*, yielding a truly concurrent semantics. Some obvious references to this approach are [20,27,19], but we suggest also the interesting overview [15]. It is worth remarking that our presentation is slightly different from usual ones, since it is presented as the minimal net generated by suitable inference rules, rather than by making explicit the chain of finite nets that approximate it.

The construction provides a distributed interpreter for PT nets. We remark that the unfolding applies only to marked nets, i.e., it requires an initial marking.

Starting from a net N, the unfolding produces a nondeterministic occurrence net $\mathcal{U}(N)$ (an acyclic net, where transition pre- and post-markings are sets instead of multisets and where each place has at most one entering arc, i.e., backward conflicts are not allowed), together with a mapping from $\mathcal{U}(N)$ to N that tells which places and transitions of the unfolding are instances of the same element of N. Hence the places of $\mathcal{U}(N)$ represent the tokens and the transitions (called *events*) the occurrences of transitions in N in all possible runs of the net. For this kind of nets the notion of *causally dependent*, of *conflicting* and of *concurrent* elements can be straightforwardly defined and are represented by the binary relations $\_ \preceq \_$, $\_\#\_$ and $\mathbf{co}(\_,\_)$, respectively. Formally, the relation $\_ \preceq \_$ is the transitive and reflexive closure of the *immediate precedence* relation $\_ \prec_0 \_$ defined as

$$\prec_0 \stackrel{\text{def}}{=} \{(a,t) \mid a \in \text{pre}(t)\} \cup \{(t,a) \mid a \in \text{post}(t)\},$$

while the binary conflict relation is defined as the minimal symmetric relation that contains $\_\#_0\_$ defined below, and that is hereditary with respect to $\_ \preceq \_$ (i.e., such that if $o_1, o_2 \in S \cup T$ and $o_1 \# o_2$ and $o_1 \preceq o$ then also $o \# o_2$).

$$t_1 \# _0 t_2 \stackrel{\text{def}}{\Leftrightarrow} t_1 \neq t_2 \wedge \text{pre}(t_1) \cap \text{pre}(t_2) \neq \emptyset$$

Since the conflict relation is required to be irreflexive, then $\_ \preceq \_$ and $\_\#\_$ have empty intersection. The concurrency relation is defined by letting $\mathbf{co}(o_1, o_2)$ if it is not the case that ($o_1 \prec o_2$ or $o_2 \prec o_1$ or $o_1 \# o_2$). In particular, the relation $\mathbf{co}$ is usually extended to generic sets of elements by writing $\mathbf{co}(X)$ if and only if for all $o_1, o_2 \in X$ we have $\mathbf{co}(o_1, o_2)$.

More concretely, the places of $\mathcal{U}(N)$ have the form $\langle a, n, H \rangle$, where $a \in S_N$, n is a positive natural number that is used to distinguish different tokens with the same history, and H is the history of the place under inspection and therefore either consists of just one event (the one that produced the token) or is empty (if the token is in the initial marking). Analogously, a generic transition of $\mathcal{U}(N)$ has the form $\langle t, H \rangle$ with $t \in T_N$, since each transition is completely identified by its history H, which in this case consists of the set of consumed tokens. The set H cannot be empty since transitions with empty pre-set are not allowed. The net $\mathcal{U}(N)$ is defined as the minimal net generated by the rules in Table 3.

$$\frac{u_{\text{in}}(a) = n, \ 1 \leq k \leq n}{\langle a, k, \emptyset \rangle \in S_{\mathcal{U}(N)}}$$

$$\frac{t : u \to \bigoplus_{j \in J} n_j b_j \in T_N, \ \Theta = \{\langle a_i, k_i, H_i \rangle \mid i \in I\} \subseteq S_{\mathcal{U}(N)}, \ \mathbf{co}(\Theta), \ u = \bigoplus_{i \in I} a_i}{e = \langle t, \Theta \rangle \in T_{\mathcal{U}(N)}, \ \Upsilon = \{\langle b_j, m, \{e\} \rangle \mid j \in J, \ 1 \leq m \leq n_j\} \subseteq S_{\mathcal{U}(N)}, \ \text{pre}(e) = \Theta, \ \text{post}(e) = \Upsilon}$$

Table3: The unfolding $\mathcal{U}(N)$.

We now give a computational interpretation of such rules. The first rule defines the initial marking of $\mathcal{U}(N)$. The second rule is the core of the unfolding: it looks for a set Θ of concurrent tokens that enables a transition t of N, atomically locks them, fires the event e (that is an occurrence of t), and produces some fresh tokens according to post(t). Notice that the condition $\mathbf{co}(\Theta)$ depends exclusively on the histories H_i for $i \in I$ and therefore cannot be altered by successive firings. In fact, as in *memoizing* for logic programming, or more generally in *dynamic programming*, the history is completely encoded in the tokens, so that it is not necessary to compute it at every firing. Also note that histories retain concurrent information rather than just sequential, therefore each token/event is generated exactly once (though it can be referred many times successively). Moreover, several occurrences of the second rule can be applied concurrently and therefore the unfolding can be implemented as a distributed algorithm.

2.2 ZS Net Unfolding

The unfolding of the underlying net N_B does not yield a faithful representation of the behavior of B. In fact, we must forbid the consumption of stable resources that were not inserted in the initial marking.

Example 2.1. Let us consider the ZS net B in Figure 5. It has three stable places a, b and c, a zero place z and two transitions t_1 and t_2, with

$\text{pre}(t_1) = a$ and $\text{post}(t_1) = b \oplus z$;
$\text{pre}(t_2) = b \oplus c \oplus z$ and $\text{post}(t_2) = 0$.

Then, if $u_{\text{in}} = a \oplus c$, the net B cannot perform any transaction, because although t_1 is enabled and thus can fire, transition t_2 must be used to end the transaction consuming the fresh token in z, but this would require also the consumption of the token in b generated by the firing of t_1, which is not available before the commit. Hence $\mathscr{R}(B, a \oplus c) = \{a \oplus c\}$. Instead, if we add an initial token in b, then such token can be used to close the transaction, by firing t_2 after t_1, and therefore $\mathscr{R}(B, a \oplus b \oplus c) = \{a \oplus b \oplus c, b\}$.

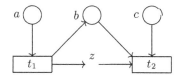

Fig.5. The ZS net of Example 2.1.

Thus, to some extent, we are unfolding on zero places only. Moreover, we must be able to apply the commit when the transaction step has consumed all the zero tokens produced so far.

The net $\mathcal{U}(B)$ is defined as the minimal net generated by the rules in Table 4. Together with the unfolding net we compute the set of reachable markings $\mathscr{R}(B, u)$ for a generic stable marking u.

$$\frac{u(a) = n, \ 1 \le k \le n}{\langle a, k, \emptyset \rangle \in S_{\mathcal{U}(B)}}$$

$$\frac{t \colon (u, x) \to (v, \bigoplus_{j \in J} n_j z_j) \in T_B, \ \Theta = \{\langle s_i, k_i, H_i \rangle \mid i \in I\} \subseteq S_{\mathcal{U}(B)}, \ \mathbf{co}(\Theta), \ u \oplus x = \bigoplus_{i \in I} a_i}{e = \langle t, \Theta \rangle \in T_{\mathcal{U}(B)}, \ \varUpsilon = \{\langle z_j, m, \{t\} \rangle \mid j \in J, \ 1 \le m \le n_j\} \subseteq S_{\mathcal{U}(B)}, \ \mathrm{pre}(e) = \Theta, \ \mathrm{post}(e) = \varUpsilon}$$

$$\frac{}{u \in \mathscr{R}(B, u)} \qquad \frac{\varGamma \subseteq T_{\mathcal{U}(B)}, \ \mathbf{co}(\varGamma), \ \mathbf{ZProd}(\varGamma) = \mathbf{ZCons}(\varGamma)}{u \ominus \mathbf{SCons}(\varGamma) \oplus \mathbf{SProd}(\varGamma) \in \mathscr{R}(B, u)}$$

Table4: The unfolding $\mathcal{U}(B)$.

The first two rules define the unfolding, which remains similar to the classical algorithm, except for the fact that stable tokens in the post-set of the fired transition are not released to the system. In fact, while the set Θ must contains enough tokens to provide both the stable and the zero resources needed by t (as expressed by the condition $u \oplus x = \bigoplus_{i \in I} a_i$), the tokens that are produced by the occurrence of t applied to Θ (i.e., tokens in the set $\varUpsilon = \mathrm{post}(e)$) just match the zero place component $\bigoplus_{j \in J} n_j z_j$ of $\mathrm{post}(t)$ and not the stable place component v. The marking v will not be released until a commit related to e will occur.

The third rule is obvious. The fourth rule defines the commit of a transaction step. To shorten the notation, we introduce the following functions that given an event e return the set of zero tokens respectively consumed and produced by the ancestors of e (and by e itself), i.e., we let

$$\mathbf{ZCons}(e) \stackrel{\mathrm{def}}{=} \bigcup_{\langle t, \Theta \rangle \preceq e} \{\langle z, k, H \rangle \in \Theta \mid z \in Z_B\}$$

$$\mathbf{ZProd}(e) \stackrel{\mathrm{def}}{=} \bigcup_{e' \preceq e} \mathrm{post}(e')$$

where $\mathbf{ZCons}(e)$ is the set of zero tokens that have been consumed by some $e' \preceq e$; similarly $\mathbf{ZProd}(e)$ represents the set of zero tokens that have been produced by some $e' \preceq e$ (note that for any $\langle z, k, H \rangle \in \mathbf{ZCons}(e)$ we have $\langle z, k, H \rangle \preceq e$, while $\mathbf{ZProd}(e)$ can also contain tokens that are concurrent with e or produced by e). We remark that $\mathbf{ZCons}(e) \subseteq \mathbf{ZProd}(e)$, because the marking u is stable and therefore does not contain zero tokens with empty histories. For stable places the situation is different, since we are just interested in knowing how many tokens have been consumed and will be produced for each places by the antecedents of e, thus

$$\mathbf{SCons}(e) \stackrel{\text{def}}{=} \bigoplus_{\langle t:(u,x)\to(v,y),\Theta\rangle \preceq e} u$$

$$\mathbf{SProd}(e) \stackrel{\text{def}}{=} \bigoplus_{\langle t:(u,x)\to(v,y),\Theta\rangle \preceq e} v.$$

The four functions that we have defined are extended to sets of events in the obvious way.

The fourth rule takes a set Γ of concurrent events and checks that the zero tokens produced by each event are consumed by an antecedent of some other event in Γ. The latter condition can be conveniently expressed as the equality $\mathbf{ZProd}(\Gamma) = \mathbf{ZCons}(\Gamma)$. In fact, if a certain token o is in $\mathbf{ZProd}(\Gamma)$, then the condition states that there exists at least an event $e \in \Gamma$ and a unique[2] $e' \preceq e$ such that $o \in \text{pre}(e)$. If these premises are satisfied, then the rule extends $\mathscr{R}(B, u)$ with the multiset obtained by subtracting from u the stable resources consumed by all the antecedents of events in Γ, but adding those that would have been produced during the step. This rule defines a commit since it synchronizes local commits, as the following result shows.

Proposition 2.1. *If $\Gamma \subseteq T_{\mathcal{U}(B)}$ such that $\mathbf{co}(\Gamma)$ and $\mathbf{ZProd}(\Gamma) = \mathbf{ZCons}(\Gamma)$, then for any $e = \langle t, \Theta \rangle \in \Gamma$ we have that t does not produce any zero token.*

Proof. We proceed by contradiction. Let us assume that $t:(u,x) \to (v,y)$ for some stable markings u, v and zero markings x and y, with $y \neq 0$. Since $e \in \Gamma \subseteq T_{\mathcal{U}(B)}$, it must have been generated by the second rule in Table 4. Hence $\Upsilon = \text{post}(e) \neq \emptyset$. Let o be a token in Υ. Since $\Upsilon \subseteq \mathbf{ZProd}(\Gamma)$ and $\mathbf{ZProd}(\Gamma) = \mathbf{ZCons}(\Gamma)$ by hypothesis, it follows that there exist an event $e' \neq e$ in Γ and an event $e'' \preceq e'$ such that $o \in \text{pre}(e'')$. But then we have $e \preceq o \preceq e'' \preceq e'$ with $e, e' \in \Gamma$, that contradicts the hypothesis $\mathbf{co}(\Gamma)$, concluding the proof.   □

The resulting algorithm is as much distributed as the classical one when applied to the abstract net of B. In fact all the useful relations are defined by just looking at the history of the elements in the premises, which, under the atomicity assumption reduce to the stable preset of the abstract step. To improve efficiency, the sets $\mathbf{ZProd}(e)$, $\mathbf{ZCons}(e)$, $\mathbf{SProd}(e)$ and $\mathbf{SCons}(e)$ could be also encoded in e more directly, although they can be easily calculated from the history component.

The main result can be formulated as the following theorem.

Theorem 2.1. $\mathscr{R}(B, u) = \{v \mid u \Rightarrow_B v\}$.

Proof. (sketch) The inclusion $\mathscr{R}(B, u) \subseteq \{v \mid u \Rightarrow_B v\}$ can be proved by rule induction. In fact, if $w \in \mathscr{R}(B, u)$, then either $u = w$ and we can apply the rules **identities**, **underlying** and **commit** to conclude that $u \in \{v \mid u \Rightarrow_B v\}$, or $w = u \ominus \mathbf{SCons}(\Gamma) \oplus \mathbf{SProd}(\Gamma)$ for some set $\Gamma \subseteq T_{\mathcal{U}(B)}$, such that $\mathbf{co}(\Gamma)$ and $\mathbf{ZProd}(\Gamma) = \mathbf{ZCons}(\Gamma)$, and therefore there exists a concurrent computation

---

[2] Otherwise a conflict would arise.

for which Γ is a slice, that starts from the marking $\mathbf{SCons}(\Gamma)$ included in u and that consumes all the zero tokens that produces. Moreover, such computation is completely described by the events in

$$C = \{e' \mid \exists e \in \Gamma, \ e' \preceq e\}.$$

By summing up all the stable tokens produced by the transitions associated to the events in C we obtain exactly $\mathbf{SProd}(\Gamma)$. Therefore we have

$$(\mathbf{SCons}(\Gamma), 0) \rightrightarrows_B (\mathbf{SProd}(\Gamma), 0).$$

By the rules **identities** and **underlying** we have also

$$(u \ominus \mathbf{SCons}(\Gamma), 0) \rightrightarrows_B (u \ominus \mathbf{SCons}(\Gamma), 0).$$

Horizontally composing the two we get

$$(u, 0) \rightrightarrows_B (u \ominus \mathbf{SCons}(\Gamma) \oplus \mathbf{SProd}(\Gamma), 0).$$

Recalling that $w = u \ominus \mathbf{SCons}(\Gamma) \oplus \mathbf{SProd}(\Gamma)$ and applying the rule **commit** we obtain $u \Rightarrow w$, that proves the inclusion.

Conversely, to prove that $\{v \mid u \Rightarrow_B v\} \subseteq \mathcal{R}(B, u)$, we proceed by induction on the proof of $u \Rightarrow_B w$, by mimicking the firings in the unfolding of the net. $\square$

3   Related Issues

3.1   Computing the Abstract Net

Since the unfolding encodes the proof of the transaction step (via the history components), it is possible to use the same scheme for computing the abstract net (whatever philosophy is preferred). However, for doing this efficiently, we must be able to recognize isomorphic processes. For example, note that given a certain computed process, any renaming of the stable tokens in the initial marking (i.e., any permutation of tokens in the same place) yields a different but isomorphic process that is also calculated during the unfolding. To solve this problem, we can either try to avoid having several isomorphic processes in the unfolding by some clever construction, or just check if a freshly computed transaction is isomorphic to some transaction already computed when executing the commit.

In general, the problem of checking the equivalence of two (finite, deterministic) processes can be exponential on the process size, the problem being given by multiple copies of the initial resources and by multiple instances of a resource produced by the same transition, making it hard to establish the correspondence between analogous elements of the two processes. However, for 1-safe computations the problem is clearly linear (the correspondence, if it exists, is the obvious one). The problem is also linear for processes of *semi-weighted nets*[3] when the initial marking is a set.

---

[3] A semi-weighted net is a net where the post-set of transitions are sets instead of multisets, whereas pre-sets can still be multisets.

3.2   Some Considerations on Computability

Since a ZS net B can contain cycles that produce an unbounded number of zero tokens, the unfolding can become infinite.

Example 3.1. Let B be the ZS net (see Figure 6) with stable places a and b, $u_{in} = a$, and three transitions t_1, t_2 and t_3, with

- $\mathrm{pre}(t_1) = a$ and $\mathrm{post}(t_1) = z$;
- $\mathrm{pre}(t_2) = z$ and $\mathrm{post}(t_2) = 2z$;
- $\mathrm{pre}(t_3) = z$ and $\mathrm{post}(t_3) = b$.

Then $\mathscr{R}(B,a) = \{a\} \cup \{nb \mid n > 0\}$. In fact any transaction can be started by a firing of t_1, then t_2 can fire any number of times and t_3 can fire as many times as t_2 plus one. Moreover, note that since the net is not semi-weighted (because of t_2), many isomorphic processes will be computed during the unfolding.

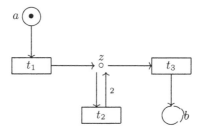

Fig.6. The ZS net of Example 3.1.

So an important question concerns the decidability of $\mathscr{R}(B,u)$. In a private communication to the authors [14], Nadia Busi proved that such set is indeed decidable. Roughly speaking the idea is to simulate the behaviors of B by a PT net with exactly one inhibitor arc,[4] for which the reachability problem has been solved in [22].

The set $\mathscr{R}(B,u)$ is recursively enumerated by the inference system, and if it is infinite we cannot do any improvement. But when $\mathscr{R}(B,u)$ is finite, it would be desirable to find some condition for halting the execution of the algorithm, that otherwise could continue computing transaction segments that cannot be completed.

---

[4] Nets with inhibitor arcs, also called with negative arcs, have been introduced in [1,2] for modeling systems where the presence of certain resources can inhibit the firing of some transitions. We recall that the reachability problem is undecidable for the class of nets with two or more inhibitor arcs.

3.3   Halting Problem for the Distributed Algorithm

Finding some general condition for halting the unfolding of ZS is an open problem that we leave for future investigations. Of course, in many cases, some very simple criterion can be applied. The problematic cases are given by cycles that do not involve stable places. If the net under inspection contain no such cycle, then the algorithm will eventually terminate. Now suppose that the net contains an increasing cycle of that kind for the zero place z, and observe that there is no decreasing cycle and that all other transitions that consumes some zero tokens from z, except the one involved in the cycle, consumes some stable token, then we can remember that the stable resources are bounded and therefore we have a bound for the tokens that can be consumed from z. Since we only consider transactions that lead to stable markings, such a bound can be put also on the zero tokens that can be created in z. It is worth noticing that these criteria are sufficient to deal with the examples of ZS nets we encountered in our previous work.

Conclusions

We have proposed the framework of zero-safe nets as a basis for modeling and implementing distributed transactions. In fact, ZS nets can provide both the refined view of the systems where actions have finer grain and an abstract view where transactions are seen just as transitions of an ordinary PT net. Working at the level of ZS nets is very important since it allows to keep smaller the size of the system description (for example the abstract net can have an infinite number of transitions also when the refined net is finite).

We have extended previous semantic investigations with the definition of a distributed interpreter for computing on ZS nets. We want to remark that the resulting implementation does not violate the locality assumptions, since it is completely analogous to the widely accepted implementation for PT nets.

We can now answer to the questions formulated at the beginning of Section 2. Thus backtracking is not necessary and the correctness of the interpreter is given by Theorem 2.1. Completeness is ensured by our inference system. We are confident to find a nice formalization of halting criteria for an expressive class of ZS nets.

References

1. T. Agerwala. A complete model for representing the coordination of asynchronous processes. Hopkins Computer Research Report 32, John Hopkins University, 1974.
2. T. Agerwala and M. Flynn. Comments on capabilities, limitations and "correctness" of Petri nets. *Computer Architecture News*, 4(2):81–86, 1973.
3. E. Best and R. Devillers. Sequential and concurrent behaviour in Petri net theory. *Theoret. Comput. Sci.*, 55:87–136, 1987.

4. E. Best, R. Devillers, and J. Esparza. General refinement and recursion for the Petri Box calculus. In *Proceedings STACS'93*, volume 665 of *Lect. Notes in Comput. Sci.*, pages 130–140. Springer-Verlag, 1993.

5. E. Best, R. Devillers, and J. Hall. The Box calculus: A new causal algebra with multi-label communication. In G. Rozenberg, editor, *Advances in Petri Nets'92*, volume 609 of *Lect. Notes in Comput. Sci.*, pages 21–69. Springer-Verlag, 1992.

6. W. Brauer, R. Gold, and W. Vogler. A survey of behaviour and equivalence preserving refinements of Petri nets. In G. Rozenberg, editor, *Advances in Petri Nets'90*, volume 483 of *Lect. Notes in Comput. Sci.*, pages 1–46. Springer-Verlag, 1991.

7. R. Bruni. *Tile Logic for Synchronized Rewriting of Concurrent Systems*. PhD thesis TD-1/99, Computer Science Department, University of Pisa, 1999.

8. R. Bruni, J. Meseguer, U. Montanari, and V. Sassone. A comparison of petri net semantics under the collective token philosophy. In J. Hsiang, A. Ohori, editors, *Proceedings ASIAN'98, 4th Asian Computing Science Conference*, volume 1538 of *Lect. Notes in Comput. Sci.*, pages 225–244. Springer Verlag, 1998.

9. R. Bruni, J. Meseguer, U. Montanari, and V. Sassone. Functorial semantics for Petri nets under the individual token philosophy. In M. Hofmann, G. Rosolini, D. Pavlovic, editors, *Proceedings CTCS'99, 8th conference on Category Theory and Computer Science*, volume 29 of *Elect. Notes in Th. Comput. Sci.*, Elsevier Science, 1999.

10. R. Bruni and U. Montanari. Zero-safe nets, or transition synchronization made simple. In C. Palamidessi and J. Parrow, editors, *Proceedings EXPRESS'97, 4th workshop on Expressiveness in Concurrency*, volume 7 of *Elect. Notes in Th. Comput. Sci.*, Elsevier Science, 1997.

11. R. Bruni and U. Montanari. Zero-safe nets: The individual token approach. In F. Parisi-Presicce, editor, *Proceedings WADT'97, 12th workshop on Recent Trends in Algebraic Development Techniques*, volume 1376 of *Lect. Notes in Comput. Sci.*, pages 122–140. Springer-Verlag, 1998.

12. R. Bruni and U. Montanari. Zero-safe nets: Comparing the collective and individual token approaches. *Inform. and Comput.*, 156:46–89. Academic Press, 2000.

13. R. Bruni and U. Montanari. Zero-safe nets: Composing nets via transition synchronization. In H. Weber, H. Ehrig, and W. Reisig, editors, *Proceedings Int. Colloquium on Petri Net Technologies for Modelling Communication Based Systems*, pages 43–80. Fraunhofer Gesellschaft ISST, 1999.

14. N. Busi. On zero safe nets, April 1999. Private communication.

15. R.J. van Glabbeek. Petri nets, configuration structures and higher dimensional automata. In J.C.M. Baeten and S. Mauw, editors, *Proceedings CONCUR'99, 10th International Conference on Concurrency Theory*, volume 1664 of *Lect. Notes in Comput. Sci.*, pages 21–27. Springer-Verlag, 1999.

16. R.J. van Glabbeek and U. Goltz. Refinement of actions and equivalence notions for concurrent systems. Hildesheimer Informatik Bericht 6/98, Institut fuer Informatik, Universitaet Hildesheim, 1998.

17. R.J. van Glabbeek and G.D. Plotkin. Configuration structures. In D. Kozen, editor, *Proceedings LICS'95, 10th Annual IEEE Symposium on Logic In Computer Science*, pages 199–209. IEEE Computer Society Press, 1995.

18. U. Goltz and W. Reisig. The non-sequential behaviour of Petri nets. *Inform. and Comput.*, 57:125–147. Academic Press, 1983.

19. J. Meseguer, U. Montanari, and V. Sassone. Process versus unfolding semantics for place/transition Petri nets. *Theoret. Comput. Sci.*, 153(1-2):171–210, 1996.

20. M. Nielsen, G. Plotkin, and G. Winskel. Petri nets, event structures and domains, part I. *Theoret. Comput. Sci.*, 13:85–108, 1981.

21. C.A. Petri. *Kommunikation mit Automaten*. PhD thesis, Institut für Instrumentelle Mathematik, Bonn, 1962.
22. K. Reinhardt. Reachability in Petri nets with inhibitor arcs. Technical Report WSI-96-30, Wilhelm Schickard Institut für Informatik, Universität Tübingen, 1996.
23. W. Reisig. *Petri Nets: An Introduction*. EACTS Monographs on Theoretical Computer Science. Springer-Verlag, 1985.
24. I. Suzuki and T. Murata. A method for stepwise refinement and abstraction of Petri nets. *J. Comput. and System Sci.*, 27:51–76, 1983.
25. R. Valette. Analysis of Petri nets by stepwise refinement. *J. Comput. and System Sci.*, 18:35–46, 1979.
26. W. Vogler. Behaviour preserving refinements of Petri nets. In G. Tinhofer and G. Schmidt, editors, *Proceedings 12th International Workshop on Graph-Theoretic Concepts in Computer Science*, volume 246 of *Lect. Notes in Comput. Sci.*, pages 82–93. Springer-Verlag, 1987.
27. G. Winskel. Event structures. In W. Brauer, editor, *Proceedings Advanced Course on Petri Nets*, volume 255 of *Lect. Notes in Comput. Sci.*, pages 325–392. Springer-Verlag, 1987.

Efficient Symbolic State-Space Construction for Asynchronous Systems*

Gianfranco Ciardo[1], Gerald Lüttgen[2], and Radu Siminiceanu[1]

[1] Department of Computer Science, College of William and Mary,
Williamsburg, VA 23187, USA, {ciardo, radu}@cs.wm.edu
[2] ICASE, NASA Langley Research Center,
Hampton, VA 23681, USA, luettgen@icase.edu

Abstract. Many techniques for the verification of reactive systems rely on the analysis of their reachable state spaces. In this paper, a new algorithm for the symbolic generation of the state spaces of *asynchronous* system models, such as Petri nets, is developed. The algorithm is based on previous work that employs *Multi-valued Decision Diagrams* for efficiently storing sets of reachable states. In contrast to related approaches, however, it fully exploits *event locality*, supports intelligent *cache management*, and achieves faster convergence via advanced *iteration control*. The algorithm is implemented in the Petri net tool SMART, and run-time results show that it often performs significantly faster than existing state-space generators.

1 Introduction

Many state-of-the-art verification techniques rely on the *automated construction of the reachable state space* of the system under consideration. Unfortunately, state spaces of real-world systems are usually very large, sometimes too large to fit in a computer's memory. One contributing problem is the concurrency inherent in reactive systems, such as those specified by *Petri nets* [18]. Consequently, many research efforts in *state-exploration techniques* concentrated on the efficient exploration and storage of large state spaces. These may be categorized according to whether sets of states are stored explicitly or symbolically.

Explicit techniques represent state spaces by trees, hash tables, or graphs, where each state corresponds to an entity of the underlying data structure. Thus, the memory needed to store the state space of a system is linear in the number of the system's states, which in practice limits these techniques to fairly small systems having at most a few million states.

Symbolic techniques allow one to store reachability sets in sublinear space. They often use *Binary Decision Diagrams* (BDDs) as a data structure for efficiently representing Boolean functions [1], into which state spaces may be

---

* This work was supported by the National Aeronautics and Space Administration under NASA Contract No. NAS1-97046 while the authors were in residence at the Institute for Computer Applications in Science and Engineering (ICASE), NASA Langley Research Center, Hampton, VA 23681, USA.

M. Nielsen, D. Simpson (Eds.): ICATPN 2000, LNCS 1825, pp. 103–122, 2000.
© Springer-Verlag Berlin Heidelberg 2000

mapped. The advent of BDD-based techniques pushed the manageable sizes of state spaces to about 10^{20} states [4]. In the Petri net community, BDDs were applied by Pastor et al. [19, 20], Varpaaniemi et al. [23], and others for the generation of the reachability sets of Petri nets. Recently, symbolic state-space generation for Petri nets has been significantly improved [17] by considering *Multi-valued Decision Diagrams* (MDDs) [15] instead of BDDs. MDDs essentially represent integer functions and allow one to efficiently encode the state of an entire subnet of a Petri net using only a single integer variable, where the state spaces of the subnets are built by employing traditional techniques. Experimental results reported in [17] show that this approach enables the representation of even larger state spaces of size 10^{60} and even 10^{600} states for particularly regular nets. However, the time needed to generate some of these state spaces ranges from minutes for the *dining philosophers* [20], with 1000 philosophers, to hours for the *Kanban system* [7], with an initial token count of 75 tokens. Thus, state-space generation shifts from a *memory-bound* to a *time-bound* problem.

The objective of this paper is to improve on the time efficiency of symbolic state-space generation techniques for a particular class of systems, namely *asynchronous systems*. Our approach aims at exploiting the concept of *event locality* inherent in such systems. In Petri nets, for example, event locality means that only those sub-markings belonging to the subnets affected by a given transition need to be updated when the transition fires. Whereas event locality was investigated in explicit state-space generation techniques [6], it has been largely ignored in symbolic techniques. Only the MDD-based approach presented in [17] touches on event locality, but it exploits this concept only superficially. In particular, this approach does not support direct jumps to and from the part of the MDD corresponding to the sub-markings that need to be updated when a transition fires. The present paper develops a new algorithm for building the reachable state spaces of asynchronous systems. Like [17], it uses MDDs for representing state spaces; unlike [17], it fully exploits event locality. Moreover, it introduces an intelligent mechanism for *cache management* and also achieves faster convergence by firing events in a specific, predefined order. The new algorithm is implemented in the tool SMART [6]. When applied to a suite of well-known Petri net models, it proves to be significantly faster than the one presented in [17], while inducing only a small overhead regarding space efficiency. The algorithm can be employed immediately for verifying *safety properties*, such as the absence of deadlocks. Moreover, the developed MDD manipulation techniques may also provide a basis for implementing MDD-based model checkers [10].

2 Structured State Spaces and MDDs

We choose to specify finite-state asynchronous systems by Petri nets [18]; however, the concepts presented here are not limited to this choice. Thus, we interchangeably use the notions *net* and *system*, *subnet* and *sub-system*, *transition* and *event*, *marking* and *(global) state*, as well as *sub-marking* and *local state*.

Consider a Petri net with a finite set $\mathcal{P}$ of places, a finite set $\mathcal{E}$ of events, and an initial marking $s_0 \in \mathbb{N}^{|\mathcal{P}|}$. The semantics of Petri nets defines how the *firing* of an event e can move the net from some state s to another state s'. We denote the set of successor states reachable from state s via event e by $\mathcal{N}(e, s)$. If $\mathcal{N}(e, s) = \emptyset$, event e is *disabled* in s; otherwise, it is *enabled*. For Petri nets, $\mathcal{N}$ is essentially a simple encoding of the input and output arcs; thus, $\mathcal{N}(e, s)$ contains at most one element. For other formalisms, however, $\mathcal{N}(e, s)$ might contain several elements. We are interested in exploring the set $\mathcal{S}$ of reachable states of the considered net. $\mathcal{S}$ is formally defined as the smallest set that (i) contains s_0 and (ii) is closed under the "one-step reachability relation," i.e., if $s \in \mathcal{S}$, then $\mathcal{N}(e, s) \subseteq \mathcal{S}$, for any event $e \in \mathcal{E}$.

As in [17], our encoding of the state space of a Petri net requires us to partition the net into K subnets by splitting its set of places $\mathcal{P}$ into K subsets. This implies a partition of a *global state* s of the net into K *local states*, i.e., s has the form $(s_K, s_{K-1}, \dots, s_1)$; the "backwards" numbering will prove to be a reasonable convention when representing global states using MDDs. The partition of $\mathcal{P}$ must satisfy a fundamental *product-form requirement* [8] which demands for function $\mathcal{N}$ to be written as the cross-product of K local functions, i.e., $\mathcal{N}(e, s) = \mathcal{N}_K(e, s_K) \times \mathcal{N}_{K-1}(e, s_{K-1}) \times \cdots \times \mathcal{N}_1(e, s_1)$, for all $e \in \mathcal{E}$ and $s \in \mathcal{S}$. Furthermore, in practice, each subnet should be small enough such that its reachable *local state space* $\mathcal{S}_k = \{s_{k,0}, s_{k,1}, \dots, s_{k,N_k-1}\}$ can be efficiently computed by traditional techniques, where $N_k \in \mathbb{N}$ is the number of reachable states in subnet k. Note that this might require the explicit insertion of additional constraints to allow for the correct computation of $\mathcal{S}_k$ in isolation, or one may use a small superset of $\mathcal{S}_k$ obtained by employing *p-invariants* [18]. For all the examples we present, the computation of the local state spaces requires negligible time. Once $\mathcal{S}_k$ has been built, we can identify it with the set $\{0, 1, \dots, N_k - 1\}$. Moreover, a set $\mathcal{S}$ of global states can then be encoded by the *characteristic function* $f_{\mathcal{S}} : \{0, \dots, N_K - 1\} \times \cdots \times \{0, \dots, N_1 - 1\} \longrightarrow \{0, 1\}$ defined by $f_{\mathcal{S}}(s_K, s_{K-1}, \dots, s_1) = 1$ if and only if $(s_K, s_{K-1}, \dots, s_1) \in \mathcal{S}$.

Multi-valued Decision Diagrams. *Multi-valued Decision Diagrams* [15], or MDDs, are data structures for representing integer functions of the form

$$f : \{0, \dots, N_K - 1\} \times \cdots \times \{0, \dots, N_1 - 1\} \longrightarrow \{0, \dots, M - 1\}$$

where $K, M \in \mathbb{N}$ and $N_k \in \mathbb{N}$, for $K \geq k \geq 1$. When $M = 2$ and $N_k = 2$, for $K \geq k \geq 1$, function f is a Boolean function, and MDDs coincide with the better known *Binary Decision Diagrams* (BDDs) [1, 2]. We use the special case $M = 2$ to store the characteristic functions of the previous section.

Traditionally, integer functions are often encoded by *value tables* or *decision trees*. Figure 1, left-hand side, shows the decision tree of the minimum function $min(a, b, c)$, where the variables a, b, and c are taken from the set $\{0, 1, 2\}$. Hence, $K = 3$ and $N_1 = N_2 = N_3 = M = 3$. Each internal node, which is depicted by an oval, is labeled by a variable and has arcs directed towards its three children. The branch labeled with i corresponds to the case where the variable of the

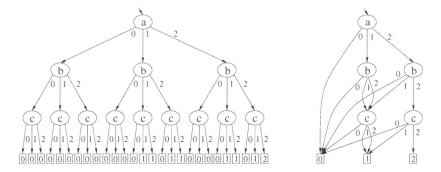

Fig. 1. Representation of $min(a,b,c)$ as decision tree (left) and as MDD (right)

node under consideration is assigned value i. Moreover, all paths through the tree have the same *variable ordering*, which in our example is $a < b < c$. Leaf nodes, depicted by squares, are labeled by either 0, 1, or 2. Each path from the root to a leaf node corresponds to an assignment of the variables to values. The value of the leaf in a given path is the value of the function with respect to the assignment for this path.

An MDD is a representation of a decision tree as *directed acyclic graph*, where identical subtrees are merged. More precisely, MDDs are *reduced* decision trees which do not contain any *non-unique* or *redundant* node: a node is non-unique, if it is a replica of another node, and redundant, if all its children are identical. Together with a fixed variable ordering, these two requirements ensure that MDDs provide a *canonical* representation of integer functions [15]. Note that the elimination of redundant nodes implies that arcs can skip levels, e.g., the arc labeled with 0 connecting node a to leaf node 0 in Fig. 1, right-hand side, skips levels b and c. Hence, the value of the function is 0, whenever a is 0. For many functions, MDD representations can be exponentially more compact than their corresponding value tables or decision trees. However, the degree of compactness depends on the considered function and the chosen variable ordering.

Data Structures for MDDs. We organize MDD nodes in levels ranging from K at the top to 1 at the bottom. Additionally, there is the special level 0 which contains either or both leaf nodes corresponding to the values 0 and 1, indicating whether a state is reachable or not. The addresses of the nodes at a given level are stored within a hash table, to provide fast access to them and to simplify detection of non-unique nodes. Hence, we have K hash tables which together represent an MDD; we also refer to this data structure as *unique table*. Note that we could as well use a single unique table for representing MDDs, but this would require us to store the level for each node; furthermore, the level-wise organization of our data structures will prove very useful below. Each node at level k consists of an array of N_k node addresses, which contains the arcs to the children of the node. Since we enforce the reducedness property, we use the value of this array to compute the hash value of the node. In the following, we let

Table 1. *Union* operation on MDDs

$Union(\text{in } p : mddAddr, \text{in } q : mddAddr) : mddAddr$
1. if $p = \langle 0, 1 \rangle$ or $q = \langle 0, 1 \rangle$ return $\langle 0, 1 \rangle$;   • deal with the base cases first
2. if $p = \langle 0, 0 \rangle$ or $p = q$ return q;
3. if $q = \langle 0, 0 \rangle$ return p;
4. $k \Leftarrow Max(p.lvl, q.lvl)$;   • maximum of the levels of p and q
5. if $LookUpInUC(k, p, q, r)$ then return r;   • found in the union cache
6. $r \Leftarrow CreateNode(k)$;   • otherwise, the union needs to be computed in r
7. for $i = 0$ to $N_k - 1$ do
8.   if $k > p.lvl$ then $u \Leftarrow Union(p, q{\rightarrow}dw[i])$;   • p is at a lower level than q
9.   else if $k > q.lvl$ then $u \Leftarrow Union(p{\rightarrow}dw[i], q)$;   • q is at a lower level than p
10.   else $u \Leftarrow Union(p{\rightarrow}dw[i], q{\rightarrow}dw[i])$;   • p and q are at the same level
11. $SetArc(r, i, u)$;   • make u the i-th child of r
12. $r \Leftarrow CheckNode(r)$;   • store r in the unique table
13. $InsertInUC(k, p, q, r)$;   • record the result in the union cache
14. return r;

mddNode denote the type of nodes and *mddAddr* the type of addresses of nodes. For convenience, we write $\langle lvl, ind \rangle$ for the node q stored in the lvl-th unique table at position ind, and $q{\rightarrow}dw[i]$ for the i-th child of q. Finally, we use nodes $\langle 0, 0 \rangle$ and $\langle 0, 1 \rangle$ to indicate the Boolean values 0 and 1 at level 0, respectively.

The *Union* Operation on MDDs. An essential operation for generating reachable state spaces is the binary *union* on sets. Since in our context all sets are represented as MDDs, an algorithm is needed which takes two MDDs as parameters and returns a new MDD, representing the union of the sets encoded by its arguments. This algorithm, which is very similar to the one used in [17], is shown in Table 1. It recursively analyzes the argument MDDs when descending from the maximum level k of the argument MDDs to the lowest level 0 and builds the result MDD when finishing the recursions by ascending from level 0 to level k. Note that the maximum of the levels of the argument MDDs is the highest level the result MDD can have.

The base cases of the recursive function *Union* are handled in Lines 1–3, where the MDDs $\langle 0, 0 \rangle$ and $\langle 0, 1 \rangle$ encode the *empty set* and the *full set*, respectively. If $k > 0$, a cache – the so-called *union cache* – is used to check whether the union of the arguments p and q has been computed previously. If so, the result stored in the cache is returned. Otherwise, a new MDD node at level k is created, whose i-th child is determined by recursively building the union of the i-th child of p and the i-th child of q, for all $0 \le i < N_k$ (cf. Lines 7–11). However, one needs to take care of the fact that some child might not be explicitly represented, namely if it is redundant (cf. Lines 8 and 9). Finally, to ensure that the resulting MDD is reduced, node r is checked by calling function *CheckNode(r)*. If r is redundant, then *CheckNode* destroys r and returns r's child, and if r is equivalent to another node r' having the same children, then *CheckNode* destroys r and returns r'. Otherwise, *CheckNode* inserts node r in the unique table and returns it. Note that the algorithm in Table 1 can be easily adapted for computing other binary operations, such as intersection, by modifying Lines 1–3.

Table 2. Iterative state-space generation

$MDDgeneration(\text{in } m : \text{array}[1, \ldots, K] \text{ of } int) : mddAddr$
1. for $k = 1$ to K do $ClearUT(k)$;   • clear unique table
2. $q \Leftarrow SetInitial(m)$;   • build the MDD representing the initial state
3. repeat   • start state-space exploration
4.   for $k = 1$ to K do $ClearUC(k)$;   • clear union cache
5.   for $k = 1$ to K do $ClearFC(k)$;   • clear firing cache
6. $mddChanged \Leftarrow false$;   • *true* if MDD changes in this iteration
7.   foreach event e do $Fire(e, q, mddChanged)$ • fire e, add newly reached states
8. until $mddChanged = false$;   • keep iterating until fixed point is reached
9. return q;   • return MDD representing the reachable state space

MDD-based State-space Construction. Table 2 shows a naive, iterative, and MDD-based algorithm to build the reachable state space of a system represented by a Petri net. As explained earlier, a global state $(s_K, s_{K-1}, \ldots, s_1)$ is stored over the K levels of the MDD, one substate per level. Recall that this requires us to partition Petri nets into subnets. While this can in principle be done automatically, it is still an open problem how to efficiently find "good" partitions, i.e., those that lead to small MDD representations of reachable state spaces; see [17] for a detailed discussion on partitioning.

The semantics of the Petri net under study is encoded in procedure *Fire* (cf. Table 2), which updates the MDD rooted at q according to the firing of event e by appropriately applying the *Union* operation. For efficiency reasons, it also makes use of another cache, which we refer to as *firing cache*. The procedure additionally updates a flag *mddChanged*, if the firing of e added any new reachable states. After first clearing the unique table, the initial marking m of the Petri net under consideration is stored as an MDD via procedure *SetInitial*. The algorithm then proceeds iteratively. In each iteration, every enabled transition is fired, and the potentially new states are added to the MDD. This is done until the MDD does not change, i.e., until no more reachable states are discovered. Finally, the root node q, representing the reachable state space of the Petri net, is returned.

3   The Concept of Event Locality

Our improvements for the MDD-based generation of reachable state spaces rely on the notion of *event locality*, which asynchronous systems inherently obey. Event locality is defined via the concept of *independence* of events from subnets. An event e is said to be *independent* of the k-th subnet of the net under consideration, or independent of level k, if $s_k = s'_k$ for all $s = (s_K, s_{K-1}, \ldots, s_1) \in \mathcal{S}$ and $s' = (s'_K, s'_{K-1}, \ldots, s'_1) \in \mathcal{N}(e, s)$. Otherwise, e depends on the k-th subnet, or on level k. If an event depends only on a single level k, it is called a *local event* for level k; otherwise, it is a *synchronizing event* [17]. We let $First(e)$ and $Last(e)$ denote the maximum and minimum levels on which e depends. Hence, e is independent of every level k satisfying $K \geq k > First(e)$ or $Last(e) > k \geq 1$, while e might or might not depend on levels strictly between

First(*e*) and *Last*(*e*). For asynchronous systems in particular, the range of affected levels is usually significantly smaller than K for most events e. We assume that all local events for level k are merged into a single *macro event* l_k satisfying $\mathcal{N}_k(l_k, s) =_{\text{df}} \bigcup_{e \in \mathcal{E}:First(e)=Last(e)=k} \mathcal{N}_k(e, s)$, for all $s \in \mathcal{S}$. This convention does not only simplify notation, but also improves the efficiency of our state-space generation algorithm.

Our aim is to define MDD manipulation algorithms that exploit the concept of event locality. Since an event e affects local states stored between levels *First*(*e*) and *Last*(*e*), firing e only causes updates of MDD nodes between these levels, plus possibly at levels higher than *First*(*e*), but only when a node at level *First*(*e*) becomes redundant, and possibly levels lower than *Last*(*e*), but only until recursive *Union* calls stop creating new nodes. To benefit from this observation, we need to be able to access MDD nodes by "jumping in the middle" of an MDD, namely to level *First*(*e*), rather than always having to start manipulating MDDs at their roots, as is done in traditional approaches and in [17]. This is the reason why we partition the unique table into a K-dimensional array of lists of nodes. However, two problems need to be addressed when accessing an MDD directly at some level *First*(*e*).

Implicit Roots. When one wants to explore an MDD from level *First*(*e*), all nodes at this level should play the role of root nodes. However, some of them might not be represented explicitly, since redundant nodes are not stored. This happens whenever there is a node p at a level higher than *First*(*e*) pointing to a node q at a level k satisfying $First(e) > k \geq Last(e)$. This situation is illustrated in Fig. 2, left-hand side. Conceptually, we have to re-insert these "implicit roots" at level *First*(*e*) when we explore and modify the MDD due to the firing of event e. There are two approaches for doing this. The first approach stores a bag (multiset) of *upstream arcs* in each node q, corresponding to the *downstream arcs* pointing to q. Hence, for each i such that $p \rightarrow dw[i] = q$, there is an occurrence of p in the bag of q's upstream arcs. Implicit roots can then be detected by scanning each node stored in the unique tables for levels *First*(*e*) + 1 through *Last*(*e*) and by checking whether the node possesses one or more upstream arcs to a node at a level above *First*(*e*). If so, an implicit root, i.e., a redundant node, is inserted at level *First*(*e*). Note that at most one implicit root needs to be inserted per node, regardless of how many arcs reach it; in our example, the arcs from both p and p' are re-routed to the same new implicit root. These redundant nodes will be deleted after firing event e, if they are still redundant. Our second approach keeps all unique redundant nodes, so that downstream arcs in the resulting MDD exist only between subsequent levels. Then, the nodes at level *First*(*e*) are exactly all the nodes from which we need to start exploring the underlying MDD when firing event e. Note that this slight variation of MDDs still possesses the fundamental property of being a *canonical* representation.

We refer to the two variants of our algorithm as *upstream-arcs approach* and *forwarding-arcs approach*. The latter approach, when compared to the former, eliminates the expensive search for implicit roots. However, both involve some memory penalty, the former for the storage of upstream arcs, which can

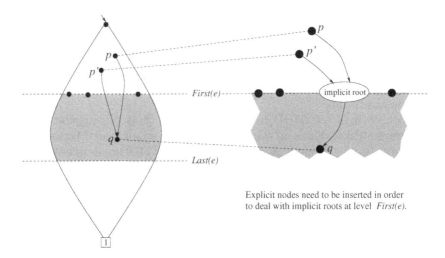

Explicit nodes need to be inserted in order
to deal with implicit roots at level *First(e)*.

Fig. 2. Illustration of the problem of *implicit roots*

in the worst case double the space requirements, and the latter because of the preservation of redundant nodes. We have implemented both approaches, and experimental results show that these memory overheads are compensated by a smaller peak number of MDD nodes when compared to [17] (cf. Sec. 6).

In-place Updates. Once all explicit and implicit nodes at level $First(e)$ are detected, one can update the MDD to reflect that the firing of event e may lead to new, reachable states. Our routine *Fire* implementing this update is described in Sec. 5. It relies on the *Union* operation, as presented in Table 1, i.e., new MDD nodes are created and appropriately inserted, as needed. However, there is one important difference with respect to existing approaches. Our *Fire* operation stops creating new MDD-nodes as soon as it reaches level $First(e)$ when backtracking from recursive calls. At this level our algorithm just links the new sub-MDDs at the appropriate positions in the original MDD, in accordance with event locality. The only difficulty with the in-place update of a node p arises when it becomes redundant or non-unique. In the former case, p must be deleted and its incoming arcs be re-directed to its unique child node q. In the latter case, p must be deleted and its incoming arcs be re-directed to replica node q.

In the upstream-arcs approach, either operation can be easily accomplished since p knows its parents. In the forwarding-arcs approach we keep redundant nodes; thus, we eliminate p only if it becomes non-unique. Instead of scanning all nodes in level $First(e) + 1$ to search for arcs to p, which is a costly operation, we mark p as deleted and set a forwarding arc from p to q. The next time a node accesses p, it will update its own pointer to p, so that it points to q instead. Since node q itself might be marked as deleted later on, forwarding chains of nodes can arise. In our implementation, the nodes in these chains are deleted after all events at level $First(e)$ have been fired and before nodes at the next higher level are explored.

It is important to note that, although these *in-place updates* change the meaning of MDD-nodes at higher levels, they do not jeopardize the correctness of our algorithm; this is due to the property of event locality. Rather than performing *in-place updates*, existing approaches reported in the literature create an MDD encoding the set of global states reachable from the current states in the state space by firing event e. This is a K-level MDD, i.e., it is expensive to build compared to our sub-MDD, especially when MDDs are tall and the effect of e is restricted to a small range of levels.

Summarizing, it is the notion of event locality that allows us to drastically improve on the time efficiency of MDD-based state-space generation techniques. Exploiting locality, we can jump in and out of the "middle" of MDDs, thereby exploring only those levels that are affected by the event under investigation. While the approach reported in [17] also exploits locality, it just considers some simplifications and improvements of MDD manipulations in the case of local events. However, it does not support localized modifications of MDDs, neither for synchronizing nor for local events.

4   Improving Cache Management and Iteration Control

The concept of event locality also paves the road towards significant improvements in *cache management* and *iteration control*.

Intelligent Cache Management. The technique of in-place updates allows us to enhance the efficiency of the union cache. In related work, including [17], the lifetime of the contents of the union cache cannot span more than one iteration, since the root of any MDD is deleted and re-created whenever additional reachable states are incorporated in the MDD.

In contrast, in our approach the "wave" of changes towards the root, caused by firing an event e, is stopped at level $First(e)$, where only a pointer is updated. This permits some union cache entries to be reused over several iterations until the referred nodes are either changed or deleted. For this purpose, MDD nodes in our implementation have two status bits attached, namely a *cached* flag and a *dirty* flag. Instead of thoroughly cleaning up the union cache after each iteration, we can now perform a *selective purging* according to the above flags. If an MDD node associated with a union cache entry is not deleted and if the copies present in the cache are not stale, the result may be kept in the union cache. Experimental studies show us that the rate of reusability of union cache entries averages about 10% and that the overall performance of our algorithm can be improved by up to 13% when employing this idea.

Additionally, we devise a second optimization technique for the union cache, which is based on *prediction*. Our prediction relies on the fact that if $Union(p,q)$ returns r, then also $Union(p,r)$ and $Union(q,r)$ will return r. Thus, these two additional results can be memorized in the cache, immediately after storing the entry for $Union(p, q)$. Experiments indicate that this heuristics accelerates our algorithm by up to 12%. The reason for such a significant improvement is the following. Assume we are exploring the firing of event e in node p at level k, and assume $j \in \mathcal{N}_k(e, i)$. Then, the set of states encoded by the MDD rooted at $p \rightarrow dw[i]$ needs to be added to the set of states encoded by the MDD rooted at $p \rightarrow dw[j]$. Let r be the result of $Union(p \rightarrow dw[i], p \rightarrow dw[j])$, which becomes the new value of $p \rightarrow dw[j]$. In the next iteration, and assuming that p has not been deleted, we explore event e in node p again and, consequently, find out that e is enabled in local state i. Hence, we need to perform the update $p \rightarrow dw[j] \Leftarrow Union(p \rightarrow dw[i], p \rightarrow dw[j])$ again. However, if p has not changed, $Union(p \rightarrow dw[i], p \rightarrow dw[j])$ is identical to $Union(p \rightarrow dw[i], r) = r$. By having cached r in the previous iteration, we can avoid computing this union.

Advanced Iteration Control. Event locality also allows us to reduce the number of iterations needed for generating state spaces. Existing MDD-based algorithms for Petri nets [17, 20] fire events in some arbitrary order within each iteration, as indicated in Line 7 of function *MDDgeneration* in Table 2. In our version of *MDDgeneration*, however, we presort events according to function $First(\cdot)$. Our algorithm then starts at level 1 and searches for the states that can be reached from the initial state by firing all events e satisfying $First(e) = 1$ and $Last(e) \geq 1$, i.e., the macro event l_1. When reaching level k, the algorithm finds the states that can be added to the current state space by firing all events e satisfying $First(e) = k$ and $Last(e) \geq 1$, i.e., the local macro event l_k at level k and all synchronizing events that affect only level k and some levels below. Moreover, in our implementation we repeatedly fire each event as long as it is enabled and as long as firing it adds new states. This specific sequence of firing events is essential for the correctness and efficiency of the implementation of our cache management. By working from the bottom to the top levels we can clear the union and firing caches selectively, thus, extending the lifetime of cache entries. Moreover, the access pattern to the caches is more regular. Our firing sequence also enables delayed node deletion which allows for the efficient collection and removal of non-unique and disconnected nodes.

In [17], repeatedly firing events is only applied for local events which are relatively inexpensive to process, while synchronizing events are still fired once and in no particular order. We stress that while the new iteration control means that our iterations are potentially more expensive than those in [17], they are also potentially fewer. More precisely, our algorithm generates state spaces in at most as many iterations as the maximum *synchronizing distance* of any reachable state s, which is defined in [17] as the minimal number of synchronizing events required to reach s from the initial state. We stress that the advanced iteration control we use implies a much finer management of MDD nodes than the one resulting from the use of breadth-first, mixed breadth-first/depth-first BDDs [25], or other techniques for reducing the intermediate sizes of BDDs [21].

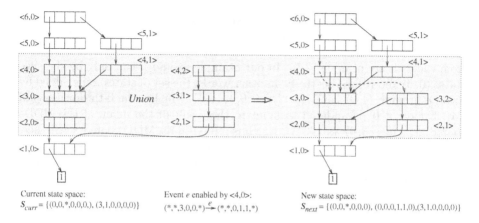

Fig. 3. Example of an MDD-modification in response to *firing an event*

5   Details of the New Algorithm

We now present some details of our new algorithm and argue for its correctness; we refer the reader to [5] for the complete pseudo-code.

Illustration of MDD-based Firing of Events. At each iteration of our algorithm, enabled events are fired to discover additional reachable states which are then added to the MDD representing the currently-known portion of the reachability set. Function $Fire(e, \cdot, \cdot)$ implements this behavior with respect to event e. Figure 3 illustrates how *Fire* works: the example net is partitioned into six subnets, each of them having four possible local states, numbered from 0 to 3. Hence, our MDD has six levels, and each MDD node has four downstream arcs; here, we do not draw node $\langle 0, 0 \rangle$, nor any arc to it. Let the current state space, depicted on the left in Fig. 3, be $\mathcal{S}_{curr} = \{(0, 0, *, 0, 0, 0), (3, 1, 0, 0, 0, 0)\}$, where "$*$" stands for any local state. Assume further that event e is enabled in every state of the form $(*, *, 3, 0, 0, *)$ and that the new state reached when firing e is $(*, *, 0, 1, 1, *)$, i.e., $First(e) = 4$ and $Last(e) = 2$. Hence, if the net is in a global state described by local state 3 at level 4 and local state 0 at levels 3 and 2, event e can fire and the local states of the affected subnets are updated to 0, 1, and 1, respectively.

Exploiting event locality, our search for enabling sequences starts directly at level $First(e) = 4$. The sub-MDDs rooted at this level are searched to match the enabling pattern of e. At level 4, only the MDD rooted at $\langle 4, 0 \rangle$ contains such a pattern, along the path $\langle 4, 0 \rangle \xrightarrow{3} \langle 3, 0 \rangle \xrightarrow{0} \langle 2, 0 \rangle \xrightarrow{0} \langle 1, 0 \rangle$. Then, our algorithm generates a new MDD rooted at node $\langle 4, 2 \rangle$, representing the set of substates for levels 4 through 1 that can be reached from $\langle 4, 0 \rangle$ via e. This MDD is depicted in Fig. 3 in the middle. Note that only nodes at levels $First(e)$ through $Last(e)$ might have to be created, since those below $Last(e)$ can simply be linked to existing nodes, such as node $\langle 1, 0 \rangle$ in our example. Indeed, in our implementation even node $\langle 4, 2 \rangle$ is actually not allocated, since we explore it one child at a time. This MDD corresponds to all states of the form $(\alpha, 0, 1, 1, \beta)$, where α is any substate leading to node $\langle 4, 0 \rangle$ and where β is a substate reachable from the 0-th arc of node $\langle 2, 0 \rangle$. In our example, α and β can only be the substates $(0, 0)$ and (0), respectively. In other words, the set of states to be added by firing e in node $\langle 4, 0 \rangle$ is $\mathcal{S}_{add} = \{(0, 0, 0, 1, 1, 0)\}$. Finally, the 0-th downstream arc of node $\langle 4, 0 \rangle$ is updated to point to the result of the union of the MDDs rooted at nodes $\langle 3, 0 \rangle$ and $\langle 3, 1 \rangle$, which is stored in an MDD rooted at the new node $\langle 3, 2 \rangle$, as depicted on the right in Fig. 3. Hence, the resulting state space $\mathcal{S}_{next}$ is $\{(0, 0, *, 0, 0, 0), (0, 0, 0, 1, 1, 0), (3, 1, 0, 0, 0, 0)\}$. Observe that our version of $Fire(e)$ is more efficient than the one in [17] since it exploits the locality of e and, thus, operates on smaller MDDs. This is important as the complexity of the *Union* operation is proportional to the sizes of its operand MDDs.

Further Implementation Details. MDD nodes store not only the addresses of their children, but also Boolean flags for garbage collection and intelligent cache management, as well as information specific to the upstream-arcs approach and to the forwarding-arcs approach.

In our implementation, nodes are stored using one *heap array* per MDD level. The pages of the heap array are created only upon request and accommodate dynamic deletion and creation of nodes. Therefore, existing nodes may not be stored contiguously in memory. For fast retrieval we maintain a doubly-linked list of nodes. Upon deletion, a node is moved to the back of the list, thereby, allowing for garbage collection (but not garbage removal) in constant time.

The unique table, the union cache, and the firing cache are organized as arrays of hash tables, i.e., one hash table per level. For the unique table, the hash key of a node is determined using the values in its dw-array. For the union cache, the addresses of the two MDD nodes involved in the union are used to determine the hash key. Together with the *cached* and *dirty* flags, this allows us to reuse union cache entries across iterations without danger of accessing stale values. Finally, the hash key for firing cache entries is determined using only the address of the MDD node to which the firing operation is applied. Note that the identity of the event is implicit, since the firing cache is cleared when moving from one event to the next. The alternative approach, i.e., allowing the co-existence of entries referring to different events, would require a larger cache with a key based on a pair of MDD node and event. However, this would not bring enough benefits as the major cost of processing the event firing lies in the *Union* operations, and these can indeed be cached across operations.

For the upstream-arcs approach, MDD nodes include the addresses of their parents, which we store in a bag. Our implementation uses a dynamic data structure for bags rather than a static data structure, since the number of parents of a node is not known in advance and may be very large, in the range of several thousand nodes. While this memory overhead is still acceptable, the approach also puts a burden on time efficiency, since each update of a downstream arc must be reflected by an update of the corresponding upstream arc. Moreover, the bag of some node q only stores the addresses of parents p, as well as the number of indexes i such that $p{\rightarrow}dw[i] = q$, but not the indexes themselves. Thus, a linear search in $p{\rightarrow}dw$ must be performed to find these indexes. The alternative of storing these indexes in q would require even more memory overhead.

Regarding the forwarding-arcs approach, time efficiency is improved by allowing redundant nodes to be represented explicitly. As a consequence, MDD nodes do not need to store bags of parents' addresses, but simply a counter indicating the number of incoming arcs [17]. When this counter reaches zero, it indicates that the node has become disconnected and can be deleted. Experiments show that the memory overhead of this approach, due to the storage of redundant nodes and the delayed deletion of non-unique nodes, is about the same as the memory overhead of the upstream-arcs approach. However, the forwarding-arcs approach is more time-efficient, as confirmed by the results in Sec. 6.

Correctness of the Algorithm. Here, we informally argue for the correctness of our algorithm since the formal proof is quite lengthy and, thus, omitted. Our comments concern three main features of the algorithm: (i) in-place updates, (ii) iteration control, and (iii) cache management.

The correctness of performing *in-place updates* is implied by the notion of event locality, i.e., by the asynchronous semantics of Petri nets. Formally, consider a snapshot of our algorithm where the MDD under construction currently encodes some state set $\mathcal{S}^* \subseteq \mathcal{S}$ and where event e will be fired next. Assume further that e is enabled in some $s = (s_K, s_{K-1}, \dots, s_1) \in \mathcal{S}^*$ and that $s' = (s'_K, s'_{K-1}, \dots, s'_1)$ is the state reached by firing it. Due to event locality we know that $s_k = s'_k$, for all k satisfying $K \geq k > First(e)$ or $Last(e) > k \geq 1$. Further, we may conclude $r' =_{df} (r_K, \dots, r_{First(e)+1}, s'_{First(e)}, \dots, s'_{Last(e)}, r_{Last(e)-1}, \dots, r_1) \in \mathcal{S}$, for all $r = (r_K, r_{K-1}, \dots, r_1) \in \mathcal{S}^*$. Hence, one may simultaneously add all these states r' to the MDD for $\mathcal{S}^*$ as is done by our in-place updates.

The exact ordering in which events are fired by the *iteration control* does not influence which MDD is returned by our algorithm. Unless an event is ignored forever during the iterations, which is not the case with our iteration control, the algorithm computes the unique least fixed point of the next-state function [11], i.e., the unique MDD encoding the reachable state space of the net under consideration. Additionally, it is obvious that our algorithm terminates for finite-state systems, since each iteration of the body of the algorithm – except the last one in which termination is detected – adds new states to the reachability set.

The correctness of our *cache management* could be hard to establish, as it is closely intertwined with iteration control and the implementation of *Fire*. For now, however, we adopt a conservative cache purging protocol to ensure that no stale entries can be accessed. Advanced protocols that achieve an even higher "hit ratio" in the union and firing caches will be the subject of further study.

6   Experimental Studies

In this section we present several performance results regarding the two variants of our algorithm and compare them with the approach most closely related to

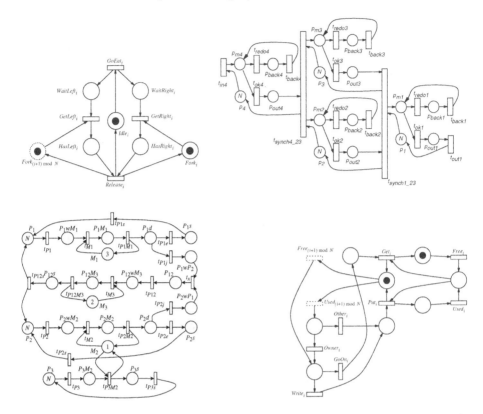

Fig. 4. Petri nets used in our experiments: dining philosophers (upper left), Kanban system (upper right), FMS (lower left), and slotted ring (lower right)

ours, namely the one reported in [17]. The variants of our algorithm are implemented in the Petri net tool SMART [6]. We apply the tool to the four Petri net models also considered in [17], i.e., the *dining philosophers*, the *slotted-ring system*, the *flexible manufacturing system* (FMS), and the *Kanban system*. The former two models, originally taken from [20], are composed of N identical *safe* subnets, i.e., each place contains at most one token at a time. The latter two models, originally taken from [7], have a fixed number of places and transitions, but are parameterized by the number N of initial tokens in certain places. The Petri nets for these systems are depicted in Fig. 4. To use MDDs, we adopt the "best" partitions found in [17]: we consider two philosophers per level and one subnet per level for the slotted-ring protocol, while we split the FMS and the Kanban system into 19 subnets (each place in a separate subnet except for $\{P_1M_1, M_1\}$, $\{P_{12}M_3, M_3\}$, and $\{P_2M_2, M_2\}$) and 4 subnets ($\{p_{mX}, p_{backX}, p_{outX}, p_X\}$ for $X = 1, 2, 3, 4$), respectively.

Table 3 presents several results for the two variants of our new algorithm, as well as the best-known existing algorithm [17], obtained when running SMART on a 500 MHz Intel Pentium II workstation with 512 MB of memory and 512 KB cache. For each model and choice of N we give the size of the state space and the final number of MDD nodes, which is of course independent of the algorithm used. Then, for each algorithm we give the peak number of MDD nodes allocated during execution and the corresponding memory consumption (in KBytes), the number of iterations, and the CPU time (in seconds). The number of iterations for the upstream-arcs and forwarding-arcs approaches coincide. The peak number of nodes and the memory consumption reported for our approach refer to the forwarding-arcs approach. For the upstream-arcs approach, the peak number of nodes reported should be decreased by one for the FMS and the Kanban system, while the memory consumption should be increased by 6–8% for the dining philosophers, 31–33% for the slotted-ring net, 41–45% for the FMS, and 1–5% for the Kanban system. This implies that, even without introducing redundant nodes, essentially all arcs already connect nodes between adjacent levels. Thus, in our examples, the only memory overhead in the forwarding arcs approach is due to postponed node deletion.

For the models we ran, our new approach is up to one order of magnitude faster and with few exceptions uses fewer MDD nodes than the one in [17]. The improvement mainly arises from the structural changes made to the core routine *Fire*, which reflect the notion of event locality. Other improvements – most importantly our cache optimizations – contribute in average about 7–13%, and up to 22% in total, to the overall improvement in time efficiency. A comparison between the run-times for the new algorithm and the ones for the algorithm in [17] indicates an increase factor in speed ranging from approximately constant for the Kanban and FMS nets to what appears to be almost linear (in N) for the slotted-ring model and the dining philosophers. Moreover, the forwarding-arcs approach is slightly faster than the upstream-arcs approach, except for the Kanban system on which we comment below. Since both variants of our new algorithm require significantly fewer peak MDD nodes, where the Kanban system is again an exception, our memory penalty is more than compensated.

The two models whose parameter N affects the height of the MDD, namely the dining philosophers and the slotted-ring model, provide a good testbed for our ideas since they give rise to tall MDDs with a high degree of event locality. For these models, the CPU times are up to 15 times faster than the ones for [17], and the gap widens as we continue to scale-up the nets. The main reason for this is that the number of explored nodes per event fired is more contained in our approach, compared to [17]. When MDD heights are small, such as for the FMS and the Kanban system, our algorithm is still faster than the one in [17], but the difference is not as impressive due to our book-keeping overhead.

The memory consumption figures for the Kanban system are poor compared to the ones for our other examples, although the number of iterations is reduced from $2 \cdot N + 1$ to 4 due to our advanced iteration control, and the solution time is still better than the one of [17]. There are several reasons for this. First, splitting the Kanban net into only four subnets leads to an MDD with a small depth, but a very large breadth, i.e., extremely large nodes. Clearly, any attempt to exploit locality in this case cannot have much pay-off. Second, our garbage

Table 3. Results: (A) dining philosophers, (B) slotted ring, (C) FMS, (D) Kanban

			Approach in [17]				Our new approach				
N	$|\mathcal{S}|$	final nodes	peak nodes	mem. (KB)	# it.	time (sec.)	peak nodes	mem. (KB)	# it.	time (sec.) upstr.	fwd.
A 10	$1.86 \cdot 10^6$	17	45	3	2	0.03	28	4	2	0.02	0.01
50	$2.23 \cdot 10^{31}$	37	285	22	2	0.74	168	26	2	0.11	0.10
100	$4.97 \cdot 10^{62}$	197	585	58	2	3.04	343	54	2	0.30	0.28
200	$2.47 \cdot 10^{125}$	397	1,185	129	2	12.23	693	109	2	1.00	0.90
300	$1.23 \cdot 10^{188}$	597	1,785	198	2	28.10	1,043	164	2	2.16	2.12
400	$6.10 \cdot 10^{250}$	797	2,385	265	2	54.16	1,393	219	2	3.95	3.80
500	$3.03 \cdot 10^{313}$	997	2,985	333	2	81.83	1,743	274	2	6.33	6.02
600	$1.51 \cdot 10^{376}$	1,197	3,585	400	2	125.74	2,093	329	2	9.19	8.77
700	$7.48 \cdot 10^{438}$	1,397	4,185	468	2	181.61	2,443	384	2	12.56	12.03
800	$3.72 \cdot 10^{501}$	1,597	4,785	535	2	247.97	2,793	439	2	16.43	15.79
900	$1.85 \cdot 10^{564}$	1,797	5,385	602	2	305.16	3,143	493	2	20.67	19.88
1,000	$9.18 \cdot 10^{626}$	1,997	5,985	669	2	386.26	3,493	548	2	25.51	24.59
B 10	$8.29 \cdot 10^9$	60	691	39	7	1.25	509	41	7	0.65	0.56
20	$2.73 \cdot 10^{20}$	220	4,546	263	12	28.64	3,197	259	12	11.04	8.75
30	$1.04 \cdot 10^{31}$	480	15,101	973	17	212.62	10,433	845	17	70.25	53.53
40	$4.16 \cdot 10^{41}$	840	37,066	2,149	22	935.75	25,374	2,055	22	282.02	210.52
50	$1.72 \cdot 10^{52}$	1,300	76,308	5,342	27	3,036.88	47,806	4,208	27	861.17	635.15
C 5	$2.90 \cdot 10^6$	149	433	21	10	0.48	240	10	10	0.22	0.18
10	$2.50 \cdot 10^9$	354	1,038	66	15	2.08	600	34	15	0.82	0.67
15	$2.17 \cdot 10^{11}$	634	1,868	145	20	5.63	1,110	78	20	2.31	1.79
20	$6.03 \cdot 10^{12}$	989	2,923	267	25	11.97	1,770	149	25	5.23	3.89
25	$8.54 \cdot 10^{13}$	1,419	4,203	441	30	23.86	2,580	253	30	10.59	7.42
50	$4.24 \cdot 10^{17}$	4,694	13,978	2,410	55	213.89	8,880	1,469	55	168.23	65.64
75	$6.98 \cdot 10^{19}$	9,844	29,378	7,035	80	863.62	18,930	4,399	80	538.29	265.03
100	$2.70 \cdot 10^{21}$	16,869	50,403	15,439	105	2,362.63	32,730	9,792	105	1,529.50	740.82
D 5	$2.55 \cdot 10^6$	7	47	3	11	0.07	56	14	4	0.05	0.04
10	$1.01 \cdot 10^9$	12	87	22	21	1.10	156	182	4	0.54	0.48
15	$4.70 \cdot 10^{10}$	17	127	86	31	6.34	306	1,005	4	3.07	2.73
20	$8.05 \cdot 10^{11}$	22	167	238	41	22.87	506	3,595	4	11.67	10.45
25	$7.68 \cdot 10^{12}$	27	207	541	51	63.70	756	9,923	4	33.88	30.37
30	$4.99 \cdot 10^{13}$	32	247	1,068	61	150.28	1,056	23,068	4	85.59	76.74
40	$9.94 \cdot 10^{14}$	42	327	3,823	81	583.01	1,806	89,189	4	378.73	343.72
50	$1.04 \cdot 10^{16}$	52	407	9,510	101	1,703.69	2,756	258,306	4	1,221.40	1,106.75

collection policy in the forwarding-arcs approach contributes to the proliferation of deleted nodes which are not truly destroyed until the end of the iteration. Combined with the reduced number of iterations in our approach, the garbage collection bin grows very rapidly. Usually, late node deletion is beneficial, since doing garbage collection in bulk reduces the number of times nodes are scanned for removal. However, in case of the Kanban system we see how this can backfire. It is worth noting that using a finer and not particularly "good" partition of the

Table 4. Results for the Kanban net with 16 levels (one place per level)

N	$\lvert S \rvert$	final nodes	Approach in [17]				Our new approach				
			peak nodes	mem. (KB)	# it.	time (sec.)	peak nodes	mem. (KB)	# it.	time (sec.) upstr.	fwd.
1	$1.60 \cdot 10^2$	32	112	31	5	0.79	56	58	6	0.20	0.19
2	$4.60 \cdot 10^3$	51	271	81	7	2,46	110	124	6	0.44	0.31
3	$5.84 \cdot 10^4$	73	521	231	10	6.56	221	230	6	0.62	0.54
4	$4.54 \cdot 10^5$	98	895	428	13	14.21	373	389	6	1.00	0.82
5	$2.55 \cdot 10^6$	126	1,414	705	16	25.46	587	613	6	1.56	1.28
8	$1.34 \cdot 10^8$	228	4,015	2,099	25	113.01	1,833	1,914	6	4.49	3.70
10	$1.01 \cdot 10^9$	311	6,838	3,629	31	237.11	3,402	3,552	6	8.01	6.56
15	$4.70 \cdot 10^{10}$	571	19,061	10,185	46	1,032.12	11,507	12,013	6	24.93	20.12
20	$8.05 \cdot 10^{11}$	906	40,741	21,902	61	3,340.99	29,137	30,419	6	62.34	48.31

Kanban net, with one place per level, drastically changes the results, as shown in Table 4. We only need to scale-up the model to $N = 20$ to see an improvement of about a factor 70 with respect to [17]. This observation testifies to the suitability of our algorithm (mostly) in cases when a good partitioning cannot be found automatically or by hand, e.g., due to insufficient heuristics.

7 Related Work

A variety of approaches for the generation of reachable state spaces of synchronous and asynchronous systems have been suggested in the literature, where state spaces are represented either in an *explicit* or in a *symbolic* fashion.

Explicit state-space generation techniques construct the reachable state space of the system under consideration by successively iterating its next-state function (see, e.g., [3, 7]). To achieve space efficiency, numerous techniques have been introduced, out of which *multi-level data structures* and *merging common bitvectors* deserve special mentioning. Multi-level data structures exploit the structure of the underlying system representation, e.g., the approach reported in [7] is based on a decomposition of a Petri net into subnets. As the name suggests, merging common bitvectors aims at compressing the storage needed for each state – a bitvector – by merging common sub-bitvectors [3, 14]; indeed, the result is somewhat analogous to the one obtained using BDDs. While explicit methods still require space linear in the number of states, they usually possess some advantages for numerical state-space analyses [16].

To avoid the problem of state-space explosion when building the explicit state space of concurrent, asynchronous systems, researchers developed three key techniques: (i) *compositional minimization techniques* build the state space of a concurrent system stepwise, i.e., parallel component by parallel component, and minimize the state space of each intermediate system according to a behavioral congruence or an interface specification [13]; (ii) *Partial-order techniques* exploit the fact that several traces of an asynchronous system may be equivalent with respect to the properties of interest [12, 22, 24]; thus, it is sufficient to explore only a single trace of each equivalence class; (iii) *techniques exploiting symmetries in systems* – such as those with repeated sub-systems – can be used to avoid the explicit construction of symmetric subgraphs of the overall state spaces [9].

Symbolic state-space generation techniques have largely concentrated on (synchronous) hardware systems rather than on (asynchronous) software systems [1, 4, 10, 15]. For safe Petri nets, Pastor et al. [20] developed a BDD-based algorithm for the generation of the reachability sets by encoding each place of a net as a Boolean variable. The algorithm is capable of generating state spaces of very large nets within hours; similar techniques were also implemented by Varpaaniemi et al. [23] in the Petri net tool PROD. In recent work, Pastor and Cortadella introduced a more efficient encoding of nets by exploiting place invariants [19]. However, the underlying logic is still based on Boolean variables. In contrast, our work uses a more general version of decision diagrams, namely MDDs [15, 17], by which the amount of information carried in each single node of a decision diagram can be increased. In particular, MDDs allow for a straightforward encoding of arbitrary, i.e., not necessarily safe, Petri nets. Since we have already compared our approach to related MDD-based techniques in the previous sections, we refrain from a repetition of the issues here.

8   Conclusions and Future Work

This paper presented a novel algorithm for constructing the reachable state spaces of asynchronous systems. As in previous work [17], state spaces are symbolically represented via Multi-valued Decision Diagrams (MDDs). In contrast to previous work, our algorithm fully exploits event locality in asynchronous systems, integrates an intelligent cache management, and achieves faster convergence via an advanced iteration control. Experimental results for examples well-known in the Petri net community show that our algorithm is often significantly faster than the one introduced in [17], with no or usually neglectable decrease in space efficiency. To the best of our knowledge, our algorithm is the first symbolic algorithm taking advantage of event locality.

Regarding future work, we would like to further explore various approaches to iteration control and to partitioning, as well as the mutual influences between partitioning and variable ordering. Moreover, we plan to employ our MDD manipulation algorithms as a basis for implementing a model checker in SMART, such that not only *safety properties* but also *liveness properties* can be automatically checked by the tool. Finally, we intend to parallelize our algorithms for shared-memory and distributed-memory architectures.

Acknowledgments. We would like to thank A.S. Miner and the anonymous referees for their valuable comments and suggestions.

References

[1] R.E. Bryant. Graph-based algorithms for Boolean function manipulation. *IEEE Trans. on Computers*, 35(8):677–691, 1986.

[2] R.E. Bryant. Symbolic Boolean manipulation with ordered binary-decision diagrams. *ACM Computing Surveys*, 24(3):393–418, 1992.

[3] P. Buchholz. Hierarchical structuring of superposed GSPNs. In *PNPM '97*, pp. 81–90. IEEE Computer Society Press, 1997.

[4] J.R. Burch, E.M. Clarke, K.L. McMillan, D.L. Dill, and L.J. Hwang. Symbolic model checking: 10^{20} states and beyond. *I&C*, 98(2):142–170, 1992.

[5] G. Ciardo, G.Lüttgen, and R. Siminiceanu. Efficient state-space construction for asynchronous systems. Techn. Rep. 99-50, Institute for Computer Applications in Science and Engineering (ICASE), 1999.

[6] G. Ciardo and A.S. Miner. SMART: Simulation and Markovian Analyzer for Reliability and Timing. In *IPDS '96*, p. 60. IEEE Computer Society Press, 1996.

[7] G. Ciardo and A.S. Miner. Storage alternatives for large structured state spaces. In *Tools '97*, vol. 1245 of *LNCS*, pp. 44–57. Springer-Verlag, 1997.

[8] G. Ciardo and M. Tilgner. On the use of Kronecker operators for the solution of generalized stochastic Petri nets. Techn. Rep. 96-35, Institute for Computer Applications in Science and Engineering (ICASE), 1996.

[9] E.M. Clarke, T. Filkorn, and S. Jha. Exploiting symmetry in model checking. In *CAV '93*, vol. 697 of *LNCS*, pp. 450–462. Springer-Verlag, 1993.

[10] E.M. Clarke, O. Grumberg, and D. Peled. *Model Checking*. MIT Press, 1999.

[11] A. Geser, J. Knoop, G. Lüttgen, B. Steffen, and O. Rüthing. Chaotic fixed point iterations. Techn. Rep. MIP-9403, Univ. of Passau, 1994.

[12] P. Godefroid. *Partial-order Methods for the Verification of Concurrent Systems – An Approach to the State-explosion Problem*, vol. 1032 of *LNCS*. Springer-Verlag, 1996.

[13] S. Graf, B. Steffen, and G. Lüttgen. Compositional minimisation of finite state systems using interface specifications. *FAC*, 8(5):607–616, 1996.

[14] G. Holzmann. The model checker Spin. *IEEE Trans. on Software Engineering*, 23(5):279–295, 1997.

[15] T. Kam, T. Villa, R.K. Brayton, and A. Sangiovanni-Vincentelli. Multi-valued decision diagrams: Theory and applications. *Multiple-Valued Logic*, 4(1–2):9–62, 1998.

[16] P. Kemper. Numerical analysis of superposed GSPNs. *IEEE Trans. on Software Engineering*, 22(4):615–628, 1996.

[17] A.S. Miner and G. Ciardo. Efficient reachability set generation and storage using decision diagrams. In *ICATPN '99*, vol. 1639 of *LNCS*, pp. 6–25. Springer-Verlag, 1999.

[18] T. Murata. Petri nets: Properties, analysis and applications. *Proc. of the IEEE*, 77(4):541–579, 1989.

[19] E. Pastor and J. Cortadella. Efficient encoding schemes for symbolic analysis of Petri nets. In *DATE '98*, pp. 790–795. IEEE Computer Society Press, 1998.

[20] E. Pastor, O. Roig, J. Cortadella, and R.M. Badia. Petri net analysis using Boolean manipulation. In *ICATPN '94*, vol. 815 of *LNCS*, pp. 416–435. Springer-Verlag, 1994.

[21] K. Ravi and F. Somenzi. High-density reachability analysis. In *ICCAD '95*, pp. 154–158. IEEE Computer Society Press, 1995.

[22] A. Valmari. A stubborn attack on the state explosion problem. In *CAV '90*, pp. 25–42. AMS, 1990.

[23] K. Varpaaniemi, J. Halme, K. Hiekkanen, and T. Pyssysalo. PROD reference manual. Techn. Rep. B13, Helsinki Univ. of Technology, 1995.

[24] F. Vernadat, P. Azema, and F. Michel. Covering step graph. In *ICATPN '96*, vol. 1091 of *LNCS*, pp. 516–535. Springer-Verlag, 1996.

[25] B. Yang and D.R. O'Hallaron. Parallel breadth-first BDD construction. In *PPOPP '97*, vol. 32, 7 of *ACM SIGPLAN Notices*, pp. 145–156, 1997.

Designing a LTL Model-Checker Based on Unfolding Graphs

Jean-Michel Couvreur[1], Sébastien Grivet[1], and Denis Poitrenaud[2]

[1] LaBRI, Université de Bordeaux I, Talence, France
`couvreur, grivet@labri.u-bordeaux.fr`
[2] LIP6, Université Pierre et Marie Curie, Paris, France
`poitrenaud@lip6.fr`

Abstract. In this paper, we present new technique designing to solve the on-the-fly model checking problem for linear temporal logic using unfolding graphs [4] and the two key algorithms presented in [2]. Our study is based on the recognition of stuttering behavior in a formula automaton and on the on-the-fly construction of an unfolding graph. Moreover, the characterization of different kinds of behaviors allows us to design efficient algorithms for the detection of accepting paths. We have extended our study to the use of the atomic proposition *dead* which holds for terminal states. Partial order techniques are not adapted to deal with this global property in the context of a LTL model checking.

Topics. Petri Nets, Model-checking, Linear temporal logic, Unfolding.

1 Introduction

Automatically checking whether a finite state program satisfies its linear specification is a problem that has gained a lot of attention during the last 15 years. Linear temporal logic is a powerful specification language for expressing safety, liveness and fairness properties. However, the model checking problem is known PSPACE-complete [22]. In practice, model checking methods face complexity related limits: the size of the state space program. Many techniques have been designed to avoid the state explosion problem. Among these ones, on-the-fly verification [2, 1, 9, 11, 13] and partial order reductions [10, 19, 23, 24] have clearly demonstrated their effectiveness for automatic verification even for fairly large-scale industrial applications [12].

Another approach consists in implementing the verification directly on a representation of the partial order. Unfoldings ([18, 5]) propose branching semantics of partial orders for Petri nets. McMillan [16] has described the construction of a finite branching process in which all reachable markings are represented. He has shown how branching processes can be used for the detection of global deadlocks and the reachability problem for finite state space Petri nets. Many improvements of this method concern the efficiency of the unfolding construction [8, 7, 14, 15] and the verification of safety properties, or the verification of branching temporal logic [6] and linear temporal logic [3, 4, 21, 26].

M. Nielsen, D. Simpson (Eds.): ICATPN 2000, LNCS 1825, pp. 123–145, 2000.

In this paper, we present new technique designing to solve on-the-fly model checking problem for linear temporal logic using unfolding graphs [4] and the two key algorithms presented in [2] (a construction of an automata for a temporal logic formula and a checking algorithm). In our work, we consider the stutter invariant linear temporal logic [20], LTL without the next-time operator ($LTL \setminus X$). The automaton of such formula may only react on a modification of the truth values of the atomic propositions. With this interpretation, behaviors of the system are recognized by finite or infinite paths in the formula automaton. Using the automata construction of [2] where each node is labelled by formulas, we deduce acceptance conditions for finite paths. Moreover, the formulas allow us to dynamically limit the reaction of the automaton to the modification of the atomic propositions of the formulas labelling the node.

The on-the-fly construction of an unfolding graph must respect constraints given by the formula automaton: the markings represented in each unfolding node must be invariant w.r.t. the atomic propositions of the formulas labelling the automaton node; if a behavior is recognized by a finite path leading to the automaton node, we need to detect terminal or infinite behavior represented locally in the unfolding node. The invariant constraint in the unfolding node is solved by considering observed transitions (transitions which may modify the truth value of the atomic propositions). The detection of infinite behavior contained locally to an unfolding node has been demonstrated to be a difficult task [4] in a general case. When it is necessary, we solve this problem using the inclusion rule for the cutoff determination.

We have extended our study to the use of the atomic proposition *dead* (*dead* holds for all terminal markings). This proposition is a global property of the system: all the transitions of the system must be observed. Hence, the use of reduction techniques, as stubborn sets, has no effect in this context. Unfolding techniques do not operate a graph reduction but propose a concise representation of all the system behaviors. To bring the *dead* proposition study to a successful conclusion, we do not take into account this proposition in the stuttering constraint. Hence, behaviors of a system are recognized by infinite paths and three kinds of finite paths in the formula automaton:

- *Live behavior*: Infinite behavior with an infinite number of modifications of the atomic propositions,
- *Divergent behavior*: Infinite behavior with a finite number of modifications of the atomic propositions,
- *Immediate dead behavior*: finite behavior (leading to a terminal state) where the last transition produces a modification of the atomic propositions,
- *Delayed dead behavior*: finite behavior (leading to a terminal state) where the last transition does not produce a modification of the atomic propositions,

The result of this study is the definition of new acceptance conditions for each of these kinds of behaviors. Moreover, we have adapted the unfolding graph construction to respect constraints binding to the new interpretation of the formula automaton.

The rest of the paper starts of preliminaries defining P/T system and linear temporal logic as well as unfoldings. Section 3 presents the interpretation of stuttering property in automaton of $LTL \setminus X$ formula. Section 4 gives the verification algorithm $LTL \setminus X$ formula with unfolding graph. Section 5 completes this study to take into account the *dead* atomic proposition. Section 6 is dedicated to the experimentations and some concluding remarks close the paper.

2 Preliminaries

2.1 P/T System and Linear Temporal Logic

Let A and B be finite or infinite set, $PowerSet(A)$ denotes the set of subsets of A and $A \to B$ denotes the set of mappings from A to B. Mappings $A \to \mathbb{N}$ are called markings and mappings $A \times B \to \mathbb{N}$ are called matrices. Mappings $S \to \{true, false\}$ are called propositions over the set S. Given a set AP of propositions over a set S, called atomic propositions, new propositions are built up using boolean operators; we denote as $Prop(AP)$ this set of propositions. We call trace function Φ_{AP} the mapping in $S \to [AP \to \{true, false\}]$ defined by $\forall m \in S, \forall p \in AP : \Phi_{AP}(m)(p) = p(m)$.

A P/T system is a pair $\langle N, m_0 \rangle$ where $N = \langle P, T, Pre, Post \rangle$ is a P/T net (P and T are disjoint sets of places and transitions, and Pre and $Post$ are incidence matrices (in $P \times T \to \mathbb{N}$), and m_0 is a initial marking (in $P \to \mathbb{N}$). The behavior of P/T system is defined by the firing rule which states that a transition t is enable for marking m (in $P \to \mathbb{N}$): $m \xrightarrow{t}$ iff $\forall p \in P : m(p) \geq Pre(p, t)$; and that its firing leads to a marking m': $m \xrightarrow{t} m'$ iff $m \xrightarrow{t} \wedge \forall p \in P : m'(p) = Post(p, t) - Pre(p, t) + m(p)$. The set of markings which can be reached from the initial marking m_0 by repeatedly firing the transitions of the net is called the reachability set and is denoted by $Reach(N, m_0)$. A P/T system is bounded iff the reachability set is finite.

A run of P/T system is an infinite sequence of marking $\rho = m_0 \cdot m_1 \cdot m_2 \ldots$ such that m_0 is the initial marking and for every $i \geq 0, \exists t \in T : m_i \xrightarrow{t} m_{i+1}$. Without loss of generality, one can add to every terminal marking a loop transition. In this case, a finite sequence leading to a terminal state induces a run by infinitely repeating it.

An atomic proposition a of a P/T system is a proposition over the set of markings $P \to \mathbb{N}$. Let AP be a set of atomic propositions, a trace of a run ρ denoted $\Phi_{AP}(\rho)$, is an infinite word over the alphabet $AP \to \{true, false\}$: $\sigma = \Phi_{AP}(\rho) = \Phi_{AP}(m_0) \cdot \Phi_{AP}(m_1) \cdot \Phi_{AP}(m_2) \ldots$

We use linear temporal logic (LTL) for our specification language. It defines a logic for the trace set of a P/T system. We will say that P/T system S fulfills a linear temporal property f iff the trace of every run of S fulfills f. The formal syntax for LTL is given below.

1. Every atomic proposition $ap \in AP$ is a LTL formula.
2. If f and g are LTL formulas, then so are $\neg f$, $f \wedge g$, Xf, fUg.

An interpretation for a LTL formula is an infinite word $\sigma = x_0.x_1.x_2 \ldots$ over the alphabet $AP \rightarrow \{true, false\}$. We use the standard notation $\sigma \models f$ to indicate the truth of a LTL formula f for an infinite word σ. We write σ_i, for the suffix of σ starting at x_i. The relation $\models$ is defined inductively as follows:

1. $\sigma \models ap$ if $x_0(ap)$ for $ap \in AP$;
2. $\sigma \models \neg f$ if $\neg(\sigma \models f)$;
3. $\sigma \models f \wedge g$ if $\sigma \models f$ and $\sigma \models g$;
4. $\sigma \models Xf$ if $\sigma_1 \models f$;
5. $\sigma \models fUg$ if $\exists i \geq 0 : (\sigma_i \models g) \wedge (\forall j < i : \sigma_j \models f)$.

The standard boolean operators and the constants $true$ and $false$ can also be used to construct LTL formulas. For a formula f, we denote by $Sub(f)$ its set of sub-formulas. This notation is extended to formula set.

2.2   Unfolding

In this subsection, unfoldings and underlying notions and notations are presented.

For a node x of a P/T net, $^\bullet x$ and $x^\bullet$ denote respectively the set of predecessors and successors of x in the net seen as a graph (this notation is extended to node sets as usual). We first define the several relations concerning the nodes of an acyclic net. Let N be an acyclic P/T net and x and y be two nodes of N. We say that

– x is causally depend to y (denoted by $x \leq y$) if there exists a path from x to y in the net seen as a graph;
– x and y are in conflict (denoted by $x \sharp y$) if there exists two transitions which are causally depend respectively to x and y and shared an input place;
– x and y are concurrent (denoted by $x \parallel y$) if there is no causally dependency between them and they are not in conflict.

Unfoldings are particular labelled P/T nets. We first introduce two important classes of nets for their definition.

Definition 1 (Occurrence and causal net). *Let $N = \langle B, E, Pre, Post \rangle$ be a P/T net. N is an* occurrence net *if*

– *$Pre, Post \in (B \times E \rightarrow \{0, 1\})$*
– *N forms an acyclic graph and every node is finitely preceded*
– *every place has at most one input transition*
– *no node is in self-conflict ($\forall x \in B \cup E, \neg x \sharp x$)*

Moreover, N is causal net *if it also satisfies that*

– *every place has at most one output transition*

In occurrence and causal nets, places and transitions are respectively called *conditions* and *events*.

It is clear that, in such nets, the relation $\leq$ forms a partial order (the net, seen as a graph, has no cycle). For an occurrence or causal net N, we denote by $Min(N)$ the set of minimal nodes in this partial order.

In an occurrence net, a subset C of events forms a *configuration* if it is left-closed by $\leq$ (i.e. $\forall x \in E, \exists y \in C : x \leq y \Rightarrow x \in C$) and is conflict-free ($\forall x, y \in C, \neg x \sharp y$). For an occurrence net N, we denoted by $Conf(N)$ its set of configurations and by $MaxConf(N)$ the set of maximal (w.r.t. set inclusion) configurations. The minimal configuration (w.r.t. set inclusion) which contains an event e is called the *local configuration* of e and is denoted $[e]$. By extension, for a condition b, we denote by $[b]$ the local configuration $[{}^\bullet b]$ if b has an input event and the empty set otherwise. Moreover, for a set $A \subseteq (B \cup E)$, $[A]$ denotes the event set obtained as the union of the local configurations of the elements in A. Note that $[A]$ is not necessarily a configuration.

Moreover, a subset of conditions forms a *cut* if it is conflict-free. Cut and configuration are closely related: to every configuration C is associated a cut denoted by $Cut(C)$ and defined by $Cut(C) = (Min(N) \cup C^\bullet) \setminus {}^\bullet C$.

Then, we state precisely the characteristics of the used labelling functions. Let h be a mapping from a set A to a set B. We extend h from $PowerSet(A)$ to $B \to \mathbb{N}$ as it follows: $\forall X \subseteq A, \forall b \in B, h(X)(b) = |\{x \in X : h(x) = b\}|$. The extension of h take a subset in A and counts how many elements in it are labelled with b (in order to deal with the multiplicity of arcs).

Definition 2 (Homomorphism). *Let $N = \langle P, T, Pre, Post \rangle$ and $N' = \langle P', T', Pre', Post' \rangle$ be two P/T nets. A mapping h from $P \cup T$ to $P' \cup T'$ is an homomorphism from N to N' if*

- $\forall p \in P, h(p) \in P'$
- $\forall t \in T, h(t) \in T'$
- $\forall t \in T, \forall p' \in P', h({}^\bullet t)(p') = Pre'(p', h(t))$ *and* $h(t^\bullet)(p') = Post'(p', h(t))$

We are now in position to define the two types of labelled nets on which the unfoldings are based.

Definition 3 (Branching Process and Process). *Let $N = \langle P, T, Pre, Post \rangle$ be a P/T net and m be a marking of N. Let $U = \langle B, E, Pr, Po \rangle$ be a P/T net and h be an homomorphism from U to N. $\langle U, h \rangle$ is a branching process of $\langle N, m \rangle$ if*

- $m = h(Min(U))$
- $\forall e, e' \in E, ({}^\bullet e = {}^\bullet e') \wedge (e^\bullet = e'^\bullet) \wedge (h(e) = h(e')) \Rightarrow (e = e')$

Moreover, if $\langle U, h \rangle$ is a causal net and satisfies these previous conditions then $\langle U, h \rangle$ is a process of $\langle N, m \rangle$.

Configurations of a branching process are closely related to processes. For a branching process $\langle U, h \rangle$ of a P/T system $\langle N, m \rangle$ and one of its configuration C, the subnet constructed from the events composing C, their surrounding conditions and the minimal conditions of U is a process of $\langle N, m \rangle$.

Configurations and maximal cuts (w.r.t. set inclusion) of a branching process characterize respectively firing sequences and reachable markings of the P/T system. Indeed, if C is a configuration then every ordering of $h(C)$ respecting $\leq$ forms a firing sequence and $h(Cut(C))$ is a reachable marking.

A process of a P/T system can be extended if the addition of a new event (and its related conditions and arcs) forms a new process. A branching process can be extended if the extension of one of its process forms a new branching process. The algorithm which extends iteratively a branching process can lead to the construction of an infinite structure even in the case of finite state P/T system. Numerous methods of verification ([16, 6, 3, 21, 17, 26, 4] and others) are based on the construction of a particular finite prefix of the complete branching process. This prefix is called an *unfolding*.

In the sequel we use the following particular notations:

- $\beta = \langle\langle B(\beta), E(\beta), Pre(\beta), Post(\beta)\rangle, h(\beta)\rangle$ is used to denote the different components of branching process β.
- $\Pi(C)$ is used to denote the process corresponding to a configuration C and $\Pi(C \setminus C_1)$ when $C_1 \subseteq C$ denotes the process from $Cut(C_1)$ composed by the events $C \setminus C_1$ (where C and C_1 are two configurations).
- $\pi_1 \leq \pi_2$ means that the process π_1 is a prefix of the process π_2.
- $\pi_1.\pi_2$ is a concatenation of two processes satisfying $h(\pi_1)(Cut(\pi_1)) = h(\pi_2)(Min(\pi_2))$.
- e_0 is the artificial initial event which produces the initial marking of any branching process.

The construction of an unfolding is based on the extension of its processes. At each step of the construction, we obtain a branching process β and a set of particular events $Cutout$ for which there are extensions that have not been explored. We call such a couple $(\beta, Cutout)$ a *stable branching process* (a more detailed construction procedure can be find in [4]).

Definition 4 (Stable branching process). *Let β be a branching process of a net N and $Cutout$ be an event set of β. The couple $(\beta, Cutout)$ is a stable branching process iff for every process π of N, one of the two following conditions holds :*

1. $\exists C \in Conf(\beta) : C \cap Cutout = \emptyset \wedge \Pi(C) = \pi$,
2. $\exists e \in Cutout, \exists C \in Conf(\pi) : \Pi([e]) = \Pi(C)$.

The set of reachable markings $Reach(\beta)$ of a stable branching process β is defined by:

$$m \in Reach(\beta) \Leftrightarrow \exists C \in Conf(\beta), C \cap Cutout = \emptyset \wedge m = h(\beta)(Cut(C)).$$

A simple rule to stop an extension of a cutout event e is to find a non cutout event $\phi(e)$ such that the cuts of $[e]$ and $[\phi(e)]$ correspond to the same state in N. The final result of the unfolding construction is called an *unfolding* and cutout events are called *cutoff* events. This simple rule is not adequate to obtain unfoldings in which all the reachable markings are represented. Cutoff rules, as in [16, 6, 8, 14], are defined to insure this reachability property. One will see two of these rules in the section 4.

3   Automata Stuttering Property for $LTL \setminus X$ Formula

In this section, we consider the stutter invariant linear temporal logic [20], $LTL \setminus X$. We demonstrate how the automaton of such formula may only react on a modification of the truth values of the atomic propositions. With this interpretation, behaviors of the system are recognized by finite or infinite paths in the formula automaton. Using the automata construction of [2] where each node are labelled by formulas, we deduce acceptance conditions for finite paths. Moreover, the formulas allow us to dynamically limit the reaction of the automaton to the modification of the atomic propositions of the formulas labelling the node. We extend this study by considering the *dead* atomic proposition.

Formally a transition Büchi automaton $\langle Q, Acc, \rightarrow, q_0 \rangle$ has the following components:

- Q is a finite set of states;
- Acc is a finite set of accepting conditions;
- $\rightarrow \subseteq Q \times Prop(AP) \times PowerSet(Acc) \times Q$ is a transition relation;
- q_0 is an initial state ($q_0 \in Q$).

An infinite word $\sigma = x_0.x_1.x_2 \ldots$ over the alphabet $AP \rightarrow \{true, false\}$ is accepted by a transition Büchi automaton iff there exists an infinite path

$$\rho = q_0 \xrightarrow{(X_0, A_0)} q_1 \xrightarrow{(X_1, A_1)} q_2 \xrightarrow{(X_2, A_2)} \ldots$$

such that $\forall i \geq 0 : ((q_i, X_i, A_i, q_{i+1}) \in \rightarrow) \wedge X_i(x_i) \wedge (\forall a \in Acc, \exists j \geq i : a \in A_j)$.

The automaton construction is based on tableau procedure. The nodes of the graph are labeled by a set of formulas and the transitions are obtained by expanding the temporal operators in order to distinguish what has to be true immediately from what has to be true from the next state on. Theorem 1 formalizes the useful properties the resulting automaton obtained in [2]. Only the atomic propositions which appear in the formulas of a node could be used in formulas of its successors.

Theorem 1. *Let f be a LTL formula. There exists a transition Büchi automaton, denoted $Bu(f)$, that accepts exactly the infinite words over the alphabet $AP \rightarrow \{true, false\}$ which satisfy f. Moreover, $Bu(f)$ fulfills the following properties:*

- $Q \subseteq PowerSet(Sub(f))$
- $\{f\}$ *is the initial state*
- $\forall G \in Q$, *the transition Büchi automaton $Bu(f)$ with G as an initial marking accepts exactly the infinite words which satisfy G*
- *if $\langle G, X, A, H \rangle \in \rightarrow$ then $Domain(H) \subseteq Domain(G)$*

where for every G in Q and when the context is clear G denotes $\bigwedge_{g \in G} g$ and Domain(G) = AP ∩ Sub(G).

Formulas in $LTL \setminus X$ fulfill the stuttering property [20]: the infinite words $\sigma = x_0 \cdot x_1 \cdot x_2 \ldots x_i \cdot x_{i+1} \ldots$ and $\sigma' = x_0 \cdot x_1 \cdot x_2 \ldots x_i \cdot x_i \cdot x_{i+1} \ldots$ satisfy exactly the same set of $LTL \setminus X$ formulas. Theorem 2 formalizes this property.

Theorem 2. *Let f be a $LTL \setminus X$ formula. Let $\sigma = x_0 \cdot x_1 \cdot x_2 \ldots$ be an infinite word over $AP \to \{true, false\}$. Let $\sigma' = x_{i_0} \cdot x_{i_1} \cdot x_{i_2} \ldots$ the infinite extracted word of σ defined by:*

- $\forall k : i_k < i_{k+1}$
- $\forall k, \forall i : i_k \le i < i_{k+1} \Rightarrow x_{i|Domain(f)} = x_{i_k|Domain(f)}$
- $\forall k : x_{i_k|Domain(f)} = x_{i_{k+1}|Domain(f)} \Rightarrow \forall i \ge i_k : x_{i|Domain(f)} = x_{i_k|Domain(f)}$

Then σ fulfills f iff σ' fulfills f.

Theorem 2 suggests a new way to verify $LTL \setminus X$ formulas on an infinite word. We consider the extracted word which change the truth value of the atomic propositions at each step (i.e. the stuttering is removed) and a path in the transition Büchi automaton recognizing this word. We have to consider two cases: if the extracted word is infinite then each accepting condition must appear infinitely often in the automaton path. If the extracted word is finite having x as the last term and the path terminates in the state F then one must have $x^\omega \models F$.

Theorem 3 characterizes for a $LTL \setminus X$ formula, the paths in its automaton which correspond to accepted stuttering words. It gives an improvement when traversal the automaton, by restricting the stuttering condition to the atomic propositions which appear in the formulas of the automaton states.

Point 1 of theorem 3 characterizes the paths of the automaton recognizing a word taking into account the local stuttering property (on a state automaton formula). Point 2 characterizes the recognizing of a finite path and states its acceptance criteria: the word is terminated by a sequence in which the values of the atomic propositions (of the last visited automaton state formula F) remain constant; the acceptance is reduced to verify $x^\omega \models F$ where x is the constant values of the atomic propositions. Point 3 resumes the acceptance criteria of an infinite path for a transition Büchi automaton.

Theorem 3. *Let f be a $LTL \setminus X$ formula. Let $Bu(f)$ be its transition Büchi automaton. Let $\sigma = x_0 \cdot x_1 \cdot x_2 \ldots$ be an infinite word over $AP \to \{true, false\}$. $\sigma \models f$ iff there exists finite or infinite path in $Bu(f)$*

$$\rho = F_0 \xrightarrow{(X_0, A_0)} F_1 \xrightarrow{(X_1, A_1)} F_2 \xrightarrow{(X_2, A_2)} \ldots$$

and a finite or infinite extracted word of σ, $\sigma' = x_{i_0} \cdot x_{i_1} \cdot x_{i_2} \ldots$ such that

1. *Stuttering characterization*
   - $|\rho| = |\sigma'|$
   - $\forall k < |\sigma'| - 1 : i_k < i_{k+1}$

$$- \forall k < |\sigma'| : X_k(x_{i_k})$$
$$- \forall k < |\sigma'| - 1, \forall i : i_k \leq i < i_{k+1} \Rightarrow x_{i|AP_k} = x_{i_k|AP_k}$$
$$- \forall k < |\sigma'| - 1 : x_{i_k|AP_k} \neq x_{i_{k+1}|AP_k}$$

2. *Finite path:* $|\sigma'| < \infty \Rightarrow$
$$- \forall i > i_{|\sigma'|-1} : x_{i|AP_{|\sigma'|-1}} = x_{i_{|\sigma'|-1}|AP_{|\sigma'|-1}}$$
$$- (x_{i_{|\sigma'|-1}})^\omega \models F_{|\sigma'|-1}$$

3. *Infinite path:* $|\sigma'| = \infty \Rightarrow (\forall a \in Acc, \forall i \geq 0, \exists j \geq i : a \in A_j)$

where $AP_i = Domain(F_i)$.

Proof. $(\Rightarrow)$ Let $\sigma = x_0 \cdot x_1 \cdot x_2 \ldots$ be an infinite word satisfying formula f. We can establish by induction on the size of $Domain(f)$ that there exists a path ρ in $Bu(f)$ and an extract word σ' which fulfills properties 1,2,3 of Theorem 3.

If $Domain(f) = \emptyset$ then $f = true$, ρ is reduced to the initial state $\{f\}$ and $\sigma' = x_0$. ρ and σ' fulfills properties 1,2,3.

If $Domain(f) \neq \emptyset$ then we consider the stutter free word $\sigma' = x_{i_0} \cdot x_{i_1} \cdot x_{i_2} \ldots$ obtains from Theorem 2. Because $\sigma' \models f$, there exists an infinite accepting path from the initial state $\{f\}$ in $Bu(f)$

$$\rho = F_0 \xrightarrow{(X_0, A_0)} F_1 \xrightarrow{(X_1, A_1)} F_2 \xrightarrow{(X_2, A_2)} \ldots$$

such that $(\forall j : X_j(x_{i_j})) \wedge (\forall a \in Acc, \forall j \geq 0, \exists k \geq j : a \in A_k)$. One can remark that each suffix word σ'_j of σ' verify formula F_j. One can consider the three following cases:

1. If $\forall j : Domain(F_j) = Domain(f) \wedge x_{i_j|Domain(f)} \neq x_{i_{j+1}|Domain(f)}$ then σ' and ρ are the looking extract word and path.

2. If $\exists n$ such that
   - $\forall j \geq n : Domain(F_j) = Domain(f)$
   - $\forall j < n : x_{i_j|Domain(f)} \neq x_{i_{j+1}|Domain(f)}$
   - $\forall j > n : x_{i_j|Domain(f)} = x_{i_n|Domain(f)}$

   Then the finite path $F_0 \ldots F_n$ and the extract word $x_{i_0} \ldots x_{i_n}$ are the looking path and word. One has to remark that the suffix word σ'_n is stuttering equivalent to $x_{i_n}{}^\omega$ with respect to $Domain(F_n)$ and then satisfies formula F_n.

3. Otherwise $\exists n$ such that
   - $\forall j < n : Domain(F_j) = Domain(f)$
   - $\forall j < n : x_{i_j|Domain(f)} \neq x_{i_{j+1}|Domain(f)}$
   - $Domain(F_n) \subset Domain(f)$

   From the automata construction theorem 1, $Bu(F_n)$ is equal to $Bu(f)$ with F_n as an initial marking. One can apply the induction property to formula F_n and the suffix word σ'_n. We obtain an new path ρ'' in $Bu(f)$ from F_n and an extract word σ'' of σ'_n which fulfills properties 1,2,3. The concatenated path $F_0 \ldots F_n \cdot \rho''$ and word $x_{i_0} \ldots x_{i_n} \cdot \sigma''$ are looking path and word.

($\Leftarrow$) Let $\sigma = x_0 \cdot x_1 \cdot x_2 \ldots$ be an infinite word. Let ρ and σ' be a path and an extract word which fulfill properties 1,2,3.

If the path ρ is finite, $(x_{i_{|\sigma'|-1}})^\omega \models F_{|\sigma'|-1}$. Because the suffix word $\sigma_{|\sigma'|-1}$ is stuttering equivalent to $(x_{i_{|\sigma'|-1}})^\omega$ with respect to $Domain(F_{|\sigma'|-1})$, it also satisfy $F_{|\sigma'|-1}$.

If the path ρ is infinite, there exists an integer n such that $\forall j > n$: $Domain(F_n) = Domain(F_j)$. The suffix word σ_{i_n} is stuttering equivalent to σ'_n and then also satisfy formula F_n.

In both case, we only have to prove the induction property :

$$\forall j : \sigma_{i_{j+1}} \models F_{j+1} \Rightarrow \sigma_{i_j} \models F_j$$

This property is deduced by the facts that if $\sigma_{i_{j+1}} \models F_{j+1}$, the word $x_{i_j}.\sigma_{i_{j+1}}$ satisfies F_j (the transition $F_j \overset{(X_j, A_j)}{\longrightarrow} F_{j+1}$ concatenated with the accepting path for $\sigma_{i_{j+1}} \models F_{j+1}$ is an accepting path for $x_{i_j}.\sigma_{i_{j+1}} \models F_j$) and that $x_{i_j}.\sigma_{i_{j+1}}$ is stuttering equivalent to σ_{i_j} with respect to $Domain(F_j)$. $\square$

It remains to solve the problem $x^\omega \models f$ which can be treated in a linear way on the size of the formula. For the temporal operator U one can apply the proposition 1. For the boolean operators, the problem is solved directly applying their definitions ($x^\omega \models \neg f \Leftrightarrow \neg(x^\omega \models f)$ and $x^\omega \models f \wedge g \Leftrightarrow (x^\omega \models f) \wedge (x^\omega \models g)$).

Proposition 1. *Let f, g be two $LTL \setminus X$ formulas. We have $\forall x \in AP \rightarrow \{true, false\}$:*

$$x^\omega \models fUg \Leftrightarrow x^\omega \models g$$

Proof. The formula to prove is the direct application of the definition of fUg. When considering all possible values of prefix and suffix of x^ω, we deduce:

$$x^\omega \models fUg \Leftrightarrow (x^\omega \models g) \vee (x^\omega \models f \wedge x^\omega \models g)$$

By simplifying this formula, we obtain the formula to prove. $\square$

We introduce the proposition *dead* over a set of markings of a P/T system: $dead(m) \Leftrightarrow$ "m is a terminal state". When considering a trace of a P/T system, it must be a legal word: when the proposition *dead* holds, it remains true in the sequel of the word as well as the values of the other atomic propositions stay constant.

Definition 5 (Legal word). *Let $\sigma = \langle x_0, d_0 \rangle \cdot \langle x_1, d_1 \rangle \ldots$ be an infinite word over $(AP \rightarrow \{true, false\}) \times \{true, false\}$. σ is a legal word if*

$$\forall i, \forall j > i : d_i \Rightarrow d_j \wedge (x_j = x_i)$$

We denote by LTL^d the linear temporal logic defined on the atomic propositions $AP \cup \{dead\}$. Theorem 4 characterizes for a $LTL^d \setminus X$ formula, the paths in its automaton which correspond to accepted stuttering words. We restrict the stuttering conditions to the atomic propositions which appear in the formulas of the automaton states without taking into account the proposition *dead*.

Point 1 of theorem 4 resumes point 1 of theorem 3 excepted that proposition *dead* is now considered for the stuttering condition. Point 2 resumes the recognizing of a finite path. One has to consider the suffix of the word where the values of the atomic propositions remain constant. The acceptance condition depends on the value of proposition *dead* in this suffix. It consists to decide the satisfaction of the last automaton state formula F by some particular words:

- "Never *dead*" (point 2.a): $\langle x, false \rangle^\omega \models F$
- "Immediately *dead*" (point 2.b): $\langle x, true \rangle^\omega \models F$
- "Sometime in strict future *dead*" (point 2.c): $\langle x, false \rangle \cdot \langle x, true \rangle^\omega \models F$

where x is the constant values of the atomic propositions. Point 3 resumes the acceptance criteria of an infinite path for a transition Büchi automaton.

Theorem 4. *Let f be a $LTL^d \setminus X$ formula. Let $Bu(f)$ be its transition Büchi automaton. Let $\sigma = \langle x_0, d_0 \rangle \cdot \langle x_1, d_1 \rangle \cdot \langle x_2, d_2 \rangle \ldots$ be a legal infinite word over $(AP \rightarrow \{true, false\}) \times \{true, false\}$. $\sigma \models f$ iff there exists finite or infinite path in $Bu(f)$*

$$\rho = F_0 \xrightarrow{(X_0, A_0)} F_1 \xrightarrow{(X_1, A_1)} F_2 \xrightarrow{(X_2, A_2)} \ldots$$

and an extracted word of σ, $\sigma' = \langle x_{i_0}, d_{i_0} \rangle \cdot \langle x_{i_1}, d_{i_1} \rangle \cdot \langle x_{i_2}, d_{i_2} \rangle \ldots$ such that

1. *Stuttering characterization*
   - $\forall k < |\sigma'| : i_k < i_{k+1}$
   - $\forall k : X_k(\langle x_{i_k}, d_{i_k} \rangle)$
   - $\forall k < |\sigma'| - 1, \forall i : i_k \leq i < i_{k+1} \Rightarrow x_{i|AP_k} = x_{i_k|AP_k}$
   - $\forall k < |\sigma'| - 1 : x_{i_k|AP_k} \neq x_{i_{k+1}|AP_k}$
2. *Finite path:* $|\sigma'| < \infty \Rightarrow \forall i > i_{|\sigma'|-1} : x_{i|AP_{|\sigma'|-1}} = x_{i_{|\sigma'|-1}|AP_{|\sigma'|-1}}$
   (a) $\forall i \geq i_{|\sigma'|-1} : \neg d_i \Rightarrow (\langle x_{i_{|\sigma'|-1}}, false \rangle)^\omega \models F_{|\sigma'|-1}$
   (b) $d_{i_{|\sigma'|-1}} \Rightarrow (\langle x_{i_{|\sigma'|-1}}, true \rangle)^\omega \models F_{|\sigma'|-1}$
   (c) $\neg d_{i_{|\sigma'|-1}} \wedge (\exists i > i_{|\sigma'|-1} : d_i) \Rightarrow \langle x_{i_{|\sigma'|-1}}, false \rangle \cdot \langle x_{i_{|\sigma'|-1}}, true \rangle^\omega \models F_{|\sigma'|-1}$
3. *Infinite path:* $|\sigma'| = \infty \Rightarrow$
   - $(\forall a \in Acc, \forall i \geq 0, \exists j \geq i : a \in A_j)$

where $AP_i = Domain(F_i) \setminus \{dead\}$.

Proof. The proof of this theorem is tedious and is almost the same as the proof of Theorem 3. In the first part of the proof ($\Rightarrow$), the induction is done on the size of $Domain(f) \setminus \{dead\}$. □

It remains to solve the problems $\langle x, false \rangle^\omega \models f$, $\langle x, true \rangle^\omega \models f$ and $\langle x, false \rangle \cdot \langle x, true \rangle^\omega \models f$. They can be treated in a linear way on the size of the formula using the proposition 2 and the definition of the boolean operators.

Proposition 2. *Let f, g be two $LTL^d \backslash X$ formulas. The following propositions hold for all $x \in AP \rightarrow \{true, false\}$:*

$$\langle x, false \rangle^\omega \models fUg \Leftrightarrow \langle x, false \rangle^\omega \models g$$
$$\langle x, true \rangle^\omega \models fUg \Leftrightarrow \langle x, true \rangle^\omega \models g$$
$$\langle x, false \rangle \cdot \langle x, true \rangle^\omega \models fUg \Leftrightarrow \langle x, false \rangle \cdot \langle x, true \rangle^\omega \models g \vee$$
$$(\langle x, false \rangle \cdot \langle x, true \rangle^\omega \models f \wedge \langle x, true \rangle^\omega \models g)$$

Proof. The two first formulas are particular application of Proposition 1. The third formula is deduced from definition of fUg. When considering all possible values of prefix and suffix of $\langle x, false \rangle \cdot \langle x, true \rangle^\omega$, we obtain the following formula:

$$\langle x, false \rangle \cdot \langle x, true \rangle^\omega \models fUg \Leftrightarrow (\langle x, false \rangle \cdot \langle x, true \rangle^\omega \models g)$$
$$\vee (\langle x, false \rangle \cdot \langle x, true \rangle^\omega \models f \wedge \langle x, true \rangle^\omega \models g)$$
$$\vee (\langle x, false \rangle \cdot \langle x, true \rangle^\omega \models f \wedge \langle x, true \rangle^\omega \models f \wedge \langle x, true \rangle^\omega \models g)$$

By simplifying this formula, we obtain the formula to prove.   □

4   Verification of $LTL \backslash X$ Formula with Unfolding Graphs

In this section, we demonstrate how the unfolding graph presented in [4] can be used for the verification of $LTL \backslash X$ formulas. First, we give briefly the definition of an unfolding graph. Then, we present how such a graph is synchronized to a transition Büchi automaton corresponding to a formula and how runs of the system which satisfy the formula can be detected on the resulting graph. Finally, we discuss on the implementation of a model-checker based on this method.

In an unfolding graph each node is a partial unfolding of the P/T net from a given marking. Such an unfolding is qualified as partial because all the reachable markings are not necessarily represented in it. Some special events called *bridges*, for which the unfolding construction has not been continued denote this incompleteness and insure the stability. In general, the cutoffs composing a partial unfolding can be defined using the size rule defined in [16].

Definition 6 (Partial size rule unfolding). *A partial size rule unfolding of a net N is a tuple $(\beta, Bridge, Cutoff, \phi)$ such that $(\beta, Bridge \cup Cutoff)$ is a stable branching process of N and ϕ is a mapping from $Cutoff$ to $E(\beta) \backslash (Bridge \cup Cutoff)$ which fulfills*
$$\forall e \in Cutoff : h(\beta)(Cut([e])) = h(\beta)(Cut([\phi(e)]) \wedge |[\phi(e)]| < |[e]|$$

Using the bridges of a partial unfolding composing a node of the graph, we can characterize its successors. For the properties related to the reachability representation in such a graph, the interested reader can confer to [4].

Definition 7 (Unfolding graph).

An unfolding graph of a net N is a graph $(G, \rightarrow, g_0)$ where

- $\forall g \in G : g$ *is a partial size rule unfolding of net* $(N, h(g)(Min(g)))$ *(N with a new initial marking $h(g)(Min(g))$) where $h(g)(Min(g)) \in Reach(N)$,*
- $\forall g \in G, \forall e \in Bridge(g), \exists! g' \in G : g \xrightarrow{e} g' \wedge h(g)(Cut([e])) = h(g')(Min(g'))$,
- $h(g_0)(Min(g_0)) = m_0$, *and*
- $\forall g \in G$, *there exists a path in $(G, \rightarrow, g_0)$ from g_0 to g.*

The partial size rule used in the definition of the unfolding graph are not adapted to the detection of local cyclic behaviors. The inclusion rule allows a simple characterization of these behaviors (the presence of a cutoff denotes a cyclic behavior, see [4]).

Definition 8 (Partial inclusion rule unfolding).

A partial size rule unfolding $(\beta, Bridge, Cutoff, \phi)$ fulfills the inclusion rule iff

$$\forall e \in Cutoff : [\phi(e)] \subset [e]$$

The definition 9 presents the synchronization of an unfolding graph and a transition Büchi automaton taking into account the stuttering. When a transition of the unfolding graph does not modify the value of the atomic propositions, it produces transitions in the synchronized product which does not change the state of the automaton. In the opposite case, it produces transitions which lead the automaton from its current state to a successor state. To be valid, the initial marking of the current unfolding graph node must satisfy the propositions labelling the transition of the automata.

Definition 9 (Synchronization). *Let $\langle N, m_0 \rangle$ be a P/T system, $\langle G, \rightarrow, g_0 \rangle$ be an unfolding graph of $\langle N, m_0 \rangle$, f be a $LTL \setminus X$ formula and $Bu(f) = \langle Q, Acc, \rightarrow, \{f\} \rangle$ be a transition Büchi automaton. The synchronized product of the unfolding graph and the transition Büchi automaton is the graph $\langle G \times Q, Acc, \rightarrow, \langle g_0, \{f\} \rangle \rangle$ where*

- $\rightarrow \subseteq (G \times Q) \times PowerSet(Acc) \times (G \times Q)$
- $\langle g, F \rangle \xrightarrow{A} \langle g', F' \rangle$ *iff*
  - $\Phi_{Domain(F)}(m_0(g)) = \Phi_{Domain(F)}(m_0(g')) \Rightarrow (\exists e : g \xrightarrow{e} g') \wedge (A = \emptyset) \wedge (F = F')$
  - $\Phi_{Domain(F)}(m_0(g)) \neq \Phi_{Domain(F)}(m_0(g')) \Rightarrow (\exists e : g \xrightarrow{e} g') \wedge (\exists X : F \xrightarrow{(X,A)} F' \wedge X(m_0(g)))$

In the construction of the synchronized product, only the initial markings of the partial unfoldings composing the nodes are considered. To be a valid support of the verification, it is essential that all the reachable markings represented in a same node satisfy the same atomic propositions. Indeed, there is no change of state in the product if a non-bridge event fires. To insure this property, we introduce the notion of observed transitions of a proposition: every transition which could potentially modify the value of the proposition. For a given state $\langle g, F \rangle$ of the synchronized product, the observed transitions related to the propositions of F can not be used in the partial unfolding g with exception for the bridges.

Definition 10 (Observed transition). *Let $\langle N, m_0 \rangle$ be a P/T system, p be an atomic proposition over $Reach(N, m_0)$. A set of observed transitions for p denoted by $Obs(p)$ must satisfy:*

$$\{t \in T : \exists m, m' \in Reach(N, m_0), m \xrightarrow{t} m' \wedge p(m) \neq p(m')\} \subseteq Obs(p)$$

Definition 11 specifies the validity of the construction. It takes into account the constraints related to the observed transitions. Moreover, we impose that the inclusion rule is used in order to detect cycles when it is necessary (i.e. when the behavior where the trace of the initial marking is infinitely repeated satisfies the formula).

Definition 11 (Compatibility). *Let $\langle N, m_0 \rangle$ be a P/T system, $\langle G, \rightarrow, g_0 \rangle$ be an unfolding graph of $\langle N, m_0 \rangle$, f be a $LTL \backslash X$ formula and $Bu(f) = \langle Q, Acc, \rightarrow, \{f\} \rangle$ be a transition Büchi automaton. A synchronized product $\langle G \times Q, Acc, \rightarrow, \langle g_0, \{f\} \rangle \rangle$ of the unfolding graph and the transition Büchi automaton is compatible iff $\forall \langle g, F \rangle \in G \times Q$*

- *$\forall e \in E(g) \setminus Bridge(g), \forall p \in Domain(F) : h(e) \notin Obs(p)$*
- *$(\Phi_{Domain(F)}(m_0(g)))^\omega \models F \Rightarrow g$ is a partial inclusion rule unfolding*

Lemma 1 exhibits how a transition of P/T system is projected in the synchronized product.

Lemma 1. *Let $\langle g, F \rangle$ be a node of a compatible synchronized product $\langle G \times Q, Acc, \rightarrow, \langle g_0, \{f\} \rangle \rangle$. Let m and m' be two markings of N :*

1. If $m = h(g)(Min(g))$ and there exists a process π from m to m' containing no observed transition of F, then there exists a path in the synchronized product from $\langle g, F \rangle$ of the form

$$\nu = \langle g, F \rangle \xrightarrow{\emptyset} \langle g_1, F \rangle \xrightarrow{\emptyset} \cdots \langle g_n, F \rangle$$

such that $m' \in Reach(g_n)$.

2. If $m \in Reach(g)$, $\Phi_{AP}(m)_{|Domain(F)} = \Phi_{AP}(m')_{|Domain(F)}$ and $m \rightarrow m'$ in N then there exists a path in the synchronized product from $\langle g, F \rangle$ of the form

$$\nu = \langle g, F \rangle \xrightarrow{\emptyset} \langle g_1, F \rangle \xrightarrow{\emptyset} \cdots \langle g_n, F \rangle$$

such that $m' \in Reach(g_n)$. Moreover if ν is reduced to node $\langle g, F \rangle$, the configuration associated to m' in g strictly include the configuration for m or g has at least a cutoff.

3. If $m \in Reach(g)$, $\Phi_{AP}(m)_{|Domain(F)} \neq \Phi_{AP}(m')_{|Domain(F)}$ and $m \rightarrow m'$ in N then for every transition $F \xrightarrow{(X,A)} F'$ from F in $Bu(f)$ such that $X(\Phi_{AP}(m))$ holds, there exists a path in the synchronized product from $\langle g, F \rangle$ of the form

$$\nu = \langle g, F \rangle \xrightarrow{A} \langle g, F' \rangle \xrightarrow{\emptyset} \cdots \langle g_n, F' \rangle$$

Proof. 1. We can establish this property by induction on the size of process π. If the process π is null, then the property is obvious ($m = m'$). Otherwise, we just have to consider the three following cases:
- The process π is in g ($\exists C \in Conf(g) : C \cap Cutout = \emptyset \wedge \Pi(C) = \pi$), then the path is reduced to the node $\langle g, F \rangle$
- The process π cuts a cutoff ($\exists e \in Cutoff, \exists C \in Conf(\pi) : \Pi([e]) = \Pi(C)$), then we can apply the induction property to the process $\pi' = \Pi(\Phi[e]).\pi \setminus \Pi([e])$ which is smaller than π and has the same initial and final state. One can note than π' contains no observed transition.
- The process π cuts a bridge ($\exists e \in Bridge, \exists C \in Conf(\pi) : \Pi([e]) = \Pi(C)$), then we can apply the induction property to the process $\pi' = \pi \setminus \Pi([e])$. π' is smaller than π, has same final state and has $h(g')Min(g')$ as initial state ,$h(g')Min(g')$ (where $g \xrightarrow{e} g'$). One can note than π' contains no observed transition. The searching path is obtained from the transition $\langle g, F \rangle \xrightarrow{\emptyset} \langle g', F \rangle$ and the path deduces from the induction.
2. The two other properties are easily deduced from the first property. We first have to consider the process π in g which have as final state m and, its extension π' with the transition $m \rightarrow m'$. If π' is in g, then m' is in $Reach(g)$ and its configuration is greater than the one of m. Otherwise π' cuts a bridge or a cutoff which is associated to the transition $m \rightarrow m'$. One can apply property 1 to the reduced process obtained from the bridge or the cutoff. One can note that this process contains no observed transition.   □

Theorem 5 shows how the existence of a trace satisfying a formula f can be directly detected in a compatible synchronized product. Theorem 3, states that a trace of the P/T system satisfying f can be recognized by a finite or infinite path in the transition Büchi automaton. In the synchronized product, the finite paths correspond to the points 1 and 2. Indeed, a finite path of the transition Büchi automaton can be represented in the synchronized product either by a finite path or by an infinite one leading to a cycle where the values of the atomic propositions stutter. The infinite paths are characterized in the point 3 of the theorem.

Theorem 5 (Checking property). *Let $\langle N, m_0 \rangle$ be a P/T system, $\langle G, \rightarrow, g_0 \rangle$ be an unfolding graph of $\langle N, m_0 \rangle$, f be a LTL $\setminus$ X formula and $Bu(f) = \langle Q, Acc, \rightarrow, \{f\} \rangle$ be a transition Büchi automaton. Let $\langle G \times Q, Acc, \rightarrow, \langle g_0, \{f\} \rangle \rangle$ be a compatible synchronized product of the unfolding graph and the transition Büchi automaton. There exists a trace σ in $\langle N, m_0 \rangle$ which fulfills f iff there exists a finite or infinite path in $\langle G \times Q, Acc, \rightarrow, \langle g_0, \{f\} \rangle \rangle$*

$$\nu = \langle g_0, \{f\} \rangle \xrightarrow{A_0} \langle g_1, F_1 \rangle \xrightarrow{A_1} \langle g_2, F_2 \rangle \xrightarrow{A_2} \dots$$

which satisfies one of the following conditions:

1. *Finite path:*
   - *$| \nu | < \infty \wedge (\Phi_{Domain(F_{|\nu|-1})}(m_0(g_{|\nu|-1})))^\omega \models F_{|\nu|-1} \wedge g_{|\nu|-1}$ contains a cutoff or a deadlock*

2. *Infinite path with stuttering cycle:*
   - $\mid \nu \mid = \infty \wedge \exists i : (\forall j > i : \Phi_{Domain(F_i)}(m_0(g_i)) = \Phi_{Domain(F_i)}(m_0(g_j)) \wedge (\Phi_{Domain(F_i)}(m_0(g_i)))^\omega \models F_i)$
3. *Infinite path without stuttering cycle:*
   - $\mid \nu \mid = \infty \wedge (\forall a \in Acc, \forall i \geq 0, \exists j \geq i : a \in A_j)$

Proof. $(\Rightarrow)$ Let $\sigma = \Phi_{AP}(m_0) \cdot \Phi_{AP}(m_1) \cdot \Phi_{AP}(m_2) \ldots$ be an infinite trace of the P/T system $\langle N, m_0 \rangle$.

From Theorem 3, there exists an stutter free extract word $\sigma\prime = \Phi_{AP}(m_0) \cdot \Phi_{AP}(m_{i_1}) \cdot \Phi_{AP}(m_{i_2}) \ldots$ of σ and a path in $Bu(f)$

$$\rho = F_0 \xrightarrow{(X_0, A_0)} F_1 \xrightarrow{(X_1, A_1)} F_2 \xrightarrow{(X_2, A_2)} \ldots$$

which fulfill the properties 1,2,3 (of Theorem 3).

When apply Lemma 1 to any transition $m \rightarrow m'$, we obtain the searching path ν :

$$\nu = \langle g_0, F_0 \rangle \xrightarrow{\emptyset} \ldots \langle g_0', F_0 \rangle \xrightarrow{A_0} \langle g_1, F_1 \rangle \xrightarrow{\emptyset} \ldots \langle g_1', F_1 \rangle \xrightarrow{A_1} \langle g_2, F_2 \rangle \ldots$$

In fact, we can conclude by considering the following case :

- if ρ is infinite (Property 3 of Theorem 3) then ν is infinite and verify the same accepting condition property (Property 3).
- if ρ is finite (Property 2 of Theorem 3) and σ is finite then ν is finite, its last node contains the terminal marking and verify the same accepting condition property (Property 1).
- if ρ is finite (Property 2 of Theorem 3) and σ is infinite then ν could be finite or infinite. If ν is finite the last node must contain a cutoff (Property 2) and if ν is infinite, it has with infinite stutter suffix from which we can deduced an stuttering cycle (Property 3). In both case the accepting condition is the same as the accepting condition of path ρ in the formula automaton.

$(\Leftarrow)$ The reciprocal proof is obvious. We just have to remark that to any transition $\langle g, F \rangle \xrightarrow{A} \langle g', F' \rangle$ in the synchronized product corresponds an sequence of the P/T system from $h(g)(Min(g))$ to $h(g')(Min(g'))$ and a transition $F \xrightarrow{\langle A, X \rangle} F'$ if there is a modification of atomic propositions of F between $h(g)(Min(g))$ and $h(g')(Min(g'))$. Then, one can construct a run of the P/T system and a path of formula automaton. For finite path in the synchronized product (Property 2), one can complete the run with a finite stuttering sequence (if the last node contains a terminal marking) or an infinite sequence (if the last node contains a cutoff). $\square$

To apply this theorem in a verification algorithm, we have to detect these three types of paths in a synchronized product. The first one consists in a simple problem of reachability. For the second type, it requires to search cycle in which the value of the atomic propositions (related to the automaton state formula) stay invariant. This problem can be solve by a depth-first-search in the synchronized product giving priority to the transitions which do not change the state of the automaton (notice that they do not modify the value of the atomic propositions). The third type of paths is detected by the existence of a strongly connected component which contains each accepting condition. We can used the algorithm presented in [2] which is based on a depth-first-search.

These three kinds of detection can be combined in a single on-the-fly algorithm based on a depth-first-search and giving the priority to the transitions which do not change the automaton state. The computation, in the synchronized product, of the successors is realized as it follows.

From a marking and a state automaton formula F, one can decide which type of partial unfolding (inclusion or size rule) must be computed by verifying if the sequence in which the trace of the marking is infinitely repeated satisfies F (that implies that the inclusion rule is used). The observed transitions are determined from F. In the case of a partial inclusion rule unfolding, the algorithm stops by the detection of a path of the first type as soon as a cutoff is added. In a case where any cutoff has been added, one has to search for a terminal state local to the partial unfolding (using the algorithm presented in [16]).

Each bridge corresponds to a successor marking in the unfolding graph. The ones which do not modify the value of the atomic propositions associated to F are considered first (and are associated to the same automaton state). The others are associated to each successor state of the transition Büchi automaton compatible with the current marking.

The consideration of each of the new transitions in the depth first search leads to the computation related to the detection of paths of types 2 and 3.

5 Verification of $LTL^d \setminus X$ Formula with Unfolding Graphs

In this section, we show how the unfolding graphs can be used for the verification of $LTL \setminus X$ formulas using the atomic proposition *dead*. We will show that the unfolding graph definition is well adapted in this context and allows to limit the set of observed transitions to a reasonable one. In the Theorem 4 which characterizes the paths stuttering with respect to a $LTL^d \setminus X$ formula, the proposition *dead* is not taken into account. One obtains this same characteristic for the synchronization of a transition Büchi automaton and an unfolding graph.

Definition 12 (Dead synchronization). *Let $\langle N, m_0 \rangle$ be a P/T system, $\langle G, \rightarrow, g_0 \rangle$ be an unfolding graph of $\langle N, m_0 \rangle$, f be a $LTL^d \setminus X$ formula and $Bu(f) = \langle Q, Acc, \rightarrow, \{f\} \rangle$ be a transition Büchi automaton. The synchronized product of the unfolding graph and the transition Büchi automaton is the graph $\langle G \times Q, Acc, \rightarrow, \langle g_0, \{f\} \rangle \rangle$ where*

- *$\rightarrow \subseteq (G \times Q) \times PowerSet(Acc) \times (G \times Q)$*
- *$\langle g, F \rangle \xrightarrow{A} \langle g', F' \rangle$ iff*
    - *$\Phi_{Domain(F) \setminus \{dead\}}(m_0(g)) = \Phi_{Domain(F) \setminus \{dead\}}(m_0(g')) \Rightarrow$*
      $(\exists e : g \xrightarrow{e} g') \wedge (A = \emptyset) \wedge (F = F')$

- $\Phi_{Domain(F)\setminus\{dead\}}(m_0(g)) \neq \Phi_{Domain(F)\setminus\{dead\}}(m_0(g')) \Rightarrow$
$$(\exists e : g \overset{e}{\longrightarrow} g') \wedge (\exists X : F \overset{(X,A)}{\longrightarrow} F' \wedge X(m_0(g)))$$

Definition 13 specifies the validity of the construction. Firstly, one obtains the constraints of Definition 11 related to the observed transition. Secondly, the inclusion rule is imposed to be used when necessary (i.e. when the trace which repeats infinitely the values of the atomic propositions of the initial marking and for which the proposition *dead* is always false satisfies the formula associated to the node).

Definition 13 (Dead compatibility). *Let $\langle N, m_0 \rangle$ be a P/T system, $\langle G, \rightarrow , g_0 \rangle$ be an unfolding graph of $\langle N, m_0 \rangle$, f be a $LTL^d \setminus X$ formula and $Bu(f) = \langle Q, Acc, \rightarrow, \{f\} \rangle$ be a transition Büchi automaton. A synchronized product $\langle G \times Q, Acc, \rightarrow, \langle g_0, \{f\} \rangle \rangle$ of the unfolding graph and the transition Büchi automaton is compatible iff $\forall \langle g, F \rangle \in G \times Q$*

- *$\forall e \in E(g) \setminus Bridge(g), \forall p \in Domain(F) : h(e) \notin Obs(p)$*
- *$\langle \Phi_{Domain(F)\setminus\{dead\}}(m_0(g)), false \rangle^\omega \models F \Rightarrow g$ is a partial inclusion rule unfolding*

Theorem 6 shows how the existence of a trace satisfying a $LTL^d \setminus X$ formula can be directly detected on a compatible synchronized product. From Theorem 4, we know that there are three types of paths corresponding to such a trace ("Never dead", "Immediately dead" and "Sometime in a strict future dead"). The paths of type "Never dead" can be distinguished in three sub-types as in Theorem 5 (finite path, infinite path with a stuttering cycle, infinite path without a stuttering cycle) and one obtains these three characterizations in respectively the point 1, 2 and 3 of Theorem 6. We can note that the proposition *dead* has the value false in this context (explicitly for the point 1 and 2 and implicitly for the point 3).

For the paths of type "Immediately dead" (point 4 of Theorem 6), it must be possible to construct a configuration of the partial unfolding composing the penultimate node such that its extension with the bridge (which leads to the last node) corresponds to a terminal marking.

For the paths of type "Sometime in a strict future dead" (point 5 of Theorem 6), one has two distinct characterizations. If the penultimate node verifies the same atomic propositions as the last one then the presence of a terminal marking in the last node unfolding is sufficient. Otherwise, the terminal marking represented in the last node unfolding must not be the initial one and consequently the unfolding must be composed by at least an event.

Theorem 6 (Dead checking property). *Let $\langle N, m_0 \rangle$ be a P/T system, $\langle G, \rightarrow , g_0 \rangle$ be an unfolding graph of $\langle N, m_0 \rangle$, f be a $LTL^d \setminus X$ formula and $Bu(f) = \langle Q, Acc, \rightarrow, \{f\} \rangle$ be a transition Büchi automaton. Let $\langle G \times Q, Acc, \rightarrow, \langle g_0, \{f\} \rangle \rangle$ be a compatible synchronized product of the unfolding graph and the transition Büchi automaton. There exists a trace σ in $\langle N, m_0 \rangle$ which fulfills f iff there exists a finite or infinite path in $\langle G \times Q, Acc, \rightarrow, \langle g_0, \{f\} \rangle \rangle$*

$$\nu = \langle g_0, \{f\} \rangle \xrightarrow{A_0} \langle g_1, F_1 \rangle \xrightarrow{A_1} \langle g_2, F_2 \rangle \xrightarrow{A_2} \ldots$$

which satisfies one of the following conditions:

1. $|\nu| < \infty \wedge \langle \Phi_{AP_{|\nu|-1}}(m_0(g_{|\nu|-1})), false \rangle^\omega \models F_{|\nu|-1} \wedge g_{|\nu|-1}$ *contains a cutoff*

2. $|\nu| = \infty \wedge$
   $\exists i : (\forall j > i : \Phi_{AP_i}(m_0(g_i)) = \Phi_{AP_i}(m_0(g_j)) \wedge \langle \Phi_{AP_i}(m_0(g_i)), false \rangle^\omega \models F_i)$

3. $|\nu| = \infty \wedge (\forall a \in Acc, \forall i \geq 0, \exists j \geq i : a \in A_j)$

4. $|\nu| < \infty \wedge \langle \Phi_{AP_{|\nu|-1}}(m_0(g_{|\nu|-1})), true \rangle^\omega \models F_{|\nu|-1} \wedge \Phi_{AP_{|\nu|-2}}(m_0(g_{|\nu|-1})) \neq$
   $\Phi_{AP_{|\nu|-2}}(m_0(g_{|\nu|-2})) \wedge (\exists e \in Bridge(g_{|\nu|-2}) : g_{|\nu|-2} \xrightarrow{e} g_{|\nu|-1} \wedge (\exists C \in$
   $Conf(g_{|\nu|-2}) : C \cap Bridge(g_{|\nu|-2}) = \{e\} \wedge C \cap Cutoff(g_{|\nu|-2}) = \emptyset \wedge$
   $Dead(h(Cut(C)))))$

5. $|\nu| < \infty \wedge \langle \Phi_{AP_{|\nu|-1}}(m_0(g_{|\nu|-1})), false \rangle \cdot \langle \Phi_{AP_{|\nu|-1}}(m_0(g_{|\nu|-1})), true \rangle^\omega \models$
   $F_{|\nu|-1} \wedge g_{|\nu|-1}$ *contains a deadlock* $\wedge$
   $(E(g_{|\nu|-1}) \neq \emptyset \vee \Phi_{AP_{|\nu|-2}}(m_0(g_{|\nu|-1})) = \Phi_{AP_{|\nu|-2}}(m_0(g_{|\nu|-2})))$

where AP_i denotes the atomic proposition set $Domain(F_i) \setminus \{dead\}$.

Proof. The proof is tedious and is almost the same as the proof of Theorem 5. $\square$

In the section 4, we have discussed on the implementation of an algorithm for the detection of the three first types of paths during the construction of the compatible synchronized product. For a $LTL^d \setminus X$ formula, we have to specify how the paths of types "Immediately dead" and "Sometime in a strict future dead" can be detected. Both correspond to a reachability problem and then their detection can be done in combination with the others.

The condition imposes by the paths of type "Sometime in a strict future dead" can be verified using the McMillan's algorithm for the detection of terminal state represented inside an unfolding (see [16]).

For the paths of type "Immediately dead", an adaptation of the McMillan's algorithm must be used. Originally, the algorithm consists in the construction of a configuration of the unfolding containing for each cutoff at least an event in conflict.

In our context, our characterization implies that the unfolding has to be extended by the events which can take place immediately at the output conditions of the considered bridge. Then, the configuration must contain the bridge and must to be in conflict with the cutoff as well as the new additional events.

6   Experimentation

Our aim is to give a first efficiency evaluation of our technique. For this purpose, we compare it to the application of partial order reduction, namely the stubborn set technique (Prod [25]). Our study concerns essentially the memory space used by both methods. Hence, the comparison criteria is the size of the visited synchronized graphs. When constructing the unfolding graph, the current unfolding node is stored in memory. For information, we give the average unfolding node size.

Our evaluation is based on two classical examples: a model of FIFO queue using slots and a model of the slotted ring protocol. Both examples are scalable in such a way that the number of reachable states is exponential with respect to a given parameter.

Model	Stubborn Sets		Unfolding Graphs		
	nodes	edges	nodes	edges	average node size
fifo3	31	39	6	6	25
fifo5	57	67	8	8	60
fifo10	157	172	13	13	232
fifo12	211	228	15	15	336
fifo15	307	327	18	18	529
fifo20	507	532	23	23	949
live3	9772	33265	43	106	185
live4	114100	436264	88	329	1075
safe3	25	25	17	17	28
safe5	73	73	30	30	40
safe10	1532	1535	82	82	72
safe12	2288	2291	110	110	85
safe15	953	953	160	160	103
safe20	7072	7075	262	262	134

Table 1. Experimental results

For the FIFO model, we consider a liveness property which is satisfied. One can note that the size of the unfolding graphs is always smaller than the ones obtained by stubborn set reductions. However, the size of the unfoldings nodes increases in a quadratic way. We believe that the stubborn set technique is better for this case. The interpretation is that the stubborn set method results in a real reduction whereas unfoldings give the complete representation of the reachable states.

The first formula studies for the slotted ring protocol ($live3 - 4$) is also a liveness property which is satisfied. It produces good experimental results for the unfolding graph against stubborn sets. The last study concerns a non satisfied safety property of the slotted ring protocol ($safe3-20$). The on-the-fly technique allows us to stop the search as soon as a counter example is detected. That explains the good results obtained by both method against the previous study. The first experimentation that we have done have given us very bad results for the unfolding graphs: the size of each unfolding node increases exponentially because of the necessary use of the inclusion rule. We have solved this problem by arbitrary limiting their sizes. It is important to note that our method is flexible and then supports this kind of heuristics. However, the detection of local infinite behavior remains an important problem: the use of inclusion is penalizing. We believe that the sufficient condition and the necessary condition presented in [4] can induce an efficient detection avoiding the use of the inclusion rule.

7   Conclusion

We have mainly presented the technical aspects for model checking of $LTL \setminus X$ formula using unfolding graphs. This study is the continuation of the unfolding verification methods presented in [4] and the adaptation of the up-to-date on-the-fly verification of LTL formula presented in [2]. In this context, we have developed:

- On-the-fly verification algorithms,
- Dynamic refinement of the stuttering constraints (which are limited to the atomic proposition labelling the formula automaton node),
- Characterization of the new accepting conditions in the case of finite path (which can be checked linearly on the size of the formula),
- Adaptation of the new verification algorithm [2] in our context,
- Efficient consideration of the *dead* atomic proposition.

The first experimental results demonstrate the practicality of our method. However, we have remarked that the use of the inclusion rule can lead to the construction of huge unfoldings. To combat this explosion, we have suggested heuristics based on sufficient condition and necessary condition presented in [4]. One needs to verify its practical pertinence to limit this problem. In some cases (illustrated by the FIFO experimentation), our approach can not compete with stubborn sets. A solution could be to combine reduction methods and unfoldings.

References

[1] C. Courcoubetis, M. Y. Vardi, P. Wolper, and M. Yannakakis. Memory efficient algorithms for the verification of temporal properties. *Formal Methods in System Design*, 1:275–288, 1992.

[2] J.-M. Couvreur. On-the-fly verification of linear temporal logic. In *Proc. of FM'99*, volume 1708 of *Lecture Notes in Computer Science*, pages 253–271. Springer Verlag, 1999.

[3] J.-M. Couvreur and D. Poitrenaud. Model checking based on occurrence net graph. In *Proc. of Formal Description Techniques IX, Theory, Applications and Tools*, pages 380–395, 1996.

[4] J.-M. Couvreur and D. Poitrenaud. Detection of illegal behaviours based on unfoldings. In *Proc. of ICATPN'99*, volume 1639 of *Lecture Notes in Computer Science*, pages 364–383. Springer Verlag, 1999.

[5] J. Engelfriet. Branching processes of Petri nets. *Acta Informatica*, 28:575–591, 1991.

[6] J. Esparza. Model checking using net unfoldings. In *Proc. of TAPSOFT'93*, volume 668 of *Lecture Notes in Computer Science*, pages 613–628. Springer Verlag, 1993.

[7] J. Esparza and S. Römer. An unfolding algorithm for synchronous products of transition system. In *Proceedings of CONCUR'99*, number 1664 in LNCS, pages 2–20. Springer, 1999.

[8] J. Esparza, S. Römer, and W. Vogler. An improvement of McMillan's unfolding algorithm. In *Proc. of TACAS'96*, volume 1055 of *Lecture Notes in Computer Science*, pages 87–106. Springer Verlag, 1996.

[9] R. Gerth, D. Peled, M. Y. Vardi, and P. Wolper. Simple on-the-fly automatic verification of linear temporal logic. In *Proc. 15th Work. Protocol Specification, Testing, and Verification*, Warsaw, June 1995. North-Holland.

[10] P. Godefroid. Partial-order methods for the verification of concurrent systems. volume 1032 of *Lecture Notes in Computer Science*. Springer Verlag, 1996.

[11] P. Godefroid and G. J. Holzmann. On the verification of temporal properties. In *Proc. 13th Int. Conf on Protocol Specification, Testing, and Verification, INWG/IFIP*, pages 109–124, Liege, Belgium, May 1993.

[12] G. J. Holzmann. *Design and Validation of Computer Protocols*. Prentice-Hall, Englewood Cliffs, New Jersey, 1991.

[13] G. J. Holzmann, D. Peled, and M. Yannakakis. On nested depth first search. In *The Spin Verification System*, pages 23–32. American Mathematical Society, 1996. Proc. of the Second Spin Workshop.

[14] A. Kondratyev, M. Kishinevsky, A. Taubin, and S. Ten. A structural approach for the analysis of Petri nets by reduced unfolding. In *Proc. of ICATPN'96*, volume 1091 of *Lecture Notes in Computer Science*, pages 346–365. Springer Verlag, 1996.

[15] R. Langerak and E. Brinksma. A complete finite prefix for process algebra. In *Proceedings of the 11th International Conference on Computer Aided Verification, Italy*, number 1633 in LNCS, pages 184–195. Springer, 1999.

[16] K.L. McMillan. Using unfoldings to avoid the state explosion problem in the verification of asynchronous circuits. In *Proc. of the 4th Conference on Computer Aided Verification*, volume 663 of *Lecture Notes in Computer Science*, pages 164–175. Springer Verlag, 1992.

[17] S. Melzer and S. Römer. Deadlock checking using net unfoldings. In *Proc. of the 9th Conference on Computer Aided Verification*, Lecture Notes in Computer Science, pages 352–363. Springer Verlag, 1997.

[18] M. Nielsen, G. Plotkin, and G. Winskel. Petri nets, events structures and domains, part I. *Theoretical Computer Science*, 13(1):85–108, 1981.

[19] D. Peled. All from one, one for all: on model checking using representatives. In *Proc. on the 5th Conference on Computer Aided Verification*, volume 697 of *Lecture Notes in Computer Science*, pages 409–423. Springer Verlag, 1993.

[20] Doron Peled and Thomas Wilke. Stutter-invariant temporal properties are expressible without the nexttime operator. *Information Processing Letters*, 63:243–246, 1997.

[21] D. Poitrenaud. *Graphes de Processus Arborescents pour la Vérification de Propriétés*. Thèse de doctorat, Université P. et M. Curie, Paris, France, 1996.

[22] A. P. Sistla and E. M. Clarke. The complexity of propositional linear temporal logic. *Journal of the Association for Computing Machinery*, 32(3):733–749, July 1985.

[23] A. Valmari. Stubborn sets for reduced state space generation. In *Advances in Petri Nets*, volume 483 of *Lecture Notes in Computer Science*, pages 491–515. Springer Verlag, 1991.

[24] A. Valmari. On-the-fly verification with stubborn sets. In *Proc. of the 5th Conference on Computer Aided Verification*, volume 697 of *Lecture Notes in Computer Science*, pages 397–408. Springer Verlag, 1993.

[25] K. Varpaaniemi and M. Rauhamaa. The stubborn set method in practice. In *Advances in Petri Nets*, volume 616 of *Lecture Notes in Computer Science*, pages 389–393. Springer Verlag, 1992.

[26] F. Wallner. Model checking LTL using net unfolding. In *Proc. on the* 10th *Conference on Computer Aided Verification*, Lecture Notes in Computer Science. Springer Verlag, 1998.

Process Semantics of Petri Nets over Partial Algebra

Jörg Desel, Gabriel Juhás, and Robert Lorenz[*]

Lehrstuhl für Angewandte Informatik
Katholische Universität Eichstätt, 85071 Eichstätt, Germany
{joerg.desel,gabriel.juhas,robert.lorenz}@ku-eichstaett.de

Abstract. "Petri nets are monoids" is the title and the central idea of the paper [7]. It provides an algebraic approach to define both nets and their processes as terms. A crucial assumption for this concept is that arbitrary concurrent composition of processes is defined, which holds true for place/transition Petri nets where places can hold arbitrarily many tokens.

A decade earlier, [10] presented a similar concept for elementary Petri nets, i.e. nets where no place can ever carry more than one token. Since markings of elementary Petri nets cannot be added arbitrarily, concurrent composition is defined as a partial operation.

The present papers provides a general approach to process term semantics. Terms are equipped with the minimal necessary information to determine if two process terms can be composed concurrently. Applying the approach to elementary nets yields a concept very similar to the one in [10].

The second result of this paper states that the semantics based on process terms agrees with the classical partial-order process semantics for elementary net systems. More precisely, we provide a syntactic equivalence notion for process terms and a bijection from according equivalence classes of process terms to isomorphism classes of partially ordered processes. This result slightly generalizes a similar observation given in [11].

1 Introduction

One of the main advantages of Petri nets is their capability to express true concurrency in a very natural way. Thus, Petri nets offer not only sequential semantics, which correspond to classical marking graphs, but also process semantics. Processes express possible runs of a system, in which independent transitions can occur concurrently. ¿From the very beginning of Petri net theory processes were based on partial order between net elements.

In [7] it is observed that place/transition nets can be understood as graphs whose vertices are multisets of places, and transitions are arcs with sources and targets given by their pre-multisets and post-multisets. Reflexive arcs represent markings. By multiset addition one can generate the concurrent marking graph

---

[*] supported by DFG: Project "SPECIMEN"

M. Nielsen, D. Simpson (Eds.): ICATPN 2000, LNCS 1825, pp. 146–165, 2000.

from a net. For example, using composition of a reflexive arc, given by a multi-set X, and an arc representing a single transiton t with pre-multiset $pre(t)$ and post-multiset $post(t)$, the arc $X + t$ changes the marking $M = X + pre(t)$ to the marking $M' = X + post(t) = M - pre(t) + post(t)$. Addition of non-reflexive arcs represents their concurrent occurrence. Concatenating graph arcs with corresponding target and source yields a representation of processes, which again can be composed concurrently or sequentially. Thus, process terms of a Petri net are obtained in a very easy way using only few production rules. Since the sum is defined for each pair of processes, this approach does not allow to express situations in which processes interfere and therefore cannot occur concurrently.

Many classes of Petri nets do not allow arbitrary concurrent composition of processes. For example, in processes of elementary net systems, no two conditions representing tokens on the same place can be concurrent. Hence, for example, no process can run concurrently with a copy of itself. A similar observation holds for nets with capacity limitations. Also inhibitor arcs and read arcs [4,8] restrict the possible concurrent composition of processes.

The aim of our work is to develop a unifying general framework to solve the problem of concurrent process composition in a conceptual way. Therefore, we employ *Petri nets over partial algebra*, defined in [5,6] as a unifying concept for Petri nets with modified occurrence rules. We claim that this approach is also suitable as a basis for process construction of different classes of Petri nets where dependencies between processes that restrict concurrent composition are taken into consideration.

In this paper, we show that Petri nets over partial algebra are suitable to define processes and their concurrent composition for elementary net systems and one-safe nets. Technically, we equip processes with the necessary information used to decide whether they are independent. We show that the minimal necessary information basically consists of the set of places associated to conditions of a process. This result coincides with the observation of [10], where also the set of involved places was used to define when two process terms can be concurrently composed.

In order to justify the algebraic approach introduced in [7] it was shown in [3] that process terms of place/transition nets from [7] are equivalent to processes based on partial order defined in [1]. In a similar fashion we show in this paper that process terms of elementary nets defined using partial algebra correspond to the usual processes of elementary nets based on partial order. Usually, processes are only defined for contact-free Petri nets, where the causal order between events is always generated by the flow of a token between the corresponding transitions. Our definition of elementary nets does not generally assume contact-freeness, i.e. we also consider situations where a transition can only occur after another transition because otherwise some place would carry two tokens at the intermediate marking. The usual way to cope with contacts is to introduce complement places in nets and according complement conditions in processes (see e.g. [9]). Our result is also based on this approach for general elementary nets, generalizing a similar result of [11] for contact-free elementary nets.

After basic definitions in Section 2, Petri nets over partial algebras (shortly PG nets) are defined in Section 3. In particular, it is shown how terms for processes are constructed within this algebra and how independency between terms is defined. An equivalence relation between such terms is introduced, which in fact is a congruence with respect to the partial operations used for the construction of the terms. In Section 4 elementary nets with corresponding PG nets are defined. Furthermore the notion of processes of elementary nets based on partial order is recalled, and a concurrent composition and concatenation of such processes is introduced. In Section 5 we present the main result of the paper, namely we show the one to one correspondence between isomorphism classes of processes of an elementary net and congruence classes of process terms of the corresponding PG net.

2   Basic Definitions

We use $\mathbb{N}$ to denote the nonnegative integers. Given two arbitrary sets A and B, the symbol B^A denotes the set of all functions from A to B. Given a function f from A to B and a subset C of A we write $f|_C$ to denote the restriction of f to the set C. The symbol 2^A denotes the power set of a set A. The set of all multi-sets over a set A is denoted by $\mathbb{N}^A$. Given a binary relation $R \subseteq A \times A$ over a set A, the symbol R^+ denotes the transitive closure of R.

Definition 1. *A partial groupoid is an ordered tuple $\mathcal{H} = (H, dom_{\dot{+}}, \dot{+})$ where H is the carrier of $\mathcal{H}$, $dom_{\dot{+}} \subseteq H \times H$ is the domain of $\dot{+}$, and $\dot{+} : dom_{\dot{+}} \to H$ is the partial operation of $\mathcal{H}$.*

Definition 2. *We say that a partial groupoid $\mathcal{H} = (H, dom_{\dot{+}}, \dot{+})$ can be embedded into a commutative monoid if there exists a commutative monoid $(H', +)$ such that $H \subseteq H'$ and the operation $+$ restricted to $dom_{\dot{+}}$ is equal to the partial operation $\dot{+}$. The monoid $(H', +)$ is called the embedding of $\mathcal{H}$.*

In the rest of the paper we will consider only partial groupoids $(H, dom_{\dot{+}}, \dot{+})$ which can be embedded into a commutative monoid, and moreover fulfil the following conditions:

- The relation $dom_{\dot{+}}$ is symmetric.
- $\forall a, b, c \in H : ((a \dot{+} b, c) \in dom_{\dot{+}} \Rightarrow (a, c), (b, c) \in dom_{\dot{+}})$.

We use the operation $\dot{+}$ to express concurrent composition of processes. As motivated in the introduction, not each pair of processes can be composed, hence $\dot{+}$ is a partial operation. $dom_{\dot{+}}$ contains the pairs of processes which are independent and can be composed. Obviously, this relation should be symmetric. The second requirement states that whenever the concurrent composition of two processes a and b is independent from c then both a and b are independent from c.

The partial groupoid $(H, dom_+, \dotplus)$ is extended to the partial groupoid $(2^H, \{dom_+\}, \{\dotplus\})$ such that

- $\{dom_+\} = \{(X, Y) \in 2^H \times 2^H | X \times Y \subseteq dom_+\}$.
- $X\{\dotplus\}Y = \{x \dotplus y | x \in X \wedge y \in Y\}$.

We will use more than one partial operations on the same carrier. Therefore the following definition: A partial algebra is a set (called carrier) together with a couple of partial operations on this set (with possibly different arity). Given a partial algebra with carrier X, an equivalence $\sim$ on X is a *congruence* if for every n-ary partial operation op ($n \in \mathbb{N}$): If $a_1 \sim b_1, \ldots, a_n \sim b_n$, $(a_1, \ldots, a_n) \in dom_{op}$ and $(b_1, \ldots, b_n) \in dom_{op}$, then $op(a_1, \ldots, a_n) \sim op(b_1, \ldots, b_n)$. If moreover $a_1 \sim b_1, \ldots, a_n \sim b_n$ and $(a_1, \ldots, a_n) \in dom_{op}$ imply $(b_1, \ldots, b_n) \in dom_{op}$ then the congruence $\sim$ is said to be *closed*. Thus, a congruence is an equivalence preserving all operations of a partial algebra, while a closed congruence moreover preserves the domains of the operations. Recall that the intersection of two congruences is again a congruence. Given a binary relation on X, there always exists the least congruence containing this relation. In general, the same does not hold for closed congruences. Given a partial algebra $\mathcal{X}$ with carrier X and a closed congruence $\sim$ on $\mathcal{X}$, we write as usual, $[x]_\sim = \{y \in X | x \sim y\}$ and $X/_\sim = \bigcup_{x \in X} [x]_\sim$. The natural homomorphism $h : X \rightarrow X/_\sim$ w.r.t. $\sim$ is given by $h(x) = [x]_\sim$. Given a subset of $A \subseteq X$, we write $[A]_\sim = \bigcup_{a \in A} [a]_\sim$. A closed congruence $\sim$ defines the partial algebra $\mathcal{X}/_\sim$ with an n-ary partial operation $op/_\sim$ defined for each n-ary partial operation $op : dom_{op} \rightarrow X$ of $\mathcal{X}$ as follows: $dom_{op/_\sim} = \{([a_1]_\sim, \ldots, [a_n]_\sim) | (a_1, \ldots a_n) \in dom_{op}\}$ and, for each $(a_1, \ldots, a_n) \in dom_{op}$, $op([a_1]_\sim, \ldots, [a_n]_\sim) = [op(a_1, \ldots a_n)]_\sim$. The partial algebra $\mathcal{X}/_\sim$ is called factor algebra of $\mathcal{X}$ with respect to the closed congruence $\sim$.

3   Process Terms of Petri Nets over Partial Algebras

Definition 3. *A graph is a quadruple* $(H, T, pre, post)$*, where* H *is a set of vertices,* T *is a set of arcs and* $pre, post : T \rightarrow H$ *are source and target functions, respectively.*

The formal definition of Petri nets over partial algebra was introduced in [5] and extended in [6].

Definition 4. *Given a partial groupoid* $(H, dom_+, \dotplus)$*, a graph* $N = (H, T, pre, post)$ *is called a* Petri net over the partial groupoid $(H, dom_+, \dotplus)$ *(shortly a PG net).*

We write $t : a \rightarrow b \in N$ to denote that $t \in T, pre(t) = a, post(t) = b$.

Elements of H are called states or markings of the net, elements of T are transitions, and $pre, post$ denote sets of pre-conditions and post-conditions.

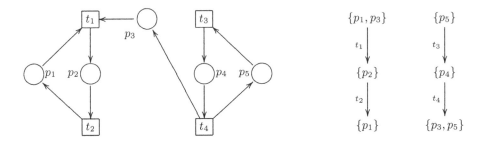

Fig. 1. An elementary net and the corresponding net over the partial groupoid $(2^{\{p_1,\dots,p_5\}}, dom_\uplus, \uplus)$ with $dom_\uplus = \{(M, M')|M \cap M' = \emptyset\}$ and $\uplus = \cup|_{dom_\uplus}$.

We consider elementary nets as elementary net systems [9] with arbitrary initial marking. Figure 1 illustrates the definition of a PG net and its relation to standard terminology of Petri nets.

In this paper we omit the definition of the enabling and firing rule for single transitions of PG nets, but rather directly define their process term semantics. The treatment of different enabling and firing rules and their relationships is discussed in [5,6].

To build process terms of a PG net, we need to have information about all states reachable in a process in order to decide whether the process is independent from another process. So we also have to consider an independence relation between states. We call two processes independent if their respective state spaces X and Y are independent, which means that every state $x \in X$ is independent from every state $y \in Y$. Given two independent processes with state spaces X and Y, the state space Z of the process derived from the concurrent composition $\dotplus$ of the two processes is defined by $Z = X\{\dotplus\}Y$.

Storing the set of all states which could be reached during a process can cause exponential growth and therefore it is not feasible. Fortunately, this exponential information is not necessary in the case of elementary nets, as will be shown in the next section. In general, for deriving a more compact information we can use any equivalence $\cong \in 2^H \times 2^H$ that is a closed congruence with respect to the operations $\{\dotplus\}$ (concurrent composition) and $\cup$ (sequential composition). Equivalence classes of the greatest (and hence coarsest) closed congruence represent the minimal information assigned to process terms necessary for concurrent composition. This congruence is unique [2].

Thus, the process semantics of a PG net is a graph generated from PG net by reflexive, additive and concatenative closure where addition respects partiality of state independence and concatenation respects equality of target and source.

Definition 5. *Let $(H, dom_{\dotplus}, \dotplus)$ be a partial groupoid, $\cong \in 2^H \times 2^H$ be the greatest closed congruence of the partial algebra $X = (2^H, dom_{\{\dotplus\}}, \{\dotplus\}, \cup)$ and $supp : X \to X/\cong$ be the natural homomorphism. Given a PG net $N = (H, T, pre, post)$ over $(H, dom_{\dotplus}, \dotplus)$, the process term semantics of N is the graph*

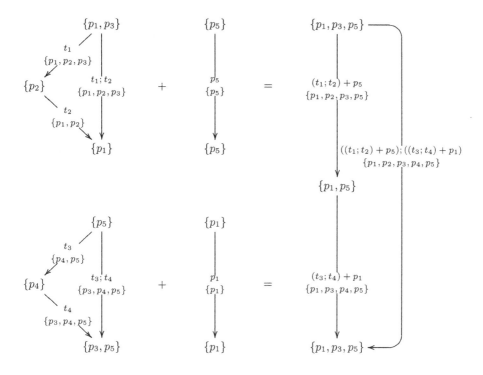

Fig. 2. Deriving a process term of the PG net from Figure 1

$\mathcal{P}(N) = (H, T_{\mathcal{P}}, pre_{\mathcal{P}}, post_{\mathcal{P}})$ *together with a function* $s : T_{\mathcal{P}} \to X/\cong$. *The elements of* $T_{\mathcal{P}}$ *(the arrows of the graph) are called process terms.* $T_{\mathcal{P}}$, $pre_{\mathcal{P}}$, $post_{\mathcal{P}}$ *and* s *are defined inductively by the following production rules, where* $\alpha : a \longrightarrow b \in \mathcal{P}(N)$ *denotes that* $\alpha \in T_{\mathcal{P}}, pre_{\mathcal{P}}(\alpha) = a, post_{\mathcal{P}}(\alpha) = b$:

$$\frac{a \in H}{a : a \longrightarrow a \in \mathcal{P}(N), s(a) = supp(\{a\})}$$

$$\frac{t \in T}{t : pre(t) \to post(t) \in \mathcal{P}(N), s(t) = supp(\{pre(t), post(t)\})}$$

$$\frac{\alpha : a \to b, A \in \mathcal{P}(N) \wedge \beta : c \to d, B \in \mathcal{P}(N) \wedge (A, B) \in dom_{\dotplus}/\cong}{(\alpha + \beta) : a \dotplus c \longrightarrow b \dotplus d \in \mathcal{P}(N), s(\alpha + \beta) = A \dotplus /\cong B}$$

$$\frac{\alpha : a \to b, A \in \mathcal{P}(N) \wedge \beta : b \to c, B \in \mathcal{P}(N)}{(\alpha; \beta) : a \longrightarrow c \in \mathcal{P}(N), s(\alpha; \beta) = A \cup /\cong B}$$

These rules define partial binary operations, called concurrent composition $(+)$ *and* concatenation $(;)$ *of process terms.*

Examples for constructing process terms are shown in Figures 2 and 3.

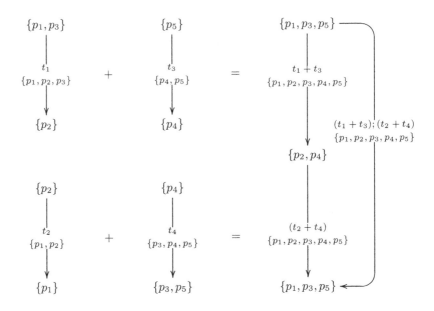

Fig. 3. Deriving another process term of PG net from Figure 1.

3.1   Equivalence of Process Terms

We now identify process terms by an equivalence relation $\sim_t$ which preserves the operations $+$ and $;$. Formally we define a congruence on $T_{\mathcal{P}}$ with respect to $+$ and $;$. Let $\sim_t$ be the least congruence on $T_{\mathcal{P}}$ with respect to $+$ and $;$ given the following axioms: Let $a, b \in H$ and $\alpha_1, \alpha_2, \alpha_3, \alpha_4$ be process terms with associated function s.

(1) $(\alpha_1 + \alpha_2) \sim_t (\alpha_2 + \alpha_1)$, whenever $+$ is defined for α_1 and α_2 .

(2) $((\alpha_1; \alpha_2); \alpha_3) \sim_t (\alpha_1; (\alpha_2; \alpha_3))$, whenever these terms are defined.

(3) $((\alpha_1 + \alpha_2) + \alpha_3) \sim_t (\alpha_1 + (\alpha_2 + \alpha_3))$, whenever these terms are defined.

(4) $\alpha = ((\alpha_1 + \alpha_2); (\alpha_3 + \alpha_4)) \sim_t \beta = ((\alpha_1; \alpha_3) + (\alpha_2; \alpha_4))$, whenever these terms are defined and $s(\alpha) = s(\beta)$.

(5) $(\alpha_1; post_{\mathcal{P}}(\alpha_1)) \sim_t \alpha_1 \sim_t (pre_{\mathcal{P}}(\alpha_1); \alpha_1)$.

(6) $a \dotplus b \sim_t (a + b)$ whenever these terms are defined.

(7) $\alpha + a \sim_t \alpha$ whenever the left term is defined, $pre_{\mathcal{P}}(\alpha) \dotplus a = pre_{\mathcal{P}}(\alpha)$ and $post_{\mathcal{P}}(\alpha) \dotplus a = post_{\mathcal{P}}(\alpha)$.

Axiom (1) represents commutativity of concurrent composition, axioms (2) and (3) associativity of concurrent composition and concatenation of process terms, axiom (4) distributivity whenever both terms have the same information about states, axiom (5) states that elements of H are partial neutral elements with respect to $;$, axiom (6) expresses that composition of these neutral elements is congruent to the neutral element constructed from their composition, and finally axiom (7) states that elements of H which are neutral to source and target of a term are neutral to the term itself.

Remark 1. Observe that for any two equivalent process terms $\alpha_1 \sim_t \alpha_2$, we have $pre_P(\alpha_1) = pre_P(\alpha_2)$ and $post_P(\alpha_1) = post_P(\alpha_2)$. Moreover $\alpha_1 \sim_t \alpha_2$ implies $s(\alpha_1) = s(\alpha_2)$. Thus, by construction of process terms, the congruence $\sim_t$ is a closed congruence.

The process term $((t_1; t_2) + p_5); ((t_3; t_4) + p_1)$ from Figure 2 and the process term $(t_1 + t_3); (t_2 + t_4)$ from Figure 3 are congruent:

$$((t_1; t_2) + p_5); ((t_3; t_4) + p_1) \quad \overset{(4),(5)}{\sim_t} \quad ((t_1 + p_5); (t_2 + p_5)); ((t_3 + p_1); (t_4 + p_1))$$

$$\overset{(1),(2),(4)}{\sim_t} \quad (t_1 + p_5); ((t_2; p1) + (p_5; t_3)); (t_4 + p_1)$$

$$\overset{(1),(5)}{\sim_t} \quad (t_1 + p_5); (t_3 + t_2); (t_4 + p_1)$$

$$\overset{(5)}{\sim_t} \quad (t_1 + p_5); ((t_3; p_4) + (p_2; t_2)); (t_4 + p_1)$$

$$\overset{(4)}{\sim_t} \quad (t_1 + p_5); ((t_3 + p_2); (t_2 + p_4)); (t_4 + p_1)$$

$$\overset{(1),(2),(4),(5)}{\sim_t} \quad (t_1 + t_3); (t_2 + t_4).$$

4   Elementary Nets

In this section we define elementary nets and some useful notations. Elementary nets can be considered as elementary net systems with arbitrary initial marking. For a given marking, we use complement places to assure contact-free behavior.

Definition 6. *(Power set with distinct union) Given a finite set P, let $(2^P, dom_{\uplus}, \uplus)$ be the partial groupoid defined by*

$$dom_{\uplus} = \{(A, B) \in 2^P \times 2^P | A \cap B = \emptyset\}$$

and $\uplus = \cup|_{dom_{\uplus}}$. Denoting $H = 2^P$, we define the mapping $supp : 2^H \to H$, $supp(A) = \bigcup_{a \in A} a$.

To define process terms for algebraic elementary nets we have to find the greatest closed congruence on $(2^H, \{\dotplus\}, dom_{\{\dotplus\}}, \cup)$.

We show that the mapping $supp$ is (isomorphic to) the natural homomorphism w.r.t. the gratest closed congruence on $(2^H, \{\dotplus\}, dom_{\{\dotplus\}}, \cup)$.

Lemma 1. *The relation $\cong \subseteq 2^H \times 2^H$ defined by $A \cong B \iff supp(A) = supp(B)$ is a closed congruence on $(2^H, dom_{\{\dotplus\}}, \{\dotplus\}, \cup)$*

Proof. Straightforward observation.

Lemma 2. *The closed congruence $\cong \subseteq 2^H \times 2^H$ is the greatest closed congruence on $(2^H, dom_{\{\dotplus\}}, \{\dotplus\}, \cup)$.*

Proof. We will show that any congruence $\approx$ such that $\cong$ is a proper subset of $\approx$ is not closed. Assume there are $A, B \in 2^H$ such that $A \approx B$ but $A \not\cong B$. Then $supp(A) \neq supp(B)$.

We construct a set $C \in 2^H$ such that $(A, C) \in dom_{\{+\}}$ but $(B, C) \notin dom_{\{+\}}$ or vice versa (which implies that $\approx$ is not closed). Denoting $supp(A) = a$ and $supp(B) = b$ we obtain $a \neq b$.

Without loss of generality we can assume $b \setminus a \neq \emptyset$. Set $C = \{c\}$ with $c = b \setminus a$. Then $c \cap a = \emptyset$, but $c \cap b \neq \emptyset$, i.e. $(A, C) \in dom_{\{+\}}$, but $(B, C) \notin dom_{\{+\}}$.

Now we are prepared to define elementary nets using our formalism.

Definition 7. *Given a finite set P (of places), a PG net $AEN = (2^P, T, pre, post)$ over a partial groupoid $(2^P, dom_{\uplus}, \uplus)$, is called an algebraic elementary net. Its process term semantics is given by $\mathcal{P}(AEN)$.*

In order to justify our approach to Petri nets and their processes defined by terms we show that the process semantics of classical elementary nets, as defined *e.g.* in [9], essentially coincides with the above formalism. Let us first recall basic definitions of elementary nets and their process semantics based on partial orders.

Definition 8. *An elementary net is a triple $EN = (P, T, F)$, where P and T are disjoint finite sets of places and transitions and $F \subseteq (P \times T) \cup (T \times P)$ is a (flow) relation such that*

(1) $\forall t \in T \exists p, q \in P : (p, t), (t, q) \in F$, and
(2) $\forall t \in T \forall p, q \in P : (p, t), (t, q) \in F \Rightarrow p \neq q$.

A marked elementary net is a tuple $MEN = (EN, M_0)$, where $EN = (P, T, F)$ is an elementary net and $M_0 \subseteq P$ is an initial marking.

Given an element $x \in P \cup T$, the set ${}^\bullet x = \{y | (y, x) \in F\}$ is called pre-set of x and the set $x^\bullet = \{y | (x, y) \in F\}$ is called post-set of x. An element x satisfying ${}^\bullet x = x^\bullet = \emptyset$ is called isolated (by definition, only places can be isolated).

Definition 9. *Given an elementary net $EN = (P, T, F)$, the corresponding algebraic elementary net $AEN = (2^P, T, pre, post)$ is defined by $pre(t) = {}^\bullet t$ and $post(t) = t^\bullet$ for each $t \in T$.*

Figure 1 shows an elementary net with corresponding algebraic elementary net.

In this paper we omit the definition of firing rule of classical elementary nets. It may be found *e.g.* in [9]. A detailed discussion about different possibilities of enabling and firing rules of elementary nets (also when understood as PG nets) can be found in [6]. Thus, we approach directly the definition of processes of classical elementary nets.

Definition 10. *A process net is an elementary net $N = (P_N, T_N, F_N)$ with unbranched places (i.e. $\forall p \in P_N : |{}^\bullet p|, |p^\bullet| \leq 1$) which is acyclic (i.e. $\forall x \in P_N \cup T_N : (x, x) \notin F_N^+$).*

Definition 11. *Given a process net* $N = (P_N, T_N, F_N)$, *the partial order* F_N^+ *generates relations* **co**, **li** $\subset (P_N \cup T_N) \times (P_N \cup T_N)$, *defined by*

(1) **co** $= \{(x, y) | (x, y), (y, x) \notin F_N^+\}$.
(2) **li** $= \{(x, y) | (x, y) \notin$ **co** $\vee x = y\}$.

A set $CO_N \subseteq P_N$ *satisfying* $\forall x, y \in CO_N : (x, y) \in$ **co** *is called co-set. A slice of* N *is a maximal co-set. The initial and final slice are given by*

(3) $^\circ N = \{p \in P_N | \nexists t \in T_N : (t, p) \in F_N\}$.
(4) $N^\circ = \{p \in P_N | \nexists t \in T_N : (p, t) \in F_N\}$.

The past and future of a slice S_N *of* N *is defined by*

(5) $^\rightarrow S_N = \{x \in P_N \cup T_N | \exists p \in S_N : (x, p) \in F_N^+ \vee x = p\}$,
(6) $S_N^\rightarrow = \{x \in P_N \cup T_N | \exists p \in S_N : (p, x) \in F_N^+ \vee x = p\}$.

Processes of elementary nets are only defined for so called contact-free marked elementary nets (for more details see e.g. [9]). In this paper, processes of a marked net which is not contact free are studied through processes of a contact-free marked net, obtained by a so called complement construction. By $\overline{MEN}$ (see [9]) we denote the net constructed from a net MEN by adding some so-called co-places. In [9] the set of co-places depends on the initial marking. Using a small simplification (which doesn't change process semantics, but only adds some unnecessary co-places) we will define a net $\overline{EN}$ by adding a co-place for each place $p \in P$. This net is contact-free for all possible initial markings of $\overline{EN}$, where an initial marking $\overline{M_0}$ of $\overline{EN}$ is constructed from an initial marking M_0 of EN by adding all co-places of places to $\overline{M_0}$ which are not in M_0.

Definition 12. *Given an elementary net* $EN = (P, T, F)$, *let* C *be a set satisfying* $|C| = |P|$ *and* $C \cap (P \cup T) = \emptyset$, *and let* $c : P \rightarrow C$ *be an arbitrary bijection. Let* $\overline{EN} = (\overline{P}, \overline{T}, \overline{F})$ *be the elementary net defined by*

- $\overline{P} = P \cup C$,
- $\overline{T} = T$ *and*
- $\overline{F} = F \cup \{((c(p), t) | (t, p) \in F\} \cup \{(t, c(p)) | (p, t) \in F\}$.

Given a marked elementary net $MEN = (EN, M_0)$, *define*

$$\overline{M_0} = M_0 \cup \{c(p) | p \in P \wedge p \notin M_0\} \text{ and } \overline{MEN} = (\overline{EN}, \overline{M_0}).$$

Note that, given an elementary net EN, the construction of $\overline{EN}$ is unique up to isomorphism.

A process of a marked elementary net MEN is now defined via the associated elementary net $\overline{MEN}$:

Definition 13. *Let* $EN = (P, T, F)$ *be an elementary net and* $M_0 \subseteq P$ *be a marking. A process* N *of* $MEN = (EN, M_0)$ *is a tuple* (P_N, T_N, F_N, Φ_N), *where* (P_N, T_N, F_N) *is a process net and* $\Phi_N : (P_N \cup T_N) \rightarrow (\overline{P} \cup \overline{T})$ *is a mapping satisfying*

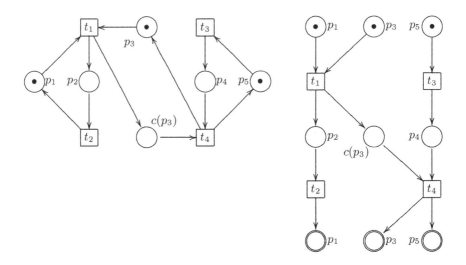

Fig. 4. A marked contact-free net where elements of the initial slice are marked and elements of the final slice are depicted by double-line circles.

(1) No isolated place of N is mapped by Φ_N to a co-place of $\overline{EN}$.
(2) $\Phi_N|_{\circ N}$ is injective.
(3) $\Phi_N(\circ N) \cap P = M_0$ and $\Phi_N(\circ N) \subseteq \overline{M_0}$.
(4) $\forall t \in T_N : \Phi_N|_{\bullet t}$ and $\Phi_N|_{t^\bullet}$ are injective, and
(5) $\forall t \in T_N : \Phi_N(\bullet t) = \bullet(\Phi_N(t))$ and $\Phi_N(t^\bullet) = (\Phi_N(t))^\bullet$,

where the $\bullet$-notation refers to $\overline{EN}$. Let $\mathcal{P}(EN, M_0)$ be the set of all processes of the marked elementary net $MEN = (EN, M_0)$. By $\mathcal{P}(EN) = \bigcup_{M_0 \subseteq P} \mathcal{P}(EN, M_0)$ we denote the set of all processes of an elementary net EN.

Note that the properties of the definition imply that $\Phi_N(P_N) \subseteq \overline{P}$ and $\Phi_N(T_N) \subseteq \overline{T}$. Moreover, Φ_N is injective on co-sets (see [9]).

We will not distinguish isomorphic processes of an elementary net.

The above definition of processes differs from the one defined in [9] since we have no isolated places which are mapped by Φ_N to co-places. Figure 4 shows a process of the elementary net from Figure 1.

We now define elementary processes according to the elementary process terms of the corresponding algebraic elementary net and the production rules.

Remark 2. Let $EN = (P, T, F)$ be an elementary net and let $\overline{EN} = (\overline{P}, \overline{T}, \overline{F})$.

(a) Let $M \subseteq P$ be a marking of EN. Then

$$N(M) := (M, \emptyset, \emptyset, id_M)$$

is a process of EN called elementary process associated to M.

(b) Let $t \in T$ be a transition of EN. Then

$$N(t) := (\,{}^{\bullet}t \cup t^{\bullet}, \{t\}, \{(p, t) : p \in {}^{\bullet}t\} \cup \{(t, p) : p \in t^{\bullet}\}, id_{\,{}^{\bullet}t \cup t^{\bullet} \cup \{t\}}),$$

where ${}^{\bullet}t, t^{\bullet}$ are defined w.r.t. $\overline{EN}$, is a process of EN, called elementary process of t.

(c) Let $N_i := (P_i, T_i, F_i, \Phi_i)$, $i = 1, 2$, be two processes of EN with disjoint sets of places and transitions, such that $\Phi_1(P_1) \cap \Phi_2(P_2) = \emptyset$. Then

$$N_1 + N_2 := (P_1 \cup P_2, T_1 \cup T_2, F_1 \cup F_2, \Phi),$$

where $\Phi|_{N_1} = \Phi_1$ and $\Phi|_{N_2} = \Phi_2$, is a process of EN, called the sum of the processes N_1 and N_2.

(d) Let $N_i := (P_i, T_i, F_i, \Phi_i)$, $i = 1, 2$, be two processes of EN with disjoint sets of places and transitions, such that $\Phi_1(N_1^{\circ}) \cap P = \Phi_2({}^{\circ}N_2) \cap P$. Define the interface $Int(N_1, N_2) \subseteq \overline{P}$ of the two processes N_1 and N_2 by

$$Int(N_1, N_2) := \Phi_1(N_1^{\circ}) \cap \Phi_2({}^{\circ}N_2).$$

Define $P_2' := P_2 \setminus \{p \in {}^{\circ}N_2 | \Phi_2(p) \in Int(N_1, N_2)\}$
and $F_2' := F_2 \cap ((T_2 \cup P_2') \times (T_2 \cup P_2'))$. Then
$N_1; N_2 := (P_1 \cup P_2', T_1 \cup T_2, F_1 \cup F_2' \cup \{(p_1, t_2) | p_1 \in N_1^{\circ} \wedge (\exists p_2 \in {}^{\circ}N_2 : \Phi_1(p_1) = \Phi_2(p_2) \wedge (p_2, t_2) \in F_2)\}, \Phi)$, where $\Phi|_{N_1} = \Phi_1$ and $\Phi|_{N_2} = \Phi_2$, is a process of EN, called the concatenation of the processes N_1 and N_2.

The Figures 5-7 illustrate the construction of the process of Figure 4 from elementary processes using the above rules (a)-(d).

Lemma 3. *Let $N_i = (P_i, T_i, F_i, \Phi_i)$, $i = 1, 2$, be processes of $EN = (P, T, F)$ satisfying $(\Phi_1(P_1) \cap \Phi_2(P_2)) \cap P = \emptyset$. Then*

$$\Phi_1(P_1) \cap \Phi_2(P_2) = \emptyset.$$

Proof. No isolated place $p_i \in P$ is mapped by Φ_i to a co-place $c(p) \in c(P)$. Hence, if $\Phi_i(p_i) = c(p) \in c(P)$ then there exists a transition $t_i \in T_i$ such that $(p_i, t_i) \in F_i \vee (t_i, p_i) \in F_i)$. Without loss of generality, let $(p_i, t_i) \in F_i$. Then, from the definition of $\overline{F}$ and the definition of processes (5), there exists a place $p_i' \in P_i$ satisfying $\Phi_i(p_i') = p$.

5 Relationship between Process Terms and Processes of Elementary Nets

This section contains the main result of the paper: The set of processes defined via process terms is identical with the set of classical processes of elementary nets.

In the sequel, let AEN be the algebraic elementary net corresponding to an elementary net $EN = (P, T, F)$. We are going to construct inductively processes $N_{\alpha} = (P_{\alpha}, T_{\alpha}, F_{\alpha}, \Phi_{\alpha})$ of EN, associated to process terms $\alpha : a \to b \in \mathcal{P}(AEN)$ with information $s(\alpha)$ (according to the 4 steps of the construction of process terms). These processes enjoy the following properties:

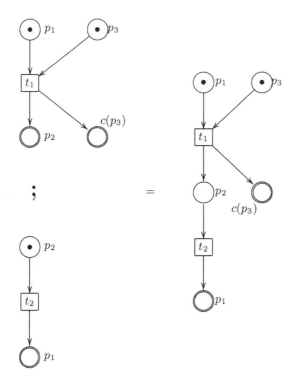

Fig. 5. Concatenation of the processes $N(t_1)$ and $N(t_2)$ for the net from Figure 1

(1) $\Phi_\alpha(°N_\alpha) \cap P = a$ and $\Phi_\alpha(N_\alpha°) \cap P = b$.
(2) $\Phi_\alpha(P_\alpha) \cap P = s(\alpha)$.

(a) Let $\alpha = M : M \to M, s(\alpha) = M$ be the reflexive process term of a marking $M \subseteq P$ of EN. Define
$$N_\alpha := N(M),$$
which is, according to Remark 2 (a), a process of EN.
Clearly properties (1) and (2) hold for $N(\alpha)$.

(b) Let $\alpha = t : pre(t) \to post(t)$ with $s(\alpha) = supp(\{pre(t), post(t)\})$ be the process term generated by a transition $t \in T$. Define
$$N_\alpha := N(t),$$
which is a process of EN according to Remark 2 (b).
Properties (1) and (2) follow from $°t \cap P = pre(t)$ and $t° \cap P = post(t)$.

(c) Let α_1, α_2 be process terms with information $s(\alpha_1), s(\alpha_2)$, such that $\alpha = \alpha_1 + \alpha_2, s(\alpha) = s(\alpha_1) \uplus s(\alpha_2)$ is a defined process term. We define a process
$$N_\alpha := N_{\alpha_1} + N_{\alpha_2}.$$

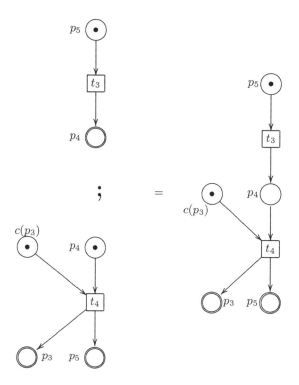

Fig. 6. Concatenation of the processes $N(t_3)$ and $N(t_4)$ for the net from Figure 1.

This is possible according to Remark 2 (c), because
(i) N_{α_1} and N_{α_2} have disjoint sets of places and transitions, and
(ii) $\Phi_{\alpha_1}(P_{\alpha_1}) \cap \Phi_{\alpha_2}(P_{\alpha_2}) = \emptyset$, where $N_{\alpha_i} = (P_{\alpha_i}, T_{\alpha_i}, F_{\alpha_i}, \Phi_{\alpha_i})$ $(i = 1, 2)$.
Condition (i) can always be achieved by appropriate renaming.
Condition (ii) follows from the fact that $\alpha_1 + \alpha_2$ is defined, $i.e.$ $s(\alpha_1) \cap s(\alpha_2) = \emptyset$. Property (2), which fulfilled for processes N_1 and N_2 according to the second condition of induction, implies $(\Phi_1(P_1) \cap \Phi_2(P_2)) \cap P = \emptyset$. Now (ii) follows from Lemma 3.
Obviously properties (1) and (2) are fulfilled.
(d) Let α_1, α_2, be process terms with information $s(\alpha_1), s(\alpha_2)$ such that $\alpha := \alpha_1; \alpha_2, s(\alpha) := s(\alpha_1) \cup s(\alpha_2)$ is a defined process term. We define a process

$$N(\alpha) = N(\alpha_1); N(\alpha_2).$$

This is possible according to Remark 2 (d), because $\Phi_{\alpha_1}(N_{\alpha_1}^\circ) \cap P = \Phi_{\alpha_2}(^\circ N_{\alpha_2}) \cap P$, by property (1) and $post_P(\alpha_1) = pre_P(\alpha_2)$.
For the new process N_α, property (1) is obvious. We have $\Phi_\alpha(P_\alpha) \cap P = \Phi_\alpha(P_{\alpha_1} \cup (P_{\alpha_2} \setminus \{p_2 \in^\circ N_{\alpha_2} | \Phi_\alpha(p_2) \in Int(N_{\alpha_1}, N_{\alpha_2})\})) \cap P$. Property (2) follows from $Int(N_{\alpha_1}, N_{\alpha_2}) \subset \Phi_\alpha(P_{\alpha_1})$.

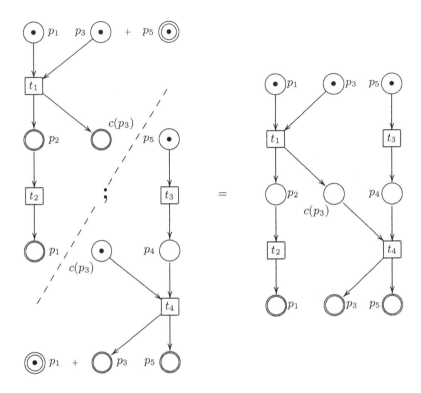

Fig. 7. The process from Figure 4 constructed from elementary processes $((N(t_1); N(t_2)) + N(p_5)); ((N(t_3); N(t_4)) + N(p_1))$.

Since the order of construction steps of N_α from elementary parts of a process term α is given by the parenthesis in α, there is a unique process net associated to a process term.

Definition 14. *Let* $\tau : T_P \to \mathcal{P}(EN)$ *be the mapping defined by* $\tau(\alpha) := N_\alpha$.

Observe that the process given in Figure 4 equals the process $\tau(\alpha)$, where α is the process term from Figure 2. The constructions of α (Figure 2) and $\tau(\alpha)$ (Figures 5-7) are analogous.

Lemma 4. *Let* $N = (P_N, T_N, F_N, \Phi_N)$ *be a process of EN and* $t_1, t_2 \in T_N$ *with* $(t_1, t_2) \in \mathbf{co} \wedge t_1 \neq t_2$. *Then* $\Phi_N(t_1) + \Phi_N(t_2)$ *is a defined process term.*

Proof. $(t_1, t_2) \in \mathbf{co}$ implies that $^\bullet t_1 \cup {}^\bullet t_2$ and $t_1^\bullet \cup t_2^\bullet$ are co-sets. Since Φ_N is injective on slices, $\Phi_N(^\bullet t_1) \cap \Phi_N(^\bullet t_2) = \Phi_N(t_1^\bullet) \cap \Phi_N(t_2^\bullet) = \emptyset$. Assume a place p in $\Phi_N(^\bullet t_1) \cap \Phi_N(t_2^\bullet)$ or $\Phi_N(^\bullet t_2) \cap \Phi_N(t_1^\bullet)$. Without loss of generality let $p \in \Phi_N(^\bullet t_1) \cap \Phi_N(t_2^\bullet)$. Then there are places $p_1 \in {}^\bullet t_1$ and $p_2 \in t_2^\bullet$ such that $\Phi_N(p_1) = \Phi_N(p_2)$. Then either $(p_1, p_2) \in \mathbf{co} \wedge p_1 \neq p_2$ (which contradicts the injectivity of Φ_N on co-sets) or $(p_2, p_1) \in F^+ \vee p_1 = p_2$ (which contradicts

$(t_1, t_2) \in \mathbf{co}$) or $(p_1, p_2) \in F^+$, which implies $(t_1, t_2) \in F^+$ because places are unbranched, what is again a contradiction..

Remark 3. (a) Given process terms α_i, $i = 1, \ldots, 4$ of AEN, whenever terms $\alpha = ((\alpha_1 + \alpha_2); (\alpha_3 + \alpha_4))$ and $\beta = ((\alpha_1; \alpha_3) + (\alpha_2; \alpha_4))$ are defined then $s(\alpha) = s(\beta)$.

(b) For any two process terms α_1 and α_2 such that $\alpha_1 + \alpha_2$ is defined we have
$\alpha_1 + \alpha_2 \sim_t (\alpha_1; post(\alpha_1)) + (pre(\alpha_2); \alpha_2) \sim_t (\alpha_1 + pre(\alpha_2)); (\alpha_2 + post(\alpha_1))$
and analogously $\alpha_1 + \alpha_2 \sim_t (\alpha_2 + pre(\alpha_1)); (\alpha_1 + post(\alpha_2))$.

(c) If $(\alpha_1; \alpha_2) + M$ is defined, M being a marking, then

$$(\alpha_1; \alpha_2) + M \sim_t (\alpha_1 + M); (\alpha_2 + M).$$

Theorem 1. *The mapping* $\tau : T_{\mathcal{P}} \to \mathcal{P}(EN)$ *is surjective.*

Proof. Let $N = (P_N, T_N, F_N, \Phi_N)$ be a process of EN. We inductively construct a process term α with $N_\alpha = N$:

(i) Set $\alpha_0 = \Phi_N(^\circ N) : \Phi_N(^\circ N) \cap P \to \Phi_N(^\circ N) \cap P$ with $s(\alpha_0) = \Phi_N(^\circ N)$.
We have $N(\alpha_0) = N|_{\circ N}$

(ii) Assume we have constructed process terms $\alpha_0, \ldots, \alpha_{n-1}$, such that
$\alpha_0; \ldots; \alpha_{n-1} : \Phi_N(^\circ N) \cap P \to \Phi_N(N_{n-1}) \cap P, s(\alpha_0) \cup \ldots \cup s(\alpha_{n-1})$ is a process term, N_{n-1} is a slice of N ($N_0 := {}^\circ N$) and $N(\alpha_0; \ldots; \alpha_{n-1}) = (P_N \cap {}^\to N_{n-1}, T_N \cap {}^\to N_{n-1}, F_N \cap ({}^\to N_{n-1} \times {}^\to N_{n-1}), \Phi_N|_{\to N_{n-1}})$. Take all transitions $t_1, \ldots, t_k \in T_N$ with ${}^\bullet t_i \subseteq N_{n-1}$, $i = 1, \ldots, k$, and define
$N' := N_{n-1} \setminus ({}^\bullet t_1 \cup \ldots \cup {}^\bullet t_k)$,
$N_n := N' \cup t_1^\bullet \cup \ldots \cup t_k^\bullet$,
$\alpha_n := (\Phi_N(N') \cap P) + \Phi_N(t_1) + \ldots + \Phi_N(t_k)$.
Clearly, α_n is well-defined, N_n is a slice of N,

$$\alpha_0; \ldots; \alpha_n : \Phi_N(^\circ N) \to \Phi_N(N_n)$$

is a process term with information $s(\alpha_0) \cup \ldots \cup s(\alpha_n)$ and

$$N(\alpha_0; \ldots; \alpha_n) = (P_N \cap {}^\to N_n, T_N \cap {}^\to N_n, F_N \cap ({}^\to N_n \times {}^\to N_n), \Phi_N|_{\to N_n}).$$

(iii) Let $m \in \mathbb{N}$, such that $N_m = N^\circ$.
Then $\alpha := \alpha_0; \ldots; \alpha_m : \Phi_N(^\circ N) \cap P \to \Phi_N(N^\circ) \cap P$ is a process term with $N(\alpha) = N$.

Corollary 1. *Every process* $N \in \mathcal{P}(EN)$ *can be inductively constructed from elementary processes using partial operations* $+$ *and* $;$ *as defined in Remark 2.*

The method of maximal steps used in the proof of Theorem 1 is illustrated in Figure 8.

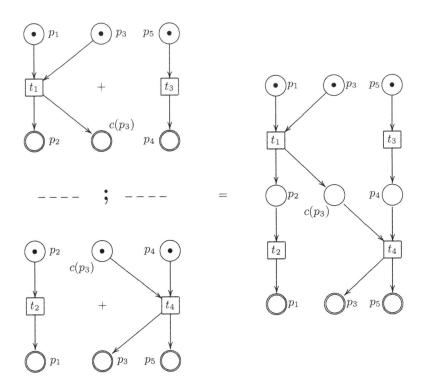

Fig. 8. Constructing the process from Figure 4 by concatenating $N(t_1) + N(t_2)$ and $N(t_3) + N(t_4)$ using the maximal step method.

Theorem 2. *For two process terms $\alpha, \beta \in \mathcal{P}(\mathcal{N})$, $\alpha \sim_t \beta$ implies that $\tau(\alpha)$ and $\tau(\beta)$ are isomorphic.*

Proof. It is sufficient to show the proposition for every (of the seven) construction rule of $\sim_t$:

(1) $N(\alpha_1 + \alpha_2) = N(\alpha_2 + \alpha_1)$ is obvious.
(2) $N(\alpha_1; \alpha_2); N(\alpha_3) = N(\alpha_1); N(\alpha_2; \alpha_3)$ is obvious.
(3) $N(\alpha_1 + \alpha_2) + N(\alpha_3) = N(\alpha_1) + N(\alpha_2 + \alpha_3)$ is obvious.
(4) We have to show, that $(N(\alpha_1) + N(\alpha_2)); (N(\alpha_3) + N(\alpha_4)) = (N(\alpha_1); N(\alpha_3)) + (N(\alpha_2); N(\alpha_4))$. Let $N_1 = (P_1, T_1, F_1, \Phi_1)$ be the process on the left side and $N_2 = (P_2, T_2, F_2, \Phi_2)$ be the process on the right side. Clearly $T_1 = T(\alpha_1) \cup \ldots \cup T(\alpha_4) = T_2$.
Further we have $P_1 = [P(\alpha_1) \cup P(\alpha_2)] \cup$
$[(P(\alpha_3) \cup P(\alpha_4)) \setminus \{p' \in {}^\circ(N(\alpha_3) + N(\alpha_4)) | \Phi_3(p') \in P \vee \Phi_4(p') \in P\})] =$
$[P(\alpha_1) \cup P(\alpha_2)] \cup$
$[(P(\alpha_3) \setminus \{p' \in {}^\circ N(\alpha_3) | \Phi_3(p') \in P\}) \cup P(\alpha_4)) \setminus \{p' \in {}^\circ N(\alpha_4) | \Phi_4(p') \in P\})].$

Because N_2 is defined, this equals
$$[P(\alpha_1) \cup (P(\alpha_3) \setminus \{p' \in^\circ N(\alpha_3) | \Phi_3(p') \in P\})] \cup$$
$$[P(\alpha_2) \cup (P(\alpha_4) \setminus \{p' \in^\circ N(\alpha_4) | \Phi_4(p') \in P\})] = P_2.$$
Since the flow relation and labeling of composed processes are constructed by restriction from the original flow relations and labelings, $F_1 = F_2$ and $\Phi_1 = \Phi_2$ follow immediately.

(5) $N(\alpha); N(post_P(\alpha)) = N(\alpha) = N(pre_P(\alpha)); N(\alpha)$ is obvious.

(6) Obvious.

(7) For elementary nets it suffices to consider the case $a = \emptyset$. Its proof is obvious.

Theorem 3. *If, for two process terms $\alpha, \beta \in T_P$, $\tau(\alpha)$ and $\tau(\beta)$ are isomorphic, then $\alpha \sim_t \beta$.*

Proof. Without loss of generality let α and β be process terms with $N(\alpha) = N(\beta) = N = (P_N, T_N, F_N, \Phi_N)$ and $\gamma = \gamma_1; \ldots; \gamma_m$ be the process term constructed from the process N in the proof of Theorem 1 by considering maximal steps. Then γ_i is of the form

$$\gamma_i = \Phi_N(t_1^i) + \ldots + \Phi_N(t_{k_i}^i) + \Phi_N(M^i),$$

$t_j^i \in T_N$ and $M_i \subseteq P_N$, $i = 1, \ldots, m$, $j = 1, \ldots, k_i$. We show that α is equivalent to γ. By symmetry, the same holds for β, and we are done.

According to Remark 3, we assume without loss of generality that α is of the form

$$\alpha = \Phi_N(t_1) + (\Phi_N(M_1) \cap P); \ldots; \Phi_N(t_k) + (\Phi_N(M_k) \cap P)$$

with transitions $t_i \in T_N$ and subsets $M_i \subseteq P_N$, $i = 1, \ldots, k$. We will use shorthands $\alpha = t_1; \ldots; t_k$, and ignore the sets M_i, because they are determined by the definition of the concatenation of process terms. Clearly, α and γ 'contain' the same transitions, i.e. $\{t_1, \ldots, t_k\} = \{t_1^1, \ldots, t_{k_1}^1, \ldots, t_1^m, \ldots, t_{k_m}^m\}$.

Assume $t_i = t_1^1$ for an $i \geq 2$. It suffices to prove

$$t_1; \ldots; t_i \sim_t t_1; \ldots; t_i; t_{i-1} \sim_t \ldots \sim_t t_i; t_1; \ldots; t_{i-1},$$

because firstly the same procedure applied to $t_2^1, \ldots, t_{k_1}^1$ provides $t_1; \ldots; t_k \sim_t \gamma_1; \ldots$ (3), and secondly this procedure applied to $\gamma_2, \ldots, \gamma_m$ finishes the theorem. In fact, it even is enough to show that we can exchange t_i and t_{i-1} in α. A sufficient condition is that $\Phi_N(t_i) + \Phi_N(t_{i-1})$ is a defined process term.

We have to distinguish two cases: If $t_{i-1} = t_k^1$ for some $k \in \{2, \ldots, k_1\}$, $\Phi_N(t_i) + \Phi_N(t_{i-1})$ is defined according to the process term γ. The other possibility is $t_{i-1} = t_k^l$ for an $l \in \{2, \ldots, m\}$ and $k \in \{1, \ldots, k_l\}$. By construction of the process N_α from α follows $(t_{i-1}, t_i) \in F_N^+$ or $t_{i-1} \mathbf{co} \, t_i$. On the other hand, by construction of γ follows either $(t_i, t_{i-1}) \in F_N^+$ or $t_i \mathbf{co} \, t_{i-1}$. It follows $t_{i-1} \mathbf{co} \, t_i$. By Lemma 4, $\Phi_N(t_i) + \Phi_N(t_{i-1})$ is defined.

Figures 7 and 8 illustrate that the equivalent terms from Figures 5 and 6 are mapped by τ onto the same process.

Finally, looking at the definition of τ, we can state our main result for elementary nets, which now follows easily from the previous theorems.

Theorem 4. *Given any elementary net EN, there exists a one-to-one correspondence between the (isomorphism classes of) processes $\mathcal{P}(EN)$ of the elementary net EN and the $\sim_t$-congruence classes of process terms $T_{\mathcal{P}}$ of the corresponding algebraic elementary net AEN. This correspondence preserves source, target and information about states of processes and process terms, as well as concurrent composition and concatenation of processes (congruence classes of process terms).*

Remark 4. Let us rephrase Theorem 4 using terminology from partial algebra [2]: Given a process term $\alpha \in T_{\mathcal{P}}$, the congruence class $[\alpha]_{\sim_t} \in [T_{\mathcal{P}}]_{\sim_t}$ corresponds to the process $\tau(\alpha) = N \in \mathcal{P}(EN)$ such that source and target are preserved, *i.e.* $\Phi(^{\circ}N) \cap P = pre_{\mathcal{P}}(\alpha)$, $\Phi(N^{\circ}) \cap P = post_{\mathcal{P}}(\alpha)$, and information about states is preserved, *i.e.* $\Phi(P_N) \cap P = s(\alpha)$. Moreover, denoting by $T_{\mathcal{P}}$ the partial algebra of process terms with concurrent composition and concatenation as defined in Definition 5, and by $\mathcal{P}(EN)$ the partial algebra of (isomorphism classes of) net processes with concurrent composition and concatenation as defined in Remark 2, the factor algebra $T_{\mathcal{P}}/\sim_t$ is isomorphic to the partial algebra $\mathcal{P}(EN)$, (*i.e.* τ is a surjective closed homomorphism between $T_{\mathcal{P}}$ and $\mathcal{P}(EN)$).

6   Conclusion

This paper has shown that concepts of partial algebra are capable to define a syntactic process semantics of elementary nets which precisely distinguishes those runs that are also obtained by partially ordered process nets. Elementary nets can be viewed as place/transition nets with a restricted occurrence rule: In case of a contact situation, a transition is not enabled. In a more general setting, we claim that partial algebra is the suitable tool to define true-concurrency semantics for arbitrary restrictions of the occurrence rule, such as capacity restrictions, inhibitor arcs, read arcs, as suggested in [5]. We are currently working on a generalization of the results of this paper to Petri nets with arbitrarily restricted occurrence rule.

References

1. E. Best and R. Devillers. Sequential and Concurrent Behaviour in Petri Net Theory. *Theoretical Computer Science*, 55, pp. 87–136, 1987.
2. P. Burmeister. Lecture Notes on Universal Algebra – Many Sorted Partial Algebras. Technical Report, TU Darmstadt, 1998.
3. E. Degano, J. Meseguer and U. Montanari. Axiomatizing the Algebra of Net Computations and Processes. *Acta Informatica*, 33(7), pp. 641–667, 1996.
4. R. Janicki and M. Koutny. Semantics of Inhibitor Nets. *Information and Computations*, 123, pp. 1–16, 1995.
5. G. Juhás. Reasoning about algebraic generalisation of Petri nets. In S. Donatelli and J. Klein (Eds.) *Proc. of 20th International Conference on Application and Theory of Petri Nets*, Springer, LNCS 1639, pp. 324-343, 1999.

6. G. Juhás. On semantics of Petri nets over partial algebra. In J. Pavelka, G. Tel and M. Bartosek (Eds.) *Proc. of 26th Seminar on Current Trends in Theory and Practice of Informatics SOFSEM'99*, Springer, LNCS 1725, pp. 408-415, 1999.
7. J. Meseguer and U. Montanari. Petri Nets are Monoids. *Information and Computation*, 88(2):105–155, October 1990.
8. U. Montanari and F. Rossi. Contextual Nets. *Acta Informatica*, 32(6), pp. 545–596, 1995.
9. G. Rozenberg and J. Engelfriet. Elementary Net Systems. In W. Reisig and G. Rozenberg (Eds.) *Lectures on Petri Nets I: Basic Models*, Springer, LNCS 1491, pp. 12-121, 1998.
10. J. Winkowski. Behaviours of Concurrent Systems. *Theoretical Computer Science*, 12, pp. 39–60, 1980.
11. J. Winkowski. An Algebraic Description of System Behaviours. *Theoretical Computer Science*, 21, pp. 315–340, 1982.

User Interface Prototyping Based on UML Scenarios and High-Level Petri Nets [1]

Mohammed Elkoutbi and Rudolf K. Keller

Département d'informatique et de recherche opérationnelle
Université de Montréal
C.P. 6128, succursale Centre-ville, Montréal, Québec H3C 3J7, Canada
voice: (514) 343-6782
fax: (514) 343-5834
e-mail: {elkoutbi, keller}@iro.umontreal.ca
http://www.iro.umontreal.ca/~{elkoutbi, keller}

Abstract: In this paper, we suggest a requirement engineering process that generates a user interface prototype from scenarios and yields a formal specification of the system in form of a high-level Petri net. Scenarios are acquired in the form of sequence diagrams as defined by the Unified Modeling Language (UML), and are enriched with user interface information. These diagrams are transformed into Petri net specifications and merged to obtain a global Petri net specification capturing the behavior of the entire system. From the global specification, a user interface prototype is generated and embedded in a user interface builder environment for further refinement. Based on end user feedback, the input scenarios and the user interface prototype may be iteratively refined. The result of the overall process is a specification consisting of a global Petri net, together with the generated and refined prototype of the user interface.

Keywords: User interface prototyping, scenario specification, high-level Petri net, Unified Modeling Language.

1   Introduction

Scenarios have been identified as an effective means for understanding requirements [16] and for analyzing human computer interaction [14]. A typical process for requirement engineering based on scenarios [7] has two main tasks. The first task consists of generating from scenarios specifications that describe system behavior. The second task concerns scenario validation with users by simulation and prototyping. These tasks remain tedious activities as long as they are not supported by automated tools.

For the purpose of validation in early development stages, rapid prototyping tools are commonly and widely used. Recently, many advances have been made in user

---

[1] This work is supported by FCAR (Fonds pour la formation des chercheurs et l'aide à la recherche au Québec) and NSERC (National Sciences and Research Council of Canada).

M. Nielsen, D. Simpson (Eds.): ICATPN 2000, LNCS 1825, pp. 166-186, 2000.

interface (UI) prototyping tools like UI builders and UI management systems. Yet, the development of UIs is still time-consuming, since every UI object has to be created and laid out explicitly. Also, specifications of dialogue controls must be added by programming (for UI builders) or via a specialized language (for UI management systems).

This paper suggests an approach for requirements engineering that is based on the Unified Modeling Language (UML) and high-level Petri nets. The approach provides an iterative, four-step process with limited manual intervention for deriving a prototype of the UI from scenarios and for generating a formal specification of the system. As a first step in the process, the use case diagram of the system as defined by the UML is elaborated, and for each use case occurring in the diagram, scenarios are acquired in the form of UML sequence diagrams and enriched with UI information. In the second step, the use case diagram and all sequence diagrams are transformed into Colored Petri Nets (CPNs). In step three, the CPNs describing one particular use case are integrated into one single CPN, and the CPNs obtained in this way are linked with the CPN derived from the use case diagram to form a global CPN capturing the behavior of the entire system. Finally, in step four, a prototype of the UI of the system is generated from the global CPN and embedded in a UI builder environment for further refinement.

In our previous work, we have investigated and implemented the generation of UI prototypes from UML scenarios using exclusively the UML, most notably UML Statecharts [5, 13, 21, 22]. In this Statechart-based approach, Statecharts are used to integrate the UML scenarios and capture object and UI behavior. In the work presented in this paper, we decided to take a Petri-net based approach, with CPNs taking the role of UML Statecharts. We opted for Petri nets because of their strong support of concurrency, their ability to capture and simulate multiple copies of scenarios in the same specification, and for the wealth of available tools for analyzing, simulating, and verifying Petri nets. A comparison of the two approaches is provided in Section 5 of the paper.

In our approach, we aim to model separately the use case and the scenario levels. We also want to keep track of scenarios after their integration. Thus, we need a PN class that supports hierarchies as well as colors or objects to distinguish between scenarios in the resulting specification. We adopted Jensen's definition of CPN [10] which is widely accepted and supported by the *designCPN* tool [3] for editing, simulating, and verifying CPNs. Object PNs could also have been used, but CPNs are largely sufficient for this work. In our current implementation, UI prototyping is embedded into the *Visual Café* environment [23] for further refinement.

Section 2 of this paper gives a brief overview of the UML diagrams relevant to our work and introduces a running example. In Section 3, the four activities leading from scenarios to executable UI prototypes are detailed. Section 4 reviews related work, and in Section 5, we discuss a number of issues related to the proposed approach. Section 6 concludes the paper and provides an outlook into future work.

2   Unified Modeling Language

The UML [19], which is emerging as a standard language for object-oriented modeling, provides a syntactic notation to describe all major views of a system using different kinds of diagrams. In this section, we discuss the three UML diagrams that are relevant for our work: Class diagram (ClassD), Use Case diagram (UsecaseD), and Sequence diagram (SequenceD). As a running example, we have chosen to study a part of an extended version of the Automatic Teller Machine (ATM) system described in [2].

2.1   Class Diagram (ClassD)

The ClassD represents the static structure of the system. It identifies all the classes for a proposed system and specifies for each class its attributes, operations, and relationships to other objects. Relationships include inheritance, association, and aggregation. The ClassD is the central diagram of UML modeling. Figure 1 depicts the ClassD for the ATM system.

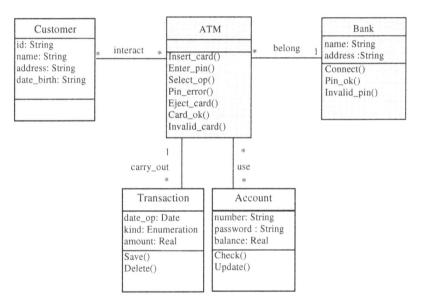

Figure 1: Class diagram of the ATM system.

2.2   Use Case Diagram (UsecaseD)

The UsecaseD is concerned with the interaction between the system and actors (objects outside the system that interact directly with it). It presents a collection of use cases and their corresponding external actors. A use case is a generic description of an entire transaction involving several objects of the system. Use cases are represented as ellipses, and actors are depicted as icons connected with solid lines to the use cases they interact with. One use case can call upon the services of another use case. Such a relation is called a *uses* relation and is represented by a directed solid line. Figure 2 shows as an example the UsecaseD corresponding to the ATM system. The *extends* relation, which is also defined in UsecaseDs, can be seen as a *uses* relation with an additional condition upon the call. A UsecaseD is helpful in visualizing the context of a system and the boundaries of the system's behavior. A given use case is typically characterized by multiple scenarios.

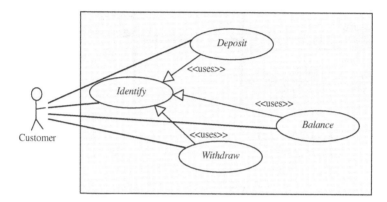

Figure 2: Use Case diagram of the ATM system.

2.3   Sequence Diagram (SequenceD)

A scenario shows a particular series of interactions among objects in a single execution (*instance*) of a use case. Scenarios can be viewed in two different ways: through SequenceDs or collaboration diagrams. Both types of diagrams rely on the same underlying semantics, and conversion from one to the other is possible. For our work, we chose to use SequenceDs for their simplicity and wide use.

A SequenceD shows the interactions among the objects participating in a scenario in temporal order. It depicts the objects by their lifelines and shows the messages they exchange in time sequence. However, it does not capture the associations among the objects. A SequenceD has two dimensions: the vertical dimension represents time, and the horizontal dimension represents the objects. Messages are shown as horizontal solid arrows from the lifeline of the object sender to the lifeline of the object receiver. A message may be guarded by a condition, annotated by iteration or concurrency information, and/or constrained by an expression. Constraints are used in our work to enrich messages with UI information.

Figures 3 and 4 depict two SequenceDs of the use case *Identify*. Figure 3 represents the scenario where the customer is correctly identified (*regularIdentify*), whereas Figure 4 shows the case where the customer entered an incorrect pin (*errorIdentify*).

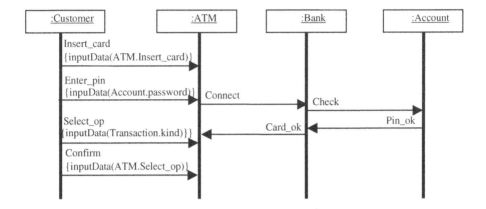

Figure 3: Scenario *regularIdentify* of the use case *Identify*.

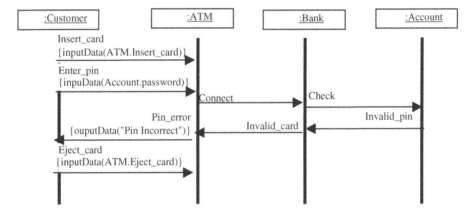

Figure 4: Scenario *errorIdentify* of the use case *Identify*.

Beyond the UML standard message constraints found in SequenceDs, we define the two additional constraints *inputData* and *outputData*. The *inputData* constraint indicates that the corresponding message holds input information from the user. The *outputData* constraint specifies that the corresponding message carries information for display. Both *inputData* and *outputData* constraints have a parameter that indicates the kind of user action. This parameter normally represents the dependency between the message and the elements of the underlying ClassD. It may be either a method name, one or several class attributes, or a string literal (see figures 3 and 4).

Once the analyst has specified the UI constraints of the messages in the SequenceD at hand, this information is used to determine the corresponding widgets that will appear in the UI prototype. Widget generation adheres to a list of rules, which is based on the terminology, heuristics and recommendations found in [8] and which includes the following eight items:

- A button widget is generated for an *inputData* constraint with a method as dependency, e.g., *Insert_card() {inputData(ATM.insert_card)}* in Figure 3.
- An enabled textfield widget is generated in case of an *inputData* constraint with a dependency to an attribute of type String, Real, or Integer, e.g., *Enter_pin() {inputData(Account.password)}* in Figure 3.
- A group of radio buttons widgets are generated in case of an *inputData* constraint with a dependency to an attribute of type Enumeration having a size less than or equal to 6, e.g., *Select_op() {inputData(Transaction.kind)}* in Figure 3.
- An enabled list widget is generated in case of an *inputData* constraint with a dependent attribute of type Enumeration having a size greater than 6 or with a dependent attribute of type collection.
- An enabled table widget is generated in case of an *inputData* constraint with multiple dependent attributes.
- A disabled textfield widget is generated for an *outputData* constraint with a dependency to an attribute of type String, Real, or Integer.
- A label widget is generated for an *outputData* constraint with no dependent attribute, e.g., *Pin_error() {outputData("Pin Incorrect")}* in Figure 4.
- A disabled list widget is generated in case of an *outputData* constraint with a dependent attribute of type Enumeration having a size greater than 6 or with a dependent attribute of type collection.
- A disabled table widget is generated in case of an *outputData* constraint with multiple dependent attributes.

3   Description of Approach

In this section, we detail the iterative process for deriving a system UI prototype from scenarios using the UML and CPNs. Figure 5 presents the sequence of activities involved in the proposed process.

In the *Scenario Acquisition* activity, the analyst elaborates the UsecaseD, and for each use case, he or she elaborates several SequenceDs corresponding to the scenarios of the use case at hand. The *Specification Building* activity consists of deriving CPNs from the acquired UsecaseD and SequenceDs. During *Scenario Integration*, CPNs corresponding to the same use case are iteratively merged to obtain an integrated CPN of the use case. Integrated CPNs serve as input to both the *CPN Verification* and the *UI Prototype Generation* activities. During *Prototype Evaluation*, the generated prototype is executed and evaluated by the end user. In the *CPN Verification* activity, existing algorithms can be used to check behavioral properties.

In the following subsections, we will discuss in detail the four activities of the UI prototyping process: scenario acquisition, specification building, scenario integration, and UI prototype generation. The CPN verification activity is discussed in [4].

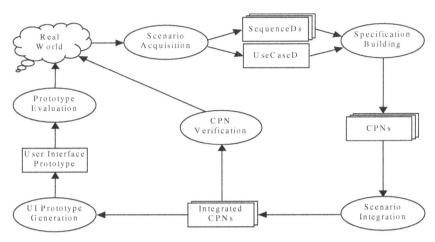

Figure 5: Activities of the proposed process.

3.1   Scenario Acquisition

In this activity, the analyst elaborates the UsecaseD capturing the system functionalities, and for each use case, he or she acquires the corresponding scenarios in form of SequenceDs. For instance, the UsecaseD of the ATM system is shown in Figure 1, and two SequenceDs of the use case *Identify* are given in figures 3 and 4.

Scenarios of a given use case are classified by type and ordered by frequency of use. We have considered two types of scenarios: normal scenarios, which are executed in normal situations, and scenarios of exception executed in case of errors and abnormal situations. The frequency of use (or the frequency of execution) of a scenario is a number between 1 and 10 assigned by the analyst to indicate how often a given scenario is likely to occur. In our example, the use case *Identify* has one normal scenario (scenario *regularIdentify* with frequency 10) and a scenario of exception (scenario *errorIdentify* with frequency 5). This classification and ordering is used for the composition of UI blocks [5].

3.2   Specification Building

This activity consists of deriving CPNs from both the acquired UsecaseD and all the SequenceDs. These derivations are explained below in the subsections *Use case specification* and *scenario specification*.

3.2.1   Use Case Specification

The CPN corresponding to the UsecaseD is derived by mapping use cases into places. The transition leading to one place (*Enter*) corresponds to the initiating action of the use case. A place *Begin* is always added to model the initial state of the system. After a use case execution, the system will return, via an *Exit* transition, back to its initial state for further use case executions. The place *Begin* may contain several tokens to model concurrent executions. Figure 6 depicts the CPN derived from the ATM system's UsecaseD (Figure 2).

In a UsecaseD, a use case can call upon the services of another one via the relation *uses*. This relation may have several meanings depending on the system being modeled. Consider a use case Uc_1 using a use case Uc_2. Figure 7(a) shows the general form of this relation. The use case Uc_1 is decomposed into three sub-use cases: Uc_{11} represents the part of Uc_1 executed before the call of Uc_2, Uc_{12} represents the part of Uc_1 that is concurrently executed with Uc_2, and Uc_{13} represents the part executed after the termination of Uc_2. Note that one or two of these three sub-use cases may be empty. The figures 7(a) through 7(g) depict the eight possible mappings.

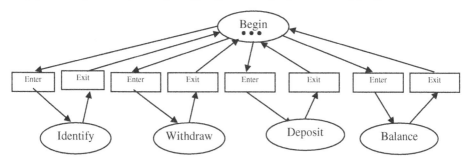

Figure 6: CPN derived from the UsecaseD of the ATM system.

A relation of type (g) between Uc_1 and Uc_2 means that Uc_2 precedes Uc_1. This implies that Uc_1 is not directly accessible from the place *Begin*. So the transitions from the place *Begin* to the place representing Uc_1 must be substituted for an *Enter* transition into Uc_2 and an *Enter* transition from Uc_1 into Uc_2. In the ATM system (Figure 1), all three *uses* relations are of type (g), and the initial CPN (Figure 7) must be updated accordingly (Figure 9(a)).

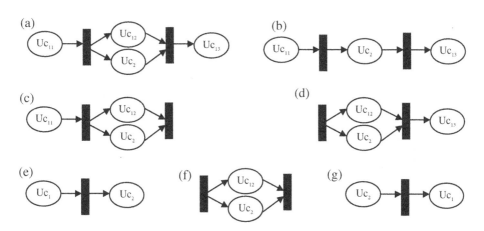

Figure 7: Possible mappings of the *uses* relation (Uc_1, Uc_2).

The designCPN tool, which we adopted in our work, allows for the refinement of transitions, but does not support the refinement of places. Therefore, in order to substitute the use cases, which are represented as places, for CPNs representing integrated scenarios (see Subsection 3.3), the CPN obtained after processing the *uses* relation (Figure 8(a)) requires adaptation: each subnet *Enter* → *place*$_i$ → *Exit* is substituted for a simple transition representing the use case underlying *place*$_i$ (cf. dashed ellipse in Figure 8(a)), and intermediary places such as *endIdentify* are inserted (see Figure 8(b)).

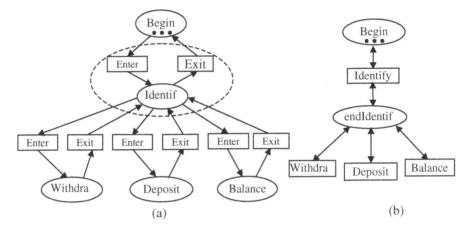

Figure 8: CPN of the UsecaseD of the ATM system after (a) processing the *uses* relation, and (b) adaptation to designCPN.

3.2.2   Scenario Specification

For each scenario of a given use case, the analyst builds an associated table of object states. This table is directly obtained from the SequenceD of the scenario by following the exchange of messages from top to bottom and identifying the changes in object states caused by the messages. For example, tables 1 and 2 show the object state tables associated with the scenarios *regularIdentify* (Figure 3) and *errorIdentify* (Figure 4). In such tables, a scenario state is represented by the state vector of the objects participating in the scenario (column *Scenario state* in Table 1 and 2, resp.).

From each object state table, a CPN is generated by transforming scenario states into places, and messages into transitions (see figures 9 and 10)[2]. Each scenario is assigned a distinct color, e.g., *rid* for the *regularIdentify* scenario, and *eid* for the *errorIdentify* scenario. All CPNs (scenarios) of the same use case will have the same initial place (state) which we call *B* in figures 9 and 10. This place will serve to link the integrated CPN (see below) with the CPN modeling the UsecaseD of the system (Figure 8(b)).

---

[2] For readability, we do not show screen dumps produced by designCPN, but replace them with CPNs redrawn by hand.

Objects Messages	Cus- tomer	ATM	Bank	Account	Scenario state
Insert_card	Present	Card_in	void	void	S1={Present, Card_in, void, void}
Enter_pin	Present	Pin-entered	void	void	S2={Present, Pin_entered, void, void}
Connect	Present	Pin-entered	Connected	void	S3={Present, Pin_entered, Connected, void}
Check	Present	Pin-entered	Connected	Checked	S4={Present, Pin_entered, Connected, Checked}
Pin_ok	Present	Pin-entered	Valid_pin	Checked	S5={Present, Pin_entered, Valid_pin, Checked}
Card_ok	Present	Valid-card	Valid_pin	Checked	S6={Present, Vaild_card, Valid_pin, Checked}
Select_op	Present	Selection	Valid_pin	Checked	S7={Present, Selection, Valid_pin, Checked}
Confirm	Present	Confir- mation	Valid_pin	Checked	S8={Present,Confirmation, Valid_pin, Checked}

Table 1: Object state table associated with the scenario *regularIdentify*.

Objects Messages	Cus- tomer	ATM	Bank	Account	Scenario state
Insert_card	Present	Card_in	void	void	S1={Present, Card_in, void, void}
Enter_pin	Present	Pin-entered	void	void	S2={Present, Pin_entered, void, void}
Connect	Present	Pin-entered	Connected	void	S3={Present, Pin_entered, Connected}
Check	Present	Pin-entered	Connected	Checked	S4={Present, Pin_entered, Connected, Checked}
Invalid_pin	Present	Pin-entered	Invalid_pin	Checked	S9={Present, Pin_entered, Invalid_pin, Checked}
Invalid_card	Present	Invalid- card	Invalid_pin	Checked	S10={Present,Invalid_card, Invalid_pin, Checked}
Pin_error	Present	Invalid_pin	Invalid_pin	Checked	S11={Present, Invalid_pin, Invalid_pin, Checked}
Eject_card	Present	Idle	Invalid_pin	Checked	S12={Present, Idle, Invalid_pin, Checked}

Table 2: Object state table associated with the scenario *errorIdentify*.

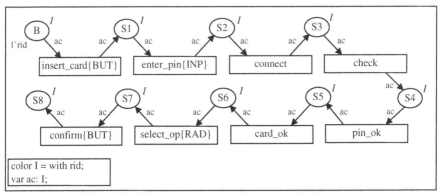

Figure 9: CPN corresponding to the *regularIdentify* scenario.

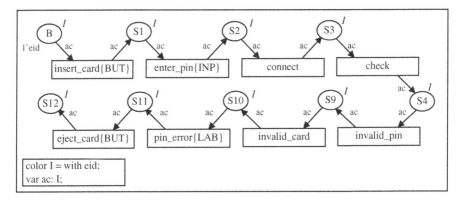

Figure 10: CPN corresponding to the *errorIdentify* scenario.

Note that in scenario specification, only the building of the object state tables is manual. The rest of the operation is fully automatic.

The next activity of the approach, scenario integration, requires as input serialized CPNs. Since designCPN uses SGML as interchange format and since conversion tools between SGML, XML, and Java are readily available, we decided to represent CPNs in XML. In our current implementation, we use the transformation scheme as depicted in Figure 11. Using the designCPN tool, the analyst will edit and then save the use case CPN and all scenario CPNs into the textual SGML format. Then, the *SX* tool [1] is used to do the conversion from SGML to rough XML (RXML), and the *XJParse* Java program [9] to convert RXML into XML.

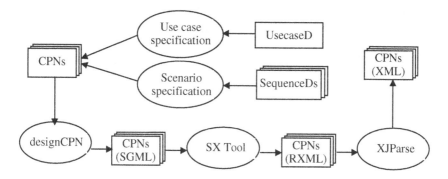

Figure 11: Generation of CPNs in XML format from UsecaseD and SequenceDs.

3.3   Scenario Integration

In this activity, we aim to merge all CPNs corresponding to the scenarios of a use case Uc_i, in order to produce an integrated CPN modeling the behavior of the use case. Our algorithm is based on a preliminary version presented in [6]. It takes an incremental approach to integration. Given two scenarios with corresponding CPNs CPN_1 and CPN_2, the algorithm merges all places in CPN_1 and CPN_2 having the same names. The merged places will have as color the union of the colors of the two scenarios. Then, the algorithm looks for transitions having the same input and output places in the two scenarios and merges them with an *OR* between their guard conditions. In the following, we describe the algorithm in pseudocode, using the "dot"-notation known from object-oriented languages.

```
Uc_i.IntergrateScenarios()
     scList = Uc_i.getListOfScenarios();
     // returns the list of scenarios of the use case Uc_i
     uc_cpn = getXML(scList[0]);
     // returns the XML file corresponding to the scenario scList[0].
     i=1;
     while (i < scList.size())
          sc_cpn = getXML(scList(i));
          sc_cpn = makeUniqueID( uc_cpn, sc_cpn);
          uc_cpn = merge(uc_cpn,sc_cpn);
          i = i + 1;
     end
end Uc_i.IntergrateScenarios
```

XML identifies each element in a CPN (place, transition, edge, etc.) by a distinct identifier (ID). Before integrating two scenarios, the method `makeUniqueID` checks if both the two input files `uc_cpn` and `sc_cpn` comprise distinct IDs. In case they share common IDs, `makeUniqueID` will modify the IDs of `sc_cpn` by adding the maximum ID of `uc_cpn` to all IDs of `sc_cpn`.

Merging (integrating) two scenarios whose CPNs have the colors *[sc₁]* and *[sc₂]*, respectively, will produce a CPN with the list *[sc₁,sc₂]* as color. The operation of merging follows the steps described below:

```
merge(uc_cpn,sc_cpn)
    uc_cpn.addPlaces(sc_cpn)
    // adds in uc_cpn places of sc_cpn that do not exist in uc_cpn
    for each t in sc_cpn.getListOfTransitions()
        t' = uc_cpn.LookForTrans(t)
        // t' is a transition of uc_cpn with •t=•t' and t•=t'•
        if (t' does not exist)
            uc_cpn.addtrans(t)
        endif
    end
    uc_cpn.addEdges(sc_cpn)
    // adds to uc_cpn edges of sc_cpn that do not exist in uc_cpn
    uc_cpn.mergeColors(sc_cpn)
    // calculates the new color of the integrated CPN (uc_cpn)
    uc_cpn.putColorsOnPlaces(sc_cpn)
    // all places of the net will have the merged color
    uc_cpn.putGuardOnTransitions(sc_cpn)
    // common transitions will be guarded by the merged color,
    // the others will be guarded by their original colors
    uc_cpn.putVariablesOnEdges(sc_cpn)
    // put on edges variables or token expressions
end merge
```

The result of applying this algorithm on the two scenarios of the use case *Identify* (figures 9 and 10) is shown in Figure 12.

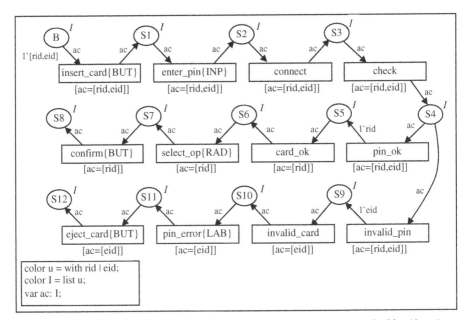

Figure 12: CPN of the use case *Identify* after merging the scenarios *regularIdentify* and *errorIdentify*.

When integrating several scenarios, the resulting specification captures in general not only the input scenarios, but perhaps even more. Figure 13 gives an example illustrating this issue: the resulting scenario Sc will capture the initial scenarios Sc_1 (T_1,T_2,T_3,T_4,T_5) and Sc_2 (T_1,T_6,T_7,T_4,T_9), as well as the two new scenarios

(T_1,T_2,T_3,T_4,T_9) and (T_1,T_6,T_7,T_4,T_5). After integrating the two scenarios, the initial place B (see Figure 13) will be shared, yet we do not know which scenario will be executed, and neither the color of Sc_1 nor the color of Sc_2 can be assigned to B. This problem was described by Koskimies and Makinen [12], and we refer to it as *interleaving problem* [6].

To solve the interleaving problem, we introduce a *chameleon token*, i.e., a token that can take on several colors [6]. Using designCPN, a chameleon token is modeled by a list of colors. Upon visiting the places of the integrated net, it will be marked by the intersection of its colors and the colors of the place being visited. When the token passes to the place S_1, it keeps the composite color $[sc_1, sc_2]$, and if it passes from S_1 to S_2, its color changes to $[sc_1]$ and will remain unchanged for the rest of its journey, or if it passes from S_1 to S_6, its color changes to $[sc_2]$ and will remain unchanged for the rest of its journey.

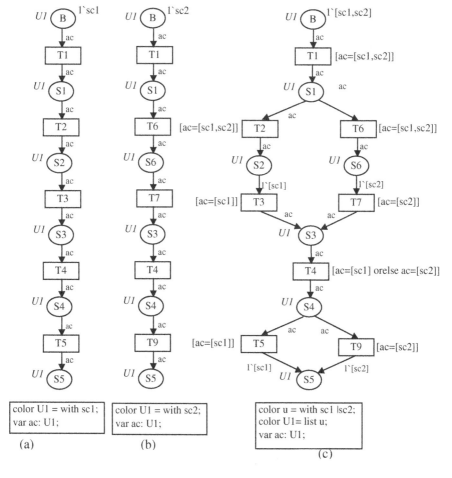

Figure 13: Interleaving problem of scenario integration: (a) scenario Sc_1, (b) scenario Sc_2 and (c) integrated scenario Sc.

Transitions that belong to only one of the scenarios Sc_1 and Sc_2 will be guarded by the color of their respective scenarios (see $T_3, T_5, T_7,$ and T_9 in Figure 13(c)). But this is not the case for the transitions T_2 and T_6 which are required to transform tokens from the composite color (list of colors) to a single color. Therefore, they must be guarded by the list of colors. For transitions that are shared by the two scenarios Sc_1 and Sc_2, they will be guarded by the composite color (see T_1 in Figure 13(c)) or by a disjunction of single colors of scenarios (see T_4 in Figure 13(c)).

An integrated CPN corresponding to a given use case can be connected to the CPN derived from the UsecaseD through a transition that is appended to the place B of the integrated CPN. This transition transforms the uncolored tokens of the CPN of the UsecaseD to the composite color of the integrated CPN.

3.4   User Interface Prototype Generation

In this activity, we derive from the CPN specifications a UI prototype of the system. The generated prototype is standalone and comprises a menu to switch between the different use cases. The various screens of the prototype represent the static aspect of the UI; the dynamic aspect of the UI, as captured in the CPN specifications, maps into the dialog control of the prototype. In our current implementation, prototypes are Java applications comprising each a number of frames and navigation functionality (see Figure 14).

The activity of prototype generation is composed of the following five operations which are described in detail in [6].

- Generating graph of transitions.
- Masking non-interactive transitions.
- Identifying UI blocks.
- Composing UI blocks.
- Generating frames from composed UI blocks.

These operations follow closely the corresponding operations in the Statechart-based approach [6], except for the operation of generating the graph of transitions. This operation consists of deriving a directed graph of transitions (GT) from the CPN of a given use case. Transitions of the CPN will represent the nodes of the GT, and edges will indicate the precedence of execution among transitions: if two transitions T_1 and T_2 are consecutively executed, there will be an edge between the nodes representing T_1 and T_2.

A GT has a list of nodes *nodeList*, a list of edges *edgeList*, and a list of initial nodes *initialNodeList* (entry nodes of the graph). The list of nodes *nodeList* of a GT is easily obtained since it corresponds to the transition list of the CPN at hand. The list of edges *edgeList* of a GT is obtained by linking the transitions in •p with the ones in p• for each place p in the CPN.

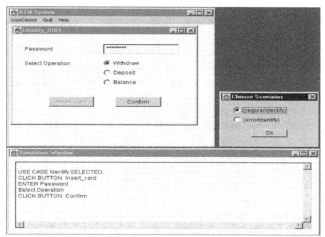

Figure 14: Prototype execution of the ATM system.

To support prototype execution, a *Simulation Window* is generated (Figure 14, bottom window), as well as a dialog box to *Choose Scenarios* (Figure 14, middle-right window). For example, after selecting the use case *Identify* from the *UseCases* menu (Figure 14, top window), a message is displayed in the simulation window that confirms the use case selection and prompts the user to click the button *Insert_card*. When this button is clicked, the fields *Password* and *Select Operation* are enabled. Then, the simulator prompts the user for information entry. When execution reaches a node in GT from which several continuation paths are possible (e.g., button *Confirm* clicked), the prototype displays the dialog box for scenario selection. In the example of Figure 14, the upper selection corresponds to the scenario *regularIdentify* and the lower one to the scenario *errorIdentify*. Once a path has been selected, the traversal of GT continues.

4   Related Work

In this section, we review some related work in the area of UI specification and automatic generation based on Petri nets. Each of the three approaches presented below suggests some high-level UI specification, yet none of them generates these specifications from scenario descriptions. For a discussion of work dealing with the simulation of specifications and with UI generation from specifications other than Petri nets refer to [5].

In the *TADEUS* approach [20] (TAsk-based DEvelopment of User interface Software), the development of the UI follows three stages: the requirements analysis and specification, the dialogue design, and the generation of the UI prototype. During the dialogue design stage, two levels of dialogues are manually built: the navigation dialogue that describes the sequencing between different task presentation units (dialogue views), and the processing dialogue that describes the changes on UI objects inside a dialogue view. The result of generation is a script file for an existing UI management system. Dialogue graphs is the formalism used to specify dialogues.

It is Petri net based, and it permits to model easily different types of dialogues (single, multi, modal, etc.). In this approach, dialogues are manually defined by the UI developer.

Palanque et al. [15] provide the Interactive Cooperative Objects (ICO) formalism for designing interactive systems. Using ICO, an object of the system has four components: a data structure, a set of operations representing the services offered by the object, an object control structure defining the object behavior with high-level Petri nets, and a presentation structured as a set of widgets. The windows of the UI and their interrelations are modeled with ICO objects. This approach is completely manual, wheras in our approach the only manual operation is the building of the object state tables (see Section 3.2.2).

De Rosis et al. [17] propose a task-based approach using Petri nets to describe interactive behavior. They represent a UI by a CPN where transitions are the tasks of the system and places represent the information display after a task execution. A logical projection function is associated to places and transitions of the CPN in order to describe in form of conditions the user actions and the information displayed as a result of performing actions. To complete the UI description, the designer links the physical aspect of the interface to the CPN by means of a physical projection function. This work focuses on specifying the UI, yet does not address UI generation.

5   Discussion of Approach

Below, we discuss our approach in respect to the following points: scenario-based approach, system and object views, visual formalisms for modeling interactive systems, and validation of approach.

5.1   Scenario-Based Approach

Our approach to UI generation exhibits the advantages of scenario-based techniques. In contrast to many data-oriented methods, UIs are generated from specifications describing dynamic system behavior, which are derived from task analysis. Once they are generated, data specifications may be used as the basis for further refinements. In line with Rosson [18] who advocates a "model-first, middle-out design approach" that interleaves the modeling of objects and scenarios, we put the emphasis on the (dynamic) modeling aspect, and generate the dynamic core of UIs rather than focus on screen design and the user-system dialog.

As scenarios are partial descriptions, there is a need to elicit all possible scenarios of the system to produce a complete specification. In our approach, colors of Petri nets are used to inhibit scenario interleaving, that is, the resulting specifications will capture exactly the input scenarios. The integration algorithm can be configured to allow scenario interleaving and to capture more than the mere input scenarios. In this way, new scenarios may be generated from already existing ones.

It is well known that scalability is an inherent problem when dealing with scenarios for large applications. Our approach eases this problem by integrating scenarios on a per use case basis, rather than treating them as an unstructured mass of scenarios.

5.2   System and Object Views

In this paper, we focus on using scenarios for the restructuring of the formal system specification (*system view*). In our previous work [5, 13], we have addressed the specification of individual objects in interactive systems (*object view*). In a scenario-based approach, the specification of the behavior of a given object can be seen as the projection of the set of acquired scenarios on that object. When projecting scenarios onto an object, only messages entering and exiting the object are considered. In this way, the sequence order of messages in the scenarios is lost, and this can lead to capture undesirable behaviors. Furthermore, in an object view, all interface objects must be explicitly identified in the underlying set of scenarios. On the other hand, in a system view, there is no loss of precision, and the UI can be generated from the system specification without the need to explicitly identify UI objects.

For the purpose of UI generation, a system view is appropriate when in all use cases and associated scenarios only one user interacts with the system. In the case of collaborative tasks (more than one user interacts with the use cases), however, an object view will be more suitable.

5.3   Visual Formalisms for Modeling Interactive Systems

Petri nets and Statecharts are among the most powerful visual formalisms used for specifying complex and interactive systems. Statecharts are an extension of state machines to include hierarchies by allowing state refinement, and concurrency by describing independent parts of a system. Concurrency in Statecharts is modeled via orthogonal states. An orthogonal state is a Cartesian product of two or more Statecharts (threads). Sending messages between threads of an orthogonal state is forbidden. Leaving a state of a thread of an orthogonal state leads to exit all threads of this orthogonal state. These restrictions do not apply to the Petri net formalism. Petri nets are known for their support of pure concurrency, all transitions having a sufficient number of tokens in their input places may concurrently be fired. Petri nets in their basic form do not support hierarchy, but the extension of Petri nets used in tools such as designCPN allow for hierarchies in the specification.

Non-deterministic choices can more easily be modeled using Petri nets than based on Statecharts. When two or more transitions share the same input places, the system chooses randomly to fire one of these transitions. In Statecharts, non-deterministic choices cannot directly be modeled because messages are ordered and broadcasted to all concurrent threads.

In many UIs, we have to model exclusive executions (modal windows). This can easily be done using the history states of Statecharts. When entering a modal window, an event must be broadcasted to all concurrent threads to enter their history states. After exiting the modal window, all concurrent threads must be re-entered in the states they were before. Modeling exclusive execution with Petri nets requires extensions of the formalism, as discussed for instance in [11].

Tokens that are specific to Petri nets can be used both in controlling and simulating system behavior, and in modeling data and resources of the system. If the place *Begin* of the Figure 6 contains only one token, the system can only execute one use case at a time. When the place contains n tokens, n concurrent executions of different use cases are possible. It may even be possible to execute *n* scenarios of the same use case (multiple instances).

Table 3 summarizes the differences between Petri nets and Statecharts based on the above discussion, indicating the strengths (+ for good and ++ for very good) and weaknesses (-) of the two formalisms. We believe that the considered criteria are all highly relevant to modern UIs. Depending on the type of UI at hand, one or the other formalism and modeling approach will be more appropriate.

Criterion	Petri net	Statechart
Concurrency	++	+
Non-determinism	+	-
History states	-	+
Multiple instances	+	-

Table 3: Differences between Petri nets and Statecharts from a UI perspective.

5.4   Validation of Approach

The scenario integration and prototype generation activities have all been implemented in Java, resulting in about 4,000 commented lines of code. The Java code generated for the UI prototype is fully compatible with the interface builder of Visual Café. For obtaining CPNs in serialized form, as required by the scenario integration algorithm, the XML tools mentioned at the end of Section 3.2.2 are used.

Our approach has been successfully applied to a number of small-sized examples. For further validation, we have started implementing the suite of examples used for validating *SUIP* [22], the implementation of our Statechart-based approach [5]. This suite includes a library system, a gas station simulator, and a filing system. The gas station simulator will be of particular interest since it involves two actors. "Extreme" examples, i.e., examples that lend themselves particularly well to the Petri net or the Statechart based approach, respectively, will also be examined, in order to further investigate the differences between the two approaches (cf. previous subsection and discussion of system versus object views).

6   Conclusion and Future Work

The work presented in this paper proposes a new approach to the generation of UI prototypes from scenarios. Scenarios are acquired as SequenceDs enriched with UI information. These SequenceDs are transformed into CPN specifications from which the UI prototype of the system is generated. Both static and dynamic aspects of the UI are derived from the CPN specifications.

The most interesting features of our approach lie in the automation brought upon by the deployed algorithms, in the use of the scenario approach addressing not only sequential scenarios but also scenarios in the sense of the UML (which supports, for instance, concurrency in scenarios), and in the derivation of executable prototypes that are embedded in a UI builder environment for refinement. The obtained prototype can be used for scenario validation with end users and can be evolved towards the target application.

As future work, we plan to move in three directions. As mentioned above, we intend to further pursue our comparison of modeling approaches. This will also include the study of the interrelationship between modeling formalism and the UI paradigm being supported. Second, we wish to investigate backward engineering, that is, allowing the automatic modification of scenarios through the UI prototype. Finally, we plan to further study the verification aspect of scenario-based modeling [4], especially the completeness of the system specification obtained from partial descriptions in form of scenarios.

Acknowledgments

We would like to thank Mazen Fahmi for helping us in the implementation part of this work. Our thanks also go to our colleagues in the *Forspec* project, Gilbert Babin, Jules Desharnais, François Lustman, and Ali Mili.

Bibliography

[1]   J. Clark. SX: An SGML System Conforming to International Standard ISO 8879, <http://www.jclark.com/sp/sx.htm>.

[2]   K. W. Derr, Applying OMT: A practical step-by-step guide to using the Object Modeling Technique, SIGS BOOKS/Prentice Hall, 1996.

[3]   designCPN: version 3, Meta Software Corp. <http://www.daimi.aau.dk/~designcpn>.

[4]   M. Elkoutbi, User Interface Engineering based on Prototyping and Formal Methods. PhD thesis, Université de Montréal, Canada, April 2000. French title: Ingénierie des interfaces usagers à l'aide du prototypage et des méthodes formelles. To appear.

[5]   M. Elkoutbi, I. Khriss, and R. K. Keller, Generating User Interface Prototypes from Scenarios, in Proceedings of the Fourth IEEE International Symposium on Requirements Engineering (RE'99), pages 150-158, Limerick, Ireland, June 1999.

[6]   M. Elkoutbi and R. K. Keller, Modeling Interactive Systems with Hierarchical Colored Petri Nets, In Proc. of 1998 Adv. Simulation Technologies Conf., pp. 432-437, Boston, MA, April 1998. Soc. for Comp. Simulation Intl. HPC98 Special session on Petri-Nets.

[7]   P. Hsia, J. Samuel, J. Gao, D. Kung, Y. Toyoshima, and C. Chen. Formal approach to scenario analysis, IEEE Software, (11)2, March 1994, pp. 33-41.

[8]   IBM, Systems Application Architecture: Common User Access – Guide to User Interface Design – Advanced Interface Design Reference, IBM, 1991.

[9]   IBM, XML Parser for Java, in Java Report's February 1999, <http://www.alphaworks.ibm.com/formula/xml>.

[10]   K. Jensen, Coloured Petri Nets, Basic concepts, Analysis methods and Pratical Use, Springer, 1995.

[11]   M. Kishinevsky, J. Cortadella, A. Kondratyev, L. Lavagno, A. Taubin, and A. Yakovlev. Coupling asynchrony and interrupts: Place chart nets. In P. Azema and G. Balbo, editors, ATPN'97, pp.328-347, Toulouse, France, June 1997. Springer-Verlag. LNCS 1248.

[12]   Koskimies K. and Makinen E.: Automatic Synthesis of State Machine from Trace Diagrams, Software Practice & Experience (1994), 643-658.

[13]   Ismail Khriss, Mohammed Elkoutbi, and R. K. Keller. Automating the synthesis of UML statechart diagrams from multiple collaboration diagrams. In J. Bezivin and P. A. Muller, ed., <<UML>>'98: Beyond the Notation, pp.132-147. Springer, '99. LNCS 1618.

[14]   B. A. Nardi, The Use of Scenarios in Design, SIGCHI Bulletin, 24(4), October 1992.

[15]   P. Palanque and R. Bastide, Modeling clients and servers in the Web using Interactive Cooperative Objects, Formals Methods in HCI,Springer, 1997, pp.175-194.

[16]   C. Potts, K. Takahashi and A. Anton, Inquiry-Based Scenario Analysis of System Requirements, Technical Report GIT-CC-94/14, Georgia Institute of Technology, 1994.

[17] F. de Rosis, S. Pizzutilo and B De Carolis: A tool to support specification and evaluation of context-customized interfaces. SIGCHI Bulletin, 28, 3, 1996.

[18] M. B. Rosson. Integrating Development of Task and Object Models, Communications of the ACM, 42(1), January 1999, pp. 49-56.

[19] J. Rumbaugh, I. Jacobson, and G. Booch, The Unified Modeling Language Reference Manual, Addison Wesley, Inc., 1999.

[20] E. Schlungbaum and T. Elwert, Modeling a Netscape-like browser Using TADEUS Dialogue graphs, In Handout of CHI'96 Workshop on Formal Methods in Computer Human Interaction: Comparison, Benefits, Open Questions, Vancouver, 1996, pp.19-24.

[21] Siegfried Schönberger, Rudolf K. Keller, and Ismail Khriss. Algorithmic Support for Model Transformation in Object-Oriented Software Development. Theory and Practice of Object Systems (TAPOS), 2000. John Wiley and Sons. to appear.

[22] SUIP. Scenario-based user interface prototyping: Website and public domain software distribution, September 1999. <http://www.iro.umontreal.ca/labs/gelo/suip/>.

[23] Symantec, Inc. Visual Café for Java : User Guide, Symantec, Inc., 1997.

Decidability of Properties of Timed-Arc Petri Nets

David de Frutos Escrig[1], Valentín Valero Ruiz[2], and Olga Marroquín Alonso[1]

[1] Dpto. Sistemas Informáticos y Programación, Fac. Matemáticas
Universidad Complutense. Madrid, SPAIN 28040
{defrutos,alonso}@sip.ucm.es
[2] Dpto. Informática, Esc. Politécnica
Univ. Castilla-La Mancha. Albacete, SPAIN 02071
valentin@info-ab.uclm.es

Abstract. Timed-arc Petri nets (TAPN's) are not Turing powerful, because, in particular, they cannot simulate a counter with zero testing. Thus, we could think that this model does not increase significantly the expressiveness of untimed Petri nets. But this is not true; in a previous paper we have shown that the differences between them are big enough to make the reachability problem undecidable. On the other hand, coverability and boundedness are proved now to be decidable. This fact is a consequence of the close interrelationship between TAPN's and transfer nets, for which similar results have been recently proved. Finally, we see that if dead tokens are defined as those that cannot be used for firing any transition in the future, we can detect these kind of tokens in an effective way.

1 Introduction

Petri nets are widely used for the modeling and analysis of concurrent systems, because of their graphical nature and the solid mathematical foundations supporting them. Several timed extensions of the basic model have been proposed to expand their application areas to those systems which exhibit a time-dependent behaviour that should be considered both in the modeling and the analysis process, such as distributed systems, communication systems and real-time systems.

A survey of the different approaches to introduce time into Petri nets is presented in [7]. We can identify a first group of models, which assign time delays to transitions, either using a fixed and deterministic value [19,20,21] or choosing it from a probability distribution [3]. Other models use time intervals to establish the enabling times of transitions [16]. Finally, we have also some models that introduce time on tokens [1,6]. In such a case, tokens become classified into two different classes: available and unavailable ones. Available tokens are those that can be immediately used for firing a transition, while unavailable tokens cannot. We have to wait for a certain period of time for these tokens to become available, although it is also possible for a token to remain unavailable forever (such tokens are said to be *dead*). More recently, Cerone and Maggiolo-Schettini

M. Nielsen, D. Simpson (Eds.): ICATPN 2000, LNCS 1825, pp. 187–206, 2000.
© Springer-Verlag Berlin Heidelberg 2000

[8] have defined a very general model (statically timed Petri nets), where timing constraints are intervals statically associated with places, transitions and arcs. Thus, models with timing constraints attached only to places, transitions or arcs can be obtained as particular subclasses of this general framework.

In this paper we analyze timed-arc Petri nets [6,23,14], a timed extension of Petri nets in which tokens have associated a natural[1] value indicating the elapsed time from its creation (its *age*), and arcs from places to transitions are also labelled by time intervals, which establish restrictions on the age of the tokens that can be used to fire the adjacent transitions.

In [6], Bolognesi *et. al* describe timed-arcs Petri nets, comparing them with Merlin's model in the framework of design and analysis of concurrent systems. The interpretation and use of timed-arcs Petri nets can be obtained from a collection of processes interacting with each other according to a rendez-vous mechanism. Each process may execute either local actions or synchronization ones. Local actions are those that the process may execute without cooperation from another process, and thus in the Petri net model of the whole system they would appear as transitions with a single precondition place, while synchronization actions would have several precondition places, which correspond to the states at which each one of the involved processes is ready to execute the action. Then, each time interval establishes some timing restrictions related to a particular process (for instance the time that a local processing may require). In consequence, the firing of a synchronization action can be done in a time window, which depends on the age of the tokens on its precondition places.

One of the applications of timed-arc Petri nets comes from the fact that it is quite easy to get a timed-arc Petri net modeling a system that has been described by means of a *Timed-LOTOS* specification [18]. Therefore, in particular, the interest of the model can be justified as a graphical tool for the design and analysis of concurrent systems.

In [5], it is proved that timed-arc Petri nets are not Turing complete, since in particular they cannot correctly simulate a 2-counter machine. Thus, we could expect that the differences with untimed Petri nets in terms of expressiveness would not be rather significant. Nevertheless, we have shown in [22] that the difference is big enough to make the reachability problem undecidable for timed-arc Petri nets.

In this paper we extend the study of decidability of properties of TAPN's proving, in particular, that coverability and boundedness are both decidable. This is a consequence of the close connection between TAPN's and transfer nets, which can simulate each other, as we show in this paper. As a consequence, the

---

[1] Although it is usual to define TAPN's taking real numbers to measure the passage of time, in this paper we will only allow natural numbers, since, for the time being, we have not been able to generalize our decidability results to the continuous time models. Instead, we considered general TAPN's when proving undecidability of reachability in [22], but it is important to note that our counterexample proving undecidability makes no use of continuous time. Reachability is also undecidable when discrete time is considered.

(un)decidability of any property preserved by these simulations can be translated from any of these classes to the other.

Transfer nets and reset nets have been thoroughly studied in [10,9,11] as particular cases of the so called Generalized Self-Modifying nets (shortly G-nets). In particular, it is proved there that coverability is decidable for both classes of nets, by applying a new general[2] *backward* technique, presented in [2,13]. On the contrary, boundedness is decidable for transfer but not for reset nets. A corollary of this decidability result is that place boundedness is undecidable in both cases.

After studying the decidability of several classic properties of TAPN's, we will concentrate on the detection of *dead* tokens. Due to the time restrictions in the nets, and also to their structure, some tokens may become *dead*, since they are too old to fire any transition in the future. Thus they stay attached to their places forever, growing and growing, and never being available. We will prove that this kind of tokens can be effectively detected, because of the fact that the firability of transitions is closely related with coverability, that it is proved to be decidable.

The paper is structured as follows. In Section 2 we present timed-arc Petri nets and their semantics; in Section 3 we recall the undecidability of reachability, and prove that, instead, it becomes decidable if we fix the duration of the considered computations. In section 4 we prove the decidability of coverability. Section 5 is the main section of the paper; there we prove the close relationship between TAPN's and transfer nets, and as a consequence we transfer to TAPN's most of the known (un)decidability results for transfer nets. In Section 6 we show that the problem of detecting dead tokens is decidable, and we also study a weaker version of that property. Finally, in Section 7 we discuss the work to be done in the future.

2   Timed-Arc Petri Nets

We deal with timed-arc Petri nets, which have their tokens annotated with an age (an integer value indicating the elapsed time from its creation), and where arcs connecting places with transitions have associated a time interval, which limits the age of the tokens consumed to fire the adjacent transition.

However, a transition is not forced to be fired when all its preconditions contain tokens with an adequate age, and the same is true even if the age of any of these tokens is about to expire. More in general, in the model there is not any kind of urgency, which can be interpreted in the sense that the model is *reactive*, as transitions will only be fired when the external context requires it. But then, it may be the case that the external context may lose the ability to fire a transition if some needed tokens become too old. Even more, it is possible that some tokens become *dead*, which means definitely useless because the increasing

---

[2] What is mainly new in these papers is their generality, which comes from the fact that the presented results can be applied to any well structured transition system. Instead, a *backward* algorithm to decide coverability for reset nets was presented in [4] as long ago as in 1976.

of their age will not allow in the future the firing of any of their postcondition transitions.

Definition 1. *(Timed-arc Petri nets)*
We define a **timed-arc Petri net** *(TAPN) as a tuple* [3] $N = (P\ T\ F\ \texttt{times})$, *where P is a finite set of* **places**, *T is a finite set of* **transitions** *($P \cap T = \emptyset$), F is the* **flow relation** *($F \subseteq (P \times T) \cup (T \times P)$), and* **times** *is a function that associates to each arc $(p\ t)$ in F a pair of natural numbers, the second of which can be infinite, i.e.:* $\texttt{times} : F|_{P \times T} \longrightarrow \mathbb{N} \times (\mathbb{N} \cup \{\infty\})$.

When $\texttt{times}(p\ t) = [t_1\ t_2]$ we write $_i(p\ t)$ to denote t_i, for $i = 1\ 2$. Since $\texttt{times}$ defines the intervals of age of the tokens to be consumed by the firing of each transition (see Def.2), we will always have $_1(p\ t) \leq\ _2(p\ t)$. Moreover, we will write $x \in \texttt{times}(p\ t)$ to denote $_1(p\ t) \leq x \leq\ _2(p\ t)$.

As we previously mentioned, tokens are annotated with natural numbers, so markings are defined by means of multisets on $\mathbb{N}$. More exactly, a marking M is a function $M : P \longrightarrow \mathcal{B}(\mathbb{N})$ where $\mathcal{B}(\mathbb{N})$ denotes the set of finite multisets of natural numbers. Thus, as usual, each place is annotated with a certain number of tokens, but each one of them has associated a natural number (its *age*). We will denote the set of markings of N by $\mathcal{M}(N)$, and using classical set notation, we denote the total number of tokens on a place p by $|M(p)|$. Finally, by some abuse of notation, sometimes we will denote the individual tokens in a marking M by pairs $(p\ i)$ with $p \in P$ and $i \in \mathbb{N}$ denoting its age, also writing $(p\ i) \in M$.

As initial markings we only allow markings M such that for all p in P, and any $x > 0$ we have $M(p)(x) = 0$ (i.e., the *initial age* of any token is 0)[4]. Then, we define *marked timed-arc Petri nets* (MTAPN) as pairs $(N\ M)$, where N is a timed-arc Petri net, and M is an initial marking on it. As usual, from this initial marking we will obtain new markings, as the net evolves, either by firing transitions, or by the passage of time. In consequence, given a non-zero marking, even if we do not fire any transition at all, starting from this marking we get an infinite reachability set of markings, due to the token aging.

A timed-arc Petri net with an arbitrary marking can be graphically represented by extending the usual representation of P/T nets with the corresponding time information. In particular, we will use the age of each token to represent it. Therefore, MTAPN's have initially a finite collection of zero values labelling each place. In Fig.1 we show a MTAPN modeling a producer/consumer system.

Let us now see how we can fire transitions, and how we model the passage of time.

Definition 2. *(Firing rule)*
Let $N = (P\ T\ F\ \texttt{times})$ be a TAPN, M a marking on it, and $t \in T$.

---

[3] To simplify some definitions we consider only arcs with weight 1, but the extension to general arcs with greater weights is straightforward.

[4] In fact, it would not be a problem to allow initial markings containing older tokens, as long as we only have a finite number of tokens. Then, the main reason for imposing this restriction is to capture the intuitive idea that initial tokens have not had yet time to become old.

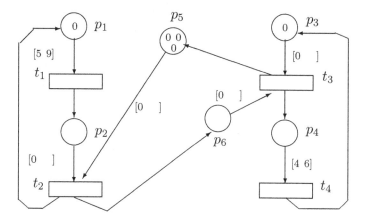

Fig. 1. Timed-arc Petri net modeling the PC-problem

(1) We say that t is **enabled** *at marking M if and only if:*

$$\forall p \in {}^{\bullet}t \;\; \exists x_p \in \mathbb{N} \;\; such \; that \;\; M(p)(x_p) > 0 \wedge x_p \in \mathtt{times}(p\;t)$$

i.e., on each precondition of t we have some token whose age belongs to $\mathtt{times}(p\;t)$.

(2) If t is enabled at M, it can be fired, and by its firing we reach any marking M′ which can be obtained as follows:

$$M'(p) = M(p) - C^{-}(p\;t) + C^{+}(t\;p) \;\; \forall p \in P$$

where both the subtraction and the addition operators work on multisets, and

$$- \; C^{-}(p\;t) \in \quad \begin{matrix} \{\{x_p\} | \; x_p \in M(p) \; and \; x_p \in \mathtt{times}(p\;t)\} \; if \; p \in {}^{\bullet}t \\ \{\varnothing\} \hspace{5.5em} otherwise \end{matrix}$$

$$- \; C^{+}(t\;p) = \quad \begin{matrix} \varnothing \;\; if \, p \notin t^{\bullet} \\ \{0\} \; otherwise \end{matrix}$$

Thus, from each precondition place of t we remove a token fulfilling (1), and we add a new token (with age 0) on each postcondition place of t.

As usual, we denote these evolutions by $M[t\rangle M'$, *but it is noteworthy that these evolutions are in general non-deterministic, because when we fire a transition t, some of its precondition places could hold several tokens with different ages, that could be used to fire it. Besides, we see that the firing of transitions does not consume any time. Therefore, to model the passage of time we need the function* **age**, *defined below. By applying it, we age all the tokens of the net by the same time:*

(3) The function $\mathtt{age} : \mathcal{M}(N) \times \mathbb{N} \longrightarrow \mathcal{M}(N)$ *is defined by:*

$$\mathtt{age}(M\;x)(p)(y) = \quad \begin{matrix} M(p)(y - x) \; if \; y \geq x \\ 0 \hspace{4em} otherwise \end{matrix}$$

The marking obtained from M after x units of time without firing any transitions will be that given by $\mathtt{age}(M\;x)$.

Although we have defined the evolution by firing single transitions, this can be easily extended to the firing of *steps* or *bags* of transitions, since those transitions that could be fired together in a single step could also be fired in sequence in any order, because no *aging* is produced by the firing of transitions. In this way, we obtain step transitions that we denote by $M[R\rangle M'$. By alternating step transitions and the passage of time we can define a timed step semantics, where timed step sequences are those sequences $= M_0[R_1\rangle_{x_1} M_1 \quad M_{n-1}[R_n\rangle_{x_n} M_n$, where $M_i's$ are markings, $R_i's$ multisets of transitions and $x_i's \in \mathbb{N}$, in such a way that $M_i[R_{i+1}\rangle M'_{i+1}$ and $M_{i+1} = \texttt{age}(M'_{i+1} \quad x_{i+1})$. Note that we admit $x_i = 0$ in order to allow null intervals of time between two causally related steps. However, note that this does not imply that both steps could be fired together in a single step. Therefore, we maintain the distinction between parallelism and sequentiality. Finally, we denote sequences $\quad$ by $M_0[\overline{R}\rangle_x M_n$, where $\overline{R} = R_1 R_2 \quad R_n$ and $x = \sum_{i=1}^{n} x_i$, whenever we do not need to keep track of the values of each x_i.

Most of the interesting properties of TAPN's, as the terminology to define them, are direct translations of properties of untimed Petri nets, although in some cases the introduction of time makes necessary to change the way in which these properties are defined. Besides, we also have some new properties for which time plays a central role.

Given a MTAPN $(N \; M_0)$, we define $[M_0\rangle$ as the set of reachable markings on N starting from M_0. Given $m \in \mathbb{N}$ we define by $[M_0\rangle_m$ the set of reachable markings by a computation of total duration m:

$$[M_0\rangle_m = \{ M \mid \exists \overline{R} \; M_0[\overline{R}\rangle_m M \}$$

When $M \in [M_0\rangle_m$ we say that M is reachable in time m. The MTAPN $(N \; M_0)$ *terminates* if there is no infinite computation. A place $p \in P$ is *bounded* if there exists $n \in \mathbb{N}$ such that for all $M \in [M_0\rangle$ we have $|M(p)| \leq n$; we say that it is *time-locally bounded* if for all $t \in \mathbb{N}$ there exists $n \in \mathbb{N}$ such that for all $M \in [M_0\rangle$ we have $|M(p)(t)| \leq n$; finally, it is *uniformly time-locally bounded* if there exists $n \in \mathbb{N}$ such that for all $t \in \mathbb{N}$ and all $M \in [M_0\rangle$ we have $|M(p)(t)| \leq n$. We say that $(N \; M_0)$ is *bounded* (resp. *time-locally bounded, uniformly time-locally bounded*) if every $p \in P$ in it is bounded (resp. time-locally bounded, uniformly time-locally bounded)[5].

As a matter of fact, we have only defined uniformly time-locally boundedness for completeness, since for TAPN's it is immediate to prove the following property:

Proposition 3. *Given a MTAPN $(N \; M_0)$ and $p \in P$, we have that p is time-locally bounded in N iff it is uniformly time-locally bounded in it.*

Proof. Since only tokens of age 0 are created, for all $M_1 \in [M_0\rangle_N$ and all $m \in \mathbb{N}$ there exists $M_2 \in [M_0\rangle_N$ such that for all $p \in P$ we have $M_1(p)(m) \leq M_2(p)(0)$.

---

[5] Note that we cannot define boundedness equivalently by saying that the reachability set is finite, since by aging a marking we can always obtain older and older different markings.

As a consequence, if p is time-locally bounded in N, the corresponding bound for age 0 is a uniform bound for any age. □

3 Decidability of Timed Reachability

In [22] we have proved the following result:

Theorem 4. *The reachability problem for MTAPN's is undecidable.*

However, timed reachability is decidable, since given (finite) time m one can simulate the behaviour of a MTAPN up to that instant by means of a plain Petri net N^m, which is defined as follows:

Definition 5. *Given a TAPN $N = (P \ T \ F \ \text{times})$, an initial marking M_0 for it [6] and $m \in \mathbb{N}$, we define the associated P/T Petri net $N^m = (P^m \ T^m \ F^m)$ as follows:*

- $P^m = (P \times 0 \ m) \cup \{ c_i \mid i \in 0 \ m \}$,
- *The set of transitions is the set*

$$T^m = \{ (t \ \text{ages} \ l) \mid \text{ages} : \{ p \mid (p \ t) \in F \} \rightarrow (\text{times}(p \ t) \cap 0 \ m) \ l \in 0 \ m \}$$
$$\cup \{ \text{tick}_i \mid i \in 1 \ m \}$$

- *The flow relation, $F^m \subseteq (P^m \times T^m) \cup (T^m \times P^m)$, is defined as the set*

$$\begin{aligned} F^m = \ & \{ (c_{i-1} \ \text{tick}_i) \ (\text{tick}_i \ c_i) \mid i \in 1 \ m \} \cup \\ & \{ (c_i \ (t \ \text{ages} \ i)) \mid i \in 0 \ m \} \cup \\ & \{ ((p \ j) \ (t \ \text{ages} \ i)) \mid (p \ t) \in F \ \ j = i - \text{ages}(p) \} \cup \\ & \{ ((t \ \text{ages} \ i) \ (p \ i)) \mid (t \ p) \in F \ \ a \in 0 \ i \} \end{aligned}$$

Finally, as initial marking we would take M_0^m defined by

$$\begin{aligned} & M_0^m(c_0) = 1 \\ & M_0^m(c_i) = 0 && \text{for all } i > 0 \\ & M_0^m((p \ 0)) = M_0(p) && \text{for all } p \in P \\ & M_0^m((p \ i)) = 0 && \text{for all } p \in P \text{ and } i > 0 \end{aligned}$$

It is not difficult to see how N^m works, simulating the behaviour of N up to the instant m. The idea is that the age of the tokens is controlled in a static way: tokens do not move when they become older; instead the *clock places* c_i do the work, since depending on the value of the global clock represented by them, the firable transitions would take the consumed tokens from different places. In this way, we capture the fact that the current age of a token on a place $(p \ j) \in P^m$ is equal to $i - j$, where c_i is the currently marked clock place. All this is formalized as follows:

---

[6] It would not be difficult to extend this construction in order to allow any finite marking as the initial marking M_0. In such a case, we take $P^m = P \times -\text{old}(M_0) \ m$, where old is defined as in Def.6. Each token $(p \ i)$ in M_0 would be represented by a token on the corresponding place $(p \ -i) \in P^m$.

Definition 6. *Let $N = (P\ T\ F\ \texttt{times})$ be a TAPN with an initial marking M_0. Given M an arbitrary marking of N and $m \in \mathbb{N}$:*

(1) We define $\texttt{old}(M) = max\{\, i \mid \exists i \in M(p)\,\}$.

(2) If $\texttt{old}(M) \leq m$ we say that the marking M^m of N^m defined by

$$M^m(c_m) = 1$$
$$M^m(c_i) = 0 \qquad\qquad for\ all\ i < m$$
$$M^m((p\ j)) = M(p)(m - j)\ for\ all\ j \in 0\ \ m$$

is the representation of M in N^m.

Theorem 7. *For each marking M of a TAPN N such that $\texttt{old}(M) \leq m$ we have that $M \in [M_0\rangle_m$ iff $M^m \in [M_0^m\rangle_{N^m}$.*

Corollary 8. *Timed reachability is decidable for TAPN's.*

4   Decidability of Coverability

In this section we will prove that coverability is decidable for timed-arc Petri nets. Nevertheless, the proof of this result cannot be obtained by generalizing the coverability tree [12] in a proper way. Instead, we have to use a *backward* technique recently presented in [2,13].

Definition 9. *([12,2] Well-structured transition systems)*
A **well-structured transition system** *(WSTS) is a structure $S = \langle Q \to \sqsubseteq\rangle$ such that:*

(1) $Q = \{M\quad\}$ is a set of **states**,

(2) $\to\, \subseteq Q \times Q$ is a set of **transitions**,

(3) $\sqsubseteq\, \subseteq Q \times Q$ is a **well-quasi-ordering** *(wqo) on the set of states, that is, a reflexive, transitive and well founded relation such that*

$$M \to M'\ and\ M_1 \sqsupseteq M\ imply\ M_1 \to M_1'\ for\ some\ M_1' \sqsupseteq M'.$$

Definition 10. *(Upward closed sets)*
Let S be a WSTS and Q its set of states.

(1) We say that $K \subseteq Q$ is **upward closed** *iff whenever $M \in K$ we also have $M' \in K$ for all $M \sqsubseteq M'$.*

(2) If $K \subseteq Q$ is upward closed and $s \in \mathbb{N}$, we define $\texttt{cover}_s(K)$ as the set of all the states from which one can reach a state in K in exactly s steps.

(3) If K is upward closed and $B \subseteq K$, we say that B is a **basis** *of K iff $K = \{\, M' \mid \exists M \in B\ \ M \sqsubseteq M'\,\}$.*

From now on, for $K \subseteq Q$, $\uparrow K$ denotes the set $\{\, M' \in Q \mid \exists M \in K\ \ M \sqsubseteq M'\,\}$.

Proposition 11. *If S is a WSTS, any upward closed set in it admits a finite basis.*

Definition 12. *(Effective WSTS)*
*Let S be a WSTS. We say that S is **effective** iff the following conditions hold:*

(1) $\sqsubseteq$ *is decidable.*

(2) *For each finite set $B \subseteq Q$ we can effectively compute a finite basis of the set $\uparrow (\mathtt{cover}_1(\uparrow B))$.*

Theorem 13. *[2,13] For effective WSTS's coverability is decidable.*

Next we recall the main ideas of the proof of this result in the case of Petri nets:

- If M can be covered from M_0 there exists $M' \in [M_0\rangle$ with $M \sqsubseteq M'$. Then we consider the last step of these computations, taking $M'' \in [M_0\rangle$ such that $M''[t\rangle M'$. We take as $\mathtt{cover}_1(M)$ the set of these markings M''.

- Iterating this construction we obtain $\mathtt{cover}_k(M)$ for all $k \in \mathbb{N}$. It can be proved that each one of these sets is upward closed.
  Then we define $\mathtt{cover}_k^*(M)$ as $\displaystyle\bigcup_{l \le k} \mathtt{cover}_l(M)$. Obviously, $k \le k'$ implies $\mathtt{cover}_k^*(M) \subseteq \mathtt{cover}_{k'}^*(M)$. But by applying the fact that $\sqsubseteq$ is a well-quasi-ordering, there must exist some $k \in \mathbb{N}$ such that $\mathtt{cover}_{k+1}^*(M) = \mathtt{cover}_k^*(M)$. If we take as $\mathtt{cover}^*(M)$ such a set, it is clear that M can be covered from M_0 iff $M \in \mathtt{cover}^*(M_0)$.

- Since the sets $\mathtt{cover}_k^*(M_0)$ are infinite we could not cope directly with them in an effective way. But applying Prop.11 we can effectively characterize these sets by their finite basis.

In order to apply this technique to prove the decidability of the coverability problem for MTAPN's, we have to find an adequate ordering relation between their markings. The natural candidate would be the multiset inclusion, as it is the immediate generalization of the adequate ordering for ordinary nets. Unfortunately, that would not work since $\mathbb{N}$ is an infinite set, and then if for any place $p \in \mathbb{N}$ we consider the singleton markings $M_i = \{(p\ i)\}$ we have $M_i \not\subseteq M_j$, whenever $i \ne j$. Nevertheless, it is possible to find an adequate ordering taking into account the fact that when a token becomes old enough its age becomes unimportant since it can be only used to fire those transitions for which $M_2(p\ t) = \infty$. This is formalized as follows:

Definition 14. *Let $N = (P\ T\ F\ \mathtt{times})$ be a TAPN.*

(1) *For each $p \in P$ we define*
$$Max(p) = Max\{\{\ _1(p\ t)\,|\,t \in p^\bullet\} \cup \{\ _2(p\ t)\,|\,t \in p^\bullet\quad _2(p\ t) < \infty\}\}$$
and $S(p) = Max(p) + 1$.

(2) *We define the **stable time** of N, $St(N)$, as $Max\{S(p)\,|\,p \in P\}$.*

Thus, we have that once the age of a token on p exceeds $Max(p)$ the only postcondition transitions $t \in p^\bullet$ that could be fired by using that token are those for which $_2(p\ t) = \infty$. Obviously, in order to fire such a transition t the age of the involved token on p is unimportant once it exceeds $S(p)$. The same is true in a uniform way for all the places of the net if we take $St(N)$ instead of the values $S(p)$.

Now we can define an adequate ordering relation $\sqsubseteq'$ between markings of a time-arc Petri net as follows:

Definition 15. *Let* $N = (P\ T\ F\ \mathtt{times})$ *be a TAPN and* M, M' *two of its markings.* $M \sqsubseteq' M'$ *iff the following conditions hold:*

(1) $M|_{St(N)-1} \subseteq M'|_{St(N)-1}$ *where for all* $p \in P$, $i \in \mathbb{N}$, *for any marking* M *and any* $k \in \mathbb{N}$ *we have*
$$M|_k(p)(i) = \begin{array}{ll} M(p)(i) & \textit{if } i \leq k \\ 0 & \textit{otherwise} \end{array}$$
(2) For all $p \in P$ $|\{i \in M(p) \mid i \geq St(N)\}| \leq |\{i \in M'(p) \mid i \geq St(N)\}|$.

Proposition 16.

(1) $\sqsubseteq'$ *is a well-quasi-ordering.*

(2) If we consider the equivalence relation $\sim'$ *induced by* $\sqsubseteq'$, *whenever* $M_1 \sim' M_2$, *for all* $t \in T$ *we have that* $M_1[t\rangle \Leftrightarrow M_2[t\rangle$, *and if* $M_1[t\rangle M_1'$ *there exists* M_2' *such that* $M_2[t\rangle M_2'$ *and* $M_1' \sim' M_2'$.

As an immediate consequence we have the following

Theorem 17.

(1) The transition system generated by the firing rule of any TAPN, N, becomes a well-structured transition system when we take as ordering relation between its markings the corresponding relation $\sqsubseteq'$.

(2) Coverability is decidable for timed-arc Petri nets.

Proof.

(1) Immediate, by application of the firing rule for TAPN's.

(2) Let M be the marking that we want to know if it is coverable from M_0. If $M = M|_{St(N)-1}$ then we can apply the procedure in the proof of Th.13 in order to decide if it is coverable. Otherwise, this procedure cannot be directly used, because of the fact that $\sqsubseteq'$ does not preserve the information about the exact age of the tokens once they become older than $St(N)$. Nevertheless, since markings are finite, we can consider $max(M) = max\{i \mid \exists(p\ i) \in M\} < \infty$, taking as ordering relation $\sqsubseteq''$ that is defined exactly as $\sqsubseteq'$ but replacing $St(N)$ by $max(M) + 1 > St(N)$. Then we have that $\sqsubseteq''$ has the same good properties as $\sqsubseteq'$, and in addition it fully preserves the information about the age of the tokens when they are younger than $max(M)$. Therefore, using this ordering relation one can decide if the given marking M is coverable from M_0.

$\square$

We could continue the study of some other properties of timed-arc Petri nets in a direct way, but in order to avoid the repetition of similar reasonings to those made in [9,10] when studying reset and transfer nets, we will prove in the next section that transfer nets and timed-arc Petri nets can simulate each other. This makes possible to translate to TAPN's most of the known results about decidability and undecidability for transfer nets.

5   Relating Timed-Arc Petri Nets and Transfer Nets

We start this section by presenting reset and transfer nets. Both are particular cases of Generalized Self-Modifying nets which have been thoroughly studied in [10].

Definition 18. *(Reset and transfer nets)*

(1) A **reset net** *is a 3-tuple* $N = \langle P\ T\ F \rangle$ *where*

- $P = \{p_1\ \ \ \ p_{|P|}\}$ *is a finite set of* **places**,
- T *is a finite set of* **transitions** *(with* $P \cap T = \varnothing$*),*
- $F : (P \times T) \cup (T \times P) \to \mathbb{N}^\infty$ *is a* **flow function** *where* $\mathbb{N}^\infty = \mathbb{N} \cup \{\infty\}$ *and for all* $(t\ p) \in (T \times P)\ \ F(t\ p) \in \mathbb{N}$.

Whenever $F(p\ t) = \infty$ *we say that* $(p\ t)$ *is a* **reset arc** *and that* t **resets** p.

(2) A **marked reset net** *is a pair* $(N\ m_0)$ *where* N *is a reset net and* $m_0 \in \mathbb{N}^{|P|}$ *is the* **initial marking**.

(3) A **transfer net** *is a 4-tuple* $N = \langle P\ T\ F\ TA \rangle$ *where* $\langle P\ T\ F \rangle$ *defines an ordinary P/T net and* TA *is a function* $TA : T \to \mathcal{P}(P \times P)$*, that defines the* **transfer arcs**, *, verifying*

$$(p\ p') \in TA(t) \Rightarrow\ (\neg\exists p'' \neq p'\ (p\ p'') \in TA(t)) \wedge\ F(p\ t) = F(t\ p') = 0$$

(4) A **marked transfer net** *is a pair* $(N\ m_0)$ *where* N *is a transfer net and* $m_0 \in \mathbb{N}^{|P|}$ *is the* **initial marking**.

Firing rules for reset and transfer nets are defined as follows:

Definition 19. *(Firing rules)*
A transition t *is* **firable** *in a reset or transfer net whenever it is firable in the corresponding plain P/T net obtained by removing reset and transfer arcs. The firing of* t *from* M *produces a new marking* M' *which is defined exactly as for ordinary nets, but whenever we have a reset or a transfer arc we have instead:*

(1) If $F(p\ t) = \infty$ *then* $M'(p) = F(t\ p)$*, which means that we take all the tokens from* p*, and we only return someone if* p *is a postcondition of* t.

(2) If $(p\ p') \in TA(t)$, we have the following (non disjoint) cases [7]:

  (a) There is no place p'' such that $(p'\ p'') \in TA(t)$: $M'(p') = M(p') + M(p)$.

  (b) There is a place p'' such that $(p'\ p'') \in TA(t)$: $M'(p') = M(p)$.

  (c) There is no place p'' such that $(p''\ p) \in TA(t)$: $M'(p) = F(t\ p)$.

  (d) If there is a place p'' such that $(p''\ p) \in TA(t)$, one can proceed as in case (b), since we have $(p''\ p)\ (p\ p') \in TA(t)$.

Next we recall some of the results about decidability of properties of reset and transfer nets which have been presented in [10,15].

Theorem 20.

(1) Coverability and termination are decidable for reset and transfer nets.

(2) Boundedness is decidable for transfer nets, but undecidable for reset nets.

(3) Structural boundedness is decidable for transfer nets.

(4) Place boundedness is undecidable for both reset and transfer nets.

In the following we will see that timed-arc Petri nets and transfer nets are closely related. First we will see how to simulate timed-arc Petri nets by transfer nets. In order to do it, we need some previous definitions:

Definition 21. *Let $N = (P\ T\ F\ \mathtt{times})$ be a TAPN and $B \in \mathbb{N}$ such that $B \geq St(N)$. We define the transfer net $N^B = (P^B\ T^B\ F^B\ TA^B)$ which simulates N by preserving the age information up to the instant B, as follows:*

- $P^B = P \times 0\ B$,
- $T^B = \{(t\ \mathtt{ages}) \mid t \in T\ \mathtt{ages} : \{p \mid (p\ t) \in F\} \to (\mathtt{times}(p\ t) \cap 0\ B)\} \cup \{\mathtt{tick}\}$,
- $F^B = \{((p\ a)\ (t\ \mathtt{ages})) \mid \mathtt{ages}(p) = a\} \cup \{((t\ \mathtt{ages})\ (p\ 0)) \mid (t\ p) \in F\}$,
- *For all $(t\ \mathtt{ages}) \in T^B$ $TA^B((t\ \mathtt{ages})) = \varnothing$,*
- $TA^B(\mathtt{tick}) = \{((p\ a)\ (p\ a+1)) \mid p \in P\ a < B\}$.

In this definition, each place $(p\ a)$ with $a < B$ represents the tokens in the original place $p \in P$ with age a, while $(p\ B)$ represents those tokens whose age is greater or equal than B. Each transition $(t\ \mathtt{ages})$ represents the firing of t by consuming tokens in its precondition places whose ages are defined by the values of $\mathtt{ages}$. So, each possible value of the function $\mathtt{ages}$ corresponds to a different selection of the ages of the consumed tokens (thus to any of the different ways to fire t in N, as defined in Def.2(2)). To be exact, we are identifying all the ages older than B, what has no negative influence, taking into account that $B \geq St(N)$. Finally, the passage of time is modelled by the transition $\mathtt{tick}$ that ages all the tokens in the net that are not older than B.

As a consequence, the markings of both nets N and N^B are related as follows:

---

[7] We have to distinguish whether we have one or more transfer arcs associated to the same transition.

Definition 22.

(1) Given a marking M of N we define the associated marking M^B of N^B by
$$M^B(p\ a) = M(p)(a) \qquad if\ a < B$$
$$M^B(p\ B) = \sum_{a \geq B} M(p)(a)$$

(2) Given a marking M' of N^B we define the set of markings represented by M', $tr(M')$, as $tr(M') = \{\ M \in \mathcal{M}(N) \mid M^B = M'\ \}$.

The following theorem formalizes the fact that N^B is a full simulation of N up to age B:

Theorem 23. *Let N be a TAPN, $B \geq St(N)$ and N^B the associated transfer net, we have:*

(1) For any marking M' of N^B, if $M_1\ M_2 \in tr(M')$ then $M_1|_{B-1} = M_2|_{B-1}$.

(2) Whenever we have $M_1[t\rangle_N M_2$, there exists some $(t\ \mathtt{ages}) \in T^B$ such that $M_1^B[(t\ \mathtt{ages})\rangle_{N^B} M_2^B$.

(3) If $M_1'[(t\ \mathtt{ages})\rangle_{N^B} M_2'$, then for all $M_1 \in tr(M_1')$ we have $M_1[t\rangle_N M_2$ for some $M_2 \in tr(M_2')$.

Therefore, TAPN's can be smoothly simulated by transfer nets; as a consequence, any decidable property for these latter preserved by the simulation is also decidable for the former. In particular, we could have used this simulation to prove decidability of coverability, but we preferred to do it in a direct way in order to emphasize the role of the bound $St(N)$, which allow us to reason about the ages of the tokens in a finitary way. We also have the following:

Corollary 24.

(1) Boundedness of MTAPN's is a decidable property.
(2) Structural boundedness of MTAPN's is also decidable.

By applying this simulation we can also conclude the decidability of termination of TAPN's. Nevertheless, this cannot be done in a straightforward way, since the computations of the nets N^B never terminate (because $\mathtt{tick}$ transition has no precondition and therefore it can be executed forever). Instead, we have to study if any infinite computation of this net does not terminate with an infinite suffix containing only $\mathtt{tick}$ transitions. But, due to the structure of these nets, whenever we iterate the execution of $\mathtt{tick}$ for at least $St(N)$ times we reach a stable marking where all the tokens are $St(N)$ old. Besides, it is easy to see that before this happens we can only reach a marking that is bigger than or equal to some previous one in the computation if we have executed some non-$\mathtt{tick}$ transition in between. Therefore, by applying the finite reachability technique in [12], and not taking into account the execution of $\mathtt{tick}$ when it does not change the current marking, we can prove the following:

Corollary 25. *Termination of TAPN's is decidable.*

In order to prove that the undecidability results for transfer nets are also preserved on TAPN's we need the converse simulation. This is a much more elaborated construction, that is somehow based on the simulation of counter machines that was used in [22] to prove undecidability of reachability for MTAPN's, and in [10,11], to prove undecidability of boundedness for reset nets. The main ideas of this simulation are to introduce transitions to represent transfer arcs, and to force the transitions to consume tokens one instant old. Together with the *good* computations also some bad but *non-dangerous* computations will be generated. Besides, older tokens become dead.

Definition 26. *Let $N = \langle P\ T\ F\ TA \rangle$ be a transfer net. We define the associated TAPN, $N^I = \langle P^I\ T^I\ F^I\ \mathtt{times}^I \rangle$, as follows:*

- $P^I = P \cup \{n\} \cup \{r_t \mid t \in T\}$, *where n and each r_t will be control states to capture the different (normal and reset) working states of the net.*
- $T^I = T \cup \{v_{t\ p} \mid p \in P\ t \in T\} \cup \{rn_t \mid t \in T\}$, *where the transitions $v_{t\ p}$ will be used to reset the age of the tokens after the firing of t and transfer them to the corresponding place when there exists some transfer arc $(p\ p') \in TA(t)$. Finally, the transitions rn_t are used to recover the normal state of the net.*
- *The flow function, $F^I \subseteq (P^I \times T^I) \cup (T^I \times P^I)$ is defined as the set*
$$
\begin{aligned}
F^I = F\ &\cup \{(p\ v_{t\ p})\ (r_t\ v_{t\ p})\ (v_{t\ p}\ r_t) \mid p \in P\ t \in T\} \\
&\cup \{(v_{t\ p}\ p') \mid p \in P\ t \in T\ (p\ p') \in TA(t)\} \\
&\cup \{(v_{t\ p}\ p) \mid p \in P\ t \in T\ \neg\exists (p\ p') \in TA(t)\} \\
&\cup \{(r_t\ rn_t)\ (rn_t\ n) \mid t \in T\} \\
&\cup \{(n\ t)\ (t\ r_t) \mid t \in T\}
\end{aligned}
$$
- $\mathtt{times}^I : F^I|_{P^I \times T^I} \to \mathbb{N} \times \mathbb{N}$ *is defined by*
$$
\mathtt{times}^I(p^I\ t^I) = \begin{array}{l} [0\ 0]\ \textit{if}\ p^I \in \{r_t \mid t \in T\} \\ [1\ 1]\ \textit{otherwise.} \end{array}
$$

It is easy to see that N^I can simulate the behaviour on N in the following way: Each marking M of N will be represented by the marking M^I given by

- For each $p \in P$ with $M(p) = k$ we take $M^I(p) = 0^k$, where 0^k denotes the multiset containing k zero values.
- $M^I(n) = 0^1$ and $M^I(r_t) = \varnothing$, for all $t \in T$.

Then, whenever we have $M[t\rangle_N M'$, we can represent in N^I the firing of t by means of the following steps:

(a) $M^I[\varnothing\rangle_1 \mathtt{ages}(M^I\ 1)$, that is, we age one unit all the tokens in M^I.
(b) $\mathtt{ages}(M^I\ 1)[t\rangle_0 M''^I$, that is, we fire (the copy of) t. As a consequence, the tokens in $^\bullet t$ are removed from these places, having age 1, while the tokens in $t^\bullet$ are added to these places, having age 0.
(c) In order to obtain the marking representing M' in N^I, M''^I, we have to transfer the tokens in each $p \in P$, for which there exists some $(p\ p') \in TA(t)$, to the corresponding p'. This is made by firing $M''^I(p)(1)$ copies of $v_{t\ p}$. Besides, for each $p \in P$ for which there is no transfer arc $(p\ p') \in TA(t)$ we have to rejuvenate the tokens in $M''^I(p)$ by firing $M''^I(p)(1)$ copies of the corresponding transition $v_{t\ p}$.
(d) Finally, we recover the "normal" state of N^I by firing rn_t.

It is easy to see that no other transition can be fired until we terminate a step of the simulation by firing the last transition rn_t. Therefore, we only have two possibilities: either we strictly follow the procedure above, thus obtaining M'^I, or we fire rn_t in advance to get some marking M''''^I. In this last case, if we denote by M^{*I} the submarking of M''''^I constituted by the tokens having age 0, we have $M^{*I} \subseteq M'^I$. However, the older tokens in M''''^I, having age 1, become dead, since the only way in which N^I can evolve is by aging M^{*I}, and in such a case these tokens become too old to be used in the future, since no transition in N^I can consume a token older than 1.

So we obtain the following:

Theorem 27. N^I *weak-simulates* N, *which means that by means of the computations of transitions in* N^I *we obtain as reachable markings, when restricting ourselves to tokens having age 0, the representations* M'^I *of the reachable markings in* $(N\ M_0)$ *and their approximations* $M^{*I} \subseteq M'^I$.

As a consequence, most of the undecidability results for transfer nets can be translated to TAPN's. In particular, we have the following:

Proposition 28. *Timed-locally boundedness is undecidable for TAPN's.*

Proof. Since the projections of the markings of the simulating net N^I on the set of tokens having age 0 are just the representations of the reachable markings in N and their approximations, for each $p \in P$ we have that p is bounded in N iff there exists some $k \in \mathbb{N}$ such that for all $M' \in [M^I\rangle_{N^I}$ we have $M'(p)(0) \leq k$. Therefore, since place boundedness is undecidable for transfer nets, we conclude that timed-locally boundedness is also undecidable for TAPN's.   □

In the following section we will present some more consequences of the mutual reductions between transfer nets and TAPN's. Before, we will note that the same construction can be used to simulate reset nets, by only including recovering transitions $v_{t\ p}$ for those places $p \in P$ with $F(p\ t) \in \mathbb{N}$. Of course, in this case, even when we exactly replicate the behaviour of the original net, something is lost; namely, the global boundedness character of the net. This is because we are not able to accurately capture the resetting character of reset arcs: instead of removing the corresponding tokens we just avoid its rejuvenation; so they become dead and thus have no other influence in the forthcoming behaviour of the net, but they remain in the reachable markings, and thus they probably alter the boundedness character of the net. As a consequence of this fact, we do not know yet if place boundedness is or not decidable for TAPN's.

6   Eliminating Dead Tokens

Let us now turn our attention to *dead tokens*. It is obvious that they can be eliminated without affecting the future behaviour of the net, although the reachability set will be affected when doing it. Moreover, the reachable markings with dead tokens are exactly those which are obtained by inserting these dead tokens into the markings that are reachable after their removal. Then, we are interested in eliminating dead tokens, because by doing so we decrease the number of reachable markings, thus slightly reducing the state explosion. At first, we have to exactly define what a dead token is.

Definition 29. *(Dead tokens)*
*Given a TAPN $N = (P\ T\ F$ times$)$ and a marking M of it, we say that a token in M is **dead** if there is no reachable marking $M' \in [M\rangle$ and no transition $t \in T$ such that t is enabled at M' and some of its firings can consume that token.*

We will see that, as a consequence of the decidability of coverability for TAPN's, dead tokens can be effectively detected. However, this is not obtained as an immediate corollary. Instead, we have to slightly change the coding of a TAPN by a transfer net, by adding a clock:

Definition 30.

*(1) We define the **clocked transfer net** N^{CB} which simulates a TAPN N by representing the global clock of N and preserving the age information up to B, by $N^{CB} = \langle P^B \cup \{$clock$\}\ T^B\ F^{CB}\ TA^B\rangle$ where $F^{CB} = F^B \cup \{($tick clock$)\}$.*

(2) We define the ordering $\sqsubseteq_B$ between markings of the net N^{CB} by $M_1 \sqsubseteq_B M_2$ iff $M_1($clock$) = M_2($clock$)$ and $M_1|_{P^B} \subseteq M_2|_{P^B}$.

We have defined in this way the covering ordering $\sqsubseteq_B$ since we need to control the passage of time and this is clearly lost if tick tokens are treated in an accumulative way. But it is clear that $\sqsubseteq_B$ is not a well-quasi-order since the number of tick-tokens is not bounded a priori. Fortunately, it is only necessary to measure time up to a known bound. This can be done by slightly changing the definition of N^{CB}:

Definition 31. *Given $B' \in \mathbb{N}$ we define $N^{B\ B}$ as N^{CB}, but taking*

- $P^{B\ B} = P^B \cup \{c_0\quad c_B\}$,
- $T^{B\ B} = T^B \cup \{$tick$_1\quad$ tick$_B$ tick$_\infty\}$,
- $F^{B\ B} = F^B \cup\{(c_{i-1}$ tick$_i)$ $($tick$_i\ c_i) \mid i \in 1\ B'\}$
  $\cup\{(c_B$ tick$_\infty)$ $($tick$_\infty\ c_B)\}$

Then we define $\sqsubseteq_{B\ B}$, taking $M_1 \sqsubseteq_{B\ B} M_2$ iff
$$M_1|_{P^B} \subseteq M_2|_{P^B} \wedge M_1|_{P^B\ B\ -P^B} = M_2|_{P^B\ B\ -P^B}$$

Now, we define *clock safe markings* as those markings M having $\overset{B}{\underset{i=0}{}} M(c_i) = 1$, so it is clear that if we take as initial marking any M_0 with $M_0(c_0) = 1$ and $M_0(c_i) = 0\ \forall i \neq 0$, all the reachable markings are clock safe. And if we restrict $\sqsubseteq_{B\ B}$ to these clock safe markings, it is easy to see that the corresponding transition system is well structured.

Let us now consider a token $(p\ i)$ in a given marking of a TAPN N that we take as its initial marking M_0. If it is not dead then we have some reachable marking $M_0[\overline{R}\rangle_m M$, where m denotes the total ellapsed time, and some firable t, $M[t\rangle$, with $(p\ t) \in F$ and $i + m \in \text{times}(p\ t)$. As a consequence, we have the following:

Proposition 32. *For each token $(p\ i)$ on the initial marking M_0 of a MTAPN N and each $t \in T$ with $(p\ t) \in F$, there exists some reachable marking $M_0[\overline{R}\rangle_m M$ with $M[t\rangle$ and $i + m \in \text{times}(p\ t)$ iff there exist some $m \in 0\ St(N)$ with $i + m \in \text{times}(p\ t)$ and some $M' \in \mathcal{M}(N^{St(N)\ St(N)})$ with $M'(c_m) = 1$ and $M'[t\rangle$, such that M' is coverable from $M_0^{B\ B}$ with respect to $\sqsubseteq_{St(N)\ St(N)}$, where $M_0^{B\ B}$ is defined as M_0^B, adding $M_0^{B\ B}(c_0) = 1$.*

Corollary 33. *We can decide when a token of a marking of a TAPN is dead.*

Proof. We only have to apply Th.13 for each one of the minimal markings of the set $\{M \in \mathcal{M}(N^{St(N)\ St(N)}) \mid M[t\rangle$ and $M(c_m) = 1$ with $i + m \in \text{times}(p\ t)\}$ □

We can also consider a weaker version of dead tokens. The reason why we introduce this new notion is because by means of dead tokens we wanted to capture the fact that they are useless, since they cannot be used to fire any transition in the future. However, when we have several identical tokens in the same place it is possible that some (but not all) of them can be used to fire such a transition. If this were the case, under our Def.29 any such token would be said to be *non-dead*. We need a weaker notion if we want to say that some of these identical tokens is indeed (weakly) dead.

Definition 34. *(Weakly dead tokens) Given a TAPN $(P\ T\ F\ \text{times})$ and a marking M of it, a token in M at the place p is* **not**[8] **weakly dead** *if there is some reachable marking $M' \in [M\rangle$ such that M' contains at p a single token as old* [9] *as the given one, and there is some enabled transitions at M' whose firing can consume that token.*

The detection of weakly dead tokens in TAPN's turns out to be a rather pathological task, as the following result shows:

Theorem 35.

(1) *It is not possible to effectively detect weakly dead tokens with age 0.*

(2) *We can effectively detect any weakly dead token having a positive age.*

Proof.

(1) We use the fact that reachability in TAPN's can be reduced to zero-reachability of a single given place. This is proved in a similar way as for classic P/T nets.

---

[8] We define this concept in a negative way in order to have a more clear definition
[9] Taking into account the fact that the given token has possibly grown during this computation.

(2) (sketch) We consider the transfer net N^B simulating N, defined as in
Def.21, but introducing into it clock places, as in Def.31, by means of which
we can control when a token from the same place and having the same
age that the given one, is consumed by the firing of a transition. Then we
introduce a counter place where we put a token each time that happens.
Then, it is clear that the given token is weakly dead iff we can put in this
counter place (at least) as many tokens as we had in the given marking at
the same place having the same age as the original one. Since coverability is
decidable for TAPN's we conclude that, in this case, weakly dead tokens are
effectively detectable.

$\square$

The reason why the situation is different depending on the age of the given
token is that whenever we create a new token its age is 0. Therefore, when the
age of the given token is positive, we know in advance how many tokens like it
have to be consumed in order that the given one will be necessary to fire some
new transition. Instead, when the age is 0, we could create some new copies of
the given token before the marking is aged, thus making impossible to control
when the last clone of the given token is consumed.

7   Conclusions, Discussion, and Future Work

We have studied timed-arc Petri nets, for which we have proved that coverability,
termination and boundedness are decidable, but timed-locally boundedness is
not. We have also seen that *dead tokens* can be effectively detected. All of this is
due to the close relationship between TAPN's and transfer nets, because enabling
of transitions is closely related to coverability.

Besides place boundedness, there are some other timed properties of TAPN
in which we are interested but unfortunately we do not know yet if they are
decidable or not. For instance, we could say that N is *non-Zeno* if it has no
infinite computation with finite duration, i.e. such that, the number of aging
steps in it is finite.

If we take into account the relationship between TAPN's and transfer nets
we see that in order to decide this kind of properties we should be able to decide
on transfer nets properties like:

*Is there any infinite computation along which only finitely many instances of
transitions having transfer arcs are executed?*

This kind of properties have been proved to be decidable for plain nets. But
the technique to do it is to characterize infinite computations by means of looping
ones. This cannot be done for transfer nets since the coverability tree does not
give enough information to characterize the behaviour of such a net.

So, in order to extend our knowledge about decidability of properties of
TAPN's, or equivalently on transfer nets, we have to study which are the logics
for which the model checking problem is decidable for this kind of nets. In
order to obtain these logics we should take into account that the corresponding

decidability proofs should be based on the use of the *backward* technique in [2,12], instead of using the coverability tree.

Concerning the possibility of extending our results to continuous time, we conjecture that it can be done in the case where the limits of the intervals defining the age of the tokens to be consumed by the firing of a transition are integer (or rational) values, but allowing arbitrary passage of time along computations to be in $\mathbb{R}$. The idea to prove such a result would be to take into account that the age of a token is only defined by the instant at which it is produced. Then, we should prove that any timed step sequence could be represented by some *equivalent* one where aging steps are defined by integer values.

If this conjecture is proved to be correct, we should look for a continuity argument by means of which these results for the rational case could be extended to arbitrary TAPN's whose firing intervals are defined by real values.

We already succeeded on the extension of decidability of coverability for timed Petri nets to the rational and real cases [21], although the ideas there applied cannot be directly translated to the case of timed-arc Petri nets. It could be the case that such a generalization would be not possible in this framework. In fact, in [21] we can also find some result for the discrete time case whose proof cannot be translated to dense time domains. But perhaps, by an alternative proof one could get such a generalization.

References

1. W.M.P. van der Aalst. *Interval Timed Coloured Petri Nets and their Analysis.* LNCS vol. 691, pp. 451-472. 1993.
2. P.A. Abdulla, K. Čerāns, B. Jonsson and T. Yih-Kuen. *General Decidability Theorems for In nite-State Systems.* Proc. 11th IEEE Symp. Logic in Computer Science (LICS'96), New Brunswick, NJ, USA, July 1996, pages 313-321. 1996.
3. M. Ajmone Marsan, G. Balbo, A. Bobbio, G. Chiola, G. Conte and A. Cumani. *On Petri Nets with Stochastic Timing.* Proc. of the International Workshop on Timed Petri Nets, IEEE Computer Society Press, pp. 80-87. 1985.
4. A. Arnold and M. Latteux. *Vector Addition Systems and Semi-Dyck Language.* Research Report 78, Laboratoire de Calcul de l'Université des Sciences et Techniques de LILLE, December 1976.
5. T. Bolognesi and P. Cremonese. *The Weakness of Some Timed Models for Concurrent Systems.* Technical Report CNUCE C89-29. CNUCE-C.N.R. October 1989.
6. T. Bolognesi, F. Lucidi and S. Trigila. *From Timed Petri Nets to Timed LOTOS.* Proc. of the 10th International IFIP WG6.1 Symposium on Protocol Specification, Testing and Verification, North-Holland. 1990.
7. F.D.J. Bowden. *Modelling Time in Petri Nets.* Proc. Second Australia-Japan Workshop on Stochastic Models. 1996.
8. A. Cerone and A. Maggiolo-Schettini. *Time-based Expressivity of Time Petri Nets for System Speci cation.* Theoretical Computer Science (216)1-2, pp. 1-53. 1999.

9. C. Dufourd, A. Finkel and Ph. Schnoebelen. *Reset Nets between Decidability and Undecidability.* Proc. 25th. Int. Coll. Automata, Languages, and Programming (ICALP'98), Aalborg, Denmark, July 1998, LNCS vol. 1443, pp:103-115. Springer-Verlag, 1998.

10. C. Dufourd. *Reseaux de Petri avec Reset/Transfert: Decidabilite et Indecidabilite.* Thése de Docteur en Sciences de l'École Normale Supérieure de Cachan. October 1998.

11. C. Dufourd, P. Jančar and Ph. Schnoebelen. *Boundedness of Reset P/T Nets.* Proc. 26th. Int. Coll. Automata, Languages, and Programming (ICALP'99), Prague, Czech Rep., July 1999, LNCS vol. 1644, pp. 301-310. Springer-Verlag, 1999.

12. A. Finkel. *Reduction and Covering of In nite Reachability Trees.* Information and Computation, 89(2):144-179, 1990.

13. A. Finkel and Ph. Schnoebelen. *Fundamental Structures in Well-Structured In nite Transition Systems.* Proc. 3rd Latin American Theoretical Informatics Symposium (LATIN'98), Campinas, Brazil, Apr. 1998, LNCS vol. 1380, pp:102-118. Springer-Verlag, 1998.

14. H.-M. Hanisch. *Analysis of Place/Transition Nets with Timed-Arcs and its Application to Batch Process Control.* Application and Theory of Petri Nets, LNCS vol. 691, pp:282-299. Springer-Verlag, 1993.

15. R. Mayr. *Lossy Counter Machines.* Technical report TUM-I9827, TU-München, 1998.

16. P. Merlin. *A Study of the Recoverability of Communication Protocols.* PhD. Thesis, Univ. of California. 1974.

17. J. L. Peterson. *Petri net Theory, and the Modeling of Systems.* Prentice-Hall. 1981.

18. J. Quemada, A. Azcorra, and D. de Frutos. *A Timed Calculus for LOTOS.* Proc. FORTE'89, pp. 245-264. 1989.

19. C. Ramchandani. *Performance Evaluation of Asynchronous Concurrent Systems by Timed Petri Nets.* PhD. Thesis, Massachusetts Institute of Technology, Cambridge. 1973.

20. J. Sifakis. *Use of Petri Nets for Performance Evaluation.* Proc. of the Third International Symposium IFIP W.G.7.3., Measuring, Modelling and Evaluating Computer Systems. Elsevier Science Publishers, pp. 75-93. 1977.

21. V. Valero, D. de Frutos Escrig, and F. Cuartero. *Decidability of the Strict Reachability Problem for TPN's with Rational and Real Durations.* Proc. 5th. International Workshop on Petri Nets and Performance Models, pp. 56-65. 1993.

22. V. Valero, D. de Frutos Escrig, and F. Cuartero. *On Non-Decidability of Reachability for Timed-Arc Petri Nets.* Proc. 8th. International Workshop on Petri Nets and Performance Models, pp. 188-196. 1999.

23. B. Walter. *Timed Petri-Nets for Modelling and Analysing Protocols with Real-Time Characteristics.* Proc. 3rd IFIP Workshop on Protocol Specification, Testing and Verification, North-Holland. 1983.

Analysing the WAP Class 2 Wireless Transaction Protocol Using Coloured Petri Nets

Steven Gordon and Jonathan Billington

Cooperative Research Centre for Satellite Systems,
University of South Australia,
Mawson Lakes SA 5095, Australia
sgordon,jb @spri.levels.unisa.edu.au

Abstract. Coloured Petri nets (CPNs) are used to specify and analyse the Class 2 Wireless Transaction Protocol (WTP). The protocol
provides a reliable request/response service to the Session layer in the
Wireless Application Protocol (WAP) architecture. When only a single
transaction is considered occurrence graph and language analysis reveals
3 inconsistencies between the protocol and service speci cation: (1) the
initiator user can receive two TR-Invoke.cnf primitives; (2) turning User
Acknowledgement on doesn't always provide the User Acknowledgement
service; and (3) a transaction can be aborted without the responder user
being noti ed. Based on the modelling and analysis, changes to WTP
have been recommended to the WAP Forum[SM].

1 Introduction

Petri nets are a proven technique for the design and verification of communication protocols [4, 15]. They provide the ability to: model at different levels of
abstraction; capture the concurrent behaviour inherent in communication protocols; and formally analyse them. This can give a high degree of confidence in the
protocol specification and design, which is important given that protocols are
an integral part of the telecommunications and computing infrastructure. This
has become more important with the emergence of e-commerce applications.

In this paper we present an initial analysis of the Class 2 Wireless Transaction
Protocol (WTP) using Coloured Petri nets. To provide context, we introduce the
Wireless Application Protocol, which includes WTP as part of its architecture.

1.1 Wireless Application Protocol

The Wireless Application Protocol (WAP) [24] defines an architecture that aims
to support the provision of Internet and advanced information services to mobile users via a wide range of predominantly hand-held devices. An example
application of WAP is to perform Web browsing on a mobile phone.

WAP is designed to take into account the limitations of the devices and the
wireless data networks they utilise. The architecture specification is defined by

M. Nielsen, D. Simpson (Eds.): ICATPN 2000, LNCS 1825, pp. 207 226, 2000.
© Springer-Verlag Berlin Heidelberg 2000

the WAP Forum^{SM}, an industry consortium of wireless service providers, device manufacturers, software companies and infrastructure and content providers.

WAP is based on the World Wide Web (WWW) programming model: requests are made to a server, via a gateway, which generates an appropriate response (e.g. the content of a Web page). This allows the experience and tools used in Web applications to be carried over to WAP applications. The gateway, which is not part of the WWW programming model, is used in WAP to perform content encoding/decoding and protocol conversion between the WAP stack and the Internet's transport protocol TCP/IP. Hence the gateway would be located at a base station – it is the interface between the wireless network (WAP) and the wired network (TCP/IP).

To provide a scalable and extensible architecture, WAP is designed in layers[1] (Fig. 1). From the bottom, the transport layer operates over a wide range of wireless bearer services. These include the GSM Short Message Service and General Packet Radio Service, CDMA Circuit Switched Data, Cellular Digital Packet Data (CDPD), and several proprietary protocols. There are three other protocol layers, security, transaction and session, and an application layer, the Wireless Application Environment (WAE). WAE defines a Wireless Markup Language (WML) and an accompanying scripting language (WMLScript) that are optimised to suit the limited display sizes of the browsing devices and low wireless link capacities.

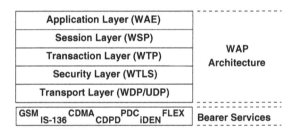

Fig. 1. WAP architecture

1.2   Previous Work

Formal methods can be applied to nearly all steps in the protocol engineering methodology – from the high level architecture design to automatic implementation and conformance testing [3]. We are interested in the analysis of the service and protocol specifications, and verification of the protocol against the service. The first phase, modelling and analysis of the Transaction service [11], included generation of the service language, i.e. the sequences of primitive events at both

---

[1] The speci cations for each layer of the WAP architecture are available via www.wapforum.org. Throughout this paper we refer to WAP Version 1.1. However for WTP, Draft Version 11-June-1999 is the basis of the analysis. This has been accepted as WTP Version 1.2.

service user interfaces. In this paper we focus on analysing the Transaction protocol and verifying its conformance to the service via a comparison of the service languages.

Transport protocols, such as TCP [20] and the OSI Transport Protocol [13], have been formally analysed using various techniques (e.g. Petri nets [2, 8, 17], Estelle [6], and systems of communicating machines [16]). However, formal analysis of transaction protocols (which can be thought of as reliable transport protocols, optimised for transactions) is limited. The only other work we are aware of is the analysis of Transaction/TCP (T/TCP) [5], a protocol designed for providing an efficient transaction service as well as stream-oriented data transfer (i.e. TCP) in the Internet. T/TCP has been specified using timed and untimed automaton models [22], and demonstrates that T/TCP doesn't provide the same service as TCP. Follow on work [23] has shown the dependence of T/TCP on accurate clocks for transaction protocols to provide efficient, reliable transactions.

Although some features are similar, there are substantial differences between the service and protocols of WTP and T/TCP to warrant the formal verification of WTP. We use Coloured Petri nets (CPNs) [14] to specify and analyse the functional properties of WTP. Previous experience and tool support (i.e. Design/CPN [18]) for modelling, simulation and (functional and performance) analysis makes CPNs a suitable choice for this task. To obtain manageable occurrence graphs, the analysis is done under restricted conditions (most notably only a single transaction is analysed). Three inconsistencies between the service and protocol specifications are detected, identifying areas for improvement of the WTP design.

1.3   Overview

The remainder of this paper is organised as follows: Section 2 introduces the Transaction service and protocol. Section 3 presents the Transaction service language. Section 4 describes the design of the protocol CPN and analysis results are given and discussed in Section 5. Section 6 concludes with a summary and areas of future work.

2   Wireless Transaction Protocol

The Wireless Transaction Protocol (WTP) [25] provides 3 classes of service:

Class 0: unreliable invoke message with no result message. This is the same datagram service as provided by the Transport layer. It is included so a datagram can be sent during a session. The User Datagram Protocol (UDP) [19] or Wireless Datagram Protocol (WDP) are recommended to be used if applications require a datagram service.

Class 1: reliable invoke message with no result message. This can be used to provide push functionality in, for example, the Wireless Session Protocol (WSP). Within the context of an existing session the server can push data to the client, which the client then acknowledges.

Class 2: reliable invoke message with one reliable result message. This is the basic transaction service.

This section briefly describes the Class 2 Transaction service and protocol.

2.1 WTP Service

The WTP service primitives and the possible types are: TR-Invoke – req (request), ind (indication), res (response), cnf (confirm); TR-Result – req, ind, res, cnf; TR-Abort – req, ind. A transaction is started by a user issuing a TR-Invoke.req primitive. This user becomes the initiator of the transaction and the destination user becomes the responder. The responder must start with a TR-Invoke.ind. Table 1 shows the primitives that may be immediately followed by a given primitive at either end point. For example, at the initiator a TR-Invoke.req can be followed by a TR-Invoke.cnf, TR-Result.ind, TR-Abort.req or TR-Abort.ind. Further details on the service can be found in [25].

Table 1. Primitive sequences for WAP Transaction Service at each end point

	TR-Invoke				TR-Result				TR-Abort	
	req	*ind*	*res*	*cnf*	*req*	*ind*	*res*	*cnf*	*req*	*ind*
TR-Invoke.req										
TR-Invoke.ind										
TR-Invoke.res		X								
TR-Invoke.cnf	X									
TR-Result.req		X*	X							
TR-Result.ind	X*			X						
TR-Result.res						X				
TR-Result.cnf					X					
TR-Abort.req	X	X	X	X	X	X	X			
TR-Abort.ind	X	X	X	X	X	X	X			

Note: the primitive in each column may be immediately followed by the primitives marked with an X. Those marked with an X* are not possible if the User Acknowledgement option is used.

Each of the primitives has several mandatory and optional parameters. The TR-Invoke request and indication must include both source and destination addresses and port numbers. Other parameters are: User Data, Class Type, Exit Info, Handle, Ack Type and Abort Code. Of special significance is Ack Type. This parameter is used to turn on or off the User Acknowledgement (User Ack) feature. When on, an explicit acknowledgement of the invoke is necessary (i.e. TR-Invoke.res and TR-Invoke.cnf). Otherwise, the result may implicitly acknowledge the invoke. Fig. 2 gives two example sequences of primitive exchanges for a successful transaction. The sequence in Fig. 2(b) is not possible if User Ack is on.

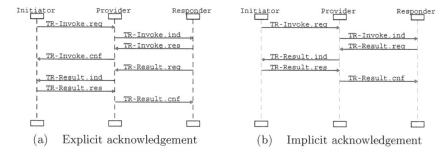

(a)   Explicit acknowledgement   (b)   Implicit acknowledgement

Fig. 2. MSC of service primitives for successful transaction

2.2   Protocol Features

Messages sent between peer protocol entities are called Protocol Data Units (PDUs) [12]. There are four primary PDUs used in the Transaction protocol: Invoke, Result, Ack and Abort. (There are 3 other PDUs available if the optional segmentation and re-assembly feature is used – this is discussed later.) Each PDU contains an integral number of octets and consists of a header, containing a fixed and variable component, and the data, if present. The relevant header fields of PDUs are given when describing the CPN model (Section 4.6).

The procedure for normal message transfer in a transaction involves 5 steps:

1. Upon receipt of a TR-Invoke.req from the user, the initiator transaction protocol entity sends an Invoke PDU to the responder. The initiator starts a retransmission timer and waits for a response.
2. Upon receipt of the Invoke PDU the responder sends a request to its user (TR-Invoke.ind) and waits for a result. The PDU includes a Transaction Identifier (TID) which is used for the remaining PDUs in the transaction. This allows several transactions to be processed concurrently. The protocol provides a mechanism for dealing with the receipt of delayed PDUs, i.e. PDUs that contain unexpected TIDs [25]. If the TID is expected, the responder can notify the user (i.e. give a TR-Invoke.ind) and proceed with the transaction. The receipt of a PDU with an unexpected TID initiates a handshake to verify if the delayed PDU should be processed or not. This involves two steps:

   (a) The responder sends an Ack PDU with a verification flag set. This asks the initiator if it has an outstanding transaction.
   (b) The initiator sends an Ack PDU with the verification flag set if the transaction is in progress, otherwise an Abort PDU is sent which indicates to the responder that the TID can be ignored.

3. While waiting for the result (from the user), the responder may send a "hold on" Ack PDU to the initiator if the responding user is taking too long to acknowledge the Invoke PDU. Then the initiator knows not to retransmit the Invoke PDU. In this case a TR-Invoke.cnf is generated at the initiator.

4. A TR-Result.req primitive from the user allows the responder to send the Result PDU. Upon receipt of the Result PDU by the initiator a TR-Invoke.cnf (if not already sent) and TR-Result.ind are passed to the user.
5. The initiator acknowledges the result by sending an Ack PDU to the responder. The initiator must either wait for a timeout before removing any transaction information or save the transaction history so it can handle a retransmission if necessary.

Reliability in WTP is provided by retransmission of PDUs until acknowledgements are received. A timer and retransmission counter are used so when the number of timeouts (and retransmissions) reaches a maximum value, the transaction is aborted. There is a maximum value for retransmitting any PDU (RCR_MAX) and for retransmitting acknowledgements (AEC_MAX). For example, assuming RCR_MAX is 1, the initiator starts a timer after sending the Invoke PDU and if no response has been received when the timer expires the PDU is retransmitted and RCR is set to 1. If again no response is received before the timeout, the initiator will abort the transaction (as RCR = RCR_MAX).

Other features of WTP include: concatenation and separation of PDUs into one service data unit, transmission of protocol parameters via transport information items (TPIs) and an optional protocol feature for segmenting and reassembling PDUs into multiple packets.

The protocol operation is described by a set of state tables in the WTP Specification [25]. The initiator and responder each have tables representing states they can be in. Each table has an event, a condition, a set of actions and the next state the entity enters. The actions have been assumed to be ordered and sequential top to bottom. In Table 2, for example, while in the LISTEN state, if the responder receives an Invoke PDU with a valid TID and User Ack is not used (U/P==false), it will generate a TR-Invoke.ind to the user, then start the timer for waiting for a response, and set the Uack variable to false. The resulting state will be INVOKE RESP WAIT.

Table 2. State table entry for responder in LISTEN state

Responder LISTEN			
Event	Condition	Action	Next State
RcvInvoke	Class == 2 1	Generate TR-Invoke.ind	INVOKE RESP
	Valid TID	Start timer, A	WAIT
	U/P==False	Uack = False	

3   Transaction Service Specification

The WAP Transaction Service has been modelled and analysed using Coloured Petri nets [11]. The model allows us to locate deficiencies in the service specification and generate it's language, the possible sequences of primitives between the service user (Session layer) and the service provider (Transaction layer). For the purpose of comparison with the protocol (see Section 5.2), the Transaction Service language from [11] is presented in Fig. 3.

The Transaction Service language has 21 nodes and 74 arcs. There are four halt states in the language: nodes 6, 16, 17 and 19 (shown in bold). Node 6 represents the case when the initiator's TR-Invoke.req is immediately followed by an abort. Node 16 represents the case when the initiator has finished and the responder has also finished or aborted. Nodes 17 and 19 represent the cases when the transaction is aborted. The primitives between the following nodes are not possible when User Ack is turned on: (2,3), (2,7), (2,13), (12,14), (8,17), (3,20). In addition, the primitives that were between 2 and 13 are now between 2 and 9.

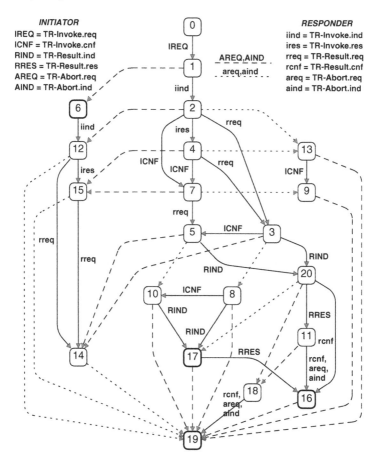

Fig. 3. Transaction Service language

4   Transaction Protocol Specification

To verify the operation of WTP, a CPN model of the protocol specification has been developed [9]. It consists of 12 pages, 55 transitions, 9 places and 21 colour sets. The necessary components for protocol verification, as shown in Fig. 4, are modelled (some at an abstract level). This enables us to gain further insight into the protocol operation, verify general properties, and compare it with the service specification. This Section describes the design decisions and assumptions made in the modelling process, using selected parts of the CPN for explanation. A complete CPN model [9] of the protocol is too large to include in this paper. The aim is to present the main ideas behind the model and to present the analysis results.

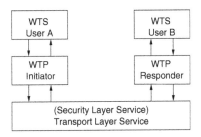

Fig. 4. Block diagram for protocol verification

4.1   Net Structure

The WTP Specification [25] defines the procedures of the protocol using state tables (as described in Section 2.2). We chose to model each table as a separate CPN page. The advantages of this approach are: it's relatively easy to transfer the table information into a CPN (see Section 4.2); the direct relationship between our CPN and WTP simplifies validation of the model by us and others; and the repetitive structure allows a manageable graphical layout of the net. A disadvantage is a lack of hierarchy in the model, and, hence, the ability to view the protocol operation from a higher level of abstraction. We feel the use of non-CPN graphics with the model (i.e. message sequences charts in Design/CPN) can compensate for this.

There are 11 top-level pages in the model as shown in Fig. 5. All pages are connected via fusion places. The first letter of the page name indicates whether it is the *I*nitiator or *R*esponder, while the remainder indicates the state the page represents. The twelfth page, I_RESULT_WAIT_RcvResult, a sub-page of I_RESULT_WAIT, is explained in Section 4.7.

4.2   Page Structure

Each CPN page is based on the following structure (see Fig. 6 for an example):

Two fusion places, InitToResp and RespToInit, to represent the medium for transporting messages between the initiator and responder, and vice versa, respectively. The medium is further described in Section 4.5.

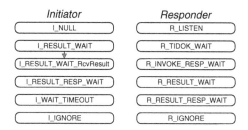

Fig. 5. Protocol CPN hierarchy page

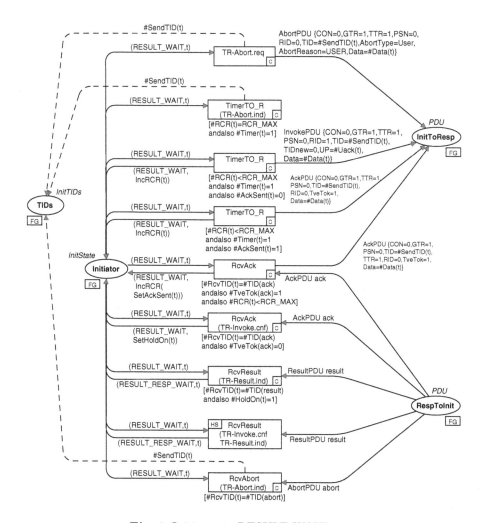

Fig. 6. Initiator in RESULT WAIT state

A fusion place that stores the current state of a protocol entity (i.e. the initiator or responder), and the context associated with that state for each transaction. The initiator therefore has a place Initiator and the responder a place Responder. The representation of the state (and transaction context) is explained in Section 4.3.

Transitions that represent events and actions from the state tables. For all top-level pages (except I_NULL – see Section 4.4), every transition has an input arc from the state place. This restricts transitions on a page to only be enabled when the entity is in the state represented by that page. Along with guards, input arcs from other places (e.g. InitToResp) can be thought of as conditions of the events. The output arcs represent an action occurring and/or state changing.

Some pages have additional places to keep track of counter values etc. For example, in Fig. 6 place TIDs stores all transaction identifiers not in use.

The WTS users (Fig. 4) are modelled at an abstract level via the primitive events. Actions in the protocol entities that represent a submit (or deliver) of a primitive from (or to) a user are indicated by the primitives as labels on the relevant transitions. For example, in Fig. 6 there are six transitions (the top two and the bottom four) that indicate communication with User A via primitives.

4.3   State of Protocol Entities

WTP supports multiple, asynchronous transactions. In the CPN, this is modelled by storing the state and context of each transaction in progress at the initiator and the responder (in places Initiator and Responder, respectively). The transactions are differentiated by their TIDs. The state of a transaction corresponds with a state table in WTP:

color States = with WAIT_TIMEOUT   RESULT_WAIT   RESULT_RESP_WAIT
    LISTEN   TIDOK_WAIT   INVOKE_RESP_WAIT;

The initiator has 4 states (its initial state (see Section 4.4) and the first 3 colours) and the responder has 5 states (the last 5 colours).

The context (TransData record) stores variables associated with each transaction. The variable t (in Figs. 6 and 8) is of this type.

color TransData = record
  SendTID:Uint16 * (* TID to send   0.. TID_MAX *)
  RcvTID:Uint16   * (* TID expected to receive   0.. TID_MAX *)
  HoldOn:Flag   * (* True if HoldOn ack received   0/1 *)
  Uack:Flag   * (* True if User Ack requested   0/1 *)
  AckSent:Flag   * (* True if Ack(TIDok/TIDve) sent   0/1 *)
  RCR:RCR_c   * (* Retransmission Counter   0..RCR_MAX *)
  AEC:AEC_c   * (* Ack Expiration Counter   0..AEC_MAX *)
  Data:Counter   * (* Data   int *)
  Timer:Flag;   (* True if Timer on   0/1 *)

The first five entries correspond to variables used by WTP at both the initiator and responder. RCR and AEC are counters used by WTP. Data is used to give each transaction a unique identifier in case of errors (see Section 4.4). As there is no time in the model, Timer has been introduced to indicate whether the timer is on (and hence a timeout can occur) or off.

We have assumed the counters can never be greater than their maximum value. This may seem obvious but an action of one entry in the Initiator RESULT WAIT state table increments RCR, but there is no condition stating RCR $<$ RCR_MAX. We have introduced this condition, as shown in Table 3 (the change is italicised).

Table 3. State table action with new condition limiting RCR

Initiator RESULT WAIT			
Event	**Condition**	**Action**	**Next State**
RcvAck	TIDve	Send Ack(TIDok)	RESP WAIT
	Class=2 1	Increment RCR	
	RCR<RCR_MAX	Start timer, R [RCR]	

The places Initiator and Responder are typed as follows:

color InitState = **product** States * TransData;
color RespState = **product** States * TransData;

Therefore, if there were two transactions in progress for example, the marking of Initiator may be:

1'(WAIT_TIMEOUT, SendTID=0, RcvTID=2, ...) +
1'(RESULT_WAIT, SendTID=1, RcvTID=3, ...)

4.4   Initialisation

The page I_NULL (Fig. 7) represents the state of the initiator before a transaction has begun. The state is modelled implicitly i.e. there is no colour NULL in the colour set States. Instead, the user can issue a TR-Invoke.req as many times as there are initial tokens in User. These tokens are parameters for each transaction. In Fig. 7 only 1 transaction can be initiated. Neither the continue (CON) option (used to identify TPIs) nor the User Ack are used.

Place TIDs stores all TIDs that are not outstanding at the initiator. This is necessary in the model to detect when old PDUs (i.e. those with a TID that is not outstanding) are received by the initiator. Place Data maintains an integer counter. This is used to generate the next TID (which is maintained as the variable GenTID in WTP), and also generate a unique data value for each transaction. Because the TID values wrap (i.e. in erroneous conditions, two transactions may have the same TID), the data value will be necessary to differentiate transactions when analysing the effect of errors on the protocol.

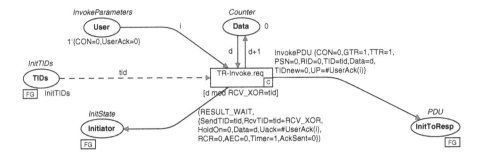

Fig. 7. Initiator in NULL state

The occurrence of TR-Invoke.req sends an InvokePDU to InitToResp and places the initiator into the RESULT_WAIT state.

Fig. 8 is the CPN page for the responder in the LISTEN state (i.e. it is waiting to receive invocations from the initiator). The three common fusion places are present (InitToResp, RespToInit and Responder), with Responder initialised to AllListen. This is a constant that says the responder can accept n asynchronous invocations, where n is double the window size at the responder. Place LastTID keeps track of the TID of the last Invoke PDU received (to store only one TID is an assumption of our current model – in practice an array of TIDs may be stored to allow Invoke PDUs to be received out-of-order). Note for the model we have set TID_MAX to 3, and for a successful receipt of an Invoke PDU, LastTID is initialised as 1. The LastTID, along with the TID of the current Invoke PDU received, is used by the windowing mechanism (implemented by TIDTest() in the guards of the two RcvInvoke transitions) to determine if Invokes are received as expected. The occurrence of the top RcvInvoke indicates the TID is expected and the transaction can proceed. The bottom RcvInvoke indicates an Invoke has been received out-of-order and a verification must occur with the initiator.

4.5   Transport Medium

From Fig. 4 the security service is provided to WTP. However, this is an optional layer, and does not alter the service (in terms of primitives) of the underlying Transport layer. Therefore, the CPN model of the transport medium must reflect the properties of the Transport layer protocols, namely WDP or UDP [19]. These datagram protocols cannot guarantee in-order delivery, removal of duplicates or loss-free delivery of messages.

Initially, the transport medium has been modelled as a single place for each direction of communication (InitToResp and RespToInit). This models the unordered characteristics of the medium. The capacity of the medium is assumed infinite. Errors in the transport medium (loss of PDUs, duplicates) have not been modelled. This is an area of future work.

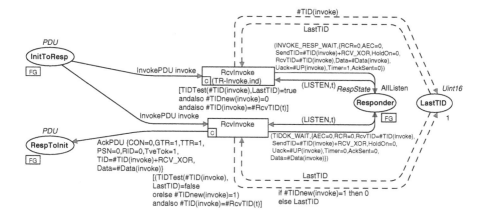

Fig. 8. Responder in LISTEN state

4.6   Protocol Data Units

PDUs are modelled as records, with entries corresponding to header fields defined in WTP. Only fields that may affect the protocol operation are included (e.g. the version field is excluded). The Invoke PDU definition is:

```
color InvokePDU_c = record
  CON:Flag      * (* Continue    0/1 *)
  GTR:Flag      * (* Group Trailer    0/1 *)
  TTR:Flag      * (* Transmission Trailer    0/1 *)
  PSN:PSN_c     * (* Packet Sequence Number    0..255 *)
  RID:Flag      * (* Retransmission Indicator    0/1 *)
  TID:Uint16    * (* Transaction Identi er   0..TID_MAX *)
  Data:Counter  * (* Data    int *)
  TIDnew:Flag   * (* Indicates wrapping of TID    0/1 *)
  UP:Flag;        (* True if User Ack on    0/1 *)
```

The first field (CON) is used in identifying TPIs and the next three are concerned with segmentation and re-assembly of PDUs. Although these protocol features (along with concatenation and separation) are not modelled, the fields are included for future use. The first 7 fields are common to all PDUs. The Result PDU has no more fields. The fields of the other PDUs are:

```
color AckPDU_c = record

. . .

  TveTok:Flag;      (* TID veri cation/TID Ok    0/1 *)
color AbortPDU_c = record

. . .
```

AbortType:AbortType_c *   (* Provider/User *)
AbortReason:AbortReason_c;   (* UNKNOWN, PROTOERR, ... *)

The colour set of the transport medium places is:

color PDU = **union** InvokePDU:InvokePDU_c + ResultPDU:ResultPDU_c +
            AckPDU:AckPDU_c + AbortPDU:AbortPDU_c;

4.7 Correspondence to Primitives

The actions specified in the WTP state tables have been modelled as atomic events. That is, for example, the generation of TR-Invoke.ind, then starting the timer and then setting Uack to False in Table 2 is modelled as the occurrence of transition RcvInvoke (TR-Invoke.ind) in Fig. 8. This allows each primitive event to be directly associated with an individual transition, and hence an arc in the OG. This is important when performing analysis because to compare the protocol and service specifications it is necessary to select arcs in the OG that correspond to primitive events (see Section 5.2). There is one exception to this case, when an action in the initiator RESULT_WAIT state table generates two primitives (Table 4). If this was modelled as a single transition, we wouldn't be able to differentiate between the two primitives when selecting arcs in the OG. Therefore, this action is decomposed into a sub-page (denoted by HS in the transition RcvResult in Fig. 6) that contains a transition for each primitive (Fig. 9).

Table 4. State table action that generates two primitives

Initiator RESULT WAIT			
Event	**Condition**	**Action**	**Next State**
RcvResult	Class == 2 HoldOn==False	Stop timer Generate TR-Invoke.cnf Generate TR-Result.ind Start timer, A	RESULT RESP WAIT

5 WTP Analysis

Occurrence graph analysis has been used to locate any undesired behaviour in the protocol specification. The OG has been reduced to a language of possible primitive events in a similar manner as the Transaction Service language [11]. A comparison of the two languages reveals errors in the protocol specification. This Section reports on these analysis results and suggests how the errors may be eliminated.

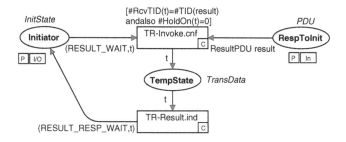

Fig. 9. Initiator in RESULT_WAIT state - sub-page for primitives

5.1   Protocol Occurrence Graph Analysis

The protocol CPN has been analysed using a single transaction. Also the maximum counter values (RCR_MAX and AEC_MAX) have been set to 1 or 2. This allows us to verify the basic operation of the protocol, as we can obtain a reasonable sized OG. The OG results with and without user acknowledgement are shown in Table 5.

Table 5. Protocol specification OG results

$OGNo$	RCR_Max	AEC_Max	$LastTID$	$UserAck$	$Nodes$	$Arcs$	$Time(s)$	TS	DTI
1	1	1	1	O	1634	6472	34	1	10
2	1	1	1	On	2321	9454	56	1	8
3	1	2	1	O	1634	6472	33	1	10
4	1	2	1	On	3350	14255	104	1	4
5	2	1	1	On	31290	165257	5950	1	4
6	2	2	1	O	19083	96356	5680	1	6
7	2	2	1	On	50491	278591	17156	1	4
8	1	1	0	On	542	2028	7	1	8

Note: TS = Terminal States, DTI = Dead Transition Instances.

The protocol terminated correctly in all cases (i.e. the single terminal state was desired). There were dead transition instances present, but they were expected. They were caused by features being modelled but not exercised, due to initial conditions. The transitions were related to the following features: User Ack; TID verification; and receiving incorrect PDUs. When User Ack is off, AEC is not used. Hence, as they only differ by the maximum AEC counter, OG number 1 and OG number 3 are identical.

OG No. 8 is the result of analysing the case of TID verification occurring. With the receipt of an Invoke PDU with TID=0 the initial value of $LastTID$ (0) forces a verification. The OG is smaller than the other cases (i.e. OG No. 2) because the receipt of the Ack PDU for verification by the initiator, implicitly acknowledges the Invoke PDU, disallowing further re-transmissions (note that essentially this means sequences leading to two TR-Invoke.cnf primitives, as discussed in the following Section, are not possible).

It should be noted that using a single transaction does not test several important features of the protocol (e.g. the windowing mechanism with sequence numbers). Analysing multiple transactions is an area for future work.

5.2   Comparison of Service and Protocol Languages

To generate the protocol language the OG is treated as a finite state automata (FSA). The sequence of primitives is of interest, so all binding elements that do not correspond to primitive events are treated as empty transitions in the FSA. A common FSA reduction technique [1] is applied to obtain the minimised deterministic FSA, or the Transaction Protocol language.

The languages have been obtained for OG No. 1, 2, 3, 4 and 8. The other OGs are too large to minimise using our current tools [7, 21]. In the following we concentrate on the languages obtained from OG No. 1 and 2 (User Ack off and on, respectively). The statistics of these languages (and for comparative purposes, the service languages from [11]) are given in Table 6.

Table 6. Transaction service and protocol FSA results

	User Ack	Nodes	Arcs	Halts	Sequences	Longest	Shortest
Service	O	21	74	4	450	9	2
Protocol	O	39	119	5	527	10	2
Service	On	21	69	4	194	9	2
Protocol	On	45	133	7	334	10	2

There are sequences possible in the protocol language, but not defined in the service. These identify errors in the protocol specification. For both User Ack cases, an error occurs when a second TR-Invoke.cnf can be generated at the initiator. More specifically, in relation to Fig. 3, TR-Invoke.cnf primitives can follow nodes 5, 7, 9 and 10. By tracing the sequences back to the OG, we have identified all possible sequences of protocol actions that result in this error (there are 174). A scenario showing the error is given in Fig. 10.

Closer investigation of some of the error sequences in the protocol CPN indicates the retransmission of an Ack PDU by the responder when it receives a retransmitted Invoke PDU while in the RESULT_WAIT state results in the second TR-Invoke.cnf primitive. The Ack PDU is retransmitted by the responder to cope with the situation when the first Ack PDU is lost. In that case the first TR-Invoke.cnf would not be generated, and everything would proceed correctly. Therefore to remedy the error, on receipt of the second Ack PDU (i.e. after a TR-Invoke.cnf has been issued) WTP should not re-issue a TR-Invoke.cnf and can discard (and log) the Ack PDU.

When User Ack is used there are two other inconsistencies with the service specification:

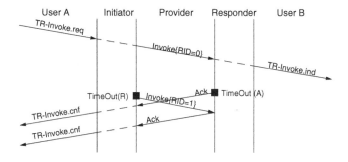

Fig. 10. Error scenario: Two TR-Invoke.cnf primitives

1. A TR-Result.req may immediately follow a TR-Invoke.ind at the responder. That is, the arcs between nodes 2 and 3, and 12 and 14 are present in the protocol language however, as stated in Section 3 and easily seen in Table 1, they are not present in the service language (Fig. 3) when User Ack is on. By restricting a TR-Result.req to only be issued when User Ack is off, as shown in Table 7, the error can be removed. With this change made to the protocol CPN, the language obtained from the OG disallows the extra TR-Result.req primitives, as required.

Table 7. State table entry with new condition restricting TR-Result.req

Responder **INVOKE RESP WAIT**			
Event	**Condition**	**Action**	**Next State**
TR-Result.req	$Uack==False$	Reset RCR Start timer R[RCR] Send Result PDU	RESULT RESP WAIT

2. There are inconsistent halt states in the protocol language. In relation to Fig. 3, node 12 would be a halt state. For example, Fig. 11 shows a sequence that constitutes a transaction. After the number of timeouts at the responder reaches the maximum, the transaction is aborted. The retransmitted Invoke PDU initiates a TID verification which fails. At the end of the sequence the initiator is in the NULL state and the responder in the LISTEN state, indicating that both entities have discarded any state information for that transaction. However, the responder user has been issued a TR-Invoke.ind but no other primitive to indicate the end of the transaction. A solution is to issue a TR-Abort.ind primitive to the responder user when a transaction is aborted due to a timeout (and the responder goes back to the LISTEN state).

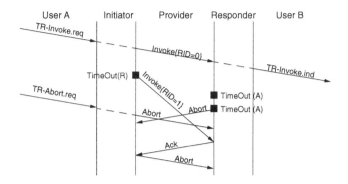

Fig. 11. Error scenario: Invalid halt state

6 Conclusions

Initial Coloured Petri net analysis of the WAP Class 2 Wireless Transaction Protocol has revealed discrepancies between the service specification and the protocol specification. This can be used for improving the specification and design of WTP.

WTP provides three classes of service, the most commonly used (Class 2) being a reliable request/response service. The service specification has been modelled using Coloured Petri nets and the Transaction Service language presented in [11]. The focus of this paper has been outlining the protocol specification CPN model and presenting the initial analysis results. In summary, we have:

> Presented a CPN model of the protocol, which was developed based on the WTP state tables. The model was developed to be able to analyse protocol features using a modular approach. Not all features are currently modelled. Using occurrence graph analysis, shown that, when a single transaction is possible and under restricted conditions on the counter limits, no deadlocks occur in the protocol and the dead transitions are as expected.
>
> Generated the protocol language, and compared it to the Transaction Service language. This identified the following discrepancies between the two:
>
> 1. Two TR-Invoke.cnf primitives could be generated at the initiator.
> 2. The User Ack service is not always provided when user acknowledgement is on (i.e. a TR-Request.req can follow a TR-Invoke.ind).
> 3. When User Ack is on a transaction can finish without the responder user being notified.
>
> Proposed the following changes to WTP:
>
> 1. For all actions when the initiator increments RCR, a condition should restrict RCR < RCR_MAX. The specific change to the state tables is given in Table 3.
> 2. To disallow two TR-Invoke.cnf primitives, the initiator should discard a second Ack PDU after already issuing a TR-Invoke.cnf, and not issue the second primitive.

3. Introduce a condition on the responder user issuing a TR-Result.req, as given in Table 7, to disallow the primitive following a TR-Invoke.ind when User Ack is on.
4. Generate a TR-Abort.ind when the the Responder aborts due to a time-out in the INVOKE_RESP_WAIT state so the user is notified of the end of the transaction.

These changes have been submitted to the WAP Forum [10].

Further analysis is required to be confident of WTPs correctness and performance. Firstly, the operation of the protocol under more complex conditions (multiple transactions, errors) needs to be verified. This work is in progress, and it is likely other analysis techniques (e.g. state space reduction) will be required. This analysis will also allow us to further evaluate the implications of the errors found and their proposed solutions. Finally, it will be necessary to analyse the performance of the protocol, either using the existing CPNs, or some other tools and techniques (e.g. OPNET).

Acknowledgements

This work was carried out with financial support from the Commonwealth of Australia through the Cooperative Research Centres Program.

References

[1] W. A. Barret and J. D. Couch. *Compiler Construction: Theory and Practice.* Science Research Associates, 1979.

[2] M. Y. Bearman, M. C. Wilbur-Ham, and J. Billington. Speci cation and analysis of the OSI Class 0 Transport Protocol. In J. M. Bennet and P. Pearcey, editors, *Proc. of the 7th Int. Conf. on Computer Communications*, pages 602 607, Sydney, Australia, 30 Oct. - 2 Nov. 1984. North Holland Publishers.

[3] J. Billington. Formal speci cation of protocols: Protocol engineering. In *Encyclopedia of Microcomputers*, pages 299 314. Marcel Dekker, New York, NY, 1991.

[4] J. Billington, M. Diaz, and G. Rozenberg, editors. *Application of Petri Nets to Communication Networks: Advances in Petri Nets.* LNCS 1605. Springer-Verlag, Berlin, 1999.

[5] R. Braden. T/TCP - TCP extensions for transactions functional speci cation. IETF RFC 1644 URL: `ftp://ftp.isi.edu/in-notes/rfc1644.txt`, July 1994.

[6] S. Budkowski, B. Alkechi, M. L. Benalycherif, P. Dembinski, M. Gardie, E. Lallet, J. P. Mouchel La Fusse, and Y. Soussi. Formal speci cation, validation and performance evaluation of the Xpress Transfer Protocol. In A. Danthine, G. Leduc, and P. Wolper, editors, *Protocol Speci cation, Testing and Veri cation, XIII*, pages 191 206, Liege, Belgium, May 1993.

[7] B. Cheong and S. Gordon. *Automata Reduction Tool (ART): System Manual.* Institute for Telecommunications Research, University of South Australia, 25 February 1999.

[8] J. C. A. de Figueiredo and L. M. Kristensen. Using Coloured Petri nets to investigate behavioural and performance issues of TCP protocols. In K. Jensen, editor,

Proc. of the 2nd Workshop on the Practical Use of Coloured Petri Nets and Design/CPN, pages 21 40, Aarhus, Denmark, 13-15 October 1999. Department of Computer Science, Aarhus University. PB-541.

[9] S. Gordon and J. Billington. Modelling and analysis of the WAP class 2 Wireless Transaction Protocol using Coloured Petri nets. Draft technical report, Institute for Telecommunications Research, University of South Australia, Adelaide, Australia, November 1999.

[10] S. Gordon and J. Billington. WAP Forum Input Document: Inconsistencies in the Wireless Transaction Protocol. Submitted to WAP Forum 19 November 1999.

[11] S. Gordon and J. Billington. Modelling the WAP Transaction Service using Coloured Petri nets. In H.-V. Leong, W.-C. Lee, B. Li, and L. Yin, editors, *Proc. of the 1st Int. Conf. on Mobile Data Access*, LNCS 1748, pages 109 118, Hong Kong, 16-17 December 1999. Springer-Verlag.

[12] ITU. Information Technology OSI Basic reference model: The basic model. ITU-T Recommendation X.200 ISO/IEC 7498, July 1994.

[13] ITU. Information Technology OSI Protocol for providing the connection-mode transport service. ITU-T Recommendation X.224 ISO/IEC 8073, November 1995.

[14] K. Jensen. *Coloured Petri Nets. Basic Concepts, Analysis Methods and Practical Use*, volume 1, Basic Concepts, *Monographs in Theoretical Computer Science*. Springer-Verlag, Berlin, 1997.

[15] K. Jensen. *Coloured Petri Nets. Basic Concepts, Analysis Methods and Practical Use*, volume 3, Practical Use, *Monographs in Theoretical Computer Science*. Springer-Verlag, Berlin, 1997.

[16] G. M. Lundy and R. C. McArthur. Formal model of a high speed transport protocol. In R. J. Linn and M. U. Uyar, editors, *Protocol Speci cation, Testing and Veri cation, XII*, pages 97 111, Lake Buena Vista, FL, June 1992.

[17] H. Mehrpour and A. E. Karbowiak. Modelling and analysis of DOD TCP/IP protocol using numerical Petri nets. In *Proc. of IEEE Region 10 Conf. on Computer Communication Systems*, pages 617 622, Hong Kong, September 1990.

[18] Meta Software Corporation. *Design/CPN Reference Manual for X-Windows, Version 2.0*. Meta Software Corporation, Cambridge, MA, 1993.

[19] J. Postel. User Datagram Protocol. IETF RFC 768 URL: `ftp://ftp.isi.edu/in-notes/rfc768.txt`, August 1980.

[20] J. Postel. Transmission Control Protocol. DARPA Internet program protocol speci cation. Information Sciences Institute, University of Southern California, Marina del Ray, CA, September 1981. IETF RFC 793 URL: `ftp://ftp.isi.edu/in-notes/rfc793.txt`.

[21] D. Raymond and D. Wood. *User's Guide to Grail*. Dept. of Computer Science, University of Western Ontario, London, Canada, March 1996. Version 2.5.

[22] M. A. Smith. Formal veri cation of communication protocols. In R. Gotzhein and J. Bredereke, editors, *Formal Description Techniques IX: Theory, Applications, and Tools*, pages 129 144. Chapman & Hall, London, UK, October 1996.

[23] M. A. Smith. Reliable message delivery and conditionally-fast transactions are not possible without accurate clocks. In *Proc. of the 17th Annual ACM Symposium on Principles of Distributed Computing*, Puerto Vallarto, Mexico, June 1998.

[24] WAP Forum. Wireless Application Protocol Architecture Speci cation. Available via `http://www.wapforum.org/`, 30 April 1998.

[25] WAP Forum. Wireless Application Protocol Wireless Transaction Protocol Speci cation. Available via `http://www.wapforum.org/`, 11 June 1999.

Liveness Veri cation of Discrete Event Systems Modeled by n-Safe Ordinary Petri Nets

Kevin X. He and Michael D. Lemmon

Dept. of Electrical Engineering
University of Notre Dame
Notre Dame, IN 46556, USA
(219)-631-8309
fax:(219)-631-4393
xhe, lemmon@maddog.ee.nd.edu

Abstract. This paper discusses liveness veri cation of discrete-event systems modeled by n-safe ordinary Petri nets. A Petri net is *live*, if it is possible to re any transition from any reachable marking. The veri cation method we propose is based on a partial order method called *network unfolding*. Network unfolding maps the original Petri net to an acyclic *occurrence net*. A nite pre x of the occurrence net is de ned to give a compact representation of the original net's reachability graph. A set of *transition cycles* is identi ed in the nite pre x. These cycles are then used to establish necessary and su cient conditions that determine the original net's liveness.

1 Introduction

Owing to its simple but powerful modeling potential, Petri nets have been used extensively in the modeling and analysis of discrete event systems. Practical examples of such systems include multiprocessor computer systems, computer networks, highway traffic control systems, and process control plants. In the design of these systems, one important thing to verify is the system's ability to transition from any subsystem to any other subsystem, or from any system state to any other system state without encountering any deadlock. Systems that satisfy this property are said to be live.

Based on the Petri net model, a variety of theoretical results [3] [15] [1] [2] and computational algorithms [4] [9] have been developed to assess the liveness of certain classes of Petri nets . Most of these results were based on the fact that the liveness of a Petri net is closely related to the satisfiability of some properties on place invariants of the net, namely siphons and traps. A *siphon* is a subset of places once being emptied, will never again obtain new tokens, while a *trap* is a subset of places once marked, will always remain marked. It was shown in [3] [2] that under certain structural constraints of the net, such as free-choiceness [3] or asymmetric-choiceness [2], the liveness property is necessary-and-sufficiently determined by checking the coverability of every siphon at every reachable marking.

M. Nielsen, D. Simpson (Eds.): ICATPN 2000, LNCS 1825, pp. 227 243, 2000.

Network unfolding originated from the notion of a *branching process* that was presented in [6]. A branching process unfolds a Petri net into an acyclic structure called an *occurrence net*. Occurrence nets were used to provide a concurrence semantics to nets [11]. McMillan in [11] used unfolding to avoid the state explosion problem in the verification of asynchronous circuits modeled by Petri nets. It was shown in [11] that network unfolding avoids the enumeration of arbitrary interleaving of concurrent transitions and thus provides a compact way to describe the net system's state space. A *cut* of the occurrence net was presented in [11] and it was proven that the resulting *finite prefix* of the occurrence net enumerates all the reachable markings of the original net system. Extensions of McMillan's work to model checking were found in [7] and [10]. In [7], algorithms were developed to verify the reachability of a marking or the liveness of transitions for 1-safe Petri nets, while in [10], other properties such as boundedness and persistence were verified for general class of nets.

This paper uses network unfolding to verify the liveness of *n*-safe ordinary Petri nets. Our work is based on a new finite prefix that is composed of the finite prefix defined in [11] and the first *tier* of *cutoff* transitions. The intuition behind it is that while the finite prefix in [11] provides a compact representation of the state space, the inclusion of cutoff transitions helps identify cyclic, concurrent and causally related *transition cycles*. These transition cycles differs from the *T-invariants* computed in [15] in that transition cycles include causality between transitions while T-invariants do not. Transition cycles can be used to verify liveness in a modular fashion. In particular, we will show that liveness can be verified by examining the behavior of each single cycle and the relation between different cycles. To give a heuristic idea of our approach, let us view the relationship between cycles and the whole Petri net as concurrent processes managed by a computer operating system and view tokens flowing from presets to postsets of transitions as system resources being used by processes. Based on this analogy, the necessary and sufficient conditions for liveness can be described by the following three conditions that should be satisfied for a live operating system. First, there should be no process being *starved*, meaning that a process cannot obtain enough system resources. Next, there should be no process either deadlocked by itself or not releasing system resources upon its completion. Finally, there should be no group of processes such that every process in the group is waiting for resource from another process in order to proceed. We define this *circular waiting* phenomena as *cyclic lock* later in the paper.

The importance of transition cycles goes far beyond the verification of liveness. Intuitively, every transition cycle can be viewed as a *steady state mode* that the net's dynamics may rest upon. The new finite prefix can therefore be seen as a *snapshot* that captures every steady state mode and all the causally related transition sequences or *transients* that lead to these modes. Following this intuition, it is therefore possible to identify the exact transition sequence that leads to the undesirable behavior, thereby provides an efficient and modular way for Petri net supervisory control [8].

The remainder of the paper is organized as follows. Section 2 reviews preliminary definitions related to ordinary Petri nets. Section 3 introduces concepts related to the network unfolding. Section 4 proves the main result of this paper. Section 5 discusses the potential of extending the verification method to liveness-enforcing supervisory control. Finally, section 6 concludes with directions of future research work.

2   Ordinary Petri Nets

This section reviews some basic notions related to Petri nets. More details can be found in [13] and [14].

An ordinary Petri net $\mathcal{N}$ is often characterized by the 3-tuple, $(S\ T\ F)$ where S is the set of *places*, T is the set of *transitions*, $F \subset (S \times T) \cup (T \times S)$ is a set of input arcs (from places to transitions) and output arcs (from transitions to places). We denote the *preset* of a transition $t \in T$ as $\bullet t$ and define it as the set of places, $s \in S$ such that $(s\ t) \in F$. In a dual manner, we introduce the postset of a transition $t \in T$ as $t\bullet$ and define it as the set of places, $s \in S$ such that $(t\ s) \in F$. We define presets and postsets of places in a similar way.

Let T_1 be a set of transitions of $\mathcal{N}$, we define the preset of T_1 as

$$\bullet T_1 = \{s \in S | s \in \bullet t \text{ for some } t \in T_1 \text{ and } s \notin t'\bullet \ \forall t' \in T_1\}$$

In a dual manner, we define the postset of T_1 as

$$T_1\bullet = \{s \in S | s \in t\bullet \text{ for some } t \in T_1 \text{ and } s \notin \bullet t' \ \forall t' \in T_1\}$$

For example, in figure 1, if we take $T_1 = \{t_5\ t_4\}$, then $\bullet T_1 = \{s_5\ s_7\}$ and $T_1\bullet = \{s_4\ s_8\}$.

The current "state" of the Petri net is represented by the *marking* of the network. The marking $\mu : S \to Z^+$ is a mapping from the places onto nonnegative integers. The marking $\mu(s)$ of place s denotes the number of *tokens* in that place. Graphically, we represent places by open circles, transitions by bars and tokens by small filled circles. Figure 1 shows an example Petri net with initial marking $\mu_0 = [\ \mu_0(s_1)\ \mu_0(s_2)\ \cdots\ \mu_0(s_{10})]^T = [1000111010]^T$.

The dynamics of ordinary Petri nets are characterized by the way in which the network marking evolves. We say that the transition t is *enabled* if $\mu(s) > 0$ for all $s \in \bullet t$. An enabled transition may *fire*. We introduce a firing function $q : T \to \{0\ 1\}$ such that $q(t) = 1$ if t is firing and is zero otherwise. If $\mu(s)$ and $\mu'(s)$ denote the marking of place p before and after the firing of enabled transition t, denoted by $\mu \xrightarrow{t} \mu'$, then

$$\mu'(s) = \begin{cases} \mu(s) + 1 & \text{if } s \in t\bullet \setminus \bullet t \\ \mu(s) - 1 & \text{if } s \in \bullet t \setminus t\bullet \\ \mu(s) & \text{otherwise} \end{cases} \qquad (1)$$

We define a *net system* as the pair $(\mathcal{N}\ \mu_0)$, where $\mathcal{N} = (S\ T\ F)$ is a Petri net and μ_0 is its initial marking. We say a sequence of transitions $\sigma = t_1 t_2 \ \cdots\ t_n$ is an *occurrence sequence*, if there exist markings $\mu_1\ \mu_2\ \cdots\ \mu_n$ such that

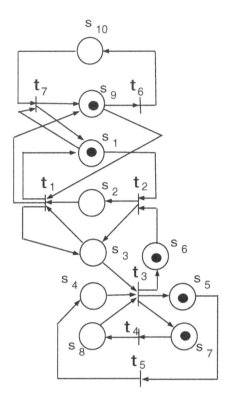

Fig. 1. An example net

$$\mu_0 \xrightarrow{t_1} \mu_1 \xrightarrow{t_2} \ldots \mu_{n-1} \xrightarrow{t_n} \mu_n$$

μ_n is the marking reached by the occurrence of σ, also denoted by $\mu_0 \xrightarrow{\sigma} \mu_n$. Given two markings μ_1 and μ_2, we say μ_2 is *reachable* from μ_1, if there exists an occurrence sequence σ' such that $\mu_1 \xrightarrow{\sigma'} \mu_2$. The *reachability graph* of network $\mathcal{N}$ is a labeled graph having the set of reachable markings of $\mathcal{N}$ as nodes and the relations $\xrightarrow{\sigma}$ between markings as edges.

We define $R(\mu_0)$ as the set of markings reachable from μ_0. We say a net system $(\mathcal{N}, \mu_0)$ is *n-safe*, if there exists a finite number n such that $\mu(s) \leq n, \forall s \in S, \forall \mu \in R(\mu_0)$. We say a net system is 1-safe, if $n = 1$. The net system in figure 1 is 1-safe.

We say a transition t is reachable from a marking μ if there exists a marking μ' and an occurrence sequence σ' such that $\mu \xrightarrow{\sigma} \mu'$ and μ' enables t. We say a place s is reachable from a marking μ if there exists a transition t such that t is reachable from μ and $s \in t\bullet$. Moreover, we say a set of transitions T_1 is reachable from a marking μ if every transition $t \in T_1$ is reachable from μ. We say a set of places S_1 is reachable from a marking μ if every place $s \in S_1$ is reachable from μ.

We denote a sequence of arcs $(x_1\ x_2), (x_2\ x_3), \cdots, (x_{N-2}\ x_{N-1}), (x_{N-1}\ x_N)$ in the net $\mathcal{N}$ as a *path*. We say that the net is *acyclic* if there is no path such that $x_1 = x_N$.

Let $\mathcal{N} = (S\ T\ F)$ be an acyclic net and $x_1\ x_2 \in S \cup T$ be two *nodes* of $\mathcal{N}$. We define the *ordering relations* between x_1 and x_2 in the following way:

We say x_1 *precedes* x_2, denoted as $x_1 < x_2$, if and only if there exists a sequence , , of arcs (also called a *path*) of the form

$$= (y_1\ y_2)\ (y_2\ y_3)\ \cdots\ (y_j\ y_{j+1})\ (y_{j+1}\ y_{j+2})\ \cdots\ (y_{N-1}\ y_N) \qquad (2)$$

such that $y_1 = x_1$ and $y_N = x_2$. In figure 2, we can see that transition t_2 precedes transition t_1, since there exists the path $(t_2\ s_2)(s_2\ t_1)$.

We say x_1 and x_2 are in *conflict*, denoted by $x_1 \# x_2$, if there exist distinct transitions $t_1, t_2 \in T$ such that $\bullet t_1 \cap \bullet t_2 \neq \emptyset$ and $t_i < x_i$ for $i = 1\ 2$. We say a node x is in *self-conflict* if $x \# x$. This means that there is a node y preceding x such that x can be reached by more than one distinct occurrence sequence from y. In figure 2, transitions $t_1\ t_7$ are in conflict, since $t_2 < t_1$ and $\bullet t_2 \cap \bullet t_7 = \{s_1\}$.

We say x_1 and x_2 are *concurrent*, denoted by $x_1 \| x_2$, if they are not in precedence and not in conflict. In figure 2, transitions t_4 and t_5 are concurrent.

A Petri net is said to be *finitary* if every node is preceded by a finite number of nodes.

3   Network Unfolding

Given a network $\mathcal{N} = (S\ T\ F)$, let $\min(\mathcal{N})$ denote the set of places

$$\min(\mathcal{N}) = \{s \in S\ :\ \bullet s = \emptyset\} \qquad (3)$$

An *occurrence net* is a finitary acyclic net $\mathcal{N}' = (S'\ T'\ F')$ with initial marking $_0$ such that

1. $|\bullet s| \leq 1$ for every $s \in S'$
2. no transition $t \in T$ is in self-conflict.
3. $_0(s) = 1$ if and only if $s \in \min(\mathcal{N}')$.

The acyclic net in figure 2 shows an example occurrence network.

Let $\mathcal{N}_1 = (S_1\ T_1\ F_1)$ and $\mathcal{N}_2 = (S_2\ T_2\ F_2)$ be two nets with initial markings $_{01}$ and $_{02}$, respectively. A *net homomorphism*, $h : S_1 \cup T_1 \rightarrow S_2 \cup T_2$ is a mapping between nodes of $\mathcal{N}_1$ and $\mathcal{N}_2$ such that

1. $h(S_1) \subseteq S_2$ and $h(T_1) \subseteq T_2$.
2. for every $t \in T_1$, the restriction of h to $\bullet t$ is a bijection between $\bullet t$ (in $\mathcal{N}_1$) and $\bullet h(t)$ (in $\mathcal{N}_2$). Similarly for the postsets $t\bullet$ and $h(t)\bullet$. In other words, a net homomorphism preserves the preset and postset of transitions.

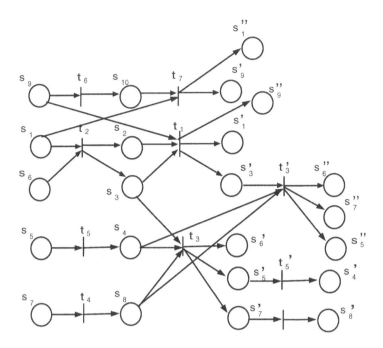

Fig. 2. The unfolding of the network in figure 1

3. the restriction of h to $\min(N_1)$ is a bijection between $\min(\mathcal{N}_1)$ and $\min(\mathcal{N}_2)$. In other words, a net homomorphism also preserves the initial marking.

A *branching process* of a net system $= (\mathcal{N}\ _0)$ is a pair $= (\mathcal{N}'\ h)$ such that

1. $\mathcal{N}' = (S'\ T'\ F')$ is an occurrence network and
2. h is a net homomorphism from $\mathcal{N}'$ to $\mathcal{N}$ such that if $\bullet t_1 = \bullet t_2$ and $h(t_1) = h(t_2)$, then $t_1 = t_2$.

Two branching processes are said to be *isomorphic* if there is a bijective homomorphism between them such that $h_2 \circ h = h_1$. Given two occurrence nets $\mathcal{N}_1 = (S_1\ T_1\ F_1)$ and $\mathcal{N}_2 = (S_2\ T_2\ F_2)$, we say $\mathcal{N}_1$ *contains* $\mathcal{N}_2$, denoted as $\mathcal{N}_2 \subseteq \mathcal{N}_1$, if $S_2 \subseteq S_1\ T_2 \subseteq T_1$ and for all $t \in T_2$, $\bullet t$ is the same in $\mathcal{N}_2$ as in $\mathcal{N}_1$, so is $t\bullet$. We say a branching process $_1 = (\mathcal{N}_1\ h_1)$ *contains* $_2 = (\mathcal{N}_2\ h_2)$ if $\mathcal{N}_2 \subseteq \mathcal{N}_1$. A branching process is *maximal* if it contains all other branching processes of a network $\mathcal{N}$. An *unfolding* of a net system   is the maximal branching process of   .

Let $\mathcal{N}' = (S'\ T'\ F')$ be the occurrence net obtained from the branching process of   . A set of transitions $C \subseteq T$ is a *configuration*, if the following holds.

1. if $t \in C$, then $t' < t$ implies $t' \in C$.
2. no two elements in C are in conflict.

A configuration is called a *maximal configuration* if it is not a subset of any other configuration. Figure 2 shows the unfolding of the network in figure 1. In the occurrence net, transitions $\{t_6 \ t_7\}$ form a maximal configuration.

From the definition, we can see that if a transition t is in a configuration, then all the transitions preceding t should be in the same configuration. In addition, transitions that are concurrent with t can also be included in that configuration. Let $t \in T'$ be a transition of $\mathcal{N}'$, we denote $[t]$ as the set $\{t' \in T' | t' < t \text{ or } t' = t\}$. We call $[t]$ the *cause* of t. This notion of *cause* is the same as the notion of *local configuration* in [11]. For example, in figure 2, $[t_7] = \{t_6 \ t_7\}$. The following two lemmas describe some characteristics of $[t]$ and the configuration.

Lemma 1. *For any transition t of an occurrence net $\mathcal{N}'$, $[t]$ forms a configuration.*

Proof: It is clear that transitions in $[t]$ satisfy the first condition in the definition of configuration. We only need to prove that no transitions in $[t]$ are in conflict. This argument is true since if there exist $t_1 \ t_2 \in [t] \ t_1 < t_2 \ t_1 \# t_2$, then $t_2 \# t_2$, meaning t_2 is in self-conflict. This contradicts the fact that no transition in $\mathcal{N}'$ is in self-conflict. •

Lemma 2. *For a configuration C, if transition $t \in C$, then $[t] \subseteq C$.*

Proof: We prove this lemma by contradiction. Assume there is a transition $t_1 < t$ and $t_1 \notin C$. By the definition of configurations, the only reason for $t_1 < t \ t_1 \notin C$ is that there exists another transition $t_2 \in C$ and $t_1 \# t_2$. Since $t_1 \# t_2$ and $t_1 < t$, then $t \# t_2$. Since $t \in C$ and $t_2 \in C$, then by the definition of configurations, t_2 and t must not be in conflict. This is a contradiction. •.

Lemma 3. *A set of transitions C of $\mathcal{N}'$ forms a configuration, if and only if there exists a set of concurrent transitions T_c of $\mathcal{N}'$ such that $C = \cup_{t \in T_c} [t]$. Furthermore, the set T_c is unique for configuration C.*

Proof: Sufficiency: Let t be a transition in T_c. From lemma 1, we know that $[t]$ forms a configuration. Let t_1 be another transition in T_c, since $t_1 \| t$, then there exists no transition in $[t_1]$ which is in conflict with some transition in $[t]$. It follows that $[t] \cup [t_1]$ forms a configuration. Pick another (if any) transition $t_2 \in T_c$. Following the same argument as above, we have $[t] \cup [t_1] \cup [t_2]$ forms a configuration. Keep doing this process till we enumerate all the transitions in T_c, we then have $C = \cup_{t \in T_c} [t]$ forms a configuration.

Necessity: Consider the set of transitions $T_c = \{t \in C | \ \nexists t' \in C \ t < t'\}$. We first want to prove that all transitions in T_c are concurrent. The statement is true because, first, by definition, there does not exist two transitions $t_1 \ t_2 \in T_c$ such that $t_1 < t_2$, meaning the precedence relationship does not exist in T_c. Secondly, by the definition of configurations, there also does not exist two transitions in T_c that are in conflict. We therefore conclude that all transitions in T_c are concurrent. Next, from lemma 2, observe that if two transitions $t_1 \| t_2$, then for any transitions $t_1' \in [t_1] \ t_2' \in [t_2]$, it is impossible that $t_1' \# t_2'$, since otherwise we will have $t_1 \# t_2$. We thus conclude that if $t_1 \ t_2 \in C$, then $[t_1] \cup [t_2] \subseteq C$, because first $[t_1]$ and $[t_2]$ satisfy the first condition in the definition of configurations, and, there is no conflict between transitions in $[t_1]$ and $[t_2]$, which satisfy the second condition in the definition. Pick another transition $t_3 \in T_c$, following the same argument, we have $[t_1] \cup [t_2] \cup [t_3] \subseteq C$. Keep picking transitions until we exhaust all transitions in T_c, we conclude that $\cup_{t \in T_c}[t] \subseteq C$. Now, we want to prove that $C \subseteq \cup_{t \in T_c}[t]$. We only need to prove that for every $t \in C$, there exists a $t' \in T_c$ such that $t \in [t']$. This is true because by definition of t_c, t has to either precede or be equal to some $t' \in T_c$, and therefore $t \in [t']$. We can thus conclude that $C = \cup_{t \in T_c}[t]$.

Finally, we need to prove that the T_c defined in the necessity proof is a minimal set satisfying $C = \cup_{t \in T_c}[t]$ and this minimal set T_c is unique for configuration C. Supposing there is another set T'_c such that $C = \cup_{t' \in T'_c}[t']$, then for every $t' \in T_c$, there should be a $t \in T_c$ such that $t' < t$ or $t' = t$. We thus have $[t'] \subseteq [t]$ and $\cup_{t' \in T'_c}[t'] \subseteq \cup_{t \in T_c}[t]$. $\cup_{t' \in T'_c}[t'] = \cup_{t \in T_c}[t]$ holds only if there is a subset of T'_c equal to T_c. This means T_c is a subset of any other set T'_c satisfying $C = \cup_{t \in T'_c}[t]$, which is enough to prove that T_c is minimal and unique. •

We define the *cut* of a configuration C as $Cut(C) = (min(\mathcal{N}') \cup C\bullet) \setminus \bullet C$. *Cut*s of configurations are used to represent reachable markings of the original net system. The link between *Cut*s and reachable markings can be defined in the following way. Recall that given the set of places S, the *multiset* S^{multi} of S is the set $\{S^m | \forall s \in S^m, s \in S, \text{and there may exist } s_1, s_2 \in S^m, \text{such that } s_1 = s_2\}$. Define $M : S^{multi} \to Z^{|S|}$ as a mapping that maps a multiset on S to a marking vector in such a way that M_i is equal to the number of place s_i's in the multiset. It is has been proven [7] that given a configuration C, $M(h(Cut(C)))$ is a reachable marking of the original net system. For example, in figure 2, $Cut(\{t_6, t_7\}) = \{s_1, s_5, s_6, s_7, s_9\} \cup \{s''_1, s''_9\} \setminus \{s_1, s_9\} = \{s''_1, s_5, s_6, s_7, s''_9\}$, and $h(\{s''_1, s_5, s_6, s_7, s''_9\}) = \{s_1, s_5, s_6, s_7, s_9\}$. We have $\mu = M(h(\{s''_1, s_5, s_6, s_7, s''_9\})) = M(\{s_1, s_5, s_6, s_7, s_9\}) = [1000111010]^T$. In other words, μ_0 is reachable from μ_0 after firing t_6, t_7.

An unfolding may be infinite, and therefore it makes sense to define a *finite prefix* of it for verification purposes. A branching process $\beta = (\mathcal{N}', h')$ is a *prefix* of the unfolding $\beta_m = (\mathcal{N}_m, h_m)$ if $\mathcal{N}'$ is contained by $\mathcal{N}_m$.

Definition A transition t of the unfolding β_m is called a *cut-off* transition if there exists a smaller cause $[t'] \subset [t]$ such that $h_m(Cut([t'])) = h_m(Cut([t]))$ or, $M(h_m(Cut([t]))) = \mu_0$. A cut-off transition t of β_m is called a *post cut-off* transition, if $\forall t' \in \bullet(\bullet t), t'$ is a cut-off transition, i.e. t's immediate predecessors are all cut-off transitions.

Definition We define $\beta_c = (\mathcal{N}_c, h_c)$ as the finite prefix obtained after removing all post cut-off transitions from β_m.

Remark: The relationship between β_c and the finite prefix β_f defined in [11] can be illustrated as follows. Recall that $\beta_f = (\mathcal{N}_f, h_f)$ is obtained after removing all cut-off transitions from β_m. It is therefore easily seen that β_c contains β_f and the first "tier" of cut-off transitions in β_m. The inclusion of cut-off transitions allows us to find transition cycles which are important for the verification of liveness. (Details will appear in section 4.) Some initial efforts that attempt to include cut-off transitions in the finite prefix to verify net properties can be found in [16].

Figure 2 shows the finite prefix $\mathcal{N}_c$ of the unfolding of the net in figure 1. Note that in figure 2, transitions t_7 t_3' are cutoff transitions and $\mathcal{N}_f$ is obtained by removing t_7 t_3' from $\mathcal{N}_c$.

In $\mathcal{N}_c$, we say that a transition t is an *end transition*, if there is no transition t' such that $t < t'$. Note that an end transition is either a cut-off transition, or a *deadlocked* transition that precedes no transition in $\mathcal{N}_m$. Consider the occurrence net shown in figure 2. In that figure, transition t_7 t_3' t_4' t_5' are end transitions. Among them, t_7 t_3' are cut-off transitions and t_4' t_5' are end transitions that precede no transition in $\mathcal{N}_m$.

The following lemmas were proven in [12].

Lemma 4. *A marking μ is a reachable marking of the original net system $\mathcal{N}$ if and only if there is a configuration C of $\mathcal{N}_f$ such that $\mu = M(h_f(Cut(C)))$.*

Proof: See [12]. •

Lemma 5. *If $\mathcal{N}$ is n-safe, then $\mathcal{N}_f$ is finite.*

Proof: See [12]. •

It is obvious to see that $\mathcal{N}_c$ enumerates all the reachable markings of the original net system, since $\mathcal{N}_c$ contains $\mathcal{N}_f$. It is also clear that $\mathcal{N}_c$ is finite if $\mathcal{N}_f$ is finite, since there are a finite number of end transitions in $\mathcal{N}_f$. The following lemma proves that $\mathcal{N}_c$ enumerates all the transitions reachable from $\mathcal{N}_0$.

Lemma 6. *For any transition t of $\mathcal{N}$ that is reachable from $\mathcal{N}_0$, there exists a transition t_c of $\mathcal{N}_c$ such that $h_c(t_c) = t$.*

Proof: Note that for a configuration C of the unfolding $\mathcal{N}_m = (\mathcal{N}_m$ $h_m)$, all the transitions in the original net $\mathcal{N}$ which are enabled by $M(h_m(Cut(C)))$ should have their image appear in $\mathcal{N}_m$. Since every reachable marking of $\mathcal{N}$ maps to a Cut of $\mathcal{N}_f$ and $\mathcal{N}_c$ contains $\mathcal{N}_f$, then any reachable transition must have their image appear in $\mathcal{N}_c$. Another version of the proof can be found in [16]. •

Lemma 7. *Let C_1 and C_2 be two configurations of $\mathcal{N}_f$ and $C_1 \subseteq C_2$, then in the original net system $\mathcal{N} = (\mathcal{N}$ $_0)$, $M(h_f(Cut(C_2)))$ is reachable from $M(h_f(Cut(C_1)))$.*

Proof: By lemma 3, there exists a minimal set of concurrent transitions T_{c_1} of $\mathcal{N}'$ such that $C_1 = \cup_{t \in T_{c_1}} [t]$ and a minimal set of concurrent transitions T_{c_2} of $\mathcal{N}'$ such that $C_2 = \cup_{t \in T_{c_2}} [t]$. Since $C_1 \subseteq C_2$, then for every $t_1 \in T_{c_1}$, there must exists a $t_2 \in T_{c_2}$ such that $t_1 < t_2$ or $t_1 = t_2$. Pick a transition $t_1 \in T_{c_1}$, and look for the associated transition $t_2 \in T_{c_2}$. We can see that t_2 is not in conflict with any transition in T_{c_1}, since otherwise t_2 and transitions in T_{c_1} are in the same configuration. Since t_2 is not in conflict with any transition in T_{c_1}, then all transitions in $[t_2]\backslash[t_1]$ can be fired one after another and it follows that $M(h_f(Cut(C_1 \cup [t_2])))$ is reachable from $M(h_f(Cut(C_1)))$. Now, pick another transition in T_{c_1} and continue the above process, by induction we conclude that $M(h_f(Cut(C_2)))$ is reachable from $M(h_f(Cut(C_1)))$. •

4   Main Result

In this section, we present necessary and sufficient conditions determining the liveness of n-safe ordinary Petri nets. As indicated in the introduction, our result is based on the understanding that liveness is closely related to the cyclic behavior of the network. We show that, for an n-safe ordinary Petri net, once we have the finite prefix of its unfolding, transition *cycles* can be identified from the occurrence network and the liveness property can be determined by investigating the liveness of each transition cycle and the existence of cyclic locks within certain group of cycles. Informally, our result proves that an n-safe net is live if and only if every transition is in a cycle, every cycle is live, and there exists no cyclic lock in any group of cycles.

We start our presentation by the definition of transition cycles. Let $\Sigma = (\mathcal{N}, \mu_0)$ be a net system and β_c be the new finite prefix of its unfolding defined earlier. Let t' be a cut-off transition of β_c and t be the transition such that $t < t'$ and $h_c(Cut([t])) = h_c(Cut(t'))$. We define a *cycle* $C_{t'}^{t'}$ as the set $C_{t'}^{t'} = [t']\backslash[t]$. Note that "$\backslash$" here means "set subtraction". We say a cut-off transition t' *defines* a cycle $C_{t_1}^{t'_1}$, if $t' = t'_1$. We say a cycle $C_t^{t'}$ is *reversible*, if $M(h_c(Cut([t']))) = \mu_0$. This means the marking goes back to the initial marking if all transitions in a reversible cycle are fired. Note that in this case we have $t = [t] = \emptyset$ and we write the reversible cycle as $C_\emptyset^{t'}$. We say a cycle $C_t^{t'}$ is *live*, if there exists a set of cycles $C_{t_1}^{t'_1}, \cdots, C_{t_k}^{t'_k}$ such that

1. $[t] \subset [t'_1]$,
2. $[t_j] \subset [t'_{j+1}]$, for $j \in \{1, \cdots, k-1\}$, and
3. either $C_{t_k}^{t'_k}$ is reversible., or $h_c(\cup_{i=1}^k C_{t_i}^{t'_i} \cup C_t^{t'}) = T$, where T is the set of all transitions in the original net.

We call this set of cycles the *live set* of $C_t^{t'}$. Intuitively, The first two conditions mean reachability between different cycles. Specifically, $[t] \subset [t'_1]$ means that we can reach cycle $C_{t_1}^{t'_1}$ from cycle $C_t^{t'}$ and $[t_j] \subset [t'_{j+1}]$ means that that we can reach cycle $C_{t_{j+1}}^{t'_{j+1}}$ from cycle $C_{t_j}^{t'_j}$. The third condition means that either the path $C_t^{t'}, C_{t_1}^{t'_1}, \cdot, C_{t_k}^{t'_k}$ is repeatable ($C_{t_k}^{t'_k}$ reversible), or it traverses all the transitions ($h_c(\cup_{i=1}^k C_{t_i}^{t'_i} \cup C_t^{t'}) = T$). The notion of a live cycle further means that if we fire the live set of cycles sequentially, we should reach every transition in the original net, or we should not encounter any deadlock.

As an example, consider the occurrence net in figure 3. There are two cut-off transitions t'_4, t'_5 in the net. The two cut-off transitions define two cycles, namely, cycle $C_{t'_1}^{t'_5}$ and cycle $C_\emptyset^{t'_4}$. For these two cycles, cycle $C_\emptyset^{t'_4}$ is reversible since $M(h_c(Cut([t'_4]))) = M(h_c(s''_1)) = \mu_0$. Cycle $C_{t'_1}^{t'_5}$ is not reversible since $M(h_c(Cut([t'_5]))) = M(h_c(s''_2)) \neq \mu_0$, but it is live since these exists a reversible cycle $C^{t'_4}$ such that $[t'_1] \subset [t'_4]$. In this case the live set of cycle $C_{t'_1}^{t'_5}$ is $\{C_\emptyset^{t'_4}\}$.

The following lemmas prove some properties of the live set.

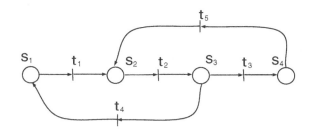

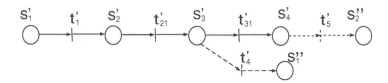

Fig. 3. Cycles in the occurrence net

Lemma 8. *Let $C_{t_1}^{t_1'}$ and $C_{t_2}^{t_2'}$ be two cycles in the finite prefix $_c$. If $[t_1] \subset [t_2']$, then every transition in $h_c(C_{t_2}^{t_2'})$ is reachable from $M(h_c(Cut([t_1])))$.*

Proof: Since $[t_1] \subset [t_2']$, then $M(h_c(Cut([t_2'])))$ is reachable from $M(h_c(Cut([t_1])))$. Since $M(h_c(Cut([t_2']))) = M(h_c(Cut([t_2])))$, then $M(h_c(Cut([t_2])))$ is reachable from $M(h_c(Cut([t_1])))$. Since every transition in $h_c(C_{t_2}^{t_2'})$ is reachable from $M(h_c(Cut([t_2])))$, it follows that every transition in $h_c(C_{t_2}^{t_2'})$ is reachable from $M(h_c(Cut([t_1])))$. •

Lemma 9. *If cycle $C_t^{t'}$ is live and the last cycle in its live set $C_{t_k}^{t_k'}$ is reversible, then $_0$ is reachable from $M(h_c(Cut([t])))$.*

Proof: Let the live set of $C_t^{t'}$ be $\{C_{t_1}^{t_1'} \cdots C_{t_k}^{t_k'}\}$. Since $[t] \subset [t_1']$, then $M(h_c(Cut([t_1])))$ is reachable from $M(h_c(Cut([t])))$. Since $[t_j] \subset [t_{j+1}']$, then $M(h_c(Cut([t_{j+1}])))$ is reachable from $M(h_c(Cut([t_j])))$. By induction, we have $M(h_c(Cut([t_k])))$ reachable from $M(h_c(Cut([t])))$ and therefore $_0 = M(h_c(Cut([t_k])))$ is reachable from $M(h_c(Cut([t])))$. •

Lemma 10. *If cycle $C_t^{t'}$ is live and $h_c(\cup_{i=1}^k C_{t_i}^{t_i'}) \cup C_t^{t'} = T$, then every transition in T is reachable from $M(h_c(Cut([t])))$.*

Proof: By lemma 8, since $[t] \subset [t_1']$ and $[t_j] \subset [t_{j+1}']$, then every transition in $h_c(C_{t_1}^{t_1'})$ is reachable from $M(h_c(Cut([t])))$ and every transition in $h_c(C_{t_{j+1}}^{t_{j+1}'})$

is reachable from $M(h_c(Cut([t_j])))$. Also, note that $M(h_c(Cut([t_1])))$ is reachable from $M(h_c(Cut([t])))$ and $M(h_c(Cut([t_{j+1}])))$ is reachable from $M(h_c(Cut([t_j])))$. It follows that every transition in $T = h_c(\cup_{i=1}^{k} C_{t_i}^{t_i'}) \cup C_t^{t'}$ is reachable from $M(h_c(Cut([t])))$. •

Lemma 11. *If cycle $C_t^{t'}$ is not live, then there must exists a transition of the original net which can not be enabled repeatedly from $M(h_c(Cut([t])))$.*

Proof: This lemma is proven by contradiction. Suppose that every transition can be enabled repeatedly from $M(h_c(Cut([t])))$ then there must exists a set of cycles $C_{t_1}^{t_1'} \cdots C_{t_k}^{t_k'}$, such that $\cup_{j=1}^{k} C_{t_j}^{t_j'} = T$ and $[t] \subset [t_j'] \; \forall j \in \{1 \cdots k\}$. Note that these cycles should also satisfy $[t_j] \subset [t']$, since otherwise transitions in $h_c(C_t^{t'} \backslash C_{t_j}^{t_j'})$ will not be reachable from $M(h_c(Cut([t_j])))$. We can now claim that $C_t^{t'}$ is live, because we can construct a live set of $C_t^{t'}$ as follows. Using $C_{t_1}^{t_1'} \cdots C_{t_k}^{t_k'}$, build a set of cycles $C_{t_1}^{t_1'} \cdots C_{t_{2k-1}}^{t_{2k-1}'}$ such that

$C_{t_{2j-1}}^{t_{2j-1}'} = C_{t_j}^{t_j'} \; \forall j \in \{1 \cdots k\}$, where the "=" means $t_{2j-1}' = t_j'$ and $t_{2j-1} = t_j$, and

$C_{t_{2j}}^{t_{2j}'} = C_t^{t'} \; \forall j \in \{1 \cdots k\}$.

Note that $C_{t_1}^{t_1'} \cdots C_{t_{2k-1}}^{t_{2k-1}'}$ is indeed a live set of $C_t^{t'}$, because we have $[\bar{t}_{2j-1}] \subset [\bar{t}_{2j}']$, since $[t_j] \subset [t']$; and $[\bar{t}_{2j}] \subset [\bar{t}_{2j+1}']$, since $[t] \subset [t_j']$. Here comes the contradiction. •

A set of cycles $C_{t_1}^{t_1'} \cdots C_{t_k}^{t_k'}$ is said to be in *cyclic lock*, if there exist two sets of transitions $t_1^c \cdots t_k^c$ and $t_1^{c'} \cdots t_k^{c'}$, denoted as *lock sets* such that

1. $t_1^c \| t_2^c \| \cdots \| t_k^c$,
2. $t_i^c \; t_i^{c'} \in [t_k']$ $j = \{1 \cdots k\}$,
3. $t_i^c < t_i^{c'}$ $j = \{1 \cdots k\}$,
4. $t_i^{c'} \# t_{i+1}^c$ $j = \{1 \cdots k-1\}$, and
5. $t_k^{c'} \# t_1^c$.

Intuitively, this notion of cyclic lock means that if we fire those cycles containing a lock set concurrently, then we should reach a (local) deadlock that prohibits the reachability of transitions succeeding the lock set.

As an example, consider the occurrence net in figure 2. Two cycles, namely $C_\emptyset^{t_7}$ and $C_\emptyset^{t_3'}$ are identified for the graph. Both of them are reversible and live. But they are in cyclic lock, because there exist $t_6 \; t_7 \in C_\emptyset^{t_7}$ and $t_2 \; t_1 \in C_\emptyset^{t_3'}$ such that $t_6 < t_7 \; t_2 < t_1 \; t_1 \# t_6 \; t_7 \# t_2$. The consequence of this cyclic lock is, if we fire t_6 and t_2 consecutively, $t_7 \; t_1$ will never be enabled.

We are now ready to present the main theorem for liveness verification. We prove that there are four necessary and sufficient conditions determining the liveness of n-safe ordinary Petri nets. The first condition basically says every transition is reachable from the initial marking; the second condition means every transition is contained in some cycle; the third condition says every cycles is live and the last condition suggests there should not exists cyclic lock in any group of transition cycles.

Theorem 1. *A n-safe net system* $(\mathcal{N}, M_0)$ *is live if and only if in* $\mathcal{N}_c = (\mathcal{N}_c, h_c)$ *of* $\mathcal{N}$*'s unfolding, all the following conditions are satisfied.*

1. $h_c(T_c) = T$, where T_c is the set of transitions in $\mathcal{N}_c$,
2. $\forall t \in T_c \backslash T^{cut}$, where T^{cut} is the set of cut-off transitions, there exists $t' \in T^{cut}$ such that $t < t'$,
3. every cycle of $\mathcal{N}_c$ is live, and
4. there does not exist a set of cycles of $\mathcal{N}_c$ which is in cyclic lock.

Proof Sufficiency: Since $h_c(T_c) = T$, then every transition of $\mathcal{N}$ is reachable from the initial marking M_0. We want to proof that every transition is reachable from any reachable marking $M \in R(M_0)$. Let M be a reachable marking of $\mathcal{N}$, then it follows from lemma 4 that we have a configuration C of $\mathcal{N}_f$ such that $h_f(Cut(C)) = M$. By lemma 3, we know that there exists a set of concurrent transitions $\{t_1^1 \cdots t_k^1\}$ such that $C = \cup_{i=1}^{k}[t_i^1]$. By the third condition of the hypothesis, every t_i^1 precedes a cut-off transition t_i'. Consider configuration C, since no set of cycles is in cyclic lock, then there exists at least one cut-off transition t_{j1}' is not in conflict with any transition in $C_{t_i}^{t_i'}$ for all $i \neq j1$. Therefore, all transitions in $[t_{j1}'] \backslash [t_{j1}^1]$ are reachable from $M(Cut(C))$. After firing all these transitions the marking of the finite prefix becomes $M(Cut(C \cup [t_{j1}'] \backslash [t_{j1}^1]))$ with the tokens inside cycle $C_{t_{j1}}^{t_{j1}'}$ reach $Cut([t_{j1}'])$. Since every cycle is live, then $Cut([t_{j1}'])$ is a cut in the next cycle in cycle $C_{t_{j1}}^{t_{j1}'}$'s live set. Now, let $C = C \cup [t_{j1}'] \backslash [t_{j1}^1]$, repeat the above process. Note that at each step, because of no cyclic lock, there is always one cycle that can fire all its transition and move to the next cycle in its live set. Eventually, there will be some cycle finishes firing all transitions in its live set. If this cycle's live set contains all the transitions in the original net, then the proof is done. If for every cycle in $\cup_{i=1}^{k} C_{t_i}^{t_i'}$, the last cycle in its live set is reversible, then the marking reached after firing all transitions in everyone of these cycle's live set is M_0. Since every transition is reachable from M_0, then it follows that every transition is reachable from M.

Necessity: If $h_c(T_c) \neq T$, then any transition in $T \backslash h_c(T_c)$ is not reachable from M_0. If there exists a transition $t \in T_c \backslash T^{cut}$ such that $\not\exists t' \in T^{cut}$ such that $t < t'$, then t must precede some "dead transition" that can not be repeatedly enabled. If there is a cycle $C_t^{t'}$ which is not live, then by lemma 11, there must exists some transition which can not be enabled repeatedly from $M(h_c(Cut([t])))$, which means the original net system is not live. If there exists a set of cycles $CY = \{C_{t_1}^{t_1'} \cdots C_{t_k}^{t_k'}\}$ in cyclic lock, assuming the two lock sets are $\{t_1^c \cdots t_k^c\}$ and $\{t_1^{c'} \cdots t_k^{c'}\}$, then no transition in $h_c(\cup_{i=1}^{k}[t_i^{c'}])$ is reachable from $M(h_c(Cut(\cup_{i=1}^{k} t_i^c)))$, which also means the original net system is not live.

∎.

Remark: The computational complexity involved in the verification of liveness depends heavily on the computation spent in obtaining the finite prefix. It can be seen that, among the four necessary and sufficient conditions, the first, second and third conditions can be easily verified on the finite prefix, while the last condition can be checked during the unfolding process. Even though there exist cases where the finite prefixes do not scale well with the size of the original net, the verification method in this paper is still worth doing, because it provides valuable information for supervisory control. The unfolding method extracts strings of causally related transitions and the causal relationship can help develop necessary and sufficient conditions for the existence of liveness-enforcing supervisory policies. Exploring the causal relationship among transition is extremely useful when the original net contains uncontrollable transitions, in which case the controller need to decide, among all the transitions preceding the uncontrollable transition, which transition to control in order to achieve liveness of the controlled net and maximum permissiveness of the supervisory policy.

5 Example

Consider the net system in figure 1 and its occurrence net in figure 2. There are two cut-off transitions t_7 and t_3' in the occurrence net. Note that transitions t_4' and t_5' are not cut-off transitions since no transition succeeds them in the unfolding. (In other words, they are "dead transitions".) The two cut-off transitions define two live cycles $C_\emptyset^{t_7}$ and $C_\emptyset^{t_3'}$. It is also known that these two cycles are in cyclic lock. The whole net system is therefore not live, since condition 2 and condition 4 in theorem 1 are not satisfied.

The result in this paper can be extended to provide necessary and sufficient conditions that determine the existence of maximally permissive supervisor that enforces liveness [8]. To illustrate this point, let us consider whether or not a liveness-enforcing supervisory policy can be easily obtained based on the liveness verification result. (Details can be found in [8].) Define the supervisory policy $\mathcal{P}$ as a mapping $\mathcal{P} : R(\mu_0) \rightarrow \{0,1\}^{|T|}$. This means $\mathcal{P}$ returns a control vector for each reachable marking. A transition t_i is *control enabled* (*control disabled*) at marking μ, if $\mathcal{P}(\mu)_i = 1$ ($\mathcal{P}(\mu)_i = 0$). A transition can not fire if it is control disabled. Instinct tells us that, to enforce liveness of the example net, we need to control disable the occurrence of both the dead transitions and the cyclic lock. This control objective can be easily achieved by examining the occurrence net. To be specific, to control disable the cyclic lock, we just have to control disable t_6 (or t_2) right after the firing of t_2 (or t_6). To disable the occurrence of t_4' and t_5', we just have to control disable the firing of t_3 right after the firing of t_2, t_4, t_5. A complete supervisory policy is shown in figure 4. The first column shows the current configuration, i.e. the set of transitions that have been fired. The second column shows the cut of the current configuration. The third column shows the marking vector corresponding to each cut and the last column shows the control vector associated with each marking vector. It can also be verified that the supervisory policy is maximally permissive.

Configuration	Cut	Marking	Control - vector
	{ s1, s5, s6, s7, s9 }	[1000111010] ^{T}	[1 1 1 1 1 11] ^{T}
{ t2 }	{ s2, s3, s5, s7, s9 }	[0110101010] ^{T}	[1 1 1 1 1 01] ^{T}
{ t6 }	{ s1, s5, s6, s7, s10 }	[1000111001] ^{T}	[1 0 1 1 1 11] ^{T}
{ t2, t4 }	{ s2, s3, s5, s8, s9 }	[0110100110] ^{T}	[1 1 1 1 1 01] ^{T}
{ t6, t4 }	{ s1, s5, s6, s8, s10 }	[1000110101] ^{T}	[1 0 1 1 1 11] ^{T}
{ t2, t5 }	{ s2, s3, s4, s8, s9 }	[0110100110] ^{T}	[1 1 1 1 1 01] ^{T}
{ t6, t5 }	{ s1, s4, s6, s8, s10 }	[1000110101] ^{T}	[1 0 1 1 1 11] ^{T}
{ t2, t4, t5 }	{ s2, s3, s4, s8, s9 }	[0111000110] ^{T}	[1 1 0 1 1 01] ^{T}
{ t 6, t4, t5 }	{ s2, s3, s4, s8, s10 }	[0111000101] ^{T}	[1 0 1 1 1 11] ^{T}
Otherwise	/	/	[1 1 1 1 1 11] ^{T}

Fig. 4. The liveness-enforcing supervisory policy for the example net system

Furthermore, let us assume that transition t_3 is not controllable. An uncontrollable transition is a transition that cannot be control disabled, i.e. it can still fire even if it is control disabled. In this case, a natural way to disable the firing of t_3 is to control disable the firing of some transition that precedes t_3. From the occurrence net, we know that we can disable the firing of t_3 by control disabling t_4 or t_5. Two liveness-enforcing supervisory policies assuming t_4 or t_5 is controllable are shown in figure 5.

6   Conclusion

In this paper, we present necessary and sufficient conditions for liveness of n-safe ordinary Petri net based on a partial order method called network unfolding. We identify cyclic, concurrent and causally related strings of transitions in the network and we prove that a net is live if and only if every transition is in a cycle, every cycle is live and there is no cyclic lock among cycles. An example is provided to illustrate the verification of liveness based on the new results and the extension of these results to liveness-enforcing supervisory control.

References

1. K. Barkaoui, Liveness of Petri nets and its relations with deadlocks, traps, and invariants, Report 92-06, Laboratoire CEDRIC-CNAM, Paris, France, 1995.
2. K. Barkaoui, J.F. Pradat-Peyre, On liveness and Controlled Siphons in Petri nets, in *Application and Theory of Petri Nets*, Springer Verlag, 1996.

Policy P_1				Policy P_2			
Configuration	Cut	Marking	Control - vector	Configuration	Cut	Marking	Control - vector
	{s1, s5, s6, s7, s9}	[1000111010]^T	[11110 11]^T		{s1, s5, s6, s7, s9}	[1000111010]^T	[11101 11]^T
{t4}	{s1, s5, s6, s8, s9}	[1000110110]^T	[11110 11]^T	{t5}	{s1, s4, s6, s7, s9}	[1001011010]^T	[11101 11]^T
{t2}	{s2, s3, s5, s7, s9}	[0110101010]^T	[11111 001]^T	{t2}	{s2, s3, s5, s7, s9}	[0110101010]^T	[11101 01]^T
{t6}	{s1, s5, s6, s7, s10}	[1000111001]^T	[1011011]^T	{t6}	{s1, s5, s6, s7, s10}	[1000111001]^T	[10101 11]^T
{t2, t4}	{s2, s3, s5, s8, s9}	[0110100110]^T	[11111 001]^T	{t2, t5}	{s2, s3, s4, s7, s9}	[0110101010]^T	[11101 01]^T
{t6, t4}	{s1, s5, s6, s7, s10}	[1000111001]^T	[1011011]^T	{t6, t5}	{s1, s4, s6, s7, s10}	[1001011001]^T	[10101 11]^T
Otherwise	/	/	[11111111]^T	Otherwise	/	/	[11111 11]^T

Fig. 5. The two supervisory policies

3. F. Commoner, Deadlocks in Petri Nets , Wake eld, Applied Data Research, Inc., Report #CA-7206-2311, 1972.

4. J.C. Corbett and G.S. Avrunin, Using integer programming to verify general safety and liveness properties, *Formal Methods in System Design: An International Journal*, vol. 6, no. 1, pp. 97-123, January 1995.

5. J. Desel and J. Esparza, Free Choice Petri Nets, *Cambridge Tracts in Theoretical Computer Science 40*, Cambridge University Press 1995.

6. Engelfriet, J., Branching processes of Petri nets. *Acta Informatica 28*, 575 591, 1991.

7. Esparza, J., Model checking using net unfoldings . In M. G. Gaudel and J. P. Jouannaud, editors, TAPSOFT'93: *Theory and Practice of Software Development. 4th Int. Joint Conference* CAAP/FASE, Volume 668 of *Lecture Notes in Computer Science*, pp 613-628. Spring-verlag, 1993.

8. Kevin X. He and Michael D. Lemmon, Liveness-enforcing supervision of *n*-safe ordinary Petri nets with uncontrollable transitions , to appear in the *proceedings of the 2000's IFAC International Conference on Control Systems Design*, special session on Petri nets, Slovakia, June 2000.

9. Kemper, P. and Bause, F., An e cient polynimial-time algorithm to decide liveness and boundedness of free choice nets, *LNCS, No. 616:263-278, 1992*.

10. A. Kondratyev, M. Kishinevsky, A. Taubin and S. Ten, Structural approach for the analysis of Petri nets by reduced unfoldings , *Proceedings of the 17th International Conference on Application and Theory of Petri Nets*, Osaka, Japan, June 24-28, 1996.

11. McMillan, K., Using unfoldings to avoid the state explosion problem in the veri cation of asynchronous circuits , in: *Computer Aided Veri cation, Fourth International Workshop, CAV'92* (B.V. Bochmann and D.K. Probst, Eds.). Vol. 663 of *Lecture Notes in Computer Science*. Springer-Verlag. pp. 164 177, 1992.

12. McMillan, K., *Symbolic Model Checking*, Kluwer Academic Publishers, 1993.

13. Murata, T., Petri nets: Properties, analysis, and applications , *Proceedings of the IEEE*, 77(4):541-580.

14. Reisig, W. (1985). *Petri Nets.* Springer-Verlag, 1985.
15. Ridder, H. and Lautenbach, K., Liveness in bounded Petri nets which are covered by t-invariants, *LNCS, No. 815:358-375, 1994.*
16. A. Semenov, Verification and Synthesis of Asynchronous Control Circuits Using Petri Net Unfoldings, Newcastle upon Tyne, 1998.(British Lending Library DSC stock location number: DXN 016059).
17. R.S. Sreenivas, On the existence of supervisory control in discrete event dynamic systems modeled by controlled Petri nets. *IEEE Trans. on Automatic Control, 42(7), July, 1997, pp. 928-945.*
18. R.S. Sreenivas, On supervisory policies that enforce liveness in complete controlled Petri nets with directed cut-places and cut-transitions, in *IEEE Trans. on Automatic Control, 44(6), June, 1999, pp. 1221-1225.*

Modelling and Analysing the SDL Description of the ISDN-DSS1 Protocol*

Nisse Husberg, Teemu Tynjälä, and Kimmo Varpaaniemi

Helsinki University of Technology,
Laboratory for Theoretical Computer Science,
P.O. Box 9700, FIN-02015 HUT, Espoo, Finland
Nisse.Husberg@hut.fi, Teemu.Tynjala@hut.fi, Kimmo.Varpaaniemi@hut.fi

Abstract. The modelling of a telecommunication protocol, ISDN-DSS1, defined using SDL, is described and the problems and methods used are discussed. The formalism used for the model is a high-level Petri net, the PROD input language which is close to a predicate/transition net. The influence of the model on the reachability analysis is also discussed with a special attention on the use of priorities and reduction methods. Finally, further development of tools which aid in modelling and analysis of this class of systems is discussed.
Keywords: ISDN, DSS1, SDL, protocol verification, reachability analysis, high-level Petri nets

1 Introduction

We present modelling and analysis of the ISDN-DSS1 [4] protocol. The project originated in problems with the implementation of DSS1 in telephone exchanges used by the telephone operator Helsinki Telephone Corporation. The task was to check whether the DSS1 protocol allowed false connections, and, if the protocol itself was correct, to test the implementations, i.e. the concrete switches, in order to find possible loopholes in the implementations of the protocol.

DSS1 is the network level protocol used in the ISDN network. It is responsible for connecting and releasing traffic channels as well as reserving and releasing the Call Reference (a unique reference tag used to identify a call in progress across the ISDN network). The ISDN network has differentiated the signalling channels and traffic channels from each other. Traffic channels are referred to as *B channels* and the signalling channels as *D channels*. The DSS1 protocol signalling is carried out only on the D channel.

A block diagram of the DSS1 protocol and the other protocol blocks to which it is related is given in Figure 1 . The operation of DSS1 is tightly related to the *Call Control Block*. The Call Control Block enables and disables accounting and initiates call setup and teardown. DSS1 offers a Service Access Point (SAP) to the Call Control Block.

---

* This work has been supported by the ETX program of The National Technology Agency of Finland and by Helsinki Telephone Corporation.

M. Nielsen, D. Simpson (Eds.): ICATPN 2000, LNCS 1825, pp. 244–260, 2000.
© Springer-Verlag Berlin Heidelberg 2000

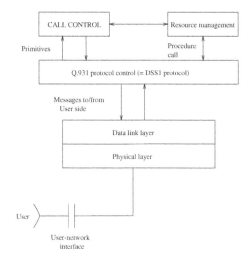

Fig. 1. DSS1 and related protocols ([4], p. 11).

The SDL [13] diagrams in the standard [4] of DSS1 were used as the source material in the project. PROD [19], a reachability analysis tool for Pr/T-nets, was used in the analysis. Modelling SDL is rather easy using high-level Petri nets, but there are a few problems, like the input queues and the timers. However, even if a system is described in a clear and formal way, it is necessary to *prepare* the model for analysis, usually abstract from a lot of implementation features. Although the DSS1 protocol specification is in itself fairly abstract, there are details which might be irrelevant for the specific analysis.

2   SDL and High-Level Nets

There are two different forms of SDL, the Graphical Representation (SDL/GR) and the (textual) Phrase Representation (SDL/PR). Here we are referring to the *graphical* form of SDL because the DSS1 description [4] is given using SDL diagrams.

2.1   The Graphical Representation of SDL

If we forget about the block structure of SDL, a system can be thought of as a collection of *SDL processes*. An SDL process can be seen as an *extended finite state machine*. It has an *input queue*, and the communication between processes is *asynchronous* and can only happen by putting signals (messages) into the queue. A process also has *timers* which have a fairly complex semantics. When a timer expires it sends a signal to the input queue of the process. However, if a timer which has expired is reset, the signal it has put into the queue should be removed.

An SDL process has a number of *SDL states* in which the process is waiting for an input from the queue. After the input (or possibly a SAVE statement), the process makes an *SDL transition* which consists of a number of *SDL statements* like OUTPUT, TASK, CREATE (a process), DECISION etc. The SDL transition always ends in a NEXTSTATE or other terminator statement.

In Figure 2, the first "box" is an SDL state (DSS1 *call state* 1: CALL INITIATED), the second an INPUT statement for the signal PROCEEDING REQUEST, the third an OUTPUT statement for signal CALL PROCEEDING with *parameter* B-CHNL, the fourth a TASK statement CONNECT B-CHNL and the last one a NEXTSTATE statement moving the control to DSS1 call state 3: OUTGOING CALL PROCEEDING.

Fig. 2. An SDL transition.

2.2   Tools for SDL Analysis

There are commercially available tools for SDL, like SDT (Telelogic Tau SDL Suite) [20] and ObjectGEODE [17], but these mainly use *simulation* for the analysis of the systems. We are interested in applying *reachability analysis* to SDL systems (but not only to these) and especially in developing methods to handle such analysis efficiently.

We also have an analyser, EMMA [7,8,9,10], which is a front-end for our reachability analyser PROD. EMMA is, however, made for TNSDL (TeleNokia SDL) which is a special dialect of SDL-88. It is furthermore a *programming language* with data types and control structures. Thus it is quite different from the SDL diagrams used in the DSS1 standard. The use of EMMA in this work would have required the rewriting of the SDL diagrams as TNSDL code. On the other hand, the model generator of EMMA is somewhat inflexible and often produces strictly unoptimal nets w.r.t. both analysis and readability. Therefore, a net model was made for the PROD analyser manually [14].

3   Modelling DSS1 Using Pr/T-Nets

For pragmatic reasons, it was decided to treat the primitives sent by the Call Control Block as "spontaneous". (The only so far obtained public description of the block is not very informative. Effectively, it only tells us that the state of the block can be fully calculated from the information that has been in the high-level net models all along.) An arbitrarily behaving Call Control Block can force the User side and the Network side to make their output queues full w.r.t. the chosen capacities and then immediately ask both sides to write again. In this way, a deadlock is reached. The phenomenon is independent of the capacities of the queues as long as all the capacities are finite.

In the chosen way to model DSS1, in order to avoid deadlocks of the above kind, it was decided that a request from the Call Control Block is taken into account when and only when the request can be realized immediately. For the same reason, it was decided that if a reading of a timeout signal has an associated write operation, the reading is synchronised with the writing. In principle, the latter decision damages the timeout mechanism, but if the timer abstraction described below is accepted, the decision becomes acceptable, too.

For the purposes in our analysis the few messages received from the Data Link layer were deemed extraneous.

The SDL description includes two resources which our Pr/T-net model must adopt. The first is the B channel state which is always exactly one of the following: *connected, committed* or *free*. In our net, there is a place for each of the three states on both the Network and the User side. The other resource to be modelled is the call reference. A call reference is used as an identifier for a single call in an ISDN network. It has two possible states, *free* and *committed*. Our model includes a place for both states on the Network and the User side. As for transitions related to the resources, a *select* transition moves a B channel from free to committed, and *connect* moves a B channel from committed to connected. A *disconnect* transition does the inverse of *connect* and *release* does the opposite of *select*. For call references the situation is similar. A *select* transition changes the state of call reference from free to committed and *release* does the opposite action.

The use of "don't care" symbols is one of the most popular abstraction techniques and an obvious way to alleviate state space explosion. At certain states of the models of DSS1, the status of a B channel was known to remain unused until any potential next setting of the status and unused in any such setting as well. At such states, a fixed "don't care" value was used for representing the status.

The SDL diagrams defining the DSS1 protocol have divided the flow of control into 17 states on the Network and 16 states on the User side. Each of these *call states* is represented by one net place. When a token moves from one such

248   Nisse Husberg, Teemu Tynjälä, and Kimmo Varpaaniemi

place to another it indicates a change in the call state. The mapping of SDL states to net places is very intuitive.

The modelling of the SDL diagrams was quite straightforward except for some problems with timers, queues, atomic and non-atomic SDL transitions which are discussed below.

3.1   Converting SDL to Pr/T-Nets

Let us see how one of the SDL transitions in the DSS1 standard [4] can be modelled. The SDL transition in Figure 2 can be modelled by the net transition in Figure 3. The places N_1 and N_3 are modelling the control points in the SDL diagrams. The places N_B_chan_committed and N_B_chan_connected are Network side resources (variables). Finally, the places U_input_empty and U_input model a fixed writing slot of the User side input queue. (One might think that we should always keep the contents of the queue in some "normal form". However, as becomes explained in Section 4.3, we have a good excuse for not doing that.) The output arc to place U_input contains either CALL_PROCEEDING or UNRECOGNIZED_MSG as the message to be transmitted. The model will generate an execution path for both possibilities. The latter represents an erroneous message transmission or message corruption. This is discussed further in Section 3.4. Figure 3 translated into PROD code is shown in Figure 4.

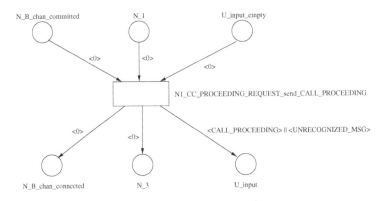

Fig. 3. A Pr/T-net transition corresponding to Figure 2.

It is justifiable to model an SDL transition by one net transition if the SDL transition is *atomary*, i.e. has no interference with the other SDL processes or global resources in the protocol, and if the SDL statements in the SDL transition are independent of each other. (See [10] for a discussion about merging of SDL statements in general.) A non-atomary SDL transition needs several net transitions. Then the protocol control can be modelled by *internal call state places*. According to the semantics of SDL no signals can be read by the process from its input queue unless the control is in the correct (call) state. Thus the internal call

```
#trans N1_CC_PROCEEDING_REQUEST_send_CALL_PROCEEDING
    in          N_1: <.0.>;   N_B_chan_committed: <.0.>;
                U_input_empty: <.0.>;
    out         N_3: <.0.>;   N_B_chan_connected: <.0.>;
                U_input: <.msg.>;
    comp        msg = CALL_PROCEEDING;   Accept();
                msg = UNRECOGNIZED_MSG; Accept();
#endtr
```

Fig. 4. A PROD representation of Figure 3.

state places only allow interleaving of the SDL transition with net transitions generated from other SDL processes.

3.2   Modelling Communication in the DSS1 Model

In SDL the input queues are infinite, but in real systems the queue length is, of course, finite. The analysis of long queues is usually impossible due to the huge state space which they generate. Restricting queue lengths, however, brings up a new problem: what to do if a process wants to write to a queue that is full? There are at least three ways of overcoming the problem. One is that the message is modelled as being lost. In some systems this may be a wise modelling choice. The second way is to overwrite the last element in the queue. This way is analogous to the first way since the only difference is in the choice of the lost message. The third way is to prevent the process from writing to the queue until there is room. Our DSS1 model follows the third way. The reason for this is that in reality very few messages are lost in a wire line telephone network.

3.3   Modelling SDL Timers

There is no time concept in the net class used in PROD. In EMMA, a timer is modelled by a net transition with a usual control place and another Lock place. The timer is started by the arrival of a control token in the control place, but the Lock place stops the timer from expiring. EMMA defines a *timer window* where the timer can expire, by using other transitions in the model to *open* and *close* the Lock. The expiry window can be moved to "interesting" SDL statements to see what happens if the timer expires concurrently to those statements. The net transitions in the window will be interleaved with the timer expiry transition in all possible ways in the analysis.

The expiry of a timer sends a timeout signal to the input queue of the process. This approach is justified by the fact that the communication between processes is completely asynchronous and the time concept has a meaning only within the same process.

In our model we have adopted a variant of this approach. Each timer is represented by two net places: one place indicates that the timer is ON and the other indicates that it is OFF. A *start* transition takes a token from timer_on and puts it to timer_off. A *timeout* transition does the reverse. The timer window in which a timeout can occur is normally small: one call state. DSS1 has been designed in such a way that a maximum of two timers can be on simultaneously in any call state.

Our model does *not* strictly follow the SDL semantics of time. Setting and resetting of timers as well as reading of timeout signals were modelled, but time itself was not modelled, partially because the timers were essentially not "competing" with each other, and partially because any serious inclusion of time would have seriously complicated all analysis. Reading of timeout signals was modelled without modelling the signals explicitly and without modelling expiries explicitly. It was assumed that whenever the latest explicitly modelled operation concerning a timer is a set-it-on operation, a timeout signal from the timer is readable. At worst this means that a timeout signal may become read before some "normal" signal that has been in the input queue for a longer time.

Even if we had decided to follow the SDL semantics, we would have had the problem that the SDL description in [4] is incomplete w.r.t. timers. For example, the timer T307 of the Network side is such that the SDL description in [4] refers to it only in a footnote on page 19: "The expiry of timer T307 is not shown in these SDLs as it runs in the Call Control Block." The project has tried to obtain an SDL description of the Call Control Block, but according to [11], the block does not have any public SDL description. However, a detailed textual description of T307 was found from [3]. The description is shown in Figure 5.

Timer number	Default time-out value	State of call	Cause for start	Normal stop	At the first expiry	At the second expiry	Cross reference
T307	3 min	Null	SUSPEND ACKNOW-LEDGE sent.	RESUME received.	Clear the network connection. Release call identity.	Timer is not restarted.	Mandatory

Fig. 5. An excerpt from "Table 9.1 (2 of 3): Timers in the network side" ([3], p. 153).

3.4   Representing Errors in the DSS1 Model

One of the purposes of the DSS1 model is to verify the correctness of the protocol in various error situations. The DSS1 model considers two erroneous behaviours of a protocol, firstly the loss of a message in the transmission channel, secondly the distortion of data in the transmission. Both errors are simulated by the transmission of a special signal UNRECOGNIZED_MSG. Each time a message is sent there are two possibilities. Either the intended message or the erroneous UNRECOGNIZED_MSG is transmitted over the line.

We used the capability of PROD to model non-deterministic behaviour in net transitions. In the analysis, this means that in addition to correct behaviour, all possible error paths related to message loss and corruption are included in the reachability graph.

3.5   Modelling Accounting in DSS1

As an end-to-end communication protocol, DSS1 plays an integral role in accounting [11]. The Call Control Block is the entity carrying out the actual accounting. Accounting starts when the remote party answers the phone and can end in two ways. One is that the Network side initiates call teardown (this can be thought of as the remote party replacing the receiver). In this case the Call Control Block (in the local exchange) disables accounting as it sends the message DISCONNECT_REQUEST to DSS1. The second manner of call teardown is initiated by the User side of the DSS1 protocol (analogous to the initiator party replacing the receiver). The local exchange of the call initiator will then receive call teardown messages and its DSS1 block shall send messages DIS-CONNECT_IND or RELEASE_IND to the Call Control Block. The reception of either of these messages will disable accounting.

The procedure described above is included in the DSS1 model. The addition of accounting means that the model is extended with two new places, Call_type and Accounting. Call_type describes whether the call is incoming or outgoing. Call charging is applicable only to outgoing calls since our model does not consider collect calls. The Accounting place contains a token indicating whether accounting is active or not.

The transitions handling the connection, disconnection and release of B channels must be augmented with the above places. Transitions that model the activation (connection) of B channels also enable accounting whereas transitions modelling disconnection or release of B channels must make active accounting inactive.

3.6   About the Sizes of the Nets

The modelling covered 145 pages of [4], all these pages consisting of SDL diagrams only. Due to parameters, options and abstractions, there are several models which are variants of each other. For example, the capacity of a queue is a parameter, whereas the inclusion/exclusion of UNRECOGNIZED_MSG is an option. Moreover, some analysis tasks imply inclusion of transitions that would otherwise be excluded.

The number of high-level places and high-level transitions strongly depends on the degree of folding. Each of the nets described in [14] has about 600 high-level places and about 1100 high-level transitions. By carrying out systematic folding, more compact nets have later been obtained, one of them having precisely 10 high-level places and precisely 70 high-level transitions. It would theoretically be possible to continue folding until there would be only one high-level place and only one high-level transition.

If we really want to measure the amount of information in a high-level net, we are more or less forced to talk about a low-level net that corresponds to the high-level net "in a reasonable way". For each analysed model of DSS1, the low-level net computed by PROD has more than 1000 places, more than 10 000 transitions, and more than 90 000 arcs.

4   Analysis of the DSS1 Model

4.1   About the Space Consumed in the Analysis

The amount of information in the full reachability graph depends very strongly on the parameters, options and abstractions. In order to measure the amount of information in a (full or reduced) reachability graph, we should take into account not only the number of vertices and edges but also the contents of the vertices and edges. In the case of PROD, the amount of hard disk space consumed by the graph can be used as a measure. When, as usual, the operating system uses 32-bit addresses for addressing a file, 2 gigabytes is an upper bound for the size of a single file that PROD can handle. In the case of the DSS1 models, this upper bound has been encountered several times. It would of course have been possible to choose an operating system that uses 64-bit addresses instead. However, hard disks of the day still have considerably limited capacities. On the other hand, mere availability of physical resources will hardly ever solve the state space explosion problem.

Due to the unbounded queues of SDL, the complete state space of an SDL system is in principle infinite. In the project being presented, the queues were abstracted to have the capacity 1 or 2, and consequently, the full reachability graphs of the net models were guaranteed to be finite. For a certain model where every queue had the capacity 2, the full reachability graph was still known to be too large to be generated in an operating system that had the above mentioned 32-bit problem. Fortunately, the stubborn set method [15,16] in PROD successfully constructed a reduced reachability graph that sufficed for showing that the full reachability graph had no terminal state. However, even that construction took more than two days. Dropping all the queue capacities from 2 to 1 changed the situation dramatically: it took less than one hour from PROD to construct the full reachability graph and to conclude that it was strongly connected.

4.2   Prioritised Transition Instances for Fast Debugging

Constructing a model of a large system by hand is inevitably error-prone. Detection of modelling errors needs practically as much automatic support as the search for the actual errors of the system. When there are many modelling errors, it is often the case that almost every branch in the reachability graph contains an indication of a modelling error. The state space explosion problem still complicates reachability analysis so much that some way to reduce the reachability graph is needed. Assigning *priorities* to transition instances is an intuitive way to reduce the reachability graph, with the risk that the intuition does not fully grasp the reality.

If an instance a has a priority over an instance b, at most a is allowed to be executed when a and b are simultaneously enabled. (We said "at most" since there can be an instance c which has a priority over a.) The user of PROD is allowed to assign the priorities quite arbitrarily. The cost of such a freedom is that the user is responsible for the quality of the reduced reachability graph. However, the quality is not a problem as long as errors are found from the reduced reachability graph. Intentionally restrictive priorities can be used for reducing the time needed for detecting some of the errors. (Randomly incomplete error detection is a research field of its own. See e.g. [6].)

It is possible to use priorities in such a way that also the absence of errors can be concluded from the reduced reachability graph. As shown by [1], such a controlled priority method is effectively the stubborn set method with strong guidance from the user. In PROD, the priority method has not been integrated with the stubborn set method.

4.3   Supporting the Stubborn Set Method

In low-level Petri net terms, a stubborn set consists of transitions. The stubborn set method computes a stubborn set in every state of the reduced reachability graph which in turn is determined by the computed stubborn sets. The reduced reachability graph is guaranteed to contain all terminal states of the full reachability graph. Additional constraints can be used in order to preserve more properties, such as the validity/invalidity of an LTL formula without a next-time operator. The reduced reachability graph can be finite even when the full reachability graph is infinite.

The stubborn set method in PROD works on P/T-nets (place/transition nets). The needed unfolding from a Pr/T-net into a P/T-net is done automatically, but the user must understand e.g. the correlation between the number of high-level transition instances and the number of low-level transitions. We now consider two difficulties which have been encountered in the DSS1 project. The first difficulty is related to the modelling of queues, whereas the second difficulty is related to redundant places. Both of the difficulties are such that a modeller can circumvent them, and they were circumvented in the DSS1 project in particular. (The difficulties could to some extent be eliminated automatically, but then there should be tools which would really do that. We shall return to this subject in Section 5.3.)

One thing the modeller should learn is that the stubborn set method does not like "simultaneity by atomicity". For example, think of a high-level transition which models reading of a single element from a queue by taking the whole contents of the queue as an input and by returning the new contents as an output.

If the capacity is low enough, unfolding succeeds, but the low-level transitions corresponding to the read operations tend not to commute with the low-level transitions that correspond to the write operations on the same queue. The reduced reachability graph then becomes much larger than necessary. What the modeller should do is to make the elements in the queue move one element per time and one slot per time. With minimal additional guidance from the user, the stubborn set method in PROD is clever enough to avoid introducing redundant interleavings that would be due to the change in atomicity. PROD also has an option which eliminates "intermediate states" [16], such as the states where the queue is not "in the normal form".

The stubborn set method does not easily recognise redundant places, and so it is best to avoid such places. For example, imagine a high-level place that keeps count of the number of tokens in another high-level place while the only use for the value of the counter is to compare the value to a known theoretical maximum. If the value of the counter is represented by a high-level token, transition instances incrementing the counter do not commute with the instances that decrement it. If the value of the counter is represented by a multiplicity of low-level tokens, the stubborn set method falsely assumes that these tokens form a critical resource. (If the counter place also has a classic complement place, the method analogously assumes that the tokens in the complement place form a critical resource, too. Such a pair of critical resource assumptions has a concrete effect, no matter which stubborn set computation algorithm we choose from the literature.) Then again, the reduced state can become much larger than necessary.

4.4   Fairness and Diagnosis of Counterexamples

Liveness properties to be verified were written as LTL formulas, without including any fairness assumption in the formulas. Since PROD has no built-in support for fairness assumptions, unfair counterexamples were obtained. Looking at the counterexamples, it was observed that the unfairness was always caused by messages interchanged during a fully established phone call. Such messages were redundant w.r.t. the properties being verified, and it was easy to abstract such messages out from the model. After sufficiently many redundant messages had been abstracted away, the liveness properties were seen to hold.

The LTL counterexamples reported by PROD are paths in the product of a reachability graph and a certain Büchi automaton. A counterexample which is found is reported in the form of a "characteristic prefix" that contains exactly one cycle, the cycle being at the end of the prefix. (Each terminal state in a reachability graph is transformed into an artificially infinite path by adding a dummy edge from the terminal state to the state itself.) In the analysis of the models of DSS1, a single characteristic prefix reported by PROD typically had hundreds of states. By carrying out separate semiautomatic searches, it was found out that considerably shorter prefixes would have been obtained if PROD had chosen some other counterexamples to the same formulas.

4.5   Potential Errors Found during the Analysis Procedure

The analysis of DSS1 protocol uncovered three problems with the protocol specification. Two are of minor importance but the third one encompasses almost
every call state. Fortunately none of the three problematic definitions is disas-
trous to the correct operation of the protocol. Each one is introduced briefly in
the subsequent paragraphs.

The first problem is on the User side of the protocol. Consider Figure 6 which
shows two branches from different call states on the User side. In the call state 6
(CALL PRESENT), it is possible to receive a PROCEEDING_REQUEST from
the Call Control Block. As the branch shows, the control moves to the call state
9 (INCOMING CALL PROCEEDING), without any timers starting. In the call
state 9, it is possible to receive a DISCONNECT message. The branch on the
right hand side of the figure shows the actions performed as a result. The first
action is to turn off the timer T313. However, that timer is not on in the execution
path described. By investigating all the other diagrams that refer to T313, it can
be shown that T313 is never on in the call state 9. So, either the resetting of
T313 in Figure 6 is a redundant operation, or a few things are missing from the
graphical representation of the standard. The project proceeded by assuming
that the resetting is redundant.

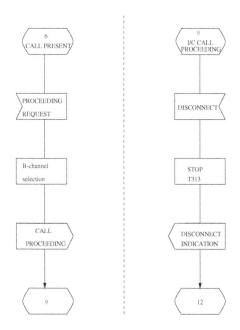

Fig. 6. A potential error on the User side.

The second problem (or inaccuracy) of the specification can be found on
the Network side. In the call state 15 (SUSPEND REQUEST), the execution

Fig. 7. A potential error on the Network side.

after Call Control message SUSPEND_RESPONSE is given in Figure 7. The control moves to the call state 0 (NULL) after the Call Reference is released. However, the B channel is left in an active state. This is a contradiction with the knowledge that the project had about accounting, i.e. accounting was supposed to be inactive during call suspension. The project decided to formally keep the B channel in a "suspended" state when the call is suspended.

The third problem, which is more serious, in the protocol description concerns timers that are possibly left on as the control moves to Call State 0. Every call state has transitions for the reception of RELEASE and RELEASE_COMP-LETE. These transitions do not turn off timers before control moves to call state 0. If a timer is active in call state 0, it may remain active in some subsequent call states too and expire at a moment it should not expire. In order to make further analysis free from the effects of this error, it was eliminated from all variants of the net model by making every reception of messages RELEASE or RELEASE_COMPLETE to reset all timers.

In principle, the same observations could have been made by using static analysis only. However, we have the opinion that some of the positively verified safety and liveness properties would never have become verified without a true reachability analysis tool. It should also be pointed out that the potential problems in the implementations of DSS1 are not necessarily caused by the standard.

4.6   About the Positively Verified Properties

Section 4.1 already mentioned two positive verification results. More generally, whenever PROD succeeded in generating a full reachability graph for a correct model of DSS1, each queue had the capacity 1 and the graph was strongly connected. On the other hand, whenever the stubborn set method in PROD succeeded in generating a reduced reachability graph for a correct model of DSS1, the graph had no terminal state. (The "largest" of these correct models is the one mentioned in Section 4.1, i.e. a model where each queue has the capacity 2.) Provided that the method has been correctly implemented in PROD, the stubborn set theory [15,16] implies that the full reachability graph of the same model has no terminal state either. Though the absence of terminal states is not necessarily any actual requirement, terminal states typically indicate poor design. Analogously, if a full reachability graph is not strongly connected, a good explanation is typically needed.

Though the stubborn set theory and its implementation in PROD supports e.g. verification of LTL formulas, the following results are limited to the cases where the generation of the full reachability graph is no problem.

The positively verified safety properties were related to the correctness of the status of the B channel and to the correctness of accounting, up to the abstraction used in the modelling of the accounting. Each of these properties can be verified by using an additional high-level transition that is enabled precisely at those states which violate the property. For example, it was shown that the accounting is correct w.r.t. those states where the input queue of the User side is empty. (The project did not find a meaningful "correctness of accounting" criterion that would have covered all states.)

The positively verified liveness properties concerned the repeatability of the call state 0. As said in Section 4.4, the liveness properties were formulated as LTL formulas. For example, it was shown that if the Network side is infinitely often in the call state 0, the User side is infinitely often in the call state 0, too.

5   Future Work

5.1   DSS1 Testing System

The modelling of the DSS1 protocol was the first stage in a subproject of the MARIA [18] analyser project. The task is to check the operability of the implementations of the DSS1 protocol using formal methods. We have, however, no access to the documentation of all these implementations. It is thus necessary to design a test system where the implementation is treated as a black box.

The test problem is not one of conformance analysis. These exchanges have been thoroughly tested already and probably work very well according to the conformance tests. The problem is that in real life, there are some users which are behaving very badly. It is thus necessary to create tests which follow the protocols to some extent only.

The problem is that the possibilities are enormous, not only in the sequences of messages but also in the timing. The strategies for such testing are not within the range of this paper but a testing system has been created which can connect this model to the ISDN line of an exchange. The model can be used to create correct sequences of communication and then distorted a little in different ways. Some heuristics can be used to create sequences which have a fair chance of fooling the implementation.

So far the project has succeeded in building a system which attempts to realize any given sequence of send/receive actions. Though the built system has not yet found errors in any exchange, it has detected a modelling error: a certain passed sequence indicated that the behaviour of a certain exchange conformed to an unmodelled branch in a certain SDL diagram.

5.2   MARIA: A Modular Reachability Analyser

During the EMMA project, some deficiencies of the implementation of PROD became obvious. There were problems with the expression power of the input language, especially missing data types. On the other hand, adding new algorithms to PROD tended to be more and more complicated in each addition. Therefore, it was decided that a new modular analyser should be designed.

The MARIA analyser is to have all the capabilities of PROD, but it is *modular* in order to make it easy to develop it further and to add new front-ends and algorithms. We hope to give a large group of users the possibility to alter the code to suit some particular problem.

One of the main tasks in the analyser project is to build a front-end for standard SDL. We also plan to add interfaces to MARIA to allow it to read the output of some largely used SDL tools. Thus the manual error-prone and time-consuming modelling work could be avoided or at least supported by efficient tools.

5.3   The Effect of the DSS1 Project on MARIA

The Partial-Order Package [5] solves the queue problem of Section 4.3 without disturbing the modeller. The solution is to make the queue a black box which can only be accessed by built-in operations of the protocol description language. In PROD, implementing the idea would mean constructing a "black box low-level subnet" which would be handled exceptionally. We plan to implement this idea in the MARIA tool.

An interesting way to solve the problem of redundant places, also discussed in Section 4.3, would be to refine the definition of concrete stubbornness towards the definition of *dynamic stubbornness* [16]. However, a more straightforward way to solve problems related to recognition of redundancy is to carry out a *net reduction* [2] before constructing any reachability graph. Section 4.3 suggests that it may be best to avoid promoting "simultaneity by atomicity" in a net reduction. If some algorithms later need the original net, a bidirectional translation between the original net and the reduced version suffices for that purpose. Unfortunately,

PROD has no automatic support for net reductions. It would thus be nice if MARIA had at least an interface to some net reduction tool. (The laboratory where MARIA is being developed actually has a net reduction tool, called NRED [12], but NRED does not very well support a modern analyser.)

Though justifiable in special cases, categorical elimination of certain messages from a model, as was done in Section 4.4, is not a good general purpose technique for elimination of unfair counterexamples. The MARIA tool is to have explicit support for verification under fairness assumptions.

6   Conclusions

In this work, we have modelled and analysed the ISDN-DSS1 protocol using high-level Petri nets. Several experiments with the stubborn set method have been performed, and the use of priorities has been critically investigated. Like many others, we have learnt something about the benefits and risks of abstraction.

References

1. Falko Bause. Analysis of Petri Nets with a Dynamic Priority Method. In P. Azéma and G. Balbo, editors, *Application and Theory of Petri Nets 1997*, volume 1248 of *LNCS*, pages 215–234. Springer-Verlag, 1997.
2. Gérard Berthelot. Transformations and Decompositions of Nets. In W. Brauer, W. Reisig, and G. Rozenberg, editors, *Petri Nets: Central Models and Their Properties (Advances in Petri Nets 1986, Part I)*, volume 254 of *LNCS*, pages 359–376. Springer-Verlag, 1987.
3. European Telecommunications Standards Institute. *Integrated Services Digital Network (ISDN); User-Network Interface Layer 3; Speci cations for Basic Call Control; Part 1, ETS 300 102-1*, December 1990.
4. European Telecommunications Standards Institute. *Integrated Services Digital Network (ISDN); Digital Subscriber Signalling System No. One (DSS1) Protocol; Signalling Network Layer for Circuit-Mode Basic Call Control; Part 2: Speci ca tion and Description Language (SDL) Diagrams, ETS 300 403-2*, November 1995.
5. Patrice Godefroid and Didier Pirottin. Refining Dependencies Improves Partial-Order Verification Methods. In C. Courcoubetis, editor, *Computer Aided Veri ca tion (CAV '93)*, volume 697 of *LNCS*, pages 438–449. Springer-Verlag, 1993.
6. Gerard J. Holzmann. An Analysis of Bitstate Hashing. *Formal Methods in System Design*, 13(3):289–307, November 1998.
7. Nisse Husberg. Verifying SDL Programs Using Petri Nets. In *1998 IEEE International Conference on Systems, Man, and Cybernetics (Volume 1)*, pages 208–213. IEEE, 1998.
8. Nisse Husberg and Tapio Manner. Emma: Developing an Industrial Reachability Analyser for SDL. In J.M. Wing, J. Woodcock, and J. Davies, editors, *FM'99 — Formal Methods (Volume I)*, volume 1708 of *LNCS*, pages 642–661. Springer-Verlag, 1999.
9. Tero Jyrinki. Dynamical Analysis of SDL Programs with Predicate/Transition Nets. Report B17, Digital Systems Laboratory, Helsinki University of Technology, 1997.

10. Markus Malmqvist. Methodology of Dynamical Analysis of SDL Programs Using Predicate/Transition Nets. Report B16, Digital Systems Laboratory, Helsinki University of Technology, 1997.
11. Sakari Pernu (Nokia Telecommunications), private communication, 1998 and 2000.
12. Marko Rauhamaa. Design and Implementation of a Reduction Tool for PrT-Nets. Report B6, Digital Systems Laboratory, Helsinki University of Technology, 1988.
13. Specification and Description Language (SDL), ITU-T Z.100, 1988.
14. Teemu Tynjälä. Reachability-Based Verification of DSS1 Protocol. Master's thesis, Department of Computer Science and Engineering, Helsinki University of Technology, 1998.
15. Antti Valmari. State Space Generation: Efficiency and Practicality. Doctoral thesis, Publications 55, Tampere University of Technology, 1988.
16. Kimmo Varpaaniemi. On the Stubborn Set Method in Reduced State Space Generation. Doctoral thesis, Report A51, Digital Systems Laboratory, Helsinki University of Technology, 1998.
17. Worldwide web, page http://www.csverilog.com/products/geode.htm.
18. Worldwide web, page http://www.tcs.hut.fi/Software/maria/index.html.
19. Worldwide web, page http://www.tcs.hut.fi/Software/prod/index.html.
20. Worldwide web, page http://www.telelogic.se/solution/tools/sdl.asp.

Process Semantics of
P/T-Nets with Inhibitor Arcs

Jetty Kleijn[1] and Maciej Koutny[2]

[1] LIACS, Leiden University, P.O.Box 9512, NL-2300 RA Leiden, The Netherlands
[2] Dept. of Comp. Sci., University of Newcastle, Newcastle upon Tyne NE1 7RU, U.K.

Abstract. In this paper, we define a process semantics of P/T-nets with inhibitor arcs (PTI-nets). For PTI-nets with bounded inhibiting places, we combine the existing approaches for ordinary P/T-nets and for elementary net systems with inhibitor arcs. To deal with unbounded inhibiting places, a new feature has to be added to the underlying occurrence nets. In either case we show how to construct a process from a step sequence and give a complete characterization of all processes which can be obtained in this way. Using these processes it is possible to express the causal relationships between events in a PTI-net behaviour.
Keywords: Causality/partial order theory of concurrency; analysis and synthesis, structure and behaviour of nets.

1   Introduction

Petri nets with inhibitor arcs have been around for quite some time now and as stated in [12], 'Petri nets with inhibitor arcs are intuitively the most direct approach to increasing the modelling power of Petri nets'. Unlike a 'normal' Petri net, a Petri net with inhibitor arcs has the possibility of testing whether a place is empty in the current marking (*zero testing*). Thus inhibitor arcs are very well suited to model situations involving testing for a specific condition, rather than producing and consuming resources. Place/Transition nets with inhibitor arcs (PTI-nets) are strictly more expressive than ordinary Place/Transition-nets (P/T-nets). They can simulate the computations of Turing machines and several important problems like reachability and liveness which are decidable for P/T-nets are undecidable for PTI-nets.

This paper is concerned with the description of the causal relationships in (concurrent) runs of PTI-nets. The research presented here is a natural continuation of the work of [8] regarding elementary net systems with inhibitor arcs. There, so-called stratified order structures are employed to provide a causality semantics which is consistent with the operational semantics in terms of step sequences. Whereas for an elementary net system, an abstract causality semantics can be given in terms of partial orders alone, the presence of inhibitor arcs requires more information on the relationships between event occurrences. As

---

This research was supported by a grant from the Netherlands Organization for Scientific Research (NWO) and the British Council.

M. Nielsen, D. Simpson (Eds.): ICATPN 2000, LNCS 1825, pp. 261–281, 2000.

an example (borrowed from [8]), consider the net with the two events, e and f, shown in figure 1.

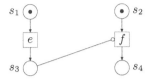

Fig. 1. An elementary net system with inhibitor arc.

In addition to the normal arcs, there is an inhibitor arc from condition s_3 to f. This implies that f can only occur if s_3 is empty (and the standard enabling conditions in an elementary net system are fulfilled). This net has three non-empty firing sequences: $\omega_1 = e$, $\omega_2 = f$ and $\omega_3 = fe$. Note that the occurrence of e is completely independent of the occurrence of f. However, f is disabled after the occurrence of e. This implies that independence of events is no longer symmetric. In the a priori semantics of [8], e and f may also be executed simultaneously, since the inhibiting condition s_3 of f does not hold *prior* to the occurrence of f. Thus also the step $\{e, f\}$ may be executed. This implies that independence and absence of ordering are no longer the same.

Stratified order structures take care of these more involved relations between event occurrences by providing next to a partial order a weak partial order. The partial order describes the strict causal relationships between event occurrences whereas the weak partial order describes weak causal relationships as the above: f *may precede* e but not vice versa and hence the step $\{e, f\}$ may be sequentialised to fe, but not to ef.

For elementary net systems (without inhibitor arcs), an abstract partial order semantics follows immediately from their process semantics (see, e.g., [14]). A process is constructed by unfolding the system according to a given run represented by a firing sequence. The result is an occurrence net: a (labelled) acyclic net with non-branching conditions, since conflicts are resolved during the run. By abstracting from the conditions of the occurrence net, one obtains a (labelled) partial order which describes precisely the causal relationships between the events in the given run: all linearisations of the partial order are firing sequences of the elementary net system and they include the firing sequence on the basis of which the process was constructed.

Also in [8], first a process semantics is given. Since in the a priori semantics not all concurrent runs of the system can be sequentialised to a firing sequence, this process semantics is based on step sequences. (Consider again the elementary net system in figure 1, with an additional inhibitor arc from s_4 to e. Now, $\omega_3 = fe$ is no longer a firing sequence, although $\sigma = \{e, f\}$ is still a legal step sequence.) Given a step sequence, the system is unfolded into a (labelled) occurrence net with additional arcs to represent the zero testing. Testing if a condition does not hold (inhibitor arc) is in the unfolding represented by testing if its complement condition does hold (activator arc). In the resulting activator occurrence net the conditions are again non-branching (with respect to the normal arcs). Moreover, it is acyclic in a sense which includes the activator arcs ($\Diamond$-acyclic) and thus allows to extract a (labelled) stratified order structure which describes precisely the causality and weak causality relationships between the events in the given run: all step sequences which obey the constraints imposed by the stratified order structure are step sequences of the system and they include the step sequence on basis of which the process was constructed.

In this paper we propose a process semantics for PTI-nets with the aim to provide a basis for their abstract causality semantics. Since the nets are no longer necessarily safe (markings may assign more than one token to a place), we combine the ideas of [8] with the definition of processes for (finite) P/T-nets as discussed in, e.g., [6] and [1]. In these processes each token in a place of the original P/T-net is represented by a distinct condition in the process net. Consequently, unfolding the net according to a step sequence in general yields more than one occurrence net. However, the same occurrence nets as employed for elementary net systems are used in the process definition of P/T-nets.

First we consider the case of PTI-nets in which the number of tokens in an inhibiting place cannot grow arbitrarily large (the inhibiting places are bounded). We refer to these nets as PTBI-nets. For them, using complementary places for the inhibiting places and activator arcs in the processes, the ideas of [8] can be combined with the approach of [1] which relates process axioms and inductively defined unfoldings. We define the processes of PTBI-nets and give an unfolding construction based on step sequences. We show that these definitions are consistent with each other, and that they can be used to extract the causal relationships between the events in a run of a PTBI-net.

Next we turn to the unbounded case. In this case, the classical place complementation can no longer be applied. Instead we introduce a new feature in the form of additional conditions (z-conditions) to the occurrence nets. A z-condition represents an empty inhibiting place and is connected by an activator arc to the events representing transitions which test that place for zero tokens. Z-conditions are introduced 'on-demand' during the construction of a process for a given step sequence, and with their introduction an up-date of the occurrence net has to take place. This differs from the standard unfolding procedures discussed above which do not refer to the past and are purely local (based on the neighbourhood of the transitions in the original net). Moreover, z-conditions may be branching (with respect to the normal arcs). Still, the resulting z-activator occurrence nets can be fully (axiomatically) characterised, and they provide us with an abstract causal semantics for the unbounded case.

Both the process semantics and the causal semantics for PTI-nets are consistent with those for PTBI-nets, which in their turn generalise the semantics of P/T-nets as defined in [1] and the semantics of elementary net systems with inhibitor arcs from [8].

This paper is largely self-contained, although it may be an advantage for the reader to be acquainted with the 'classical' process theory as presented in [1,6] and [14]. Due to the page limit, some proofs are either only sketched or omitted.

2   Preliminaries

$\mathbf{N}$ denotes the set $\{0, 1, 2, \ldots\}$ of natural numbers. All functions considered in this paper are total. For a finite set X, we denote by $|X|$ its cardinality.

Let X be a set. A *multiset* (over X) is a function $m : X \to \mathbf{N}$. The sum of two multisets m_1 and m_2 over X is denoted by $m_1 + m_2$ and is defined by $(m_1 + m_2)(x) = m_1(x) + m_2(x)$, for all $x \in X$. The *empty multiset*, denoted by $\underline{0}$, is defined by $\underline{0}(x) = 0$, for all $x \in X$. Note that a multiset m over X may be seen as the subset $\{x \in X \mid m(x) \geq 1\}$ of X, the elements of which are equipped with multiplicities. Conversely, every subset of X may be viewed through its characteristic function as a multiset over X. We denote $x \in m$ if $m(x) \geq 1$.

A *step sequence* (over X) is a finite sequence $m_1 \ldots m_n$ of non-empty multisets m_i (over X). The empty sequence is denoted by λ. If each of the multisets m_i in a step sequence $\sigma = m_1 \ldots m_n$ is a singleton set $\{x_i\}$ (i.e., $m_i(x_i) = 1$ and $m_i(y) = 0$, for all $y \neq x_i$), then σ may be written as $x_1 \ldots x_n$. Thus X^*, the set of all finite sequences of occurrences of elements from X, is a subset of the set of all step sequences over X.

Now assume that X is finite. A *labelling* of X is a function $l : X \to A$, where A is some set of labels (the labelling alphabet). It is extended to step sequences over X in the following way: For $m : X \to \mathbf{N}$, we define $l(m) : A \to \mathbf{N}$ by $l(m)(a) = \sum_{\{x \mid l(x) = a\}} m(x)$, for all $a \in A$. For $\sigma = m_1 \ldots m_n$, we set $l(\sigma) = l(m_1) \ldots l(m_n)$. In particular, $l(\lambda) = \lambda$. Hence step sequences over X are mapped to step sequences over A. Observe that $l(\sigma)$ is in A^*, whenever $\sigma = x_1 \ldots x_n$ is in X^*. In general, however, a set is mapped to a multiset.

For two relations $P, Q \subseteq X \times X$, their composition $P \circ Q$ is also a binary relation over X, defined by $P \circ Q = \{(x, z) \mid \exists y \in X : (x, y) \in P \wedge (y, z) \in Q\}$. Let $id_X = \{(x, x) \mid x \in X\}$ be the identity relation in X. A binary relation P over X is reflexive if $id_X \subseteq P$; it is irreflexive if $id_X \cap P = \emptyset$; and it is transitive if $P \circ P \subseteq P$. The transitive closure of P is denoted by P^+, and its transitive and reflexive closure by P^*.

2.1   Partially Ordered Sets

A *partial order* on X is an irreflexive and transitive binary relation over X. If $\prec \subseteq X \times X$ is a partial order, then the pair $(X, \prec)$ is referred to as a *partially ordered set*, or *poset* for short. In this paper we will only consider *finite* posets (X is finite).

A *labelled poset* is a triple $(X, \prec, l)$ such that $(X, \prec)$ is a poset and $l : X \to A$ is a labelling of X. As we will be mainly dealing with labelled posets, all terminology is introduced directly for labelled posets. If need be, it can be carried over to posets by identifying the poset $(X, \prec)$ with the labelled poset $(X, \prec, id_X)$.

Let $(X, \prec, l)$ be a labelled poset. As usual, for $x, y \in X$, we write $x \prec y$ rather than $(x, y) \in \prec$ and we use $x \preceq y$ to denote that $x = y$ or $x \prec y$. The notation $x \nleftrightarrow y$ indicates that x and y are distinct incomparable elements ($x \neq y \wedge x \nprec y \wedge y \nprec x$).

The labelled poset $(X, \prec, l)$ is *linear* (or *total*), if every two distinct elements are comparable (the relation $\not\smile$ is empty). It is *stratified* [4] if $x \not\smile y$ and $y \not\smile z$ imply that $x \not\smile z$ whenever $x \neq z$. Thus a linear labelled poset is always stratified. Note that $(X, \prec, l)$ is stratified if and only if $\not\smile \cup\, id_X$ is an equivalence relation. If $(X, \prec, l)$ is stratified it defines a unique (ordered) sequence of subsets $X_1 \ldots X_k$ of X, the equivalence classes of $\not\smile \cup\, id_X$, with the property: $\prec = \bigcup_{i<j} X_i \times X_j$, and $\not\smile = (\bigcup_{i=1}^{k} X_i \times X_i) \setminus id_X$. Hence each labelled stratified poset $po = (X, \prec, l)$ as above defines a unique step sequence $\mho_{po} = l(X_1) \ldots l(X_k)$. Conversely, if $po = (X, \prec, l)$ is such that X can be partitioned into non-empty sets $X_1, \ldots, X_k$ satisfying the above conditions, then it is stratified and $\mho_{po} = l(X_1) \ldots l(X_k)$.

Two labelled posets $po_1 = (X_1, \prec_1, l_1)$ and $po_2 = (X_2, \prec_2, l_2)$ are *isomorphic* if there is a bijection $f : X_1 \to X_2$ such that for all $x, y \in X_1$, $l_1(x) = l_2(f(x))$, and $x \prec_1 y$ if and only if $f(x) \prec_2 f(y)$.

Note that every step sequence σ defines an isomorphism class of labelled stratified posets po with the property that $\mho_{po} = \sigma$. In the sequel, however, we are not really interested in the underlying set which is only used to carry labels and we will simply use po_σ to denote any labelled stratified poset po such that $\mho_{po} = \sigma$.

2.2   Stratified Order Structures

A *relational structure* is a triple $\mathcal{S} = (X, \prec, \sqsubset)$, where $\prec$ and $\sqsubset$ are two binary relations over a finite set X. $\mathcal{S}$ is called a *stratified order structure* [5,7], or an *so-structure* for short, if for all $x, y, z \in X$ the following hold (again using the infix notation):

$$
\begin{array}{rcll}
 & x \not\sqsubset x & & \text{C1} \\
x \prec y & \implies & x \sqsubset y & \text{C2} \\
x \sqsubset y \sqsubset z \;\wedge\; x \neq z & \implies & x \sqsubset z & \text{C3} \\
x \sqsubset y \prec z \;\vee\; x \prec y \sqsubset z & \implies & x \prec z & \text{C4.}
\end{array}
$$

It is easily seen that $(X, \prec)$ is a poset and, furthermore, that $x \prec y$ implies $y \not\sqsubset x$. Furthermore, if $(X, \prec)$ is a poset, then $(X, \prec, \prec)$ is an so-structure. In diagrams, $\prec$ is represented by solid arcs, and $\sqsubset$ by dashed arcs. We can omit arcs that can be deduced using C1-C4.

The elements of a relational structure $(X, \prec, \sqsubset)$ will usually be labelled. Thus we consider structures $\mathcal{S} = (X, \prec, \sqsubset, l)$, such that $(X, \prec, \sqsubset)$ is a relational structure and $l : X \to A$ is a labelling of X. All remaining terminology is now introduced directly for labelled relational structures. (It can be carried over to the non-labelled case by identifying $(X, \prec, \sqsubset)$ with $(X, \prec, \sqsubset, id_X)$.) In diagrams, we do not name the nodes but only give their labels.

Concurrency theory employs partial orders $\prec$ to model both specifications and observations of behaviours. On the level of observations, they are used to define operational semantics; $\prec$ is then interpreted as the *earlier than* relation, and $\not\smile$ as (potential) *simultaneity*. On the level of behaviour specifications, $\prec$ is usually interpreted as *causality*, and $\not\smile$ as *independence*. The first relation $\prec_\mathcal{S}$ in an so-structure $\mathcal{S}$, should be interpreted as the standard *causality*, and the second relation, $\sqsubset_S$, as a *weak causality*. While causality is an abstraction of the 'earlier than' relation, weak causality is a similar abstraction of the 'not later than' relation (this should be clearer if one looks at the formula (1) where $\prec_{po}$ represents the former, and $\prec_{po} \cup \not\leftrightarrow$ the latter relation). For a detailed discussion of so-structures the reader is referred to [7].

When used as a tool for representing concurrent behaviours, so-structures are derived from locally defined information involving events which directly interact with one another. This local information then needs to be combined into a global relationship involving all the event occurrences. For this a closure operation is applied which builds an so-structure from representative local relations. The $\Diamond$-closure of a relational structure was introduced in [8] to serve such a purpose.

Let $S = (X, \prec, \sqsubset, l)$ be a labelled relational structure. The $\Diamond$-*closure* of S is the labelled relational structure $S^\Diamond = (X, \prec_{S\Diamond}, \sqsubset_{S\Diamond}, l)$, where

$$\prec_{S\Diamond} = (\prec \cup \sqsubset)^\star \circ \prec \circ (\prec \cup \sqsubset)^\star \quad \text{and} \quad \sqsubset_{S\Diamond} = (\prec \cup \sqsubset)^\star \setminus id_X.$$

We also say that a labelled relational structure S is $\Diamond$-*acyclic* if $\prec_{S\Diamond}$ is irreflexive. The property of $\prec_{S\Diamond}$ being irreflexive, which holds when the structure $S^\Diamond$ obtained from S is an so-structure, has a straightforward interpretation in operational terms. Basically, it means that in any single system history as described by S, there are no event occurrences $e_1, e_2, \ldots, e_k$ such that each e_i has occurred *before or simultaneously with* e_{i+1}, while e_k has occurred *before* e_1.

Proposition 1. *[8] Let $S = (X, \prec, \sqsubset, l)$ be a labelled relational structure.*

1. *$S^\Diamond$ is a labelled so-structure if and only if $\prec_{S\Diamond}$ is irreflexive.*
2. *If S is an so-structure, then $S^\Diamond = S$.* $\qquad\qquad\qquad\qquad\qquad\qquad$ □

We now turn to the relationship between so-structures and stratified posets which resembles that between partial orders and their linear extensions.

A labelled stratified poset $po = (X_{po}, \prec_{po}, l_{po})$ is an *extension* of a labelled so-structure $S = (X_S, \prec_S, \sqsubset_S, l_S)$ if they have the same domain $X_{po} = X_S$ and the same labelling $l_{po} = l_S$, and moreover, $\prec_S \subseteq \prec_{po}$ and $\sqsubset_S \subseteq \prec_{po} \cup \not\leftrightarrow_{po}$. We denote this by $po \in strat(S)$. If $S = (X, \prec, \sqsubset, l)$ is a labelled so-structure then we have [8]:

$$S = \left(X, \bigcap_{po \in strat(S)} \prec_{po}, \bigcap_{po \in strat(S)} (\prec_{po} \cup \not\leftrightarrow_{po}), l \right). \tag{1}$$

Thus S can be derived from its poset extensions. Recall that Szpilrajn's theorem [13] states that each poset is unambiguously identified by its linear extensions. A similar result does not hold for so-structures since these do not necessarily have total order extensions, e.g., $S = (\{a, b\}, \emptyset, \{(a, b), (b, a)\})$. For them one needs to consider stratified poset extensions [9].

Again, we are not interested in the actual carriers of the labels in a poset and so in the sequel we will use the notation $strat(S)$ to denote the set of all isomorphic copies of the labelled stratified poset extensions of S.

We say that two labelled relational structures, $\mathcal{S}_1 = (X_1, \prec_1, \sqsubset_1, l_1)$ and $\mathcal{S}_2 = (X_2, \prec_2, \sqsubset_2, l_2)$, are *isomorphic* if there is a bijection $f : X_1 \to X_2$ such that for all $x, y \in X_1$, $l_1(x) = l_2(f(x))$, and $x \prec_1 y$ if and only if $f(x) \prec_2 f(y)$, and $x \sqsubset_1 y$ if and only if $f(x) \sqsubset_2 f(y)$.

3   Place/Transition Nets with Inhibitor Arcs

This section introduces the notation and terminology for P/T-nets with inhibitor arcs (PTI-nets, for short) and discusses their operationally defined a priori step sequence semantics. PTI-nets have an underlying structure consisting of a net augmented with inhibitor arcs.

A *net* is a triple $N = (S, T, F)$ such that S and T are disjoint finite sets, and $F \subseteq (T \times S) \cup (S \times T)$. The elements of S and T are respectively called *places* and *transitions*, and F is called the *flow relation*. We assume that, for every $t \in T$, $\{s \mid (s, t) \in F\} \neq \emptyset$ and $\{s \mid (t, s) \in F\} \neq \emptyset$ (nets are *T-restricted*).

An *inhibitor net* is a net together with a (possibly empty) set of *inhibitor arcs* leading from places to transitions. (In diagrams, inhibitor arcs have small circles as arrowheads.) Thus an inhibitor net N is specified as a tuple (S, T, F, I) such that (S, T, F) is a net (the underlying net of N) and $I \subseteq S \times T$ is its set of inhibitor arcs. A net (S, T, F) (without inhibitor arcs) is considered as a special instance of an inhibitor net and identified with the inhibitor net $(S, T, F, \emptyset)$.

Given an inhibitor net $N = (S, T, F, I)$ and $x \in S \cup T$, the *post-set* of x, denoted by $x^\bullet$, is defined by $x^\bullet = \{y \mid (x, y) \in F\}$ and the *pre-set* of x, denoted by $^\bullet x$, is defined by $^\bullet x = \{y \mid (y, x) \in F\}$. In addition, for all $t \in T$, $^\circ t = \{s \in S \mid (s, t) \in I\}$ denotes the set of *inhibiting places* of t. These notations are extended to multisets over $S \cup T$ in the following way: For a multiset $U : S \cup T \to \mathbf{N}$, $U^\bullet = \{y \mid \exists x \in U : (x, y) \in F\}$ and $^\bullet U = \{y \mid \exists x \in U : (y, x) \in F\}$; and for a multiset $U : T \to \mathbf{N}$, $^\circ U = \{s \in S \mid \exists t \in U : (s, t) \in I\}$.

A PTI-net is an inhibitor net equipped with an initial state. The states of an inhibitor net are given in the form of markings.

Let $N = (S, T, F, I)$ be an inhibitor net. A *marking* of N is a multiset of places. Following standard terminology, given a marking M of N and a place $s \in S$, we say that s is marked (under M) if $M(s) \geq 1$ and that $M(s)$ is the number of tokens in s under M.

Transitions represent actions which may occur at a given marking and then lead to a new marking. Here we define this dynamics in the more general terms of multisets of (concurrently occurring) transitions. A *step* is a multiset of transitions, $U : T \to \mathbf{N}$. It is *enabled* at a marking M if, for all $s \in S$:

$$M(s) \geq \sum_{t \in s^\bullet} U(t) \quad \text{and} \quad [s \in {}^\circ U \implies M(s) = 0].$$

Thus, by the first condition, in order for U to be enabled at M, for each place s, the number of tokens in s under M should be at least equal to the total number of occurrences of transitions in U that have s as an input place. By the second condition, if a place s is an inhibiting place of some transition occurring in U, then s should be empty in M. Note that the enabledness of a step is based on an *a priori* condition: the inhibiting places of transitions occurring in that step should be empty before it occurs.[1]

If U is enabled at M, then it can be *executed*, which leads to the marking M' defined, for all $s \in S$, by:

$$M'(s) = M(s) - \sum_{t \in s^\bullet} U(t) + \sum_{t \in {}^\bullet s} U(t).$$

This means that the execution of U 'consumes' from each place s a token for each occurrence of a transition in U that has s as an input place, and 'produces' in each place s a token for each occurrence of a transition in U with s as an output place. If the execution of U leads from M to M' we write $M[U\rangle M'$. Note that the empty step $\underline{0}$ is enabled at every marking of N and that its execution has no effect: $M[\underline{0}\rangle M$ for all markings M of N.

A *step sequence* from a marking M to marking M' is a sequence $U_1 \ldots U_n$ of non-empty steps U_i, $1 \leq i \leq n$ with $n \geq 0$, such that

$$M = M_0\,[U_1\rangle\,M_1\,[U_2\rangle M_2 \cdots M_{n-1}\,[U_n\rangle\,M_n = M'$$

for some markings $M_1, \ldots, M_{n-1}$ of N. If τ is a step sequence from M to M' we write $M\,[\tau\rangle\,M'$ and M' is said to be *reachable* (in N) from M. Note that every marking is reachable from itself by the empty step sequence.

In case we want to make clear which (inhibitor) net we are dealing with, we may add a subscript N and write $[\cdot\rangle_N$ rather than $[\cdot\rangle$.

A *Place/Transition net with inhibitor arcs* (or PTI-net) is a tuple $N = (S, T, F, I, M_0)$, where $N' = (S, T, F, I)$ is its underlying inhibitor net, and M_0 is a marking of (S, T, F, I).[2] A *step sequence* of $N = (S, T, F, I, M_0)$ is a step sequence starting from M_0 in its underlying inhibitor net N'. The set of all step sequences of N is the set $steps(N) = \{\tau \mid \exists M : M_0[\tau\rangle_{N'} M\}$.

As the last point of this section, we look at the boundedness of places in N. A place $s \in S$ is *n-bounded* in N, where n is a positive integer, if $M(s) \leq n$ for every marking M reachable from M_0; it is *bounded* if it is n-bounded for some n, otherwise it is *unbounded*. N is *safe* if all of its places are 1-bounded. If s_1 is a bounded place of N, then $s_2 \in S$ is a *complement* place of s_1, if ${}^\bullet s_1 = s_2{}^\bullet$ and $s_1{}^\bullet = {}^\bullet s_2$. Then $bound(s_1) = M_0(s_1) + M_0(s_2)$ is a bound for both s_1 and s_2, and $bound(s_1) = M(s_1) + M(s_2)$, for every marking M reachable from M_0.

We call N a *PTBI-net* if all inhibiting places of all its transitions are bounded.

---

[1] In the *a posteriori* approach [3], the second condition for enabledness is strengthened: for all $s \in S$, $[s \in {}^\circ U \implies (M(s) = 0 \land s \notin U^\bullet)]$. Thus no inhibiting place of a transition in U is also an output place of any transition occurring in U.

[2] Note that I may be empty, in which case we are actually dealing with a P/T-net, and then N may also be specified in the form (S, T, F, M_0).

4   Processes

4.1   Occurrence Nets

For safe P/T-nets and elementary net systems, *processes* can be used as a non-sequential representation of runs of the net (see, e.g., [2,11,14]). Processes are based on occurrence nets and may be viewed as (partial) acyclic unfoldings of the net. Each transition represents an occurrence of a transition in the original net, while each place corresponds to a token. Conflicts between transitions are resolved and thus places do not branch. An occurrence net defines a partial order on its transitions which in turn provides a partial order description of transition occurrences in the original net.

Definition 1. *A* (labelled) occurrence net *is a labelled net* $ON = (B, E, R, l)$ *such that:* $|{}^\bullet b| \leq 1 \geq |b^\bullet|$, *for every* $b \in B$; *the relation* $(R \circ R)|_{E \times E}$ *is acyclic; and* l *is a labelling function for* $B \cup E$. *The elements of* B *and* E — *the places and transitions of* ON — *are respectively called* conditions *and* events.   □

The *minimal* and *maximal* conditions of ON are respectively $Min(ON) = \{b \in B \mid {}^\bullet b = \emptyset\}$ and $Max(ON) = \{b \in B \mid b^\bullet = \emptyset\}$. ON defines a set of step sequences which start from an implicit marking formed by $Min(ON)$ and lead to $Max(ON)$. (Note that the steps in these sequences are sets and that ON with initial marking $Min(ON)$ is safe.) Applying the labelling l to such step sequences yields the set $lsteps(ON) = \{l(\sigma) \mid Min(ON)[\sigma\rangle_{ON} Max(ON)\}$.

Since $(R \circ R)|_{E \times E}$ is acyclic, its transitive closure $\prec_{ON} = ((R \circ R)|_{E \times E})^+$ is irreflexive and we can associate with ON a labelled poset $po_{ON} = (E, \prec_{ON}, l|_E)$.

For EN-systems [14], the notion of occurrence nets provides a causality (partial order) semantics which can be defined in two different ways: (i) axiomatic, from the structure of the net; and (ii) operational, through unfolding based on step sequences. In both cases, the processes and hence also the associated partial orders are the same.

The above approach is not directly applicable to non-safe nets. For these, [6] and [1] propose to represent each of the multiple tokens in a place by a separate condition of an occurrence net. We now provide a rephrasing of the definitions of [1] for the case of general (possibly non-safe) finite P/T-nets.

Definition 2. *Let* $N = (S, T, F, M_0)$ *be a P/T-net. A* process *of* N *is an occurrence net* $ON = (B, E, R, l)$ *such that the following conditions are satisfied:*

1. $l : B \cup E \to S \cup T$ *is such that* $l(B) \subseteq S$ *and* $l(E) \subseteq T$.
2. *For all* $s \in S$: $M_0(s) = |Min(ON) \cap l^{-1}(s)|$.
3. *For all* $s \in S$ *and* $e \in E$:
   (a) $|\{s\} \cap {}^\bullet l(e)| = |\{b \in l^{-1}(s) \mid (b, e) \in R\}|$
   (b) $|\{s\} \cap l(e)^\bullet| = |\{b \in l^{-1}(s) \mid (e, b) \in R\}|$.

We will use $on(N)$ *to denote the set of all processes of* N. □

The above is the axiomatic definition. Alternatively, we can start from a step sequence and construct a corresponding process.

Definition 3. *Let* $N = (S, T, F, M_0)$ *be a P/T-net and let* $\tau = U_1 \ldots U_n$ *be a step sequence of* N. *A process generated by* τ *is the last labelled net* N_n *in a series* $N_0, \ldots, N_n$ *with* $N_k = (B_k, E_k, R_k, l_k)$, *for* $0 \leq k \leq n$, *constructed thus.*

- *Step 0:* $N_0 = (B_0, E_0, R_0, l_0)$ *where*
  - $E_0 = R_0 = \emptyset$ *and* $B_0 = \{b_{s,i,0} \mid 1 \leq i \leq M_0(s)\}$.
  - $l_0 : B_0 \to S$ *is such that* $l(b_{s,i,0}) = s$, *for all* $b_{s,i,0} \in B_0$.
  Let $Max_0 = B_0$.
- *Step* $m = k + 1$: *Let* $N_k = (B_k, E_k, R_k, l_k)$. *Then* N_m *is defined thus:*
  - $B_m = B_k \cup \{b_{s,t,i,m} \mid 1 \leq i \leq U_m(t) \wedge s \in t^\bullet\}$.
  - $E_m = E_k \cup \{e_{t,i,m} \mid 1 \leq i \leq U_m(t)\}$. *Moreover, for each* $e_{t,i,m} \in E_m$ *and each* $s \in {}^\bullet t$ *we choose[3] a distinct* $\widehat{b}_{\langle s,t,i,m \rangle} \in Max_k \cap l^{-1}(s)$.
  - $R_m = R_k \cup \left(\begin{array}{l} \{(\widehat{b}_{\langle s,t,i,m \rangle}, e_{t,i,m}) \mid e_{t,i,m} \in E_m \wedge s \in {}^\bullet t\} \cup \\ \{(e_{t,i,m}, b_{s,t,i,m}) \mid e_{t,i,m} \in E_m \wedge s \in t^\bullet\} \end{array} \right)$.
  - $l_m(b_{s,t,i,m}) = s$ *and* $l_m(e_{t,i,m}) = t$, *for all* $b_{s,t,i,m} \in B_m \setminus B_k$ *and* $e_{t,i,m} \in E_m \setminus E_k$. *Moreover,* $l_m(x) = l_k(x)$, *for all* $x \in B_k \cup E_k$.
  Let $Max_m = \{b \in B_m \mid \neg \exists e \in E_m : (b, e) \in R_m\}$.

We will use $proc_\tau$ *to denote the set of all isomorphic copies[4] of all processes generated by* τ. □

4.2 Activator Occurrence Nets

The presence of inhibitor arcs makes the unfolding procedure more complicated, due to the fact that local information regarding the lack of tokens in a place cannot be explicitly represented in an occurrence net. In [8] this problem is solved by using complement places and representing inhibitor arcs by activator arcs connected to conditions representing complement places. The notion of an occurrence net is replaced by that of an activator occurrence net.

Definition 4. *A* (labelled) *activator occurrence net* (ao-net) *is a tuple* $AON = (B, E, R, Act, l)$ *such that:* $ON = (B, E, R, l)$ *is an occurrence net;* $Act \subseteq B \times E$ *are* activators *arcs; and the relational structure* $\mathcal{S}_{aux}(AON) = (E, \prec_{aux}, \sqsubset_{aux}) = (E, (R \circ R)|_{E \times E} \cup (R \circ Act), (Act^{-1} \circ R) \setminus id_E)$ *is* $\Diamond$-*acyclic.* □

In the diagrams, activator arcs have black dots as arrowheads; see, e.g., figure 4 where (b_2, e) is an activator arc. Figure 2 shows how $\prec_{aux}$ and $\sqsubset_{aux}$ are constructed from ordinary arcs and activator arcs.

Notice that the $\Diamond$-acyclicity of $\mathcal{S}_{aux}(AON)$ implies that $(R \circ R)|_{E \times E}$ is acyclic in the usual sense. Since $\mathcal{S}_{aux}(AON)$ is $\Diamond$-acyclic, we can associate with AON the labelled so-structure $\mathcal{S}(AON) = \mathcal{S}_{aux}(AON)^\Diamond$, see proposition 1. Figure 3 shows the labelled so-structures $\mathcal{S}(AON_i)$ for the ao-nets AON_i in figure 5.

---

[3] This is the only difference with the safe case, where there is only one candidate condition $\widehat{b}_{\langle s,t,i,m \rangle}$, and so the process associated with τ is unique (up to isomorphism).

[4] The construction of a process from step sequences in this and the next sections is based on concrete nodes which carry the labels. This provides us immediately with a fully specified representative of an isomorphism class which is both intuitive and useful in proofs.

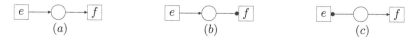

Fig. 2. (a,b) Two cases defining $e \prec_{aux} f$, and (c) one case defining $e \sqsubset_{aux} f$.

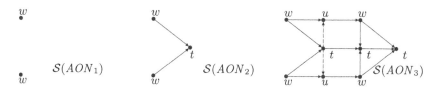

Fig. 3. Stratified order structures generated by ao-nets in figure 5.

Intuitively, an activator arc between a condition b and an event e means that the occurrence of e requires the holding of b, but the occurrence of e will not make b cease to hold. Formally, a step U of events is enabled at a marking M of AON if U is enabled in the underlying occurrence net ON at marking M and, furthermore, for all e in U and $b \in B$, $(b, e) \in Act$ implies that b is marked in M. The resulting marking M' is the same as the marking resulting from the execution of U in ON.[5] As before, we will write $M[\sigma\rangle_{AON} M'$ if executing a step sequence σ in AON leads from M to M'.

The *minimal* and *maximal* conditions of AON are respectively $Min(AON) = Min(ON)(= Min)$ and $Max(AON) = Max(ON)(= Max)$. The step sequences and the reachable markings of AON from the marking Min are also step sequences and reachable markings of ON with initial marking Min. Thus, in particular, (AON, Min) is safe, since (ON, Min) always is. As for occurrence nets, we consider those step sequences which lead from the minimal conditions to the maximal conditions. Applying the labelling l to such step sequences yields the set $lsteps(AON) = \{l(\sigma) \mid Min[\sigma\rangle_{AON} Max\}$. The following result states the correspondence between the (labelled) step sequences of an ao-net AON and the stratified extensions of its associated labelled so-structure $\mathcal{S}(AON)$.

Theorem 1. $strat(\mathcal{S}(AON)) = \{po_\sigma \mid \sigma \in lsteps(AON)\}$.

Proof. Let AON and ON be as in definition 4, and Min and Max be (safe) markings as above. It suffices to show the result assuming that l is the identity labelling for E.

Suppose that $Min[\sigma\rangle_{AON} Max$ and $\sigma = E_1 \dots E_n$. Then also $Min[\sigma\rangle_{ON} Max$. Thus, due to the standard properties of occurrence nets, each E_i is a set and E is the disjoint union of $E_1, \dots, E_n$. Moreover, there are sets of conditions $B_0, \dots, B_k$ of B (cuts of ON, see [1]) such that

$$Min = B_0[E_1\rangle_{ON} B_1 \dots B_{n-1}[E_n\rangle_{ON} B_n = Max \qquad (2)$$

---

[5] Thus an activator arc does not interfere with normal arcs, unlike *read arcs*, [15,3].

and, for every $b \in B$, there are $0 \leq k_b \leq l_b \leq n$ such that

$$b \in B_i \text{ if and only if } k_b \leq i \leq l_b. \qquad (3)$$

In the above, k_b is the index of the first cut B_i in the sequence $B_0, \ldots, B_n$ in which condition b is marked, and l_b is the index of last such cut. Clearly,

$$Min = B_0[E_1\rangle_{AON} B_1 \ldots B_{n-1}[E_n\rangle_{AON} B_n = Max \qquad (4)$$

also holds. To show that $po_\sigma \in strat(\mathcal{S}(AON))$, it suffices to prove that if $e \in E_i$ and $f \in E_j$ then:

$$(\exists b \in B : (e,b) \in R \land (b,f) \in R \cup Act) \Rightarrow i < j. \qquad (5)$$

$$(\exists b \in B : (b,e) \in Act \land (b,f) \in R) \Rightarrow i \leq j. \qquad (6)$$

From $(2,3,4)$ and $E = E_1 \uplus \ldots \uplus E_n$ and $|{}^\bullet b| \leq 1 \geq |b^\bullet|$ it follows that: $(e,b) \in R \Rightarrow i = k_b$; $(b,e) \in R \Rightarrow i-1 = l_b$; and $(b,e) \in Act \Rightarrow k_b \leq i-1 \leq l_b$. Thus (5) holds since $(e,b) \in R \land (b,f) \in R \cup Act$ implies $i = k_b$ and $l_b = j-1 \lor k_b \leq j-1$. And (6) holds since $(b,e) \in Act \land (b,f) \in R$ implies $i-1 \leq l_b = j-1$.

We have shown the $(\supseteq)$ inclusion. To prove the reverse one, suppose that $po_\sigma \in strat(\mathcal{S}(AON))$ and $\sigma = E_1 \ldots E_n$ which means that $E = E_1 \uplus \ldots \uplus E_n$ and $(5,6)$ hold. From (5) (without the $(b,f) \in Act$ part), it follows that $Min[\sigma\rangle_{ON} Max$. Hence there are $B_0, \ldots, B_n$ such that $(2,3)$ hold. To show that $Min[\sigma\rangle_{AON} Max$ also holds, it suffices to observe that if $e \in E_i$ and $(b,e) \in Act$ then $b \in B_{i-1}$. Indeed, if this was not true, then $l_b < i-1$ or $k_b \geq i$. In the former case, there is $f \in E_{l_b+1}$ such that $(b,f) \in R$, a contradiction with (6). And, in the latter case, there is $f \in E_{k_b}$ such that $(f,b) \in R$, a contradiction with (5). $\qquad \square$

The labelled step sequences of AON have a causality interpretation in terms of the partial order and the weak partial order provided by $\mathcal{S}(AON)$. In fact, a single partial order (as defined by an occurrence net) is insufficient, as it cannot fully express the relationship between simultaneous events (in a step) if they cannot be sequentialized. For example, in figure 4 we have that $\sigma_1 = \{e,f\}$ and $\sigma_2 = \{e\}\{f\}$ are step sequences leading from Min to Max, but $\{f\}\{e\}$ cannot be executed, despite the fact that e and f are independent as far as the usual partial ordering is concerned.

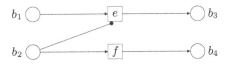

Fig. 4. An activator occurrence net where $Min = \{b_1, b_2\}$ and $Max = \{b_3, b_4\}$.

In the next section, we will combine the approaches of [1] and [8] in order to obtain a causal semantics for PTI-nets in case the inhibiting places have known bounds. The treatment of unbounded inhibiting places will require a further extension of occurrence nets.

5   The Bounded Case

In this section $N = (S, T, F, M_0, I)$ is a fixed PTBI-net and $N' = (S, T, F, M_0)$ is its underlying P/T-net. We assume here that every inhibiting place $s \in S$ has a unique complement place $s^{cpl} \in S$ with $M_0(s) + M_0(s^{cpl}) = bound(s)$ where $bound(s) > 0$ is a bound of s in N. The processes of N are defined as follows.

Definition 5. *An* activator process *of N is an ao-net $AON = (B, E, R, Act, l)$ such that $ON = (B, E, R, l) \in on(N')$ and, for all $s \in S$ and $e \in E$:*

$$|\{s\} \cap {}^\circ l(e)| \cdot bound(s) = |\{b \in l^{-1}(s^{cpl}) \mid (b, e) \in Act\}|. \tag{7}$$

We will use $aon(N)$ to denote the set of all activator processes of N.   □

Figure 5 shows an example of a PTBI-net and its three activator processes.

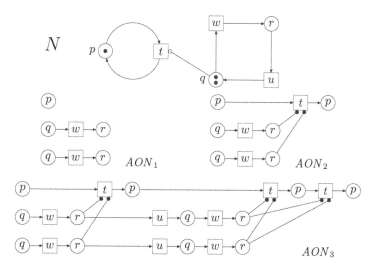

Fig. 5. Three activator processes AON_i of a PTBI-net N.

The first result we show states that an activator process of a P/T-net describes a set of valid step sequences of the original net.

Lemma 1. *If $AON \in aon(N)$, then $lsteps(AON) \subseteq steps(N)$.*

Proof. Let AON and ON be as in definition 5, and $\tau \in lsteps(AON)$. Then, by theorem 1, there is a step sequence $\sigma = E_1 \ldots E_n$ such that $\tau = l(\sigma)$ and $Min(AON)[\sigma\rangle_{AON} Max(AON)$ and $E = E_1 \uplus \ldots \uplus E_n$ and (5,6) in the proof of theorem 1 hold. Since $ON \in on(N')$, we have, by the standard theory [1], that $\tau \in steps(N')$. Moreover, there are sets of conditions $B_0, \ldots, B_n$ such that (2,3) in the proof of theorem 1 hold and:

$$M_0 = l(B_0)[l(E_1)\rangle_{N'}l(B_1) \ldots [l(E_n)\rangle_{N'}l(B_n)$$

Thus, to prove $\tau \in steps(N)$, it suffices to show that if $e \in E_i$ and $s \in {}^\circ l(e)$, then $l(B_{i-1})(s) = 0$. The latter is equivalent to $l(B_{i-1})(s^{cpl}) = bound(s)$. If this does not hold then, by (7), there is $b \in B$ such that $(b, e) \in Act$ and $l_b < i - 1$ or $i \leq k_b$. We then obtain a contradiction with (5,6), similarly as in the last part of the proof of theorem 1. □

Definition 5 can be made operational through the following net unfolding which takes a step sequence and constructs an ao-net corresponding to it.

Definition 6. *Let* $\tau = U_1 \ldots U_n$ *be a step sequence of N. An* activator pro-cess *generated by τ is the last labelled net N_n with activator arcs in a series $N_0, \ldots, N_n$ with $N_k = (B_k, E_k, R_k, Act_k, l_k)$, for $0 \leq k \leq n$, constructed thus.*

- *Step 0: $N_0 = (B_0, E_0, R_0, Act_0, l_0)$ where $Act_0 = \emptyset$, and all other components are as in Step 0 of definition 3, including Max_0.*
- *Step $m = k + 1$: Let $N_k = (B_k, E_k, R_k, Act_k, l_k)$. Then N_m is defined thus:*
  - *B_m, E_m, R_m, l_m and Max_m are as in Step m of definition 3.*
  - *$Act_m = Act_k \cup \{(b, e) \in Max_k \times (E_m \setminus E_k) \mid (l_m(b)^{cpl}, l(e)) \in I\}.$*

We will use $proc_\tau^{ao}$ to denote the set of all isomorphic copies of all activator processes generated by τ. □

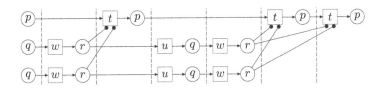

Fig. 6.   Deriving   an   activator   process   in $proc_\tau^{ao}$ for τ = $\{w, w\}\{t\}\{u, u\}\{w, w\}\{t\}\{t\}$.

Figure 6 illustrates the construction of an activator process for the PTBI-net in figure 5. The vertical lines indicate the stages (from left to right) in which the net has been derived. Notice that it is an activator process of N in figure 5 as it is isomorphic to AON_3 shown there. The next results states that this is not a mere chance, since every unfolding of a PTBI-net satisfies the axiomatic definition of an activator process.

Lemma 2. *For τ and N_n in definition 6, $N_n \in aon(N)$ and $\tau \in lsteps(N_n)$.*

Proof. Assume the notation from definition 6. That $ON = (B_n, E_n, R_n, l_n) \in proc_\tau$ for N' follows directly from the definitions and thus, by the standard results for P/T-nets [1], $ON \in on(N')$. Moreover, the construction is such that, for $k = 1, \ldots, n$, $M_0 = l(Max_0)[U_1 \ldots U_k\rangle_{N'} l(Max_k)$ and so also $M_0 = l(Max_0)[U_1 \ldots U_k\rangle_N l(Max_k)$. Thus, if $e \in E_k \setminus E_{k-1}$ and $s \in {}^\circ l(e)$, then we have $l(Max_{k-1})(s) = 0$ and so $l(Max_{k-1})(s^{cpl}) = bound(s)$. Hence

$$|\{b \in l^{-1}(s^{cpl}) \mid (b, e) \in Act_k\}| = |\{b \in Max_{k-1} \mid l(b) = s^{cpl}\}| = bound(s).$$

As a result, (7) is satisfied. To complete the proof of $N_n \in aon(N)$, we still need to show that $\mathcal{S}_{aux}(N_n)$ is $\Diamond$-acyclic. This, however, follows from an easy observation that the conditions (5,6) from the proof of theorem 1 (suitably re-interpreted by setting each E_i to be the set of events added in step i of the construction described in definition 6), hold here by construction.

That $\tau \in lsteps(N_n)$ follows immediately from the construction of N_n and a simple inductive argument. □

Corollary 1. *If $\tau \in steps(N)$ and $AON \in proc_\tau^{ao}$, then $\tau \in lsteps(AON)$.* □

Similarly as it is the case for processes of ordinary P/T-nets, the axiomatic and operational definitions of processes of a PTBI-net coincide.

Theorem 2. $aon(N) = \bigcup_{\tau \in steps(N)} proc_\tau^{ao}$.

Proof. The ($\supseteq$) inclusion follows from lemma 2. To show the reverse one, we take AON and ON as in definition 5. Then, by $strat(\mathcal{S}(AON)) \neq \emptyset$ which always holds [7], there is at least one τ such that $po_\tau \in strat(\mathcal{S}(AON))$. By lemma 1 and theorem 1, $\tau \in steps(N)$ and so $\tau \in steps(N')$. Thus, by the standard properties of processes of P/T-nets, there is a way in which the construction described in definition 3 generates a net $N_n = (B_n, E_n, R_n, l_n)$ which is isomorphic to ON. One can then re-run the construction of ON, adding at each stage the sets Act_k, as prescribed in definition 6. This is a deterministic procedure which results in an activator net which is isomorphic to AON. In proving the latter, one takes advantage of theorem 1, which guarantees that $\tau \in lsteps(AON)$. □

We now can establish that activator processes of a PTBI-net generate exactly the same step sequences as the original net.

Theorem 3. $steps(N) = \bigcup_{AON \in aon(N)} lsteps(AON)$.

Proof. The ($\supseteq$) inclusion has been proved in lemma 1. The reverse inclusion follows from corollary 1 and theorem 2. □

The last result can be re-stated in terms of labelled stratified posets and thus shows that the activator processes of a PTBI-net correctly describe causality in the runs of the net.

Corollary 2. $\{po_\tau \mid \tau \in steps(N)\} = \bigcup_{AON \in aon(N)} strat(\mathcal{S}(AON))$.

Proof. Follows from theorems 1 and 3. □

6   Unboundedness

In this section, we deal with PTI-nets whose inhibiting places can be unbounded. Thus we cannot use complement places to represent the emptiness of places, and therefore need to introduce another device. It will be provided by z-places that will play a role similar to that of the complement places in activator process. However, z-places will represent *logical conditions* rather than tokens (*resources*), and will admit branching. Let $N = (S, T, F, M_0, I)$ be a PTI-net fixed throughout this section.

Definition 7. *A (labelled) z-activator occurrence net (zao-net) is a tuple AON^z = (B, E, R, Act, l) such that: $ON = (B^n, E, R', l|_{B^n \cup E})$ is an occurrence net, where $B^n = B \setminus B^z$ and $B^z = \{b \in B \mid (b, e) \in Act\}$; $R \subseteq (B \times E) \cup (E \times B)$ and $R' = R|_{(B^n \times E) \cup (E \times B^n)}$; $Act \subseteq B^z \times E$ is a set of activator arcs; l is a labelling function for $B \cup E$; and the relational structure $S_{aux}(AON^z) = (E, \prec_{aux}, \sqsubset_{aux})$ = $(E, (R \circ R)|_{E \times E} \cup (R \circ Act), (Act^{-1} \circ R) \setminus id_E)$ is $\Diamond$-acyclic.* ☐

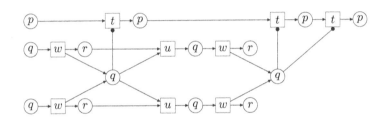

Fig. 7. A z-activator occurrence net.

Figure 7 shows an example of a zao-net. The semantics of a zao-net can be understood in two ways. First, we can take the underlying order structure $S(AON^z) = S_{aux}(AON^z)^{\Diamond}$, as we did for ao-nets, and derive all stratified order extensions, or step sequences corresponding to these. The alternative view, involving step sequences executed from the initial marking, $Min(AON^z) = Min(ON)$, to the final marking, $Max(AON^z) = Max(ON)$, is not directly applicable since z-conditions allow branching. However, it is possible to replace the z-conditions by sets of ordinary conditions for each pair of pre- and post-event of a given z-condition, as described below.

Definition 8. *Let $AON^z = (B, E, R, Act, l)$ be as in definition 7, and:*

- *$B' = B^n \cup B''$ where*
  $B'' = \{b_{x,y} \mid b \in B^z \wedge (x \in {}^\bullet b \vee x = \emptyset = {}^\bullet b) \wedge (y \in b^\bullet \vee y = \emptyset = b^\bullet)\}$.
- *$R' = R|_{(B^n \times E) \cup (E \times B^n)} \cup \{(e, b_{e,y}) \mid e \in E\} \cup \{(b_{x,e}, e) \mid e \in E\}$.*
- *$Act' = \{(b_{x,y}, e) \mid (b, e) \in Act\}$.*
- *$l'|_{B^n \cup E} = l|_{B^n \cup E}$ and $l'(b_{x,y}) = l(b)$, for all $b \in B''$.*

We then call $\zeta(AON^z) = (B', E, R', Act', l')$ the z-pruning of AON^z. ☐

It is not difficult to see that the z-pruning of AON^z is an ao-net. It is used to give the activator arcs an operational semantics which corresponds to the intuition behind the z-conditions. We define the (labelled) step sequences of AON^z by $lsteps(AON^z) = lsteps(\zeta(AON^z))$. Observe that $\mathcal{S}(\zeta(AON^z)) = \mathcal{S}(AON^z)$. Consequently, $strat(\mathcal{S}(AON^z)) = \{po_\sigma \mid \sigma \in lsteps(AON^z)\}$, by theorem 1. We now give an axiomatisation of the notion of process for the PTI-net N.

Definition 9. *A z-activator process of N is a zao-net $AON^z = (B, E, R, Act, l)$ such that:*

1. *$l : B \cup E \to S \cup T$ is such that $l(B) \subseteq S$ and $l(E) \subseteq T$.*
2. *For all $s \in S$: $M_0(s) = |Min(AON^z) \cap l^{-1}(s) \cap B^n|$.*
3. *For all $s \in S$ and $e \in E$:*
   (a) $|\{s\} \cap {}^\bullet l(e)| = |\{b \in l^{-1}(s) \cap B^n \mid (b, e) \in R\}|$.
   (b) $|\{s\} \cap l(e)^\bullet| = |\{b \in l^{-1}(s) \cap B^n \mid (e, b) \in R\}|$.
   (c) $|\{s\} \cap {}^\circ l(e)| = |\{b \in l^{-1}(s) \mid (b, e) \in Act\}|$.
4. *For all $b^z \in B^z$ and $e \in E$:*
   (a) If $(b^z, e) \in R$, then $(l(e), l(b^z)) \in F$.
   (b) If $(b^z, e) \in R^*$ and $(l(e), l(b^z)) \in F$, then there is a unique $b \in B^z$ such that $l(b) = l(b^z)$ and $(b, e) \in R$ and $(b^z, b) \in R^*$.
   (c) If $(e, b^z) \in R$, then $(l(b^z), l(e)) \in F$.
   (d) If $(e, b^z) \in R^*$ and $(l(b^z), l(e)) \in F$, then there is a unique $b \in B^z$ such that $l(b) = l(b^z)$ and $(e, b) \in R$ and $(b, b^z) \in R^*$.
5. *For all $b^z \in B^z$ and $b \in B$, if $l(b) = l(b^z)$, then $(b^z, b) \in R^*$ or $(b, b^z) \in R^*$.*

We will use $aon^z(N)$ to denote the set of z-activator processes of N. □

Note the absence of place bounds in the above definition. Instead, we have an explicit 'record' of the fact that a place was empty in the form of a z-condition. By points 4(a) and 4(c) above, if a z-condition b^z is input (output) to an event e, then the inhibiting place $l(b^z)$ of N is output (input) to the transition $l(e)$. Requirement 4(b) prescribes that whenever transition $l(e)$ adds a token to the inhibiting place $l(b^z)$, only the most recent record b of $l(b^z)$ being empty in the past of the occurrence e of $l(e)$ is input to e. Similarly, 4(d) stipulates that whenever transition $l(e)$ removes a token from the inhibiting place $l(b^z)$, while sometime in the future of this occurrence $l(b^z)$ is successfully tested for emptiness, the occurrence e of $l(e)$ is only connected to the earliest future record b of $l(b^z)$ being empty. Note that by definition 9(5), all records of the emptiness of an inhibiting place are linearly ordered by R^*. Moreover, according to R^* an inhibiting place is never recorded to be empty while it contains a token.

Figure 7 shows a z-activator processes for the net shown in figure 5. It corresponds to AON_3 in figure 5 in the sense that they generate isomorphic labelled so-structures. The last definition is also illustrated for a non-PTBI-net, in figure 8. We finally define an unfolding procedure for PTI-nets.

Definition 10. *Let $\tau = U_1 \ldots U_n$ be a step sequence of N. A z-activator process generated by τ is the last labelled net N_n with activator arcs in a series $N_0, \ldots, N_n$ with $N_k = (B_k, E_k, R_k, Act_k, l_k)$, for $0 \leq k \leq n$, constructed thus:*

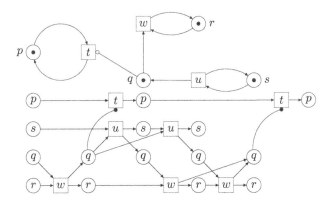

Fig. 8. A z-activator process of a PTI-net which is not a PTBI-net.

- *Step 0:* $N_0 = (B_0, E_0, R_0, Act_0, l_0)$ *where:*
  - $E_0 = R_0 = Act_0 = B_0^z = \emptyset$.
  - $B_0 = B_0^n = \{b_{s,i,0} \mid 1 \leq i \leq M_0(s)\}$.
  - $l_0 : B_0 \to S$ *is such that* $l(b_{s,i,0}) = s$, *for all* $b_{s,i,0} \in B_0$.

  Let $Max_0 = B_0$.
- *Step* $m = k+1$: *Let* $N_k = (B_k, E_k, R_k, Act_k, l_k)$. *Then* N_m *is defined thus:*
  - $B_m^n = B_k^n \cup \{b_{s,t,i,m} \mid 1 \leq i \leq U_m(t) \wedge s \in t^\bullet\}$ *and*
    $B_m^z = B_k^z \cup \{b_{s,m} \mid \exists t \in U_m : s \in {}^\circ t \setminus l_k(Max_k \cap B_k^z)\}$.
  - $E_m = E_k \cup \{e_{t,i,m} \mid 1 \leq i \leq U_m(t)\}$.
    Moreover, for each $e_{t,i,m} \in E_m$ *and for each* $s \in {}^\bullet t$ *we choose a distinct*
    $\widehat{b}_{\langle s,t,i,m\rangle} \in Max_k \cap B_k^n \cap l^{-1}(s)$.
  - $l_m(b_{s,t,i,m}) = s$ *and* $l_m(b_{s,m}) = s$ *and* $l_m(e_{t,i,m}) = t$,
    for all $b_{s,t,i,m} \in B_m^n \setminus B_k^n$ *and* $b_{s,m} \in B_m^z \setminus B_k^z$ *and* $e_{t,i,m} \in E_m \setminus E_k$.
    $l_m(x) = l_k(x)$, *for all* $x \in B_k \cup E_k$.
  - $R_m = R_k \cup \left(\begin{array}{l} \{(\widehat{b}_{\langle s,t,i,m\rangle}, e_{t,i,m}) \mid e_{t,i,m} \in E_m \wedge s \in {}^\bullet t\} \cup \\ \{(e_{t,i,m}, b_{s,t,i,m}) \mid e_{t,i,m} \in E_m \wedge s \in t^\bullet\} \end{array} \right) \cup R_m' \cup R_m''$

    where

    $$R_m' = \left\{ (e, b_{s,m}) \in E_k \times (B_m^z \setminus B_k^z) \left| \begin{array}{l} (s, l_k(e)) \in F \wedge \neg\exists b' \in B_k^z : \\ l_k(b') = s \wedge (e, b') \in R_k \end{array} \right. \right\}$$

    $$R_m'' = \left\{ (b_{s,i}, e) \in B_m^z \times (E_m \setminus E_k) \left| \begin{array}{l} (l_m(e), s) \in F \wedge \\ \forall b_{s,j} \in B_m^z : j \leq i \end{array} \right. \right\}.$$

  - $Act_m = Act_k \cup \{(b, e) \in (Max_m \cap B_m^n) \times (E_m \setminus E_k) \mid (l_m(b), l_m(e)) \in I\}$,
    where $Max_m = \{b \in B_m \mid \neg\exists e \in E_m : (b, e) \in R_m\}$.

We will use $proc_\tau^{zao}$ *to denote the set of all isomorphic copies of all z-activator processes generated by* τ. $\qquad\square$

The above definition is illustrated for the PTBI-net of figure 5 and its step sequence $\tau = \{w, w\}\{t\}\{u, u\}\{w, w\}\{t\}\{t\}$. As before, figure 9 shows stages in which the nodes and connections were generated.

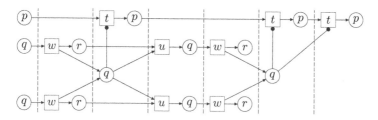

Fig. 9. A z-activator process generated for a step sequence of a PTBI-net.

The z-conditions are generated 'on-demand', when it is necessary to 'legitimise' transition occurrences. In general, this excludes undesirable orderings between events. For consider the net N in figure 5 and its step sequence $\tau = \{w, w\}\{u, u\}$. If we were to add a z-condition each time q becomes empty, then we would generate an occurrence net as shown in figure 10. Intuitively, such a net would introduce artificial causal relationships between some of the event occurrences.

As in the case of processes of P/T-nets and PTBI-nets, the axiomatic and operational definitions coincide. Moreover, we have the desired consistency between step sequences of a PTI-net and its zao-processes.

Theorem 4. *The following are satisfied.*

1. $aon^z(N) = \bigcup_{\tau \in steps(N)} proc_\tau^{zao}$.
2. $steps(N) = \bigcup_{AON^z \in aon^z(N)} lsteps(AON^z)$.
3. $\{po_\tau \mid \tau \in steps(N)\} = \bigcup_{AON^z \in aon^z(N)} strat(\mathcal{S}(AON^z))$. $\square$

The proofs of the various parts of theorem 4 follow those of similar results presented in the previous section. A main change is that we no longer can use complement places to establish the emptiness of an inhibiting place, and instead need to refer to the corresponding z-conditions.

It can be seen that both the process semantics and the causal semantics for PTI-nets developed in this section are consistent with those developed for PTBI-nets in the previous section. The latter, in turn generalises the semantics of P/T-nets [1] and elementary net systems with inhibitor arcs from [8].

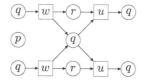

Fig. 10. Generating z-conditions may not be desirable.

7 Concluding Remarks

The basic contribution of this paper is a proposal for a process semantics for P/T-nets with inhibitor arcs while assuming an a priori operational semantics. This contrasts with the approach of [3] where transitions can occur in a step if and only if they can occur in either order. First we generalised the existing process notions for ordinary P/T-nets ([6,1,14]) and for safe nets with inhibitor arcs ([8]) to the case of P/T-nets with bounded and complemented inhibiting places. In order to obtain a process semantics for general PTI-nets, z-activator occurrence nets were introduced. Given the processes, their associated stratified order structures provide a specification of the net behaviours in terms of causality and weak causality. Thus the results in this paper form a basis for a further investigation of the abstract causal relations within the behaviours of a PTI-net. There are at least two potential applications of these results: first, they can be useful in the development of model checking algorithms for PTI-nets based on unfoldings; second, they can be used as a basis for obtaining a causality semantics for P/T-nets with priorities, extending the results obtained for the elementary net systems with priorities in [10]. Finally, the approach presented in this paper can easily be generalised to nets with weighted arcs; an extension to weighted inhibitor arcs is a matter for future research.

Acknowledgment We are grateful to the referees for their constructive criticism and many suggestions to improve our presentation.

References

1. E. Best and R. Devillers: Sequential and Concurrent Behaviour in Petri Net Theory. *Theoretical Computer Science* 55 (1988) 87-136.
2. E. Best and C. Fernández: *Nonsequential Processes. A Petri Net View.* EATCS Monographs on Theoretical Computer Science, Springer-Verlag (1988).
3. N. Busi and G. M. Pinna: Process Semantics for Place/Transition Nets with Inhibitor and Read Arcs. *Fundamenta Informaticae* 40 (1999) 165-197.
4. R. Fräisse: *Theory of Relations.* North Holland (1986).
5. H. Gaifman and V. Pratt: Partial Order Models of Concurrency and the Computation of Function. Proc. of *LICS'87*, IEEE Computer Society Press (1987) 72-85.
6. U. Goltz and W. Reisig: The Non-sequential Behaviour of Petri Nets. *Information and Control* 57 (1983) 125-147.
7. R. Janicki and M. Koutny: Structure of Concurrency. *Theoretical Computer Science* 112 (1993) 5-52.
8. R. Janicki and M. Koutny: Semantics of Inhibitor Nets. *Information and Computation* 123 (1995) 1-16.
9. R. Janicki and M. Koutny: Order Structures and Generalisations of Szpilrajn's Theorem. *Acta Informatica* 34 (1997) 367-388.
10. R. Janicki and M. Koutny: On Causality Semantics of Nets with Priorities. *Fundamenta Informaticae* 38 (1999) 1-33.
11. M. Nielsen, G. Rozenberg and P. S. Thiagarajan: Behavioural Notions for Elementary Net Systems. *Distributed Computing* 4 (1990) 45-57.

12. J. L. Peterson: *Petri Net Theory and the Modeling of Systems.* Prentice Hall (1981).
13. E. Szpilrajn: Sur l'extension de l'ordre partiel. *Fundamenta Mathematicae* 16 (1930) 386-389.
14. G. Rozenberg and J. Engelfriet: Elementary Net Systems. In: *Advances in Petri Nets. Lectures on Petri Nets I: Basic Models,* W. Reisig and G. Rozenberg (Eds.). Springer-Verlag, Lecture Notes in Computer Science 1491 (1998) 12-121.
15. W. Vogler: Partial Order Semantics and Read Arcs. Proc. of *MFCS'97,* I. Privara and P. Ruzicka (Eds.). Springer-Verlag, Lecture Notes in Computer Science 1295 (1997) 508-517.

Improved Question-Guided Stubborn Set Methods for State Properties

Lars Michael Kristensen[1] and Antti Valmari[2]

[1] University of Aarhus, Department of Computer Science
Aabogade 34, DK-8200 Aarhus N., Denmark
kris@daimi.au.dk
[2] Tampere University of Technology, Software Systems Laboratory
PO Box 553, FIN-33101 Tampere, Finland
ava@cs.tut.fi

Abstract. We present two new question-guided stubborn set methods for state properties. The first method makes it possible to determine whether a marking is reachable in which a given state property holds. It generalises the results on stubborn sets for state properties recently suggested by Schmidt in the sense that his method can be seen as an implementation of our more general method. We propose also alternative, more powerful implementations that have the potential of leading to better reduction results. This potential is demonstrated on some practical case studies. As an extension of the first method, we present a second method which makes it possible to determine if from all reachable markings it is possible to reach a marking where a given state property holds. The novelty of this method is that it does not rely on ensuring that no transition is ignored in the reduced state space. Again, the benefit is in the potential for better reduction results.

1 Introduction

State space methods have proven powerful in the analysis and verification of concurrent systems. Unfortunately, the state spaces of systems tend to grow very rapidly when systems become bigger. This well-known phenomenon is referred to as *state explosion*, and it is a serious problem for the use of state space methods in the analysis of real-life systems.

Many techniques for alleviating the state explosion problem have been suggested, such as the *stubborn set methods* [7,10]. The stubborn set methods range from methods for deadlock detection to methods preserving the properties which can be expressed in the temporal logic CTL^*_{-X}. They comprise a subgroup of rather similar methods first suggested in the late 80's and early 90's [3, 4, 5], and are based on the fact that the total effect of a set of concurrent transitions is independent of the order in which the transitions occur. Therefore, it often suffices to investigate only one or some orderings in order to reason about the behaviour of the system.

This paper presents two new stubborn set methods which make it possible to reason about *state properties*. A state property is a property that talks about

M. Nielsen, D. Simpson (Eds.): ICATPN 2000, LNCS 1825, pp. 282–302, 2000.

only one marking. For instance, $M(p) \leq 10$ is a state property, whereas $\exists M' \in [M\rangle : M'(p) > M(p)$ is not.

The first stubborn set method makes it possible to answer the following question: "is it possible to reach a marking where a given state property holds?" The method is *question-guided*, i.e., it takes a state property as input and generates a reduced state space. This reduced state space will contain a marking where the property holds if and only if there exists a reachable marking in which the state property holds. This method is important, because with it one can, e.g., find place bounds, and check reachability of a (perhaps incompletely specified) marking, more efficiently than with existing stubborn set methods [9, 8, 10, 6]. The method presented is based on the ideas in [6], but tries to compute better stubborn sets. This can potentially lead to better reduction results.

The second question-guided method makes it possible to answer the question: "is it possible from all reachable markings to reach a marking where a given state property holds?" This method can for instance be used to check liveness of a single transition with better reduction results than an earlier method [9] that check liveness of all transitions simultaneously. It can also be used to check whether a given (perhaps incompletely specified) marking is a home marking more efficiently than with the technique described in [6].

The paper is organised as follows. Section 2 recalls the basic facts of Place/-Transition Nets (PT-nets), state spaces, and stubborn sets used in the rest of this paper. Section 3 gives an informal introduction to the first stubborn set method by means of a small example. Sections 4-7 formally develop the new stubborn set methods, and Sect. 8 considers their implementation. Section 9 discusses applications of the first method to boundedness properties of PT-nets. Section 10 gives some numerical data on the performance of the first method on some case studies. Finally, we sum up the conclusions in Sect. 11.

2   Background

This section briefly summarises the basic facts and notation of PT-nets, state spaces, and stubborn sets used in the rest of the paper. We assume that the reader is familiar with the dynamic behaviour of PT-nets and the basic ideas of state spaces (also called occurrence graphs or reachability graphs/trees).

Definition 1. *A **Place/Transition Net** is tuple $PTN = (P, T, A, W, M_I)$, where P is a finite set of places, T is a finite set of transitions such that $P \cap T = \emptyset$, $A \subseteq (P \times T) \cup (T \times P)$ is a set of arcs, $W : A \to \mathbb{N}_+$ is an arc weight function, and $M_I : P \to \mathbb{N}_0$ is the initial marking.*   □

We use M_I as the initial marking instead of the more conventional M_0. This allows us to use M_0 as the first marking of *occurrence sequences* which do not necessarily start in the initial marking. If a transition t is *enabled* in a marking M_1 (denoted $M_1[t\rangle$), then t may *occur* in M_1 yielding some marking M_2. This is written $M_1[t\rangle M_2$. Extending this notation, an occurrence sequence is denoted $M_0[t_1\rangle M_1 \cdots M_{n-1}[t_n\rangle M_n$ and satisfies $M_{i-1}[t_i\rangle M_i$ for $1 \leq i \leq n$. When the intermediate markings in an occurrence sequence are not important we will write it as $M_0[t_1t_2 \cdots t_n\rangle M_n$. A *reachable marking* is a marking which can be obtained (reached) by an occurrence sequence starting in the initial marking. By $[M\rangle$ we denote the set of markings reachable from a marking M. For a place (transition) x, $\bullet x$ denotes the set of input transitions (places) of x, and $x\bullet$ is a similar notation for output transitions (places). The notation is extended to sets by taking the union of $\bullet x$ ($x\bullet$) over each member x of the set. In a marking M, the marking of a place p is denoted $M(p)$.

Definition 2. *The* **Full State Space** *of a PT-net is a directed graph* $SG = (V, E)$, *where* $V = [M_I\rangle$ *and* $E = \{ (M_1, t, M_2) \in V \times T \times V \mid M_1[t\rangle M_2 \}$.   □

In the rest of this paper we assume that a PT-net (P, T, A, W, M_I) with a *finite* full state space $SG = (V, E)$ is given. For some of the stubborn set algorithms presented in this paper we will exploit the *strongly connected components*. A strongly connected component (SCC) is a non-empty set C of reachable markings such that if $M \in C$ then $C = \{ M' \mid M' \in [M\rangle \wedge M \in [M'\rangle \}$. An SCC is said to be a *terminal strongly connected component* iff $M \in C$ implies $[M\rangle \subseteq C$.

State space construction with stubborn sets follows the same procedure as the construction of the full state space of a PT-net, with one exception. When processing a marking, a set of transitions, the so-called *stubborn set*, is constructed. Only the enabled transitions in the stubborn set are used to construct successor markings. This means that only a subset of the relation $M[t\rangle M'$ is used for the construction of the reduced state space. We denote this subset by $M[t\rangle_{SSG}M'$, and define $M[t_1 \cdots t_n\rangle_{SSG}M'$ and $[M\rangle_{SSG}$ as for the full state space but now based on the relation $M[t\rangle_{SSG}M'$. The stubborn set reduced state space (from now on called the *SS state space*) can be defined as a directed graph $SSG = (V_{SSG}, E_{SSG})$ based on the relation $M[t\rangle_{SSG}M'$ in a similar way as the full state space. We define the (terminal) SCCs for the SS state space analogously to the case for the full state space.

The choice of stubborn sets depends on the properties that are being analysed or verified of the system. Many stubborn set algorithms are surveyed in [10]. They all assume that the stubborn sets used in each marking satisfy certain conditions, and stubborn set methods for different properties are obtained by using different conditions. However, it is common to almost all of them that the conditions listed below should hold. Below $T_s(M)$ denotes the stubborn set used in the marking M.

D1 If $t \in T_s(M_0)$, $t_1, \ldots, t_n \notin T_s(M_0)$, $M_0[t_1t_2 \cdots t_n\rangle M_n$, and $M_n[t\rangle M'_n$, then there is M'_0 such that $M_0[t\rangle M'_0$ and $M'_0[t_1t_2 \cdots t_n\rangle M'_n$.
D2 If M_0 has an enabled transition, then there is at least one transition $t_k \in T_s(M_0)$ such that if $t_1, \ldots, t_n \notin T_s(M_0)$ and $M_0[t_1t_2 \cdots t_n\rangle M_n$, then $M_n[t_k\rangle$. Any transition with this property is called a *key transition* of $T_s(M_0)$.

The conditions D1 and D2 as such are not suited for constructing stubborn sets since they refer to occurrence sequences. Therefore, the construction of stubborn sets is in practice implemented by relying on rules that refer only to the structure of the PT-net and the current marking, and which express sufficient conditions to make D1 and D2 hold. The tutorial [10] lists a number of such. Below we give a simple proposition which guarantees that D1 and D2 hold. The proposition analyses the dependencies between transitions at a rather coarse level, and it is not optimal in the sense of yielding smallest possible stubborn sets and smallest SS state spaces. We will use it only for illustration purposes.

Proposition 1. *The conditions D1 and D2 hold if the following hold for every* $t \in T_s(M)$:

1. *If* $\exists t_1 \in T : M[t_1\rangle$, *then* $\exists t_2 \in T_s(M) : M[t_2\rangle$.
2. *If* $\neg M[t\rangle$, *then* $\exists p \in \bullet t : M(p) < W(p,t) \wedge \bullet p \subseteq T_s(M)$.
3. *If* $M[t\rangle$, *then* $(\bullet t)\bullet \subseteq T_s(M)$.   □

The important aspect of Prop. 1 is that the three items can be read as rules. Item 1 specifies that if there is an enabled transition, then an enabled transition has to be in the stubborn set. Item 2 specifies that if a disabled transition t has been included in the stubborn set, some place p in the preset of t which does not contain enough tokens for t to be enabled must be chosen and its preset included. Finally, item 3 specifies that if an enabled transition t has been included then the postset of the preset of t must be included. A number of algorithms for constructing stubborn sets based on propositions like Prop. 1 are given in [10].

3   An Example

In this section we introduce the first of our improved stubborn set methods in an informal way using the simple PT-net shown in Fig. 1. Figure 2 shows the full state space of this PT-net. Node 1 corresponds to the initial marking. Each arc has an associated label giving the name of the transition to which it corresponds. For a node n we denote the corresponding marking by M_n.

Suppose that we want to check that there exists a reachable marking in which the place p_{10} contains at least two tokens. This can be expressed as the *state property* $\phi \equiv M(p_{10}) \geq 2$. M_9 is the only such marking.

The stubborn set method in [6], in the following referred to as the *attractor set method*, would define an *attractor set* in M_1, denoted $A_\phi(M_1)$, for the *atomic state proposition* $M(p_{10}) \geq 2$. The role of the attractor set is to ensure that in each step of the SS state space construction, progress is made towards a marking where the property holds. The attractor set in M_1 would consist of the transitions which can add tokens to p_{10}. Hence $A_\phi(M_1) = \{t_5, t_6\}$. The attractor set method requires the attractor set to be a subset of the stubborn set in each marking. If we apply Prop. 1, then the stubborn set in M_1 will be $\{t_1, t_2, t_3, t_4, t_5, t_6\}$. Hence both enabled transitions (t_1 and t_2) are in the stubborn set in M_1.

If we consider the marking M_2 then the attractor set remains the same as in M_1 and Prop. 1 gives us $\{t_2, t_4, t_5, t_6\}$ as the stubborn set. Again, all enabled transitions are included in the stubborn set. The situation in M_3 is symmetric to M_2. In M_4, the transition t_2 will be in the stubborn set. The situation in M_6

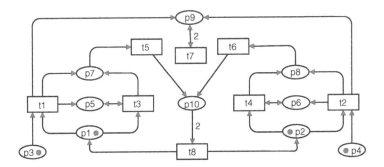

Fig. 1. Example PT-net.

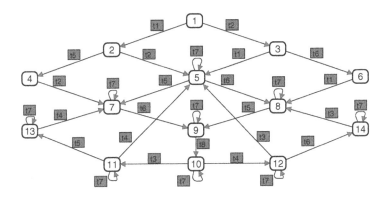

Fig. 2. Full state space for the PT-net in Fig. 1.

is symmetric to M_4, and in M_5, the transitions t_5 and t_6 will be in the stubborn set. In M_7 and M_8, the transition t_6 and t_5, respectively, will be in the stubborn set. In conclusion this means that the attractor set method yields an SS state space consisting of markings M_1 to M_9.

It can however be observed that it is possible to select stubborn sets during the construction of an SS state space with fewer enabled transitions than those required by the attractor set method. This could potentially lead to more reduction. The basic idea in our new method is to relax the requirement that the attractor set must *always* be contained in the stubborn set.

Suppose that the requirement imposed by the attractor set was totally removed. From Prop. 1 it follows that in M_1 we can select $\{t_2, t_4\}$ or $\{t_1, t_3\}$ as the stubborn set. Suppose that we select $\{t_1, t_3\}$. Proposition 1 implies that it is possible to select $\{t_5\}$ or $\{t_2, t_4\}$ as the stubborn set in M_2. If in M_2 we select the latter, then in M_5 we can select $\{t_5\}$, $\{t_6\}$, or $\{t_7\}$ as the stubborn set.

If in M_5 we select the stubborn set consisting of $\{t_7\}$ only, then the construction of the SS state space will terminate at this point, since M_5 is already included in it. This means that we would wrongly conclude that there does not exist a marking in which ϕ holds.

The problem is that we have not ensured *progress* towards such a marking. The attractor set method ensures progress in each marking of the SS state space by *always* including the attractor set in the stubborn set. Instead of this strong requirement we will ensure that from each marking in the SS state space *eventually* progress is made, i.e., a marking is reachable in the SS state space in which progress is made. For this purpose we introduce the notion of *up sets*. An up set is a set of transitions chosen such that at least one transition in it has to occur in order to make the state property hold. Hence an up set is similar to an attractor set. However, unlike the attractor set method, we will not require that the up set is always contained in the stubborn set. Moreover, we will additionally exploit that the state properties which we consider are growing Boolean functions. This makes it possible to ensure progress towards the property by either reducing the length of an occurrence sequence leading to a marking where the state property holds, or by increasing the number of atomic state propositions which are satisfied. This requirement will ensure that t_5 or t_6 is in the stubborn set in M_5. Similarly, it will ensure that t_6 is in the stubborn set in M_7, and that t_5 is in the stubborn set in M_8.

Ensuring eventual progress is however not sufficient for preserving state properties. As a simple example, suppose that we want to show that a marking is reachable in which $M(p_3) = 0$ and $M(p_4) = 1$. This corresponds to showing that M_2 or M_4 is reachable. If $\{t_2, t_4\}$ is selected as the stubborn set in M_1, then neither of M_2 and M_4 will be in the SS state space. The problem is that in M_1 the only enabled transition in the stubborn set is t_2, and an occurrence of this transition can change the value of the state property from True to False. To account for this we introduce the notion of *down sets*. A down set is a set of transitions chosen such that a transition in the down set has to occur in order to make the property not hold. We will ensure that if an enabled transition which is in the down set is in the stubborn set, then the transitions in the up set are also in the stubborn set. This will ensure that if t_2 is in the stubborn set in M_1 then also t_1 is.

4 State Properties

We consider state properties expressed as formulas that are composed of so-called *atomic state propositions* using only the logical operators "$\wedge$" and "$\vee$" and parentheses "(" and ")". For a state property ϕ we denote its atomic state propositions by $\varphi_1, \varphi_2, \ldots, \varphi_n$, and let $I = \{1, 2, \ldots, n\}$ denote the set of indices of the atomic state propositions. The atomic state propositions and state properties are interpreted on the markings of the PT-net, and the resulting truth values are denoted by $\varphi_i(M)$ and $\phi(M)$. The atomic state propositions are defined according to the following syntax, where p, p_1, and p_2 denote arbitrary places and k is an integer constant.

$$\varphi_i ::= M(p) \geq k \mid M(p_1) \geq M(p_2) \mid M(p) = k \mid M(p_1) = M(p_2) \mid$$
$$M(p) \leq k \mid M(p_1) > M(p_2) \mid M(p) \neq k \mid M(p_1) \neq M(p_2)$$

We have not included $M(p) > k$ and $M(p) < k$ as atomic state propositions since they can be expressed as $M(p) \geq k + 1$ and $M(p) \leq k - 1$, respectively. The set of atomic state propositions could be extended provided that the corresponding up and down sets to be defined in Sect. 5 are implemented properly.

Above only conjunction and disjunction were allowed as the Boolean operators. However, the atomic state propositions are closed under negation (p_1 and p_2 may be swapped when needed), so formulas which use negation can always be re-written to a form allowed by the above syntax using *De Morgan's equivalences* (i.e., $\neg(\phi_1 \vee \phi_2) = \neg\phi_1 \wedge \neg\phi_2$ and $\neg(\phi_1 \wedge \phi_2) = \neg\phi_1 \vee \neg\phi_2$). Therefore, the syntax does not restrict generality. It is however important for the correctness of the later algorithms that formulas are given in a negation-free form, i.e., that they have been preprocessed before being provided as input to the algorithms.

Definition 3. *Let M be a marking and ϕ a state property constructed from the atomic state propositions $\{ \varphi_i \mid i \in I \}$. The set of the indices of the atomic state propositions which are satisfied in M is denoted $\mathrm{on}_\phi(M)$. The set of the indices of the atomic state propositions which are not satisfied in M is denoted $\mathrm{off}_\phi(M)$. Formally:*

$$\mathrm{on}_\phi(M) = \{ i \in I \mid \varphi_i(M) \} \text{ and } \mathrm{off}_\phi(M) = \{ i \in I \mid \neg\varphi_i(M) \} \qquad \square$$

If we let $\mathbb{B} = \{\mathsf{True}, \mathsf{False}\}$, treat the φ_is as argument symbols, and define $\mathsf{False} \leq \mathsf{True}$, then a state property formula ϕ determines a monotonically increasing Boolean function from $\mathbb{B}^n$ to $\mathbb{B}$.

The following proposition lists important properties of the state property formulas which will be exploited later.

Proposition 2. *Let M and M' be markings and ϕ a state property constructed from the atomic state propositions $\{ \varphi_i \mid i \in I \}$. Then the following holds:*

1. $\forall i \in I : \varphi_i(M) \leq \varphi_i(M') \Rightarrow \phi(M) \leq \phi(M')$.
2. $\phi(M) \wedge \neg\phi(M') \Rightarrow \exists i \in I : \varphi_i(M) \wedge \neg\varphi_i(M')$. $\qquad \square$

Item 1 states that ϕ is a monotonically increasing Boolean function. Item 2 states that if ϕ is satisfied in M but not in M', then there exists at least one atomic state proposition which is satisfied in M but not satisfied in M'. Item 2 is a consequence of item 1.

5   Up/Down and Satisfiability Sets

To describe the required properties of the stubborn sets, we define two sets of transitions related to a state property ϕ: an *up set* and a *down set*. The *up set* of ϕ in a marking M is a set of transitions chosen such that if ϕ does not hold in M then at least one transition in the up set must occur before ϕ can start to hold. The *down set* of ϕ is a set of transitions chosen such that it contains at least all transitions whose occurrence can change the value of some atomic state proposition φ_i of ϕ from True to False. In addition to these two sets we define the *satisfiability set* of ϕ in M as a set of indices of the atomic state propositions such that at least one atomic state proposition that has its index in the set has to change its value from False to True in order to make the state property hold.

The implementation of concrete up and down sets will be determined from the atomic state propositions and Boolean combinators. However, the properties of up and down sets are general concepts and not tied to the specific set of state properties considered in this paper. Therefore, we define up and down sets as properties of a set of transitions. A similar remark applies to satisfiability sets.

Definition 4. *Let ϕ be a state property constructed from the atomic state propositions $\{\,\varphi_i \mid i \in I\,\}$ and let $M_0 \in [M_I\rangle$. A set of transitions $T' \subseteq T$ has the **up set property** in M_0 with respect to ϕ iff the following holds for all occurrence sequences $M_0[t_1 t_2 \cdots t_n\rangle M_n$ starting in M_0:*

$$\neg\phi(M_0) \wedge \phi(M_n) \Rightarrow \exists j : 1 \leq j \leq n \wedge t_j \in T'$$

*A set of transitions $T' \subseteq T$ has the **down set property** with respect to ϕ iff the following holds for all markings $M, M' \in [M_I\rangle$, all $t \in T$, and all $i \in I$:*

$$M\,[t\rangle\,M' \wedge \varphi_i(M) \wedge \neg\varphi_i(M') \Rightarrow t \in T'$$

*A set of indices $J \subseteq I$ has the **satisfiability set property** in M_0 with respect to ϕ iff the following holds for all occurrence sequences $M_0[t_1 t_2 \cdots t_n\rangle M_n$:*

$$\neg\phi(M_0) \wedge \phi(M_n) \Rightarrow \exists i \in J : \neg\varphi_i(M_0) \wedge \varphi_i(M_n) \qquad \square$$

The up set property and the satisfiability set property are relative to the current marking whereas the down set property is not. This is deliberate and due to the way our methods will later use these sets. It is worth observing that the definition of up (down) set property allows approximations of the up (down) sets to be used: if $T' \subseteq T''$ and T' has the up (down) set property then also T'' has the up (down) set property. A similar remark applies to indices and the satisfiability set property. This will be exploited later once we show how to construct such sets. Moreover, if a marking in which ϕ holds is reachable from a marking M where ϕ does not hold, then a satisfiability set in M exists and it is non-empty because of Prop. 2. From now on we will assume that we have an algorithm that given a state property ϕ produces some down set down_ϕ, and additionally given a marking produces some up set $\mathsf{up}_\phi(M)$ and some satisfiability set $\mathsf{sat}_\phi(M)$. We will give such an algorithm in Sect. 8.

6   Preserving Reachability of State Properties

This section presents the new stubborn set method for determining whether a reachable marking exists in which a given state property holds. The method consists of obeying the D1 condition from Sect. 2 and two additional conditions formulated in the following definition. An explanation of the definition will be given below.

Definition 5. *Let M be a marking and ϕ a state property constructed from the atomic state propositions $\{\,\varphi_i \mid i \in I\,\}$. A set $T_s(M) \subseteq T$ is **Reachability of a State Property Preserving (RSPP) stubborn** in M, iff the following hold:*

D1 *If $t_1, \ldots, t_n \notin T_s(M)$, $t \in T_s(M)$, $M[t_1 t_2 \cdots t_n\rangle M_n$, and $M_n[t\rangle M_n'$, then there is M' such that $M[t\rangle M'$ and $M'[t_1 t_2 \cdots t_n\rangle M_n'$.*
SPP1 *If $\neg\phi(M)$ and $\exists t : M[t\rangle \wedge t \in \mathsf{down}_\phi \wedge t \in T_s(M)$ then $\mathsf{up}_\phi(M) \subseteq T_s(M)$.*
SPP2 *For every $i \in \mathsf{sat}_\phi(M)$ there is an occurrence sequence $M_0[t_1\rangle M_1[t_2\rangle \cdots [t_n\rangle M_n$ such that $M = M_0$, t_j is a key transition of $T_s(M_{j-1})$ for $1 \leq j \leq n$, and $\phi(M_n) \vee \mathsf{up}_{\varphi_i}(M_n) \subseteq \bigcup_{j=0}^n T_s(M_j)$.* □

The intuitive purpose of SPP1 is to ensure that a next step in the SS state space can be taken in such a way that we do not get further away from a marking where ϕ holds. SPP1 requires that if we have taken an enabled transition in the down set, then we have also included the transitions in the up set. The latter transitions represent a step towards a marking where ϕ holds, since we know that a transition in the up set has to occur in order to make ϕ hold. Therefore, if one transition makes regress then there is another transition that makes progress. It is also possible that no enabled transition makes regress or progress.

SPP2, on the other hand, is there to *ensure* progress – to ensure that we will eventually get closer to a marking where ϕ holds. If ϕ holds in M then SPP2 holds trivially since we can then choose $n = 0$. If SPP2 does not take us directly to a marking where ϕ holds, then it ensures that there is a path in the SS state space where we have eventually included every transition in the up set of some atomic state proposition which has to change its value. This represents progress, since such an additional atomic state proposition has to be satisfied in order to make the state property hold. SPP2 states its requirement to every element $i \in \mathsf{sat}_\phi(M)$ because $\mathsf{sat}_\phi(M)$ is an upper approximation and we do not necessarily know which member is important.

We now turn to the correctness of the RSPP-stubborn set method. The key to establishing correctness is the following lemma.

Lemma 1. *Let ϕ be a state property, $SG = (V, E)$ the full state space, and $SSG = (V_{SSG}, E_{SSG})$ an SS state space constructed using RSPP-stubborn sets. Let $M_0 \in V_{SSG}$ be a marking such that $\neg\phi(M_0)$ and for which there exists an occurrence sequence $M_0[t_1\rangle M_1[t_2\rangle \cdots [t_n\rangle M_n$ such that $\phi(M_n)$ holds. Then there is a marking $M_0' \in [M_0\rangle_{SSG}$ such that the following holds.*

1. *There are transitions $t_1', t_2', \ldots, t_m'$ and markings $M_1', M_2', \ldots, M_m'$ such that $M_0'[t_1'\rangle M_1'[t_2'\rangle \cdots [t_m'\rangle M_m'$ and $\phi(M_m')$ holds.*
2. *The occurrence sequence in item 1 leading to a marking where ϕ holds is no longer than the original occurrence sequence, i.e., $m \leq n$.*

3. *The length of the occurrence sequence in item 1 has decreased, i.e., $m < n$, or the set of the atomic state propositions of ϕ which are satisfied has grown, i.e., $\mathrm{on}_\phi(M_0) \subset \mathrm{on}_\phi(M_0')$.* □

Proof. Since $\neg\phi(M_0)$ and $\phi(M_n)$ then $\mathrm{sat}_\phi(M_0)$ contains an i such that $\neg\varphi_i(M_0)$ and $\varphi_i(M_n)$. Since $\neg\phi(M_0)$ and $T_\mathrm{s}(M_0)$ (the RSPP-stubborn set in M_0) satisfies the condition SPP2 there exist key transitions $\bar{t}_1, \ldots, \bar{t}_k$ and markings $\bar{M}_0, \ldots, \bar{M}_k \in V_{SSG}$ such that $M_0 = \bar{M}_0[\bar{t}_1\rangle_{SSG}\bar{M}_1[\bar{t}_2\rangle_{SSG}\cdots[\bar{t}_k\rangle_{SSG}\bar{M}_k$, and $\phi(\bar{M}_k) \vee \mathrm{up}_{\varphi_i}(\bar{M}_k) \subseteq \bigcup_{j=0}^k T_\mathrm{s}(\bar{M}_j)$.

If $\phi(\bar{M}_j)$ holds for some $0 \leq j \leq k$ then the claim holds by choosing $M_0' = \bar{M}_j$ and $m = 0$. In this case $m < n$ since $\neg\phi(M_0)$ and $\phi(M_n)$. From now on we may therefore assume that $\forall j, 0 \leq j \leq k : \neg\phi(\bar{M}_j)$. In the rest of the proof the following fact is needed:

(1) If $\{t_1, \ldots, t_n\} \cap T_\mathrm{s}(\bar{M}_h) = \emptyset$ for every $0 \leq h < l \leq k$, then, due to the key transition property of $\bar{t}_1, \ldots, \bar{t}_k$ and D1, there are markings $\hat{M}_0, \ldots, \hat{M}_l$ such that $M_n = \hat{M}_0[\bar{t}_1 \cdots \bar{t}_l\rangle\hat{M}_l$ and $\bar{M}_l[t_1 \cdots t_n\rangle\hat{M}_l$. Furthermore, if we assume that there exists a smallest index h such that $\bar{t}_{h+1} \in \mathrm{down}_\phi$ then $\phi(\hat{M}_0), \ldots, \phi(\hat{M}_h)$ hold due to the down set property. Thus, $\neg\phi(\bar{M}_h)$ and $\phi(\hat{M}_h)$ and SPP1 implies that $\emptyset \neq \{t_1, \ldots, t_n\} \cap \mathrm{up}_\phi(\bar{M}_h) \subseteq T_\mathrm{s}(\bar{M}_h)$ contrary to our assumption. Consequently, $\bar{t}_1, \ldots, \bar{t}_l \notin \mathrm{down}_\phi$ and we therefore have $\mathrm{on}_\phi(\bar{M}_0) \subseteq \mathrm{on}_\phi(\bar{M}_1) \subseteq \cdots \subseteq \mathrm{on}_\phi(\bar{M}_l)$, $\mathrm{on}_\phi(\hat{M}_0) \subseteq \mathrm{on}_\phi(\hat{M}_1) \subseteq \cdots \subseteq \mathrm{on}_\phi(\hat{M}_l)$, and $\phi(\hat{M}_0), \ldots, \phi(\hat{M}_l)$ hold.

We now split the proof in two cases.

Case A: $\{t_1, \ldots, t_n\} \cap T_\mathrm{s}(\bar{M}_j) \neq \emptyset$ for some $0 \leq j \leq k$. In this case we can pick the smallest such j and apply (1) for $l = j$. Since $T_\mathrm{s}(\bar{M}_j)$ contains at least one of the transitions $t_1, t_2, \ldots, t_n$, then we can pick the first such transition t_h and apply D1 on $\bar{M}_j[t_1 \cdots t_{h-1}t_ht_{h+1} \cdots t_n\rangle\hat{M}_j$ to obtain a marking M'' such that $\bar{M}_j[t_h\rangle M''[t_1 \cdots t_{h-1}t_{h+1} \cdots t_n\rangle\hat{M}_j$. The claim now holds with $M_0' = M''$ and $m = n - 1$.

Case B: $\{t_1, \ldots, t_n\} \cap T_\mathrm{s}(\bar{M}_j) = \emptyset$ for every $0 \leq j \leq k$. In this case (1) gives us that $\mathrm{on}_\phi(\bar{M}_0) \subseteq \mathrm{on}_\phi(\bar{M}_k)$ and $\mathrm{on}_\phi(\hat{M}_0) \subseteq \mathrm{on}_\phi(\hat{M}_k)$. If $\mathrm{on}_\phi(\bar{M}_0) = \mathrm{on}_\phi(\bar{M}_k)$ then we must have one of the $t_1, t_2, \ldots, t_n$ in $\mathrm{up}_{\varphi_i}(\bar{M}_k)$ and by SPP2 also in $\bigcup_{j=0}^k T_\mathrm{s}(\bar{M}_j)$ which contradicts the assumption that $\{t_1, \ldots, t_n\} \cap T_\mathrm{s}(\bar{M}_j) = \emptyset$ for every $0 \leq j \leq k$. Therefore $\mathrm{on}_\phi(\bar{M}_0) \subset \mathrm{on}_\phi(\bar{M}_k)$ and the claim holds with $M_0' = \bar{M}_k$ and $m = n$.

□

The following theorem states that if there exists a marking in the SS state space from which it is possible to reach in the full state space a marking where the state property holds, then the SS state space also contains a marking in which the state property holds. The correctness of the RSPP stubborn set method follows immediately from the theorem by letting $M_0 = M_I$.

Theorem 1. *Let ϕ be a state property, $SG = (V, E)$ be the full state space, $SSG = (V_{SSG}, E_{SSG})$ an SS state space constructed using RSPP-stubborn sets, and let $M_0 \in V_{SSG}$. Then:*

$$\exists M \in [M_0\rangle : \phi(M) \Leftrightarrow \exists M' \in [M_0\rangle_{SSG} : \phi(M') \qquad \square$$

Proof. The $\Leftarrow$ direction follows from the fact that $V_{SSG} \subseteq V$ and $E_{SSG} \subseteq E$. For establishing the $\Rightarrow$ direction we apply Lemma 1 inductively to obtain $M_1 \in [M_0\rangle_{SSG}, M_2 \in [M_0\rangle_{SSG}, \ldots$, until we find an $M_n \in [M_0\rangle_{SSG}$ such that $\phi(M_n)$ holds. The induction hypothesis is that there is a marking M_ϕ^i and an occurrence sequence σ_i such that $M_i[\sigma_i\rangle M_\phi^i$ and $\phi(M_\phi^i)$ holds. When $i = 0$ this holds with $M_\phi^0 = M$.

Define the *distance* $\Delta(M', \sigma, \phi)$ between a marking $M' \in V_{SSG}$ and a marking $M_\phi \in V$ which satisfies ϕ and which can be reached from M' by the occurrence sequence σ as follows ($|\sigma|$ denotes the length of the occurrence sequence σ):

$$\Delta(M', \sigma, \phi) = (|I| + 1) \cdot |\sigma| + |\mathsf{off}_\phi(M')|$$

The reason for the rather complicated definition of distance is that if at a marking we choose a transition which decreases the length of the occurrence sequence leading to a marking where ϕ holds we may at the same time switch some of the atomic state propositions off.

If ϕ does not hold in M_{i-1}, then Lemma 1 gives a marking $M_i \in [M_{i-1}\rangle_{SSG}$, an occurrence sequence σ_i, and a marking M_ϕ^i such that $M_i[\sigma_i\rangle M_\phi^i$ and ϕ holds in M_ϕ^i. Items 2 and 3 of Lemma 1 ensure that $\Delta(M_i, \sigma_i, \phi) < \Delta(M_{i-1}, \sigma_{i-1}, \phi)$. Clearly $0 \leq \Delta(M_i, \sigma_i, \phi) < \infty$, so eventually this process terminates in a marking $M_n \in [M_0\rangle_{SSG}$ in which ϕ holds. $\qquad \square$

7   Preserving Home State Properties

This section presents a new stubborn set method for determining whether from all reachable markings it is possible to reach a marking where a given state property holds. This can be formally expressed as determining whether $\forall M \in [M_I\rangle : \exists M' \in [M\rangle : \phi(M')$. The method presented is based on the observation that by negation this is the same as determining whether a reachable marking exists from which it is *not* possible to make the given state property hold. This can be expressed as $\exists M \in [M_I\rangle : \forall M' \in [M\rangle : \neg\phi(M')$. We will use "$\phi \in [M\rangle$" as an abbreviation of $\exists M' \in [M\rangle : \phi(M')$ (from M a marking M' can be reached where ϕ holds), and "$\phi \in [M\rangle_{SSG}$" as an abbreviation of $\exists M' \in [M\rangle_{SSG} : \phi(M')$.

The method consists of obeying the conditions of the RSPP-stubborn set method from Sect. 6 and two additional conditions formulated in the following definition.

Definition 6. *Let M be a marking and ϕ a state property. A set $T_s(M) \subseteq T$ is* **Home State Property Preserving (HSPP) Stubborn** *in M, iff $T_s(M)$ is RSPP stubborn in M and the following hold:*

D2' *If* $t_1, \ldots, t_n \notin T_s(M)$, $t \in T_s(M)$, $M[t_1 t_2 \cdots t_n \rangle M_n$, *and* $M[t\rangle$, *then* $M_n[t\rangle$.

SPP3 *For every* $t \in \mathsf{down}_\phi$ *there is an occurrence sequence* $M_0[t_1\rangle \cdots [t_n\rangle M_n$ *such that* $M = M_0$, $t_j \in T_s(M_{j-1})$ *for* $1 \leq j \leq n$, *and* $t \in T_s(M_n)$.   □

The intuitive purpose of $T_s(M)$ being RSPP stubborn in the context of this method is to ensure that we from any marking in the SS state space always attempt to make ϕ hold (recall that we are trying to show that there exists a reachable marking from which ϕ cannot be made to hold). The D2' condition is like the D2 condition from Sect. 2, except that it requires all enabled transitions in the stubborn set to be key transitions and allows $T_s(M) = \emptyset$. Together with the D1 condition inherited from RSPP this implies that HSPP stubborn sets are *strong stubborn sets* [10]. SPP3 is there to *ensure* progress, to ensure that we eventually get closer to a marking from which ϕ cannot be made to hold. This is formulated in terms of transitions in the down set since such a transition has to occur in order to make ϕ not hold.

The correctness of the HSPP stubborn set method follows immediately from the following theorem by letting $M_0 = M_I$.

Theorem 2. *Let* ϕ *be a state property,* $SG = (V, E)$ *be the full state space,* $SSG = (V_{SSG}, E_{SSG})$ *an SS state space constructed using HSPP stubborn sets, and let* $M_0 \in V_{SSG}$. *Then:*

$$\exists M \in [M_0\rangle : \phi \notin [M\rangle \Leftrightarrow \exists M_{SSG} \in [M_0\rangle_{SSG} : \phi \notin [M_{SSG}\rangle_{SSG} \qquad \square$$

Proof. The $\Leftarrow$ direction follows immediately from Thm. 1. We prove the $\Rightarrow$ direction by showing the following for a strictly decreasing sequence of values of n:

(1) There are $M_0^n, t_1^n, \ldots, t_n^n$ and M^n such that $M_0^n \in [M_0\rangle_{SSG}$, $M_0^n[t_1^n t_2^n \cdots t_n^n\rangle$ M^n and $\phi \notin [M^n\rangle$.

Initially we get (1) for some finite non-negative value of n from the left hand side of the theorem if we choose $M_0^n = M_0$. If $\phi \notin [M_0^n\rangle$, then $\phi \notin [M_0^n\rangle_{SSG}$, so M_0^n can be chosen as the M_{SSG} and the right hand side of the theorem holds. This happens at the latest when $n = 0$. Therefore, we get our result by showing that if (1) holds for some n and $\phi \in [M_0^n\rangle$, then (1) holds also for some m such that $0 \leq m < n$.

So we assume that $\phi \in [M_0^n\rangle$ holds. Thm. 1 asserts the existence of $t_1'', \ldots, t_h''$ and $M_0'', \ldots, M_h''$ such that $M_0^n = M_0''$, $M_0''[t_1''\rangle_{SSG} M_1''[t_2''\rangle_{SSG} \cdots [t_h''\rangle_{SSG} M_h''$ and $\phi(M_h'')$. We first establish the existence of an occurrence sequence $\bar{M}_0[\bar{t}_1\rangle_{SSG} \bar{M}_1$ $[\bar{t}_2\rangle_{SSG} \cdots [\bar{t}_k\rangle_{SSG} \bar{M}_k$ such that $M_0^n = \bar{M}_0$ and $\{t_1^n, \ldots, t_n^n\} \cap T_s(\bar{M}_j) \neq \emptyset$ for some $1 \leq j \leq k$. We split the proof in two cases.

Case A: If at least one $t_1^n, \ldots, t_n^n$ belongs to $T_s(M_0'') \cup \cdots \cup T_s(M_h'')$, then we just choose $k = h$ and $\bar{t}_i = t_i''$ and $\bar{M}_i = M_i''$ for $1 \leq i \leq h$.

Case B: If none of $t_1^n, \ldots, t_n^n$ is in $T_s(M_0'') \cup \cdots \cup T_s(M_h'')$, then due to D1 and D2' there is M_h' such that $M_h''[t_1^n \cdots t_n^n\rangle M_h'$ and $M^n[t_1'' \cdots t_h''\rangle M_h'$. Because of $\phi \notin [M^n\rangle$ we know that $\neg\phi(M_h')$. Because $\phi(M_h'')$ holds, there is $t_i^n \in \{t_1^n, \ldots, t_n^n\}$ such that $t_i^n \in \mathsf{down}_\phi$. Therefore, SPP3 gives the desired occurrence sequence.

We can now choose the smallest j such that $\{t_1^n, \ldots, t_n^n\} \cap T_s(\bar{M}_j) \neq \emptyset$. Let i be the smallest number such that $t_i^n \in T_s(\bar{M}_j)$. Due to D1 and D2' there is M^{n-1} such that $\bar{M}_j[t_1^n \cdots t_n^n\rangle M^{n-1}$ and $M^n[\bar{t}_1 \cdots \bar{t}_j\rangle M^{n-1}$. We have $\phi \notin [M^{n-1}\rangle$, because otherwise $\phi \notin [M^n\rangle$ would not hold. Due to D1 there is M_0^{n-1} such that $\bar{M}_j[t_i^n\rangle_{SSG} M_0^{n-1}[t_1^n \cdots t_{i-1}^n t_{i+1}^n \cdots t_n^n\rangle M^{n-1}$. We thus have (1) for $n-1$. $\square$

8   Implementation

We now consider the implementation of the RSPP and HSPP stubborn set methods presented in the previous sections. In Sect. 8.1 we show how to construct up, down, and satisfiability sets. In Sect. 8.2 we discuss different ways of implementing the conditions D1, D2', and SPP1-3.

8.1   Implementation of Up/Down and Satisfiability Sets

In this section we show how to define up, down and satisfiability sets for the state properties considered in this paper. The construction is in all three cases specified inductively using the syntactical structure of the state properties. We end the section with a proposition which states that the defined up, down, and satisfiability sets posses the up, down, and satisfiability set properties as defined in Def. 4. First we give the definition of up sets.

Definition 7. *Let M be a marking and ϕ a state property. The up set $\mathsf{up}_\phi(M)$ in M is defined as follows. If ϕ holds in M we define $\mathsf{up}_\phi(M) = \emptyset$. If ϕ does not hold in M we define $\mathsf{up}_\phi(M)$ according to the following cases:*

Case $\phi \equiv M(p) \leq k$: *$\mathsf{up}_\phi(M)$ consists of the transitions which can remove tokens from p and which add at most k tokens to p:*

$$\mathsf{up}_\phi(M) = \{ t \in T \mid W(p,t) > W(t,p) \wedge W(t,p) \leq k \}$$

Case $\phi \equiv M(p) = k$: *If p contains too few tokens then $\mathsf{up}_\phi(M)$ consists of the transitions which can add tokens and do not require additional tokens to be present on p, and if p contains too many tokens then $\mathsf{up}_\phi(M)$ consists of the transitions which can remove tokens from p and which add at most k tokens:*

$$\mathsf{up}_\phi(M) = \begin{cases} \{ t \in T \mid W(p,t) < W(t,p) \wedge W(t,p) \leq M(p) \} & \text{if } M(p) < k \\ \{ t \in T \mid W(p,t) > W(t,p) \wedge W(t,p) \leq k \} & \text{if } M(p) > k \end{cases}$$

Case $\phi \equiv M(p) \geq k$: *$\mathsf{up}_\phi(M)$ consists of the transitions which can add tokens and which do not require additional tokens to be present on p:*

$$\mathsf{up}_\phi(M) = \{ t \in T \mid W(p,t) < W(t,p) \wedge W(t,p) \leq M(p) \}$$

Case $\phi \equiv M(p) \neq k$: $\mathsf{up}_\phi(M)$ *consists of the transitions which can change the marking of p and which do not require additional tokens to be present on p:*

$$\mathsf{up}_\phi(M) = \{\, t \in T \mid W(p,t) \neq W(t,p) \wedge W(p,t) \leq k \,\}$$

Case $\phi \equiv M(p_1) > M(p_2)$ **or** $\phi \equiv M(p_1) \geq M(p_2)$:

$$\mathsf{up}_\phi(M) = \{\, t \in T \mid W(t,p_1) - W(p_1,t) > W(t,p_2) - W(p_2,t) \,\}$$

Case $\phi \equiv M(p_1) = M(p_2)$: $\mathsf{up}_\phi(M) =$

$$\begin{cases} \{\, t \in T \mid W(t,p_1) - W(p_1,t) > W(t,p_2) - W(p_2,t) \,\} \text{ if } M(p_1) < M(p_2) \\ \{\, t \in T \mid W(t,p_1) - W(p_1,t) < W(t,p_2) - W(p_2,t) \,\} \text{ if } M(p_1) > M(p_2) \end{cases}$$

Case $\phi \equiv M(p_1) \neq M(p_2)$:

$$\mathsf{up}_\phi(M) = \{\, t \in T \mid W(t,p_1) - W(p_1,t) \neq W(t,p_2) - W(p_2,t) \,\}$$

Case $\phi \equiv \phi_1 \wedge \phi_2$: $\mathsf{up}_\phi(M)$ *is the up set of one of the ϕ_is which does not hold in M:*

$$\left(\neg\phi_1(M) \wedge \mathsf{up}_\phi(M) = \mathsf{up}_{\phi_1}(M)\right) \vee \left(\neg\phi_2(M) \wedge \mathsf{up}_\phi(M) = \mathsf{up}_{\phi_2}(M)\right)$$

Case $\phi \equiv \phi_1 \vee \phi_2$: $\mathsf{up}_\phi(M) = \mathsf{up}_{\phi_1}(M) \cup \mathsf{up}_{\phi_2}(M)$ $\qquad\qquad\square$

Definition 8. *Let ϕ be a state property. The down set down_ϕ is defined as follows:*

Case $\phi \equiv M(p) \leq k$: $\mathsf{down}_\phi = \{\, t \in T \mid W(p,t) < W(t,p) \wedge W(p,t) \leq k \,\}$
Case $\phi \equiv M(p) = k$: $\mathsf{down}_\phi = \{\, t \in T \mid W(p,t) \neq W(t,p) \wedge W(p,t) \leq k \,\}$
Case $\phi \equiv M(p) \neq k$: $\mathsf{down}_\phi = \{\, t \in T \mid W(p,t) \neq W(t,p) \wedge W(t,p) \leq k \,\}$
Case $\phi \equiv M(p) \geq k$: $\mathsf{down}_\phi = \{\, t \in T \mid W(p,t) > W(t,p) \wedge W(t,p) < k \,\}$
Case $\phi \equiv M(p_1) > M(p_2)$ **or** $\phi \equiv M(p_1) \geq M(p_2)$:

$$\mathsf{down}_\phi = \{\, t \in T \mid W(p_1,t) - W(t,p_1) > W(p_2,t) - W(t,p_2) \,\}$$

Case $\phi \equiv M(p_1) = M(p_2)$ **or** $\phi \equiv M(p_1) \neq M(p_2)$:

$$\mathsf{down}_\phi = \{\, t \in T \mid W(t,p_1) - W(p_1,t) \neq W(t,p_2) - W(p_2,t) \,\}$$

Case $\phi \equiv \phi_1 \wedge \phi_2$ **or** $\phi \equiv \phi_1 \vee \phi_2$: $\mathsf{down}_\phi = \mathsf{down}_{\phi_1} \cup \mathsf{down}_{\phi_2}$ $\qquad\square$

Definition 9. *Let M be a marking and ϕ a state property constructed from the atomic state propositions $\{\, \varphi_i \mid i \in I \,\}$. The satisfiability set $\mathsf{sat}_\phi(M)$ in M is defined as follows. If $\phi(M)$ holds then $\mathsf{sat}_\phi(M) = \emptyset$. Otherwise it is defined as follows.*

Case $\phi \equiv \varphi_i$: $\mathsf{sat}_\phi(M) = \{\, i \,\}$
Case $\phi \equiv \phi_1 \vee \phi_2$: $\mathsf{sat}_\phi(M) = \mathsf{sat}_{\phi_1}(M) \cup \mathsf{sat}_{\phi_2}(M)$
Case $\phi \equiv \phi_1 \wedge \phi_2$: $\mathsf{sat}_\phi(M)$ *is the satisfiability set of one of the ϕ_is which does not hold in M:*

$$\left(\neg\phi_1(M) \wedge \mathsf{sat}_\phi(M) = \mathsf{sat}_{\phi_1}(M)\right) \vee \left(\neg\phi_2(M) \wedge \mathsf{sat}_\phi(M) = \mathsf{sat}_{\phi_2}(M)\right)$$

$\qquad\qquad\qquad\qquad\qquad\qquad\qquad\qquad\qquad\qquad\qquad\qquad\square$

The following proposition states that the up, down, and satisfiability sets defined above have the required properties. The proof of the proposition is based on structural induction on the state properties and is not contained in this paper.

Proposition 3. *Let M be a marking and assume that $\mathsf{up}_\phi(M) \subseteq T$, $\mathsf{down}_\phi \subseteq T$, and $\mathsf{sat}_\phi(M) \subseteq I$ are constructed according to Def. 7, Def. 8, and Def. 9, respectively. Then the following hold:*

1. $\mathsf{up}_\phi(M)$ *has the up set property in M with respect to ϕ.*
2. down_ϕ *has the down set property with respect to ϕ.*
3. $\mathsf{sat}_\phi(M)$ *has the satisfiability set property in M with respect to ϕ.*   □

8.2   Implementation of RSPP and HSPP Stubborn

We now consider the implementation of D1, D2′, and SPP1-3. The implementation of D1, D2′ and SPP1 is rather straightforward. Techniques for ensuring D1 and D2′ are well-established (see, e.g., [10] for a survey), and SPP1 can be handled with similar techniques. Below we suggest three implementations of SPP2 and SPP3. The more complex implementations have the potential of leading to better reductions of the state space.

Attractor Set. A simple way to implement SPP2 is to ensure that in each marking M encountered during the SS state space generation we have $\mathsf{up}_{\varphi_i}(M) \subseteq T_{\mathsf{s}}(M)$ for every $i \in \mathsf{sat}_\phi(M)$. In the case of SPP3 we also ensure that $\mathsf{down}_\phi \subseteq T_{\mathsf{s}}(M)$. This guarantees that the n in the formulation of SPP2 and SPP3 can be chosen to be zero. SPP1 is automatically guaranteed since $\bigcup_{\{i \in \mathsf{sat}_\phi(M)\}} \mathsf{up}_{\varphi_i}(M) \subseteq T_{\mathsf{s}}(M)$ has the up set property in M. This implementation of SPP2 coincides with the attractor set method suggested in [6], and thus tends to produce unnecessarily large stubborn sets as was described in Sect. 3.

Terminal SCC Detection. A more powerful implementation of SPP2 can be obtained by exploiting strongly connected components (SCCs) and the fact that for a directed graph it is always possible to reach the nodes belonging to some terminal SCC. This fact implies that if all enabled transitions in the stubborn sets used are key transitions, then a sufficient condition for SPP2 and SPP3 to hold is that for every i there exists an occurrence sequence satisfying SPP2 and SPP3 in each of the terminal SCCs of the SS state space. Stubborn sets in which all enabled transitions are key transitions are also referred to as *strong stubborn sets*. Strong stubborn sets are already guaranteed in case of HSPP due to D2′. In case of RSPP, we can obtain strong stubborn sets by simply ensuring also D2′ in addition to D1, SPP1, and SPP2.

Checking that the terminal SCCs satisfy the requirement formulated above can be done on-the-fly when combining a depth-first generation of the SS state space with generation of SCCs by means of TARJAN's algorithm [2]. If a terminal SCC C is about to be completed and the construction of the SS state space is about to backtrack from the marking M_0 then we check that either

$\phi(M)$ holds in some $M \in C$ or for each atomic state proposition φ_i we have that $\mathsf{up}_{\varphi_i}(M_0) \subseteq \bigcup_{\{ M \in C \}} T_{\mathsf{s}}(M)$. If we find an atomic state proposition i such that 1) $\mathsf{up}_{\varphi_i}(M_0) \not\subseteq \bigcup_{\{ M \in C \}} T_{\mathsf{s}}(M)$ and 2) the stubborn set in M_0, $T_{\mathsf{s}}^{\mathsf{up}_i}(M_0)$ containing $\mathsf{up}_{\varphi_i}(M_0) - \bigcup_{\{ M \in C \}} T_{\mathsf{s}}(M)$ contains enabled transitions which are not in $T_{\mathsf{s}}(M_0)$, then we extend the stubborn set used in M_0 with $T_{\mathsf{s}}^{\mathsf{up}_i}(M_0)$. The extension of the stubborn set is simple to implement since the union of two stubborn sets is again a stubborn set. SPP3 requires also the check that for every $t \in \mathsf{down}_\phi$, there is a $M \in C$ such that $t \in T_{\mathsf{s}}(M)$.

The use of terminal SCCs was first suggested in [9] for a condition which from an implementation point of view resembles SPP2 and SPP3. The condition was later called "S" in [10]. We refer to [9] for additional details about its implementation.

Cycle Detection. An approximation to ensuring that an occurrence sequence exists satisfying SPP2 and SPP3 in each of the terminal SCCs is to ensure the stronger requirement that such an occurrence sequence exists in each of the SCCs. This can be implemented without the use of TARJAN's algorithm, and we can rely on depth-first generation and strong stubborn sets only. The algorithm operates as follows.

Whenever we reach a marking M_1 during the SS state space generation which is on the depth-first search stack, we have detected a cycle. We search backwards in the stack through markings $M_n, M_{n-1}, \ldots, M_1$ and check whether for all i we have that $\mathsf{up}_{\varphi_i}(M_1) \subseteq \bigcup_{j=1}^{n} T_{\mathsf{s}}(M_j)$. For all atomic state proposition i such that $\mathsf{up}_{\varphi_i}(M_1) \not\subseteq \bigcup_{j=1}^{n} T_{\mathsf{s}}(M_j)$ we compute a new stubborn set in M_1, $T_{\mathsf{s}}^{\mathsf{up}_i}(M_1)$ containing $\mathsf{up}_{\varphi_i}(M_1) - \bigcup_{j=1}^{n} T_{\mathsf{s}}(M_j)$, and extend the stubborn set used in M_1 with $T_{\mathsf{s}}^{\mathsf{up}_i}(M)$. Again, SPP3 requires taking also down_ϕ into account in the check.

9   Applications

In this section we develop stubborn set methods for *boundedness properties* based on the RSPP stubborn set method. The considered boundedness properties are inspired from how boundedness properties of High-level Petri Nets are interpreted at the level of the equivalent PT-net. The purpose of this section is twofold. Firstly, to develop methods for a general set of boundedness properties as such, and secondly to illustrate how the results of this paper can be applied as a tool for developing stubborn set methods for state properties composed of atomic state propositions beyond those considered in this paper.

Best Upper Bounds. An integer k is an *upper bound* for a set of places $P' \subseteq P$ iff $\forall M \in [M_I\rangle : \sum_{p \in P'} M(p) \leq k$. We are interested in finding the minimal such k, denoted the *best upper bound* of P'. One approach is to check the state properties $\phi^k(M) \equiv \sum_{p \in P'} M(p) \geq k$ starting with $k = 0$ and incrementing k until a k_0 is found for which a marking with $\sum_{p \in P'} M(p) \geq k_0$ is not reachable. $k_0 - 1$ is then the best upper bound. A problem which has to be solved before this approach works is that we have not allowed $\sum_{p \in P'} M(p) \geq k$ as an atomic state proposition. However, all that is needed to make our stubborn set algorithm work in this case is to define proper up and down sets for this "new" atomic state proposition. The up set for ϕ^k can be defined as the set of transitions which adds tokens to P' and which does not require additional tokens to be present on P'. The down set can be defined as the transitions which can remove tokens from P' and which produces less than k token on P'. Formally:

$$\mathsf{up}_{\phi^k}(M) = \{\, t \in T \mid \sum_{p \in P'} W(p,t) < \sum_{p \in P'} W(t,p) \wedge \sum_{p \in P'} W(p,t) \leq \sum_{p \in P'} M(p) \,\}$$

$$\mathsf{down}_{\phi^k} = \{\, t \in T \mid \sum_{p \in P'} W(p,t) > \sum_{p \in P'} W(t,p) \wedge \sum_{p \in P'} W(t,p) < k \,\}$$

An alternative is to observe that up_{ϕ^k} is independent of k and down_{ϕ^k} can be approximated from above by removing "$\sum_{p \in P'} W(t,p) < k$" from its definition. This means that the stubborn sets used are then independent of k. It therefore suffices to generate just a single SS state space.

Best Lower Bounds. An integer k is a *lower bound* for a set of places $P' \subseteq P$ iff $\forall M \in [M_I] : \sum_{p \in P'} M(p) \geq k$. We are interested in finding the maximal such k, referred to as the *best lower bound* of P'. Similarly to the upper bound case we consider state properties of the form : $\phi^k(M) \equiv \sum_{p \in P'} M(p) \leq k$ starting with $k = 0$ and continuing until the first k_0 is found for which a marking with $\sum_{p \in P'} M(p) \leq k_0$ is reachable. k_0 is then the best lower bound. The up and down sets for the atomic state proposition $\sum_{p \in P'} M(p) \leq k$ can be defined as shown below. If one is interested in generating only a single SS state space when finding the best lower bound of a set of places, then the dependency of k can be eliminated like for the best upper bound case by approximating the up and down sets to become independent of k.

$$\mathsf{up}_{\phi^k}(M) = \{\, t \mid \sum_{p \in P'} W(p,t) > \sum_{p \in P'} W(t,p) \wedge \sum_{p \in P'} W(t,p) \leq k \,\}$$

$$\mathsf{down}_{\phi^k} = \{\, t \mid \sum_{p \in P'} W(p,t) < \sum_{p \in P'} W(t,p) \wedge \sum_{p \in P'} W(p,t) \leq k \,\}$$

10   Experimental Results

We have implemented the RSPP stubborn set method on top of the state space tool of Design/CPN [1]. The prototype implements the *Attractor Set* and *Cycle Detection* algorithms given in Sect. 8.2. The construction of stubborn sets is based on the *strong component algorithm* described in [10] adapted to take the condition SPP1 into account.

Tables 1 and 2 give numerical data on the reduction obtained with the two implemented algorithms on some examples. For PETERSON's and HYMAN's mutual exclusion algorithms we consider mutual exclusion (*Mutual Excl.*), and that each of the two processes can reach the critical section (*Reach. of CS*). For the Reader/Writer protocol we consider three state properties: the writers can get write access (*Reach. of Write*), the readers can get read access (*Reach. of Read*), and the protocol guarantees exclusive write (*Excl. Write*). We consider a configuration with 2 writers and 3 readers. For the Master/Slave protocol we consider two properties: a marking is reachable in which the master has received a response from all slaves which in turn have returned to their idle state (*DoneIdle*), and the master never continues before having received a response from all slaves (*DoneWIdle*). We consider configurations with 3, 5 and 6 slaves.

Table 1 gives information about the performance of the *Cycle Detection* algorithm. The table contains two main parts. In the *Up set Driven* part the construction of the stubborn set is initiated from the transitions in the up set and it favours stubborn sets containing transitions in the up set. In the *Up set/Enabling Driven* part the construction of the stubborn set is initiated from the transitions in the up set but it does not favour stubborn sets containing transitions in the up set. The *DFG* columns represent a depth-first generation of the state space with *early termination*, i.e., as soon as a marking has been found where the state property holds the generation stops. The *CG* columns represent a complete generation, i.e., the generation continues even though a marking has been found where the state property holds. This gives information about how large a state space the corresponding algorithm considers in the worst-case. For those properties where no marking is reachable where the property holds, depth-first and complete generation coincide, and only the numbers for the complete generation are given. The entries in the *Min Length* columns are of the form x/y, where x gives the number of nodes in a shortest path leading to a marking where the property holds for the depth-first generation (if such one exists), and y gives the corresponding number for the complete generation. This gives information about how good the algorithm is at providing short witness paths.

Table 2 gives information about the performance of the *Attractor Set* algorithm and the full state space. The table contains two main parts. The *Full State Space* part lists the size of the full state space. In the *Attractor Set Method* part, the *DFG* and *CG* columns represent depth-first generation with early termination, and complete generation, respectively. The *BFG* column represents a breadth-first generation with early termination. The entries in the *Min Length* are of the form x/y, where x gives the number of nodes in a shortest path leading to a marking where the property holds for the depth-first generation (if such one exists), and y gives the corresponding number for the *BFG* generation. It was proved in [6] that the latter equals the number of nodes in one of the shortest paths of the full state space. All state spaces reported on in this section were generated in less than 2 minutes on a 166 Mhz PII PC.

If we first compare the numbers for the complete generation (*CG*) in Tables 1 and 2 then in all cases the *Up set/Enabling Driven* implementation gives much better reduction than the *Attractor Set Method*. The *Up set Driven* implementation gives approximately the same reduction as the *Attractor Set Method*. As a consequence of this the *Up set/Enabling Driven* implementation outperforms

Table 1. Experimental results – Cycle detection algorithm.

Model/ Property	Up set Driven					Up set/Enabling Driven				
	DFG		CG		Min	DFG		CG		Min
	Nodes	Arcs	Nodes	Arcs	Length	Nodes	Arcs	Nodes	Arcs	Length
Peterson										
Reach. of CS	9	8	36	59	9/6	6	5	34	51	6/6
Mutual Excl.	-	-	48	84	-	-	-	47	79	-
Hyman										
Reach. of CS	5	4	60	95	5/5	8	7	38	46	8/8
Mutual Excl.	19	19	64	106	17/12	18	20	45	57	12/12
Reader/Writer										
Reach. of Write	3	2	77	221	3/3	7	6	14	17	7/7
Reach. of Read	3	2	85	197	3/3	14	17	14	17	7/7
Excl. Write	-	-	46	111	-	-	-	24	37	-
Master/Slave										
DoneIdle-3	60	67	229	548	60/15	30	30	130	152	30/15
DoneIdle-5	230	277	7,837	32,412	230/23	691	790	1,654	2,172	483/23
DoneIdle-6	516	622	46,781	233,276	513/27	1,744	2,084	5,600	7,658	1053/27
DoneWIdle-3	-	-	231	562	-	-	-	185	272	-
DoneWIdle-5	-	-	7,839	32,494	-	-	-	3,745	6,592	-
DoneWIdle-6	-	-	46,783	233,470	-	-	-	16,769	31,168	-

the *Attractor Set Method* in the cases where the state property does not hold. If we consider the set of state properties which holds then for the first three examples the *Attractor Set Method* seems slightly better than the cycle detection algorithm in terms of yielding small state spaces and generating short witness paths. However, when we turn to the larger *Master/Slave* example, then the *Up set/Enabling Driven* implementation again outperforms the *Attractor Set Method* in terms of reduction and it is still able to generate a short witness path. The intuitive reason for the *Up set/Enabling Driven* implementation to be better in these cases is that if the state property is located "far" from the initial marking (as is the case for the *Master/Slave* example), then the *Attractor Set Method* and to some extent also the *Up Set Driven* implementation have a high risk of investigating wrong branches of the state space first. For the cases where the state property holds the *Up set/Enabling Driven* implementation therefore seems to represent a good solution to the trade-off between generating short witness paths and considering large state spaces.

11   Conclusions

We have presented two new stubborn set methods for reasoning about state properties. The method for determining whether a reachable marking exists in which a given state property holds was based on ideas first presented in [6]. The main difference between our new method and [6] is in how progress towards the state property is ensured. We have replaced the *always progress* condition of [6] with the weaker *eventual progress* condition, which has the potential of leading to better reduction results, and which contains the always progress condition as a special case. We have demonstrated the potential on some practical case studies by means of an implementation of the new method. The case studies showed that the new stubborn set method is significantly better when the state property does

Table 2. Experimental results – Full state space and attractor set algorithm.

Model/ Property	Full State Space		Attractor Set Method							
			DFG		BFG		CG		Min.
	Nodes	Arcs	Nodes	Arcs	Nodes	Arcs	Nodes	Arcs	Length
Peterson	58	116							
Reach. of CS			9	8	10	11	39	67	9/5
Mutual Excl.			-	-	-	-	50	90	-
Hyman	80	160							
Reach. of CS			7	6	14	17	60	95	7/5
Mutual Excl.			36	42	49	76	64	106	30/12
Reader/Writer	136	532							
Reach. of Write			3	2	3	2	77	221	3/3
Reach. of Read			3	2	3	2	85	197	3/3
Excl. Write			-	-	-	-	118	419	-
Master/Slave									
DoneIdle-3	232	588	61	79	212	520	229	548	45/15
DoneIdle-5	7,840	32,656	1,623	4,144	7,700	31,966	7,837	32,412	575/23
DoneIdle-6	46,784	233,856	9,819	37,155	33,092	156,317	46,701	233,276	1566/27
DoneWIdle-3	232	588	-	-	-	-	231	562	-
DoneWIdle-5	7,840	32,656	-	-	-	-	7,839	32,494	-
DoneWIdle-6	46,784	233,856	-	-	-	-	46,783	233,470	-

not hold in any reachable marking. When a reachable marking exists in which the state property does hold, then it represents a good solution to the trade-off between short witness paths and large state spaces. From an implementation point of view the more powerful implementations which we have suggested for the eventual progress condition require *strong stubborn sets*, whereas for the always progress implementation it suffices to use *weak stubborn sets*.

We have extended the first stubborn set method to obtain a second stubborn set method representing a novel technique for determining, e.g., whether a marking is a home marking, and for checking liveness of only a single transition. Like existing methods for checking liveness of transitions it relies on strong stubborn sets, but it does not require *ignoring* [9] to be eliminated.

As an application to boundedness properties we have illustrated the use of the results presented in this paper as a tool for developing stubborn set methods for state properties beyond those considered in the paper. In fact, it can be observed that we only directly referred to PT-nets in the implementation of up and down sets, and hence the suggested methods can be transferred to other modelling formalisms – provided that they allow for the definition of sets of transitions satisfying the properties of up and down sets.

References

[1] S. Christensen, J. B. Jørgensen, and L. M. Kristensen. Design/CPN - A Computer Tool for Coloured Petri Nets. In *Proceedings of TACAS'97*, volume 1217 of *Lecture Notes in Computer Science*, pages 209–223. Springer-Verlag, 1997.

[2] A. Gibbons. *Algorithmic Graph Theory*. Cambridge University Press, 1985.

[3] P. Godefroid. Using Partial Orders to Improve Automatic Verification Methods. In *Proceedings of Computer-Aided Verification '90*, volume 531 of *Lecture Notes in Computer Science*, pages 175–186. Springer-Verlag, 1990.

[4] P. Godefroid. *Partial-Order Methods for the Verification of Concurrent Systems, An Approach to the State-Explosion Problem*, volume 1032 of *Lecture Notes in Computer Science*. Springer-Verlag, 1996.

[5] D. Peled. All from One, One for All: On Model Checking Using Representatives. In *Proceedings of Computer-Aided Verification '93*, volume 697 of *Lecture Notes in Computer Science*, pages 409–423. Springer-Verlag, 1993.

[6] K. Schmidt. Stubborn Sets for Standard Properties. In *Proceedings of ICATPN'99*, volume 1639 of *Lecture Notes in Computer Science*, pages 46–65. Springer-Verlag, 1999.

[7] A. Valmari. Error Detection by Reduced Reachability Graph Generation. In *Proceedings of the 9th European Workshop on Application and Theory of Petri Nets*, pages 95–112, 1988.

[8] A. Valmari. A Stubborn Attack on State Explosion. In *Proceedings of Computer-Aided Verification '90*, volume 531 of *Lecture Notes in Computer Science*, pages 156–165. Springer-Verlag, 1990.

[9] A. Valmari. Stubborn Sets for Reduced State Space Generation. In G. Rozenberg, editor, *Advances in Petri Nets '90*, volume 483 of *Lecture Notes in Computer Science*, pages 491–515. Springer-Verlag, 1990.

[10] A. Valmari. The State Explosion Problem. In W. Reisig and G. Rozenberg, editors, *Lectures on Petri Nets I: Basic Models*, volume 1491 of *Lecture Notes in Computer Science*, pages 429–528. Springer-Verlag, 1998.

A Compositional Model of Time Petri Nets

Maciej Koutny

Department of Computing Science, University of Newcastle
Newcastle upon Tyne NE1 7RU, U.K.

Abstract. This paper presents two related algebras which can be used
to specify and analyse concurrent systems with explicit timing information. The first algebra is based on process expressions, called t-expressions,
and a system of SOS rules providing their operational semantics. The
second algebra is based on a class of time Petri nets, called ct-boxes,
and their transition firing rule. The two algebras are related through a
mapping which, for a t-expression, returns a corresponding ct-box with
behaviourally equivalent transition system. The resulting model, called
the Time Petri Box Calculus (tPBC), extends the existing approach of
the Petri Box Calculus (PBC).

Keywords: Net-based algebraic calculi; time Petri nets; relationships
between net theory and other approaches; process algebras; box algebra;
SOS semantics.

1 Introduction

Process algebras deal with large and complex concurrent computing systems
by employing specific operators corresponding to commonly used programming
constructs [1,10,11,15]. Petri nets, on the other hand, provide both a graphical
representation of such systems and, through their being based on a theory of
partial orders (capturing explicit asynchrony), an additional means of verifying their correctness efficiently, and a way of expressing properties related to
causality and concurrency in system behaviour [16,19].

The standard treatment of the structure and semantics of concurrent systems provided by process algebras and Petri nets is different, making it virtually
impossible to take full advantage of their relative strengths (i.e., *compositionality* and *explicit asynchrony*) when used in isolation. This problem has been
addressed by the Box Algebra [5,6,7,13] and its precursor, the Petri Box Calculus (PBC) [3,4]. It is a generic model that embodies both Petri nets and process
algebras, and thus provides a bridge between these two approaches. However,
the existing framework does not allow one to specify timing restrictions for the
actions employed by a concurrent system.

This paper introduces and investigates two different models for the specifications of concurrent systems including explicit timing information. The models
have an algebraic structure based on operators present in the standard PBC.

The first algebra is based on process expressions, called t-expressions, and
a system of rewriting rules providing a structural operational semantics of t-
expressions in the style of [17]. The second algebra is based on a class of time

M. Nielsen, D. Simpson (Eds.): ICATPN 2000, LNCS 1825, pp. 303–322, 2000.

Petri nets, called ct-boxes, and their execution rules [2,8,14,18]. This means, in particular, that: (i) each transition is provided with two time bounds, e and l, representing the *earliest firing time* and the *latest firing time*, respectively; (ii) the local clock of a transition is started at the very moment it becomes enabled; and (iii) time is discrete. The two algebras are related through a compositionally defined mapping which, for every t-expression, returns a corresponding ct-box (its denotational semantics). The main result is that the denotational and operational semantics of a t-expression are equivalent. The resulting framework consisting of two consistent algebras is called the Time Petri Box Calculus, or simply tPBC.

The paper is organised as follows. Section 2 defines the syntax of t-expressions, and section 3 defines the rules of operational semantics. Section 4 develops a compositional net model based on time Petri nets. The next section gives a net semantics of t-expressions, and states the consistency result between the operational and denotational semantics of t-expressions.

Throughout the paper, we assume that the reader is familiar with the basic concepts of PBC and the Box Algebra and, in particular, with the notion of net refinement [3,5] on which the compositional treatment of nets is based.

2　Syntax

In this section, we introduce process expressions used throughout the paper.

Static Expressions The class of process expressions we will consider is derived from the standard recursion-free PBC [4]. For brevity of presentation, we omitted the relabelling operator which is a syntactic sugar in this context, and assumed a simple form of basic actions which is, essentially, that of CCS [15]. More precisely, we assume that there is a fixed set of *communication* actions $\mathcal{A}$ and that, for every $a \in \mathcal{A}$, there exists its *conjugate*, $\widehat{a} \in \mathcal{A}$, such that $a \neq \widehat{a}$ and $\widehat{\widehat{a}} = a$. In addition, there is a *silent* (or internal) action $\imath \notin \mathcal{A}$. It is assumed that a *synchronisation* of two conjugate communication actions gives rise to the silent action $\imath$.

The following is the syntax for *static time box expressions* (or static t-expressions), E.

$$E ::= C \quad \mid \quad E \text{ sy } A \quad \mid \quad E \text{ rs } A \quad \mid \quad E \,\square\, E \quad \mid \quad E \| E$$
$$C ::= \alpha[e,l] \quad \mid \quad C \text{ sy } A \quad \mid \quad C \text{ rs } A \quad \mid \quad C \,\square\, C \quad \mid \quad E;E \quad \mid \quad [E * C * E].$$

The only significant modification, when compared with the PBC syntax, is that a different type of constant expression is used, viz. $\alpha[e,l]$ where: $\alpha \in \mathcal{A} \cup \{\imath\}$ is a basic action; e is a non-negative integer; and $l \geq e$ is a non-negative integer or ∞. Intuitively, e denotes the *earliest* and l the *latest* execution time for α, measured from the moment it has become enabled.

In the above, C defines an auxiliary class of *ex-exclusive* expressions [6,7], for whom the corresponding nets are such that no marking reachable from the initial one can mark simultaneously an entry and an exit place. As a result, in the translation from t-expressions to safe Petri nets, when modelling iteration we will be able to use a simple operator box Ω_* shown in figure 1.

Static expressions describe the structural characteristics of concurrent systems. Their operational semantics will be modelled using dynamic expressions, defined next.

Dynamic Expressions The syntax of dynamic PBC expressions is changed by adding time annotations to the over- and underbars. Each such annotation is a non-negative integer intuitively corresponding to the *age* of an over- or underbar. For example, $\overline{E}^3$ is an expression E which has been in its initial state for 3 time units, and $\underline{E}_{23}; F$ a sequential composition where the first component terminated 23 time units ago.

The *dynamic t-expressions* are defined below, where E denotes a static t-expression, C denotes an ex-exclusive static t-expression, and $\tau \in \mathbf{N} = \{0, 1, \ldots\}$.

$$G ::= \overline{E}^\tau \mid \underline{E}_\tau \mid G \text{ sy } A \mid G \text{ rs } A \mid G \,\square\, E \mid E \,\square\, G \mid G \| G \mid K$$

$$K ::= \overline{C}^\tau \mid \underline{C}_\tau \mid K \text{ sy } A \mid K \text{ rs } A \mid K \,\square\, C \mid C \,\square\, K \mid G; E \mid E; G \mid$$
$$[G * C * E] \mid [E * K * E] \mid [E * C * G] .$$

As the C before, K is an auxiliary class of ex-exclusive dynamic t-expressions.

The above syntax may appear to be too permissive; e.g., it admits an expression $\overline{\alpha[0, 2]}^6$ with an inconsistent timing information (the action has to be executed within 2 time units after being enabled, yet it has been overbarred for much longer than that). However, such an expression may be a part of another time consistent expression, e.g., $(\overline{\alpha[0, 2]}^6)$ rs $\{a\}$, and thus cannot be excluded. In the approach adopted in this paper, $\overline{\alpha[0, 2]}^6$ will not be a correct expression representing a complete system, but it still may model a subsystem within a larger expression.

3 Operational Semantics of t-Expressions

In this section, we present an operational semantics for dynamic t-expressions, by following the steps through which the semantics of PBC was defined [5,13]. We first define a structural equivalence relation on expressions, and then introduce rules of the structural operational semantics (SOS) in the style of [17].

3.1 Structural Equivalence

The structural equivalence relation on expressions aims to capture the most fundamental correspondence between expressions. For example, $\underline{E}_\tau; F \equiv E; \overline{F}^\tau$

states that a sequential system in which the first component terminated τ units of time ago is the same as that in which the second component has been in its initial state for τ time units.

The equations of the structural equivalence relation defined for the standard PBC need to be adjusted in order to reflect the time annotations of over- and underbars. Formally, $\equiv$ is the least equivalence relation on dynamic t-expressions such that the rules in table 1 are satisfied.

$$\overline{E\|F}^{\,\tau} \equiv \overline{E}^{\,\tau}\|\overline{F}^{\,\tau} \qquad\qquad \underline{E}_{\tau}\|\underline{F}_{\tau'} \equiv \underline{E\|F}_{\min\{\tau,\tau'\}}$$

$$\overline{E\,\square\,F}^{\,\tau} \equiv \overline{E}^{\,\tau}\,\square\,F \qquad\qquad \underline{E}_{\tau}\,\square\,F \equiv \underline{E\,\square\,F}_{\tau}$$

$$\overline{E\,\square\,F}^{\,\tau} \equiv E\,\square\,\overline{F}^{\,\tau} \qquad\qquad E\,\square\,\underline{F}_{\tau} \equiv \underline{E\,\square\,F}_{\tau}$$

$$\overline{E\,\mathsf{rs}\,A}^{\,\tau} \equiv \overline{E}^{\,\tau}\,\mathsf{rs}\,A \qquad\qquad \underline{E}_{\tau}\,\mathsf{rs}\,A \equiv \underline{E\,\mathsf{rs}\,A}_{\tau}$$

$$\overline{E\,\mathsf{sy}\,A}^{\,\tau} \equiv \overline{E}^{\,\tau}\,\mathsf{sy}\,A \qquad\qquad \underline{E}_{\tau}\,\mathsf{sy}\,A \equiv \underline{E\,\mathsf{sy}\,A}_{\tau}$$

$$\overline{E;F}^{\,\tau} \equiv \overline{E}^{\,\tau};F \qquad\qquad E;\underline{F}_{\tau} \equiv \underline{E;F}_{\tau}$$

$$\underline{E}_{\tau};F \equiv E;\overline{F}^{\,\tau} \qquad\qquad [D*\underline{E}_{\tau}*F] \equiv [D*E*\overline{F}^{\,\tau}]$$

$$\overline{[D*E*F]}^{\,\tau} \equiv [\overline{D}^{\,\tau}*E*F] \qquad\qquad [D*E*\underline{F}_{\tau}] \equiv \underline{[D*E*F]}_{\tau}$$

$$[\underline{D}_{\tau}*E*F] \equiv [D*\overline{E}^{\,\tau}*F] \qquad\qquad [D*\overline{E}^{\,\tau}*F] \equiv [D*\underline{E}_{\tau}*F]$$

Table 1. Rules of the structural equivalence.

We do not give any rule for $\overline{E}^{\,\tau}\|\overline{F}^{\,\tau'}$ with $\tau \neq \tau'$ as such an expression can never be derived from initially marked static expressions. Expressions of this form are excluded in order to obtain a crucial consistency result between t-expressions and time Petri nets. For consider the expression

$$G = \alpha[0,\infty]\,\square\,(\overline{\beta[0,1]}^{\,1} \| \overline{\gamma[0,1]}^{\,0})\,.$$

In the corresponding net α is enabled, and so in order to have a corresponding move of the operational semantics, we need to be able to make $\alpha[0,\infty]$ over-barred. Suppose that

$$G \equiv H = \overline{\alpha[0,\infty]}^{\,\tau}\,\square\,(\beta[0,1] \| \gamma[0,1])\,,$$

for some τ. By a symmetric argument, we need to be able to transfer the τ-labelled overbar back, e.g.,

$$H \equiv J = \alpha[0,\infty]\,\square\,(\overline{\beta[0,1]}^{\,\tau} \| \overline{\gamma[0,1]}^{\,\tau})\,.$$

Thus $G \equiv J$, yet the corresponding time Petri nets would have different seman-tics. A possible solution would be to 'remember' somehow within the overbar of G that it was created from two different overbars of age 0 and 1, respectively. However, such an approach would be much more complex and, since no t-expression in its initial state will ever yield sub-expressions like G, we simply exclude them from further considerations. More precisely, we exclude from the set of dynamic t-expressions any expression which contains a subexpression H such that $H \equiv \overline{E}^\tau \| \overline{F}^{\tau'}$ where $\tau \neq \tau'$.

Proposition 1. *Assuming that we treat the rules in table 1 as term rewriting rules, if $G \equiv H$ and G is a time expression, then so is H.*

3.2 SOS Rules

There are two kinds of operational semantics moves, namely *action* moves and *time* moves.

A time move has the form $G \xrightarrow{\checkmark} H$ where $\checkmark$ is a special symbol indicating the passage of one unit of time.

An action move has the form $G \xrightarrow{\Gamma} H$ where $\Gamma = \{\gamma_1, \ldots, \gamma_k\}$ is a finite multiset ($k \geq 0$). Each γ_i is an *action occurrence* of the form $\alpha[e, \tau, l]$ where: $\alpha \in \mathcal{A} \cup \{\imath\}$ is a communication or silent action; $e \leq \tau$ are non-negative integers; and l is a non-negative integer greater or equal to e, or ∞. The action occurrence $\gamma_i = \alpha[e, \tau, l]$ is *legal* if $\tau \leq l$, and *illegal* if $\tau > l$. Intuitively, τ is the age of the (most recent) overbar consumed during the execution of γ_i. Legal action occurrences are created when all timing constraints are satisfied at the moment of the execution of an action. Illegal action occurrences should not be generated if an expression is supposed to represent a complete system, but it may happen that they are generated within by a subexpression. An illegal action occurrence $a[e, \tau, l]$ represents an action which cannot occur on its own (it is restricted at a higher level of the syntax tree), but if synchronised (before being restricted) with a suitable legal action occurrence $\hat{a}[e', \tau', l']$, it can contribute 'half' of a composite action occurrence $\imath[\max\{e, e'\}, \min\{\tau, \tau'\}, \min\{l, l'\}]$. Thus, in particular, we will allow the execution of $a[e, l]$ with an overbar labelled by a time value greater than l. In a synchronisation with a legal action occurrence this may not matter since it is the joint enabledness of illegal and legal action which triggers the local clock of the synchronised action. We will return to this issue in section 3.4.

We now formally define the various types of moves of the structural operational semantics of dynamic t-expressions.

Empty Moves The following rules deal with the empty action moves.

$$\frac{G \equiv H}{G \xrightarrow{\emptyset} H} \qquad \frac{G \xrightarrow{\emptyset} J \xrightarrow{\Gamma} H}{G \xrightarrow{\Gamma} H} \qquad \frac{G \xrightarrow{\Gamma} J \xrightarrow{\emptyset} H}{G \xrightarrow{\Gamma} H}$$

Basic Action A basic action can yield a legal action occurrence if its timing restrictions are satisfied by the age of an overbar, and it can yield an illegal action occurrence if the age of its overbar exceeds the latest firing time:

$$\overline{\alpha[e,l]}^{\tau} \xrightarrow{\{\alpha[e,\tau,l]\}} \underline{\alpha[e,l]}_0 \quad \text{where } e \leq \tau$$

Note that the age a newly created underbar is always set to 0.

Restriction and Synchronisation There is a single rule for restriction:

$$\frac{G \xrightarrow{\Gamma} H}{G \text{ rs } A \xrightarrow{\Gamma} H \text{ rs } A}$$

$$\text{where } \alpha \notin A \cup \widehat{A} \text{ for every } \alpha[e,\tau,l] \in \Gamma$$

There are two synchronisation rules. The first one states that a synchronised expression can do all that the original one could, while the second captures the essential meaning of the standard PBC synchronisation, after taking into account the timing restrictions imposed on actions.

$$\frac{G \xrightarrow{\Gamma} H}{G \text{ sy } A \xrightarrow{\Gamma} H \text{ sy } A}$$

$$\frac{G \text{ sy } A \xrightarrow{\Gamma + \{a[e,\tau,l], \widehat{a}[e',\tau',l']\}} H \text{ sy } A}{G \text{ sy } A \xrightarrow{\Gamma + \{i[\max\{e,e'\}, \min\{\tau,\tau'\}, \min\{l,l'\}]\}} H \text{ sy } A}$$

$$\text{where } a \in A \text{ and } \max\{e,e'\} \leq \min\{l,l'\} \text{ and } \max\{e,e'\} \leq \min\{\tau,\tau'\}$$

The second rule essentially means that two conjugate action occurrences, $a[e,\tau,l]$ and $\widehat{a}[e',\tau',l']$, can be synchronised if their timing restrictions lead to a well-defined action occurrence, in the sense that $\max\{e,e'\} \leq \min\{l,l'\}$ and $\max\{e,e'\} \leq \min\{\tau,\tau'\}$. Such a synchronisation may or may not lead to a legal action occurrence:

– If both $a[e,\tau,l]$ and $\widehat{a}[e',\tau',l']$ are legal action occurrences, then the synchronisation is also legal since $\min\{\tau,\tau'\} \leq \min\{l,l'\}$.
– If both $a[e,\tau,l]$ and $\widehat{a}[e',\tau',l']$ are illegal action occurrences, then the synchronisation is also illegal since $\min\{\tau,\tau'\} > \min\{l,l'\}$.
– If, say, $a[e,\tau,l]$ is a legal and $\widehat{a}[e',\tau',l']$ illegal action occurrence, then the synchronisation is legal if and only if $\tau \leq l'$.

Illegal synchronisations are undesirable and one of the aims of the development of the theory is to ensure that they are not generated by what is to be considered as a well formed expression.

Other tPBC Operators There is no real difference in the rules for the remaining operators when compared with the standard PBC [5,13]. These rules are given in table 2.

$$\frac{G \xrightarrow{\Gamma} G' \, , \, H \xrightarrow{\Gamma'} H'}{G \| H \xrightarrow{\Gamma+\Gamma'} G' \| H'} \qquad \frac{G \xrightarrow{\Gamma} H}{G \,\square\, E \xrightarrow{\Gamma} H \,\square\, E}$$

$$\frac{G \xrightarrow{\Gamma} H}{E \,\square\, G \xrightarrow{\Gamma} E \,\square\, H} \qquad \frac{G \xrightarrow{\Gamma} H}{G; E \xrightarrow{\Gamma} H; E}$$

$$\frac{G \xrightarrow{\Gamma} H}{E; G \xrightarrow{\Gamma} E; H} \qquad \frac{G \xrightarrow{\Gamma} H}{[G * E * F] \xrightarrow{\Gamma} [H * E * F]}$$

$$\frac{G \xrightarrow{\Gamma} H}{[E * G * F] \xrightarrow{\Gamma} [E * H * F]} \qquad \frac{G \xrightarrow{\Gamma} H}{[E * F * G] \xrightarrow{\Gamma} [E * F * H]}$$

Table 2. Rules of the operational semantics.

Time Moves There is a single time rule:

$$\frac{\neg \exists \, G \xrightarrow{\{\alpha[e,\tau,l]\}} H \; : \; l \leq \tau}{G \xrightarrow{\sqrt{}} \sqrt{}(G)}$$

where $\sqrt{}(G)$ is G with each time annotation τ at an over- or underbar changed to $\tau + 1$. Although the above rule is a rule with negative premise, the inference rule system is well defined since time moves are not used in the premise of any rule for the action moves. Notice that a time move can only be applied at the topmost level of an expression as it cannot be 'propagated up' through the expression using action rules. This ensures that time progresses uniformly.

It can be seen that the rules of the operational semantics do not lead outside the set of dynamic t-expressions (c.f. proposition 1).

Reachable t-Expressions We are ultimately interested in those t-expressions that can be reached, through the rules of the structural operational semantics, from static t-expressions started at zero time. In what follows, we will call a dynamic t-expression G *reachable* if there is a static t-expression E such that G can be derived from $\overline{E}^{0}$ using the operational semantics rules defined in this section. A crucial property of reachable t-expressions is that they never generate illegal action occurrences.

Proposition 2. *Let G be a reachable t-expression and $G \xrightarrow{\Gamma} H$. Then, every action occurrence in Γ is legal.*

Transition Systems of Reachable t-Expressions Let G be a reachable t-expression. We will use $[G\rangle$ to denote the least set of expressions containing G such that if $H \in [G\rangle$ and $H \xrightarrow{x} J$, for some x, then $J \in [G\rangle$.
The *transition system* of G is $\mathsf{ts}_G = (V, A, v_0)$ where: $V = \{[H]_\equiv \mid H \in [G\rangle\}$ is the set of states (note that $[H]_\equiv$ is the equivalence class of $\equiv$ containing H); $v_0 = [G]_\equiv$ is the initial state; and the set of arcs is given by:

$$A = \{([H]_\equiv, x, [J]_\equiv) \mid H \in [G\rangle \wedge H \xrightarrow{x} J\} \ .$$

3.3   Examples

In the four examples below, we consider derivations from static expressions marked initially with a zero-overbar.

The first example shows a derivation where the execution of an action happens at the last possible moment.

$$\overline{a[0,2] \; \Box \; b[1,2]}^{\,0} \qquad \equiv \qquad a[0,2] \; \Box \; \overline{b[1,2]}^{\,0} \qquad \xrightarrow{\checkmark}$$

$$a[0,2] \; \Box \; \overline{b[1,2]}^{\,1} \qquad \xrightarrow{\checkmark} \qquad a[0,2] \; \Box \; \overline{b[1,2]}^{\,2} \qquad \xrightarrow{\{b[1,2,2]\}}$$

$$a[0,2] \; \Box \; \underline{b[1,2]}_{\,0} \qquad \equiv \qquad a[0,2] \; \Box \; \underline{b[1,2]}_{\,0} \ .$$

Notice that no time move is possible for $a[0,2] \; \Box \; \overline{b[1,2]}^{\,2}$ as this expression can generate an action occurrence $b[1,2,2]$ which cannot be delayed.

The second example demonstrates that time annotations in different parts of a parallel composition can have different values.

$$\overline{(a[0,2] \parallel b[1,2]); c[0,0]}^{\,0} \qquad\qquad \equiv$$

$$(\overline{a[0,2]}^{\,0} \parallel \overline{b[1,2]}^{\,0}); c[0,0] \qquad \xrightarrow{\{a[0,0,2]\}}$$

$$(\underline{a[0,2]}_{\,0} \parallel \overline{b[1,2]}^{\,0}); c[0,0] \qquad \xrightarrow{\checkmark}$$

$$(\underline{a[0,2]}_{\,1} \parallel \overline{b[1,2]}^{\,1}); c[0,0] \qquad \xrightarrow{\checkmark}$$

$$(\underline{a[0,2]}_{\,2} \parallel \overline{b[1,2]}^{\,2}); c[0,0] \qquad \xrightarrow{\{b[1,2,2]\}}$$

$$(\underline{a[0,2]}_{\,2} \parallel \underline{b[1,2]}_{\,0}); c[0,0] \qquad \equiv$$

$$\underline{(a[0,2] \parallel b[1,2])}_{\,0}; c[0,0] \qquad \equiv$$

$$(a[0,2] \parallel b[1,2]); \overline{c[0,0]}^{\,0} \qquad \xrightarrow{\{c[0,0,0]\}}$$

$$(a[0,2] \parallel b[1,2]); \underline{c[0,0]}_{\,0} \ .$$

The next example demonstrates that a potential synchronisation (i.e., one which would be allowed in PBC after ignoring time annotations) can be made impossible by the timing constraints.

$$\overline{(a[0,0] \parallel (b[1,1];\widehat{a}[0,1]))\text{ sy }\{a\}}^{\,0} \qquad \equiv$$

$$(\overline{a[0,0]}^{\,0} \parallel \overline{(b[1,1];\widehat{a}[0,1])}^{\,0})\text{ sy }\{a\} \qquad \xrightarrow{\{a[0,0,0]\}}$$

$$(\underline{a[0,0]}_{\,0} \parallel (\overline{b[1,1]}^{\,0};\widehat{a}[0,1]))\text{ sy }\{a\} \qquad \xrightarrow{\quad\checkmark\quad}$$

$$(\underline{a[0,0]}_{\,1} \parallel (\overline{b[1,1]}^{\,1};\widehat{a}[0,1]))\text{ sy }\{a\} \qquad \xrightarrow{\{b[1,1,1]\}}$$

$$(\underline{a[0,0]}_{\,1} \parallel (b[1,1]_{\,0};\widehat{a}[0,1]))\text{ sy }\{a\} \qquad \equiv$$

$$(\underline{a[0,0]}_{\,1} \parallel (b[1,1];\overline{\widehat{a}[0,1]}^{\,0}))\text{ sy }\{a\} \qquad \xrightarrow{\{\widehat{a}[0,0,1]\}}$$

$$(\underline{a[0,0]}_{\,1} \parallel (b[1,1];\widehat{a}[0,1]_{\,0}))\text{ sy }\{a\} \qquad \equiv$$

$$\underline{(a[0,0] \parallel (b[1,1];\widehat{a}[0,1]))\text{ sy }\{a\}}_{\,0}\,.$$

We observe that it is not possible to generate a synchronisation action since $a[0,0]$ has to be executed immediately, whereas $\widehat{a}[0,1]$ must wait for the completion of $b[1,1]$ which will take exactly one time unit. We shall return to this example in section 5.2.

The last example shows a (legal) synchronisation of two legal action occurrences.

$$\overline{((a[0,2];c[0,4]) \parallel (b[0,1];\widehat{c}[1,3]))\text{ sy }\{c\}}^{\,0} \qquad \xrightarrow{\{a[0,0,2]\}}$$

$$((\underline{a[0,2]}_{\,0};c[0,4]) \parallel (\overline{b[0,1]}^{\,0};\widehat{c}[1,3]))\text{ sy }\{c\} \qquad \xrightarrow{\quad\checkmark\quad}$$

$$((\underline{a[0,2]}_{\,1};c[0,4]) \parallel (\overline{b[0,1]}^{\,1};\widehat{c}[1,3]))\text{ sy }\{c\} \qquad \xrightarrow{\{b[0,1,1]\}}$$

$$((\underline{a[0,2]}_{\,1};c[0,4]) \parallel (b[0,1]_{\,0};\widehat{c}[1,3]))\text{ sy }\{c\} \qquad \xrightarrow{\quad\checkmark\quad}$$

$$((\underline{a[0,2]}_{\,2};c[0,4]) \parallel (b[0,1]_{\,1};\widehat{c}[1,3]))\text{ sy }\{c\} \qquad \xrightarrow{\quad\checkmark\quad}$$

$$((\underline{a[0,2]}_{\,3};c[0,4]) \parallel (b[0,1]_{\,2};\widehat{c}[1,3]))\text{ sy }\{c\} \qquad \equiv$$

$$((a[0,2];\overline{c[0,4]}^{\,3}) \parallel (b[0,1];\overline{\widehat{c}[1,3]}^{\,2}))\text{ sy }\{c\} \qquad \xrightarrow{\{\iota[1,2,3]\}}$$

$$\underline{((a[0,2];c[0,4]) \parallel (b[0,1];\widehat{c}[1,3]))\text{ sy }\{c\}}_{\,0}\,.$$

3.4   Operational Semantics of Subexpression

We now discuss the way in which subexpressions of a reachable t-expression can evolve as it is executed. This issue is crucial for compositional proofs of the correctness of the proposed operational semantics (such as proposition 2), as well as for establishing the consistency results between the algebra of expressions and an algebra of time Petri nets introduced in section 4. Our main observation will be that one needs to look closely at the way illegal action occurrences are generated (for example, proposition 2 will not be true for subexpressions of reachable t-expressions), and that the way time moves are applied needs to be re-considered.

Consider the following reachable t-expression $G = (\overline{a[0,0]}^0)$ rs $\{a\}$. It can easily be seen that

$$G \xrightarrow{\checkmark} (\overline{a[0,0]}^1) \text{ rs } \{a\} .$$

Now, if we were to look at this evolution from the point of view of the subexpression $H = \overline{a[0,0]}^0$ then the time move amounts to executing

$$\overline{a[0,0]}^0 \xrightarrow{\checkmark} \overline{a[0,0]}^1 .$$

However, this is not allowed by the time rule introduced above. Therefore, there may be situations when an evolution of a subexpression should be considered after taking into account what communication actions are restricted at higher levels of the syntax tree. Let us (tentatively) denote

$$H \xrightarrow{\checkmark_{\{a\}}} \overline{a[0,0]}^1$$

to indicate that this time move is a valid one provided that the context in which the expression is embedded restricts a. However, as we argue next, looking at the restricted actions may not be enough as the enclosing synchronisations also play an important role.

Consider a subexpression $H' = \overline{a[0,0]}^0 \| \overline{\widehat{a}[0,0]}^0$. When considered in the context of executing a reachable t-expression,

$$H \text{ rs } \{a\} \xrightarrow{\checkmark} (\overline{a[0,0]}^1 \| \overline{\widehat{a}[0,0]}^1) \text{ rs } \{a\}$$

this expression can be seen as performing a time move

$$H' \xrightarrow{\checkmark_{\{a\}}} \overline{a[0,0]}^1 \| \overline{\widehat{a}[0,0]}^1 .$$

Thus the device used above for H is quite sufficient (recall that rs $\{a\}$ restricts $\widehat{a}$-labelled actions as well). Next consider the same expression H' within a different context $G' = H'$ sy $\{a\}$ rs $\{a\}$. Then, using the same principle as above, one might expect that a move is possible for G' such that

$$G' \xrightarrow{\checkmark} (\overline{a[0,0]}^1 \| \overline{\widehat{a}[0,0]}^1) \text{ sy } \{a\} \text{ rs } \{a\} .$$

This, however, is impossible since G' can generate a (legal) action occurrence $\imath[0,0,0]$. We therefore have a discrepancy between what is possible within G' and what we thought is a legitimate move for H'. The reason why we encountered a problem is that we did not take into account the possibility that a pair of actions which will be restricted may create an internal $\imath$-labelled synchronisation which cannot be restricted. We will therefore annotate the 'internal' time moves not only with the actions which are restricted but also with those which are synchronised at higher levels of the syntax tree. In this way, we will have that

$$H' \xrightarrow{\;\sqrt{}_{\emptyset,\{a\}}\;} \overline{a[0,0]}^{\,1} \| \overline{\widehat{a}[0,0]}^{\,1}$$

to indicate that H' can admit a time move in a context in which no action is synchronised and a is restricted, but

$$H' \xrightarrow{\;\sqrt{}_{\{a\},\{a\}}\;} \overline{a[0,0]}^{\,1} \| \overline{\widehat{a}[0,0]}^{\,1}$$

will not be a valid time move.

We now formalise the above observations. For every pair of sets of communication actions $A, B \subseteq \mathcal{A}$, we define $\mathcal{E}_{A,B}$ to be the set of all dynamic t-expressions G such that no illegal action occurrence is enabled by the expression G sy A rs B. Intuitively, such an expression can be safely inserted into a larger expression (wherever a dynamic expression is allowed) provided that B is the set of synchronisations, and A is the set of restrictions it is subjected to.

More precisely, given a subexpression H of an expression J, let B be the union of all restriction sets of restrictions encompassing H, and let A comprise all actions a such that $J = \ldots(\ldots H \ldots)$ sy $A \ldots$ and $a \in A$ and there is no restriction on a or $\widehat{a}$ applied in-between H and sy A. We will call such a pair (A, B) the *sy/re-context* of H in J. Then, for an expression $G \in \mathcal{E}_{A,B}$, we introduce an *internal* timing rule

$$\frac{G \text{ sy } A \text{ rs } B \xrightarrow{\;\sqrt{}\;} \sqrt{}(G) \text{ sy } A \text{ rs } B}{G \xrightarrow{\;\sqrt{}_{A,B}\;} \sqrt{}(G)}$$

With the last set of definitions, one can see that given a reachable t-expression G and any of its dynamic subexpressions H, it is the case that $H \in \mathcal{E}_{A,B}$ where (A, B) is the sy/re-context of H within G. Moreover, if G admits a $\sqrt{}$-move then H admits a $\sqrt{}_{A,B}$-move. Therefore the time behaviour of a reachable t-expression G can be projected onto its subexpressions, facilitating a compositional analysis of the behaviour of G.

4   An Algebra of Time Boxes

In this section, we introduce a class of Petri nets which are a cross between Time Petri Nets [2,14,18] and Petri Boxes [4,5]. The nets inherit their structure from the latter and their time annotations and execution semantics from the former. We also introduce operators on time boxes corresponding to those defined for t-expressions.

4.1   Time Boxes

A *time box* (or t-box) is a triple $\Theta = (\Sigma, \mathit{eft}, \mathit{lft})$ such that $\Sigma = (S, T, W, \lambda, M)$ is a static or dynamic ex-directed box with transition labels in $\{\imath\} \cup \mathcal{A}$, as defined in [4,5,6,7], and $\mathit{eft}, \mathit{lft} : T \rightarrow \mathbf{N} \cup \{\infty\}$ are mappings which return, for every transition t, its *earliest firing time* and *latest firing time*, respectively. It is assumed that $\mathit{eft}(t) \leq \mathit{lft}(t)$, for every $t \in T$. The t-box Θ inherits the usual notation and terminology from the box Σ. In particular, Θ is *static* if the marking M of Σ is empty, and *dynamic* if the marking of Σ is non-empty. Notions that are necessary to deal with the timing aspects of its behaviour, are introduced below.

To achieve a concise notation, we shall employ two auxiliary transitions, skip and redo, and denote $T_{\mathsf{sr}} = T \cup \{\mathsf{skip}, \mathsf{redo}\}$. It is assumed (see [5]) that $^\bullet\mathsf{skip} = {}^\circ\Sigma$ and $^\bullet\mathsf{redo} = \Sigma^\circ$, and so the two transitions can be used to test, through their enabledness, whether the net is in its entry or exit marking, respectively. We should stress, however, that we will not consider in this paper any behaviours involving the two auxiliary transitions.

Clock Vectors   A *clock vector* for Θ is a mapping $I : T_{\mathsf{sr}} \rightarrow \mathbf{N} \cup \{-\infty\}$. Intuitively, $I(\mathsf{skip})$ and $I(\mathsf{redo})$ denote the age of the entry and exit marking of Θ, respectively.

A clock vector is *semi-consistent* if $I^{-1}(\mathbf{N}) = \mathsf{enabled}(\Theta)$, and a semi-consistent clock vector is *consistent* if $I(t) \leq \mathit{lft}(t)$, for every $t \in T$.

For $\tau \in \mathbf{N}$, the clock vector $I \oplus \tau$ is defined by $(I \oplus \tau)(t) = I(t) + \tau$, for every $t \in T_{\mathsf{sr}}$. Moreover, $^\circ\tau$ and τ° are clock vectors defined, for every $t \in T_{\mathsf{sr}}$, by:

$$
{}^\circ\tau(t) = \begin{cases} \tau & \text{if } t \in \mathsf{enabled}(\overline{\Theta}) \\ -\infty & \text{otherwise} \end{cases} \qquad \tau^\circ(t) = \begin{cases} \tau & \text{if } t \in \mathsf{enabled}(\underline{\Theta}) \\ -\infty & \text{otherwise} . \end{cases}
$$

A clock vector which returns only $-\infty$ will be denoted by $I^{-\infty}$. Note that it is the only consistent clock vector for a static t-box.

A *clocked t-box* (or ct-box) is a pair $\Delta = (\Theta, I)$ where Θ is a t-box and I is a semi-consistent clock vector for Θ. A *consistently clocked t-box* (or cct-box) is a ct-box whose clock vector is consistent.

In general, we will only be interested in ct-boxes generated from reachable t-expressions which turn out to be consistently clocked. However, in the process of constructing such cct-boxes, it may happen that some of the intermediate ct-boxes do not have consistent clock vector values for some of the transitions which are later removed through restriction. This corresponds to an observation made in section 3.4 that a reachable t-expression may contain basic actions overbarred by 'too old' time values.

Execution Semantics of ct-Boxes   There are two kinds of steps possible for a ct-box $\Delta = (\Theta, I)$, namely time steps and action steps.

A *time step* is executable if $I(t) < lft(t)$, for every $t \in T$. In this case, the *successor* ct-box is $\Delta' = (\Theta, I \oplus 1)$. We denote this by

$$\Delta \xrightarrow{\checkmark} \Delta' .$$

Note that if Δ is a ct-box (cct-box) then Δ' is also a ct-box (respectively, cct-box).

An *action step* is executable for a set of transitions $U = \{t_1, \ldots, t_k\}$ if U is enabled in Θ and $eft(t) \le I(t)$, for every $t \in U$. In this case, the *successor* ct-box is $\Delta' = (\Theta', I')$ where Θ' is such that $\Theta[U\rangle\Theta'$ and, for every $t \in T_{sr}$,

$$I'(t) = \begin{cases} -\infty & \text{if } t \notin \mathsf{enabled}(\Theta') \\ 0 & \text{if } t \in \mathsf{enabled}(\Theta') \wedge {}^\bullet t \cap {}^\bullet U \ne \emptyset \\ I(t) & \text{otherwise} . \end{cases}$$

We denote this by $\Delta \xrightarrow{\Gamma} \Delta'$ where Γ is the multiset

$$\Gamma = \{\lambda(t_1)[eft(t_1), I(t_1), lft(t_1)], \ldots, \lambda(t_k)[eft(t_k), I(t_k), lft(t_k)]\} .$$

Note that if $t \in \mathsf{enabled}(\Theta')$ and ${}^\bullet t \cap {}^\bullet U \ne \emptyset$ then, due to the ex-directedness of Θ, $t \notin \{\mathsf{skip}, \mathsf{redo}\}$. Thus the clocks of two auxiliary transitions are never reset to 0. Moreover, if Δ is a ct-box (cct-box) then Δ' is also a ct-box (respectively, cct-box).

Transition Systems of cct-Boxes Let Δ be a cct-box. We will use $[\Delta\rangle$ to denote the least set of cct-boxes containing Δ such that if $\Delta' \in [\Delta\rangle$ and $\Delta' \xrightarrow{x} \Delta''$, for some x, then $\Delta'' \in [\Delta\rangle$.
The *transition system* of Δ is $\mathsf{ts}_\Sigma = (V, A, v_0)$ where $V = [\Delta\rangle$ is the set of states, $v_0 = \Delta$ is the initial state, and the set of arcs is given by:

$$A = \{(\Delta', x, \Delta'') \mid \Delta' \in [\Delta\rangle \wedge \Delta' \xrightarrow{x} \Delta''\} .$$

4.2 Composing Static and Dynamic t-Boxes

For static and dynamic t-boxes, the composition operators are defined similarly as for the standard boxes, using the operation of net refinement [3,5,13], and the six operator boxes shown in figure 1. The only important addition is that, for each transition $t = (v, \alpha) \lhd \{t_1, \ldots, t_k\}$ ($k \ge 2$) of a newly constructed net, it is required that

$$\max\{eft(t_1), \ldots, eft(t_k)\} \le \min\{lft(t_1), \ldots, lft(t_k)\}$$

and the timing constraints of the new transition are given by:

$$eft(t) = \max\{eft(t_1), \ldots, eft(t_k)\}$$
$$lft(t) = \min\{lft(t_1), \ldots, lft(t_k)\} .$$

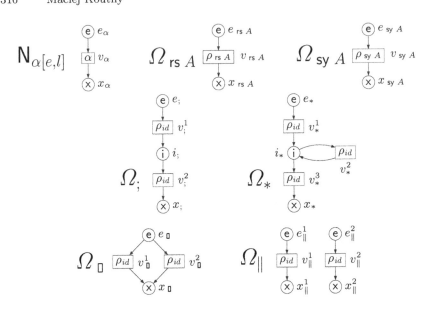

Fig. 1. A t-box $N_{\alpha[e,l]}$ (where $eft(v_\alpha) = e$ and $lft(v_\alpha) = l$) and six operator boxes.

We will now do consider in turn all the composition operators defined for static and dynamic t-boxes. To start with, composing static ct-boxes amounts to composing the underlying t-boxes.

Let $\Delta_i = (\Theta_i, I^{-\infty})$, for $i = 1, 2, 3$, be static ct-boxes and $A \subseteq \mathcal{A}$. Then:

$$\Delta_1 \text{ sy } A = (\Theta_1 \text{ sy } A, I^{-\infty}) \qquad \Delta_1 \text{ rs } A \qquad = (\Theta_1 \text{ rs } A, I^{-\infty})$$

$$\Delta_1 \,\square\, \Delta_2 = (\Theta_1 \,\square\, \Theta_2, I^{-\infty}) \qquad \Delta_1 \| \Delta_2 \qquad = (\Theta_1 \| \Theta_2, I^{-\infty})$$

$$\Delta_1; \Delta_2 \; = (\Theta_1; \Theta_2, I^{-\infty}) \qquad [\Delta_1 * \Delta_2 * \Delta_3] = ([\Theta_1 * \Theta_2 * \Theta_3], I^{-\infty}) \,.$$

When composing dynamic ct-boxes, one needs to ensure that the operations faithfully propagate the age of markings from the dynamic to static operands. It is here that we will need to take into account the age of the entry and exit markings encoded as clock values for skip and redo.

Initial and Final State Let $\Delta = (\Theta, I^{-\infty})$ be a static ct-box and $\tau \in \mathbf{N}$. Then

$$\overline{\Delta}^\tau = (\overline{\Theta}, {}^\circ\tau) \quad \text{ and } \quad \underline{\Delta}_\tau = (\underline{\Theta}, \tau^\circ) \,.$$

Note that $\overline{\Delta}^\tau$ is a cct-box if and only if $\tau \le lft(t)$, for every transition t enabled in the entry marking of Θ. Since all the t-boxes considered in this paper are ex-directed, no transition is enabled in the exit marking of Θ, and the second of the two operations always yields a cct-box.

Synchronisation and Restriction Let $\Delta = (\Theta, I)$ be a dynamic ct-box and $A \subseteq \mathcal{A}$. The synchronisation of Δ w.r.t. A is defined as

$$\Delta \text{ sy } A = (\Theta \text{ sy } A, I')$$

where $I'(\text{skip}) = I(\text{skip})$ and $I'(\text{redo}) = I(\text{redo})$. Moreover, for every transition $t = (v \text{ sy } A, \alpha) \lhd \{t_1, \ldots, t_k\}$ in $\Theta \text{ sy } A$, $I'(t) = \min\{I(t_1), \ldots, I(t_k)\}$. Note that t is a synchronisation of transitions $t_1, \ldots, t_k$ in Θ ($k \geq 1$). In our case, due to the simple form of basic actions, it is always the case that $k \leq 2$, and if $k = 2$ then t_1 and t_2 have conjugate communication labels and $\alpha = \iota$.

The restriction of Δ w.r.t. A is defined as

$$\Delta \text{ rs } A = (\Theta \text{ rs } A, I')$$

where $I'(\text{skip}) = I(\text{skip})$ and $I'(\text{redo}) = I(\text{redo})$. Moreover, for every transition $t = (v \text{ rs } A, \alpha) \lhd \{u\}$ in $\Theta \text{ rs } A$, $I'(t) = I(u)$. Note that t is based on a transition u in Θ with the label α not belonging to $A \cup \widehat{A}$.

Choice and Sequential Composition Let $\Delta_i = (\Theta_i, I_i)$, for $i = 1, 2$, be ct-boxes, Δ_1 being dynamic and Δ_2 static. The choice composition of Δ_1 and Δ_2 is defined as

$$\Delta_1 \,\square\, \Delta_2 = (\Theta_1 \,\square\, \Theta_2, I)$$

where $I(\text{skip}) = I_1(\text{skip})$ and $I(\text{redo}) = I_1(\text{redo})$. Moreover, for every transition $t = (v_\square^i, \alpha) \lhd \{u\}$ in $\Theta_1 \,\square\, \Theta_2$ (note that t is based on a transition u in Θ_i),

$$I(t) = \begin{cases} I_1(\text{skip}) & \text{if } t \in \text{enabled}(\Theta_1 \,\square\, \Theta_2) \\ I_i(u) & \text{otherwise .} \end{cases}$$

The above implies that if Δ_1 has been in the entry marking for τ time units, then all transitions enabled in the entry marking of Θ_2 have their clocks set to τ. The composition $\Delta_2 \,\square\, \Delta_1$ is defined similarly.

The sequential composition of Δ_1 and Δ_2 is defined as

$$\Delta_1; \Delta_2 = (\Theta_1; \Theta_2, I)$$

where $I(\text{skip}) = I_1(\text{skip})$ and $I(\text{redo}) = I_2(\text{redo})$. Moreover, for every transition $t = (v_;^i, \alpha) \lhd \{u\}$ in $\Theta_1; \Theta_2$ (note that t is based on a transition u in Θ_i),

$$I(t) = \begin{cases} I_1(\text{redo}) & \text{if } i = 2 \wedge t \in \text{enabled}(\Theta_1; \Theta_2) \\ I_i(u) & \text{otherwise .} \end{cases}$$

Thus, if Δ_1 has been in the exit marking for τ time units, then all transitions enabled in the entry marking of Θ_2 have their clocks set to τ. The sequential composition of Δ_2 and Δ_1 is defined as

$$\Delta_2; \Delta_1 = (\Theta_2; \Theta_1, I)$$

where $I(\text{skip}) = I_2(\text{skip})$ and $I(\text{redo}) = I_1(\text{redo})$. Moreover, for every transition $t = (v_;^i, \alpha) \lhd \{u\}$ in $\Theta_2; \Theta_1$ (note that t is based on a transition u in Θ_2 if $i = 1$, and in Θ_1 if $i = 2$),

$$I(t) = \begin{cases} I_1(u) & \text{if } i = 2 \\ I_2(u) & \text{if } i = 1 . \end{cases}$$

Parallel Composition Let $\Delta_i = (\Theta_i, I_i)$, for $i = 1, 2$, be two dynamic ct-boxes. The parallel composition of Δ_1 and Δ_2 is defined as

$$\Delta_1 \| \Delta_2 = (\Theta_1 \| \Theta_2, I)$$

where $I(\mathsf{skip}) = \min\{I_1(\mathsf{skip}), I_2(\mathsf{skip})\}$ and $I(\mathsf{redo}) = \min\{I_1(\mathsf{redo}), I_2(\mathsf{redo})\}$. Moreover, for every transition $t = (v_\|^i, \alpha) \lhd \{u\}$ in $\Theta_1 \| \Theta_2$, $I(t) = I_i(u)$. Note that t is based on a transition u in Θ_i.

Iteration Let $\Delta_i = (\Theta_i, I_i)$, for $i = 1, 2, 3$, be ct-boxes (one of them being dynamic, the other two static). The iteration of Δ_1, Δ_2 and Δ_3 is defined as

$$[\Delta_1 * \Delta_2 * \Delta_3] = ([\Theta_1 * \Theta_2 * \Theta_3], I)$$

where $I(\mathsf{skip}) = I_1(\mathsf{skip})$ and $I(\mathsf{redo}) = I_3(\mathsf{redo})$. Moreover, for every transition $t = (v_*^i, \alpha) \lhd \{u\}$ in $[\Theta_1 * \Theta_2 * \Theta_3]$ (note that t is based on a transition u in Θ_i),

$$I(t) = \begin{cases} \tau & \text{if} \quad u \notin \mathsf{enabled}(\Theta_i) \wedge t \in \mathsf{enabled}([\Theta_1 * \Theta_2 * \Theta_3]) \\ I_i(u) & \text{otherwise} \end{cases}$$

where $\tau = \max\{I_1(\mathsf{redo}), I_2(\mathsf{skip}), I_2(\mathsf{redo}), I_3(\mathsf{skip})\}$. For instance, suppose that Δ_2 is a dynamic ct-box which has been in the entry marking for τ time units. The above formula implies that in such a case, all transitions enabled in the entry markings of Θ_2 and Θ_3 have their clocks set to τ.

5   Denotational Semantics of t-Expressions

In this section, we introduce a denotational semantics of t-expressions. The semantical mapping ctbox from t-expressions to ct-boxes is defined compositionally. First, $\mathsf{ctbox}(\alpha[e, l]) = (\mathsf{N}_{\alpha[e,l]}, I^{-\infty})$ where $\mathsf{N}_{\alpha[e,l]}$ is shown in figure 1. For other static and dynamic t-expressions, we define:

$$\mathsf{ctbox}(\overline{E}^\tau) \;\; = \overline{\mathsf{ctbox}(E)}^\tau \qquad\qquad \mathsf{ctbox}(\underline{E}_\tau) \;\; = \underline{\mathsf{ctbox}(E)}_\tau$$

$$\mathsf{ctbox}(J \text{ sy } A) = \mathsf{ctbox}(J) \text{ sy } A \qquad \mathsf{ctbox}(J \text{ rs } A) = \mathsf{ctbox}(J) \text{ rs } A$$

$$\mathsf{ctbox}(J \,\square\, L) \;\; = \mathsf{ctbox}(J) \,\square\, \mathsf{ctbox}(L) \qquad \mathsf{ctbox}(J \| L) \;\; = \mathsf{ctbox}(J) \| \mathsf{ctbox}(L)$$

$$\mathsf{ctbox}(J; L) \;\; = \mathsf{ctbox}(J); \mathsf{ctbox}(L)$$

$$\mathsf{ctbox}([J * L * M]) \;\; = [\mathsf{ctbox}(J) * \mathsf{ctbox}(L) * \mathsf{ctbox}(M)] .$$

Proposition 3. *The mapping* ctbox *always returns a ct-box and, for every reachable t-expression, it returns a cct-box. Moreover, if $G \equiv H$ are dynamic t-expressions, then $\mathsf{ctbox}(G) = \mathsf{ctbox}(H)$.*

5.1   Consistency between Denotational and Operational Semantics

We now formulate a central result of this paper which states that the two semantics of reachable t-expressions are equivalent. Recall that a reachable t-expression can be obtained through the rules of operational semantics from a static expression initially marked with a zero-overbar.

Theorem 1. *For every reachable t-expression G,*

$$R = \{([H]_\equiv, \mathsf{ctbox}(H)) \mid [H]_\equiv \text{ is a node of } \mathsf{ts}_G\}$$

is a strong bisimulation between the transition systems ts_G and $\mathsf{ts}_{\mathsf{ctbox}(G)}$. The latter means that $([G]_\equiv, \mathsf{ctbox}(G)) \in R$, and if $(v, v') \in R$ then the following hold:

- *If $v \xrightarrow{x} w$ then there is w' such that $v' \xrightarrow{x} w'$ and $(w, w') \in R$.*
- *If $v' \xrightarrow{x} w'$ then there is w such that $v \xrightarrow{x} w$ and $(w, w') \in R$.*

In other words, ts_G and $\mathsf{ts}_{\mathsf{ctbox}(G)}$ are strongly equivalent transition systems.

The above result extends that for the standard PBC where the transition systems of corresponding expressions and boxes are isomorphic [6,7]. It is worth observing that the transition systems in theorem 1 are not, in general, isomorphic. For consider reachable t-expressions

$$G_\tau = (a[0, \infty]; \overline{b[0, 0]}^\tau) \; \mathsf{rs} \; \{b\}$$

for $\tau \geq 0$. In the operational semantics, we have

$$G_0 \xrightarrow{\sqrt{}} G_1 \xrightarrow{\sqrt{}} G_2 \xrightarrow{\sqrt{}} \cdots$$

and $G_i \not\equiv G_j$, for $i \neq j$. However, $\mathsf{ctbox}(G_i) = \mathsf{ctbox}(G_j)$, for all i and j. The reason for this discrepancy is that for a dynamic cct-box $\Delta = (\Theta, I^{-\infty})$,

$$\Delta \xrightarrow{\sqrt{}} \Delta \,.$$

On the other hand, a time move always results in a non-equivalent t-expression. Thus the net based and process based semantics treat deadlocks states differently. However, such a difference is not really significant.

Remarks on the Proof of Theorem 1   The proof of theorem 1 follows that developed for the standard PBC, and proceeds by induction on the structure of t-expressions. However, the technical details are more involved since, as already discussed, for the subexpressions of G one needs to take into account not only legal but also illegal action occurrences. To address this problem, for every pair of sets of communication actions $A, B \subseteq \mathcal{A}$, we define a set $\mathcal{B}_{A,B}$ comprising all ct-boxes Δ such that $\Delta \; \mathsf{sy} \; A \; \mathsf{rs} \; B$ is a cct-box. Moreover,

$$\Delta \xrightarrow{\sqrt{}_{A,B}} \Delta' \quad \text{if and only of} \quad \Delta \; \mathsf{sy} \; A \; \mathsf{rs} \; B \xrightarrow{\sqrt{}} \Delta' \; \mathsf{sy} \; A \; \mathsf{rs} \; B \,.$$

Perhaps not surprisingly, for every t-expression $G \in \mathcal{E}_{A,B}$, it is the case that $\mathsf{ctbox}(G) \in \mathcal{B}_{A,B}$. Next, both for t-expressions in $\mathcal{E}_{A,B}$, and ct-boxes in $\mathcal{B}_{A,B}$, we introduce transition systems based on the usual action steps and $\sqrt{}_{A,B}$-labelled time moves. One can then show that the transition systems of $G \in \mathcal{E}_{A,B}$ and $\mathsf{ctbox}(G) \in \mathcal{B}_{A,B}$ are strongly equivalent. And theorem 1 follows since each reachable t-expression belongs to $\mathcal{E}_{\emptyset,\emptyset}$, and $\sqrt{}_{\emptyset,\emptyset}$ is another way of writing $\sqrt{}$.

5.2 An Example Motivating Illegal Actions

We will now explain how illegal action occurrences can be used to ensure behavioural consistency between reachable t-expressions and their translations into cct-boxes. Consider again the expression $E = (a[0,0] \parallel (b[1,1]; \widehat{a}[0,1]))$ sy $\{a\}$ from section 3.3. The net corresponding to E is shown in figure 2. As already observed, $\overline{E}^{0}$ cannot execute the synchronisation action which is consistent with the net semantics. However, the situation changes for the net corresponding to $\overline{E \text{ rs } \{a\}}^{0}$ (see again figure 2). Now, it is possible to execute $b[1,1,1]$ followed by $\imath[0,0,0]$. That the same can be done in the operational semantics follows from the following argument. First, we obtain a derivation using only legal action occurrences:

$$\overline{(a[0,0] \parallel (b[1,1]; \widehat{a}[0,1])) \text{ sy } \{a\} \text{ rs } \{a\}}^{0} \qquad \equiv$$

$$(\overline{a[0,0]}^{0} \parallel \overline{(b[1,1]; \widehat{a}[0,1])}^{0}) \text{ sy } \{a\} \text{ rs } \{a\} \qquad \xrightarrow{\sqrt{}}$$

$$(\overline{a[0,0]}^{1} \parallel (\overline{b[1,1]}^{1}; \widehat{a}[0,1])) \text{ sy } \{a\} \text{ rs } \{a\} \qquad \xrightarrow{\{b[1,1,1]\}}$$

$$(\overline{a[0,0]}^{1} \parallel (\underline{b[1,1]}_{0}; \widehat{a}[0,1])) \text{ sy } \{a\} \text{ rs } \{a\} \qquad \equiv$$

$$(\overline{a[0,0]}^{1} \parallel (b[1,1]; \overline{\widehat{a}[0,1]}^{0})) \text{ sy } \{a\} \text{ rs } \{a\} .$$

We now observe that after leaving out the synchronisation and restriction operators, we can derive a step comprising a legal action occurrence $\widehat{a}[0,0,1]$ and a conjugate illegal action occurrence $a[0,1,0]$:

$$\overline{a[0,0]}^{1} \parallel (b[1,1]; \overline{\widehat{a}[0,1]}^{0}) \qquad \xrightarrow{\{a[0,1,0],\widehat{a}[0,0,1]\}} \qquad \underline{a[0,0]}_{0} \parallel (b[1,1]; \underline{\widehat{a}[0,1]}_{0}) .$$

Thus, by the second synchronisation rule, we have:

$$(\overline{a[0,0]}^{1} \parallel (b[1,1]; \overline{\widehat{a}[0,1]}^{0}) \text{ sy } \{a\} \qquad \xrightarrow{\{\imath[0,0,1]\}} \qquad (\underline{a[0,0]}_{0} \parallel (b[1,1]; \underline{\widehat{a}[0,1]}_{0}) \text{ sy } \{a\} .$$

This allows us to complete our initial derivation, using the rule for restriction:

$$(\overline{a[0,0]}^{1} \parallel (b[1,1]; \overline{\widehat{a}[0,1]}^{0})) \text{ sy } \{a\} \text{ rs } \{a\} \qquad \xrightarrow{\{\imath[0,0,1]\}}$$

$$(\underline{a[0,0]}_{0} \parallel (b[1,1]; \underline{\widehat{a}[0,1]}_{0}) \text{ sy } \{a\} \text{ rs } \{a\} \qquad \equiv$$

$$\underline{(a[0,0] \parallel (b[1,1]; \widehat{a}[0,1])) \text{ sy } \{a\} \text{ rs } \{a\}}_{0} .$$

Rules based only on legal actions could not produce a similar result since no legal action occurrence is possible for $\overline{a[0,0]}^{1}$.

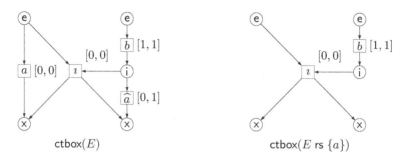

Fig. 2. Two ct-boxes.

5.3 Time Divergence

In [8], an additional assumption was made to exclude infinite sequences of action occurrences. This assumption has been removed in [9], and so we did not use it in this paper. However, in a compositional context, time divergence can easily be excluded syntactically, in the following way. First, we say that a static t-expression E is *time guarded* if:

- $E = \alpha[e, l]$ and $e > 0$.
- $E = F$ sy A or $E = F$ rs A, and F is time guarded.
- $E = F \, \Box \, F'$, and both F and F' are time guarded.
- $E = F; F'$ or $E = F \| F'$ or $E = [F * F'' * F']$, and F or F' is time guarded.

Then a static t-expression is *non-time-divergent* if, for its every sub-expression $[E * C * F]$, it is the case that C is time guarded.

6 Concluding Remarks

We introduced a model which provides two consistent semantics for a class of process expressions with explicit timing information. Since ct-boxes are essentially time Petri nets, we can apply to them model checking techniques developed in [8,9].

The model presented in this paper can easily be extended to allow more complicated basic actions, as in the original PBC, as well as action relabelling. The price to pay is an increase in the complexity of the operational semantics rules, although no entirely new devices or constructs are needed. A more challenging goal would be to apply the approach proposed in this paper to the full Box Algebra.

Acknowledgements

I would like to thank the participants of the Workshop on Timed Petri Net Algebras, organised at the University of Paris XII in October 1999, for their valuable comments on the previous version of this paper. The work described in this paper has been supported the Anglo-German Foundation ARC Project 1032 BAT (Box Algebra with Time).

References

1. J. Baeten and W. P. Weijland: *Process Algebra*. Cambridge Tracts in Theoretical Computer Science 18, Cambridge University Press (1990).
2. B. Berthomieu and M. Diaz: Modelling and verification of Time Dependent Systems Using Time Petri Nets. *IEEE Trans. on Soft. Eng.* 17 (1991) 259-273.
3. E. Best, R. Devillers and J. Esparza: General Refinement and Recursion Operators for the Petri Box Calculus. Proc. of *STACS'93*, P. Enjalbert, A. Finkel and K. W. Wagner (Eds.). Springer-Verlag, Lecture Notes in Computer Science 665 (1993) 130-140.
4. E. Best, R. Devillers and J. Hall: The Petri Box Calculus: a New Causal Algebra with Multilabel Communication. In: *Advances in Petri Nets 1992*, G. Rozenberg (Ed.). Springer-Verlag, Lecture Notes in Computer Science 609 (1992) 21-69.
5. E. Best, R. Devillers and M. Koutny: Petri Nets, Process Algebras and Concurrent Programming Languages. In: *Advances in Petri Nets. Lectures on Petri Nets II: Applications*, W. Reisig and G. Rozenberg (Eds.). Springer-Verlag, Lecture Notes in Computer Science 1492 (1998) 1-84.
6. E. Best, R. Devillers and M. Koutny: The Box Algebra - a Model of Nets and Process Expressions. Proc. of *ICATPN'99*, S. Donatelli and J. Kleijn (Eds.). Springer-Verlag, Lecture Notes in Computer Science 1639 (1999) 344-363.
7. E. Best, R. Devillers and M. Koutny: *Petri Net Algebra*. EATCS Monographs on Theoretical Computer Science, Springer-Verlag (to be published in 2000).
8. B. Bieber and H. Fleischhack: Model Checking of Time Petri Nets Based on Partial Order Semantics. Proc. of *CONCUR'99*, J. Baeten and S. Mauw (Eds.). Springer-Verlag, Lecture Notes in Computer Science 1664 (1999) 210-225.
9. H. Fleischhack: Computing a Complete Finite Prefix of a Time Petri Net. Draft paper (1999).
10. M. B. Hennessy: *Algebraic Theory of Processes*. The MIT Press (1988).
11. C. A. R. Hoare: *Communicating Sequential Processes*. Prentice Hall (1985).
12. R. Janicki and P. E. Lauer: *Specification and Analysis of Concurrent Systems - the COSY Approach*. EATCS Monographs on Theoretical Computer Science, Springer-Verlag (1992).
13. M. Koutny and E. Best: Fundamental Study: Operational and Denotational Semantics for the Box Algebra. *Theoretical Computer Science* 211 (1999) 1-83.
14. P. Merlin and D. Farber: Recoverability of Communication Protocols - Implication of a Theoretical Study. *IEEE Trans. on Soft. Comm.* 24 (1976) 1036-1043.
15. R. Milner: *Communication and Concurrency*. Prentice Hall (1989).
16. T. Murata: Petri Nets: Properties, Analysis and Applications. *Proc. of IEEE* 77 (1989) 541-580.
17. G. D. Plotkin: A Structural Approach to Operational Semantics. Teachnical Report FN-19, Computer Science Department, University of Aarhus (1981).
18. L. Popova: Time Petri Nets. *Journal of Information Processing and Cybernetics EIK* 27 (1991) 227-244.
19. W. Reisig: *Petri Nets. An Introduction*. EATCS Monographs on Theoretical Computer Science, Springer-Verlag (1985).

Composing Abstractions of Coloured Petri Nets

Charles Lakos

Computer Science Department
University of Adelaide
Adelaide, SA, 5005, Australia.
Charles.Lakos@adelaide.edu.au

Abstract: An earlier paper considered appropriate properties for abstract net components (or nodes) in the Coloured Petri Net formalism. This paper augments that earlier work in three main areas — it proposes general canonical forms for such node refinements, it identifies two other forms of refinement which will be used in concert with node refinement, and it considers the compositionality of these refinements. All of them maintain behavioural compatibility between refined and abstract nets, which is captured by the notion of a system morphism.

Keywords: Theory of High-Level Petri Nets, Abstraction, Refinement

1 Introduction

The abstraction (and refinement) of Petri Nets is not new and dates back to Carl Adam Petri himself (e.g. [16]). One approach has been to use abstraction to focus on the structural relationship between the abstract and refined nets in terms of net morphisms [1, 3, 4, 6, 17]. In the case of Free Choice nets, the structural compatibility has behavioural implications, namely the maintenance of traps and syphons. In the context of High-Level Petri Nets (such as CPNs [9] and Predicate-Transition Nets [8]), the approach adopted has typically been to use abstraction to aid the process of developing a Petri Net model, with the abstraction subsequently being discarded when the model is simulated or analysed [17]. Thus CPNs, as formalised by Jensen [9] and implemented in the Design/CPN tool [10], provide substitution transitions together with place fusion for building Hierarchical Coloured Petri Nets (HCPNs). While substitution transitions maintain structural compatibility, there is no concept of abstract behaviour, since the semantics of the construct are defined in terms of the underlying CPN — in fact, the inscriptions on the arcs incident on substitution transitions are simply ignored.

By contrast, we prefer a view of abstraction in CPNs which respects behavioural constraints (as in [15, 18]), so that the abstraction can contribute to the understanding of the net and also to its analysis. In discussing behaviour-respecting morphisms for Petri Nets, Winskel [18] raised the question of the appropriate form of morphisms for CPNs. This paper advocates three forms which respect the colour structures — *type*, *subnet* and *node* refinement. Each of these maintains behavioural compatibility between refined and abstract nets. For type refinement, this means that the refined type values can be projected onto the abstract type values; for subnet refinement (or extension), the refined behaviour can be restricted to yield the corresponding abstract

M. Nielsen, D. Simpson (Eds.): ICATPN 2000, LNCS 1825, pp. 323-345, 2000.

behaviour; for node refinement, a subnet which is abstracted by a place should maintain the notion of an abstract marking, and a subnet which is abstracted by a transition should have the notion of abstract firing.

An earlier paper focussed on the notion of abstraction for Coloured Petri Nets (CPNs) [12], in the sense of mapping place (or transition) bordered subnets into a single place (or transition). That paper sought to be as general as possible, while maintaining behavioural compatibility between the abstract and refined nets. Thus, if a place-bordered subnet were to be abstracted to a place, then there needed to be some notion of an abstract marking for the subnet, which would only be modified by interaction of the subnet with its environment (and not by internal actions of the subnet). Similarly, if a transition-bordered subnet were to be abstracted to a transition, then there ought to be some notion of abstract firing for the subnet, where a firing sequence of the subnet included the firing of border transitions with matching firing modes.

The generality of the above proposals led to difficulties in proving compositionality results [13] — it was necessary to constrain the previous definitions in order to prove that the composition of two node refinements was also a node refinement. Thus, a place refinement was required to be resettable, i.e. a return to the same abstract marking could be matched by a return to the same refined marking (of the subnet). Similarly, a transition refinement should be able to return to the initial marking (of the subnet) after each abstract firing.

The above raised a dilemma as to whether the more general or the more restricted definitions were appropriate. This paper resolves the dilemma by asserting paradoxically that both are appropriate. This is achieved by recognising that three forms of refinement — type, subnet, and node refinement — are normally used in concert. This means that we can propose canonical bases for node refinements which satisfy the more constrained definitions, but in conjunction with the other forms of refinement we can achieve the generality of the original formulation.

The current paper is intended to be self-contained, but some aspects from the earlier paper are reiterated in abbreviated form. It is therefore desirable that the reader has access to the earlier paper. An informal introduction together with a brief motivating example is presented in §2. The definitions of CPNs and their refinement are found in §3. In §4, we prove the generality of the canonical node refinements. Finally, further comments on the use of these refinements and the conclusions are given in §5.

2 Example

This section presents a simple motivating example for the refinement proposals. The example is that of a library loans system and is adapted from an example found in [5].

A book in the library will have some associated information, such as the author(s), title, publisher, etc. In a CPN, each book can be represented by a token, and this information can be captured in a token type called *Book*. Each book will be in one of several states such as available, on loan, and overdue. These states can be represented by places in the Petri Net, with the presence of a book token in a place indicating that the book is currently in that state. Thus a CPN for the library books could be as shown in fig 2.1.

In this case, we have not shown all the details of the net. Thus, the colour for the token type is not indicated in full, nor is the relation between the transition firing modes and the variables inscribing the arcs. Clearly, however, the firing mode for transition *Borrow* will include sufficient information to identify the book b and the user u.

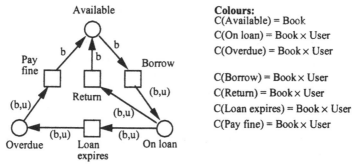

Colours:
C(Available) = Book
C(On loan) = Book × User
C(Overdue) = Book × User

C(Borrow) = Book × User
C(Return) = Book × User
C(Loan expires) = Book × User
C(Pay fine) = Book × User

Fig 2.1: CPN Books — the lifecycle of library books

A borrower or user of the library will have some associated information, such as their name, contact details, classification of membership, etc. Again, each user can be represented by a token type called *User*, and the various states of the user can be indicated by places. In this case, it is convenient to have only one place to indicate the state of the user with transitions indicating the possible actions such as borrowing or returning a book and paying a fine. Thus, a CPN for the library users could be as shown in fig 2.2.

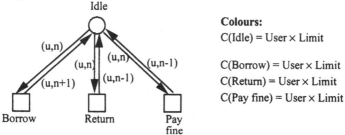

Colours:
C(Idle) = User × Limit

C(Borrow) = User × Limit
C(Return) = User × Limit
C(Pay fine) = User × Limit

Fig 2.2: CPN Users — the lifecycle of library users

The two CPNs above could be combined into a composite system by fusing the similarly-named transitions in the two nets. Clearly, many more details could be specifed, but the example above will be adequate for our purposes of illustrating the forms of refinement which we wish to support.

The first and simplest form of refinement, which we call *type refinement*, is to incorporate additional information in the net in the tokens and firing modes. However, each value of the refined type can be projected onto a value of the abstract type. For example, it may be desirable to introduce a further classification of books to vary the loan period. As far as the subnet for the books is concerned, this will simply involve extending the token type for the places, and extending the corresponding type for the various transition firing modes. The projection from refined to abstract type values is obvious. The above changes will affect the firing of the *Loan expires* transition, especially in the composite system. It is certainly the case, however, that if there is a behaviour of the refined system, then there would be a corresponding behaviour of the abstract system. This is the generic behavioural constraint which we require for acceptable refinements.

The second form of refinement, which we call *subnet refinement*, is to augment a subnet with additional places, transitions and arcs. (We also classify as subnet refinement the extension of a token type or mode type to include extra values which are independent of previous processing. Here, these values of the extended type are not projected onto values of the abstract type but are ignored in the abstraction.) For example, it may be appropriate to cater for the reservation of books. In this case, we would add a place to hold the reservation status for each book, and the transition *Borrow* would only fire if there were a compatible reservation status on the book for the given user. (We use the value *(b,-)* to indicate that no-one has a reservation for book *b*, and the value *(b,u)* to indicate that user *u* has a reservation on book *b*.) The modified part of the *Books* subnet is given in fig 2.3. Again we satisfy the constraint that if there is a refined behaviour, then there is a corresponding abstract behaviour (which ignores reservations), but not necessarily vice versa.

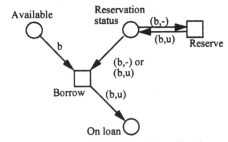

Fig 2.3: Subnet indicating the processing of book reservations

The third form of refinement, which we call *node refinement*, is to replace a place (or transition) by a place (or transition) bordered subnet. In this paper, we advocate the use of canonical forms of such refinements. The basis for a canonical place refinement is given in fig 2.4.

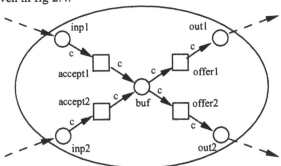

Fig 2.4: Canonical place refinement

It has separate input and output border places — in this case there are two of each. Each input (output) border place may have more than one incident input (output) arc from (to) the environment. Each input border place has an associated *accept* transition which will transfer tokens from the border place to an internal place, here called *buf*. Similarly, each output border place has an associated *offer* transition which will transfer tokens from place *buf* to the output border place. All the border places and the place *buf* have the same token type, which is also the mode type shared by the *accept* and *offer* transitions. None of these transitions constrains the flow of tokens, e.g. by using a guard. Clearly, the abstract marking of such a canonical place refinement is given by the sum of tokens in the border places and the internal place *buf*.

An arbitrary place refinement will be of the form of the basis of a canonical refinement (as above) augmented by subnet refinement which extends the *accept* and *offer* transitions. It is a contribution of this paper to observe how the combination of different refinements extends the generality of the basic constructs.

In our running example, such an incremental change might be the identification of the details of processing a book once it has been returned. In other words, the place *Available* might be replaced by a subnet which takes into account the delay in reshelving a book, the possibility of repairs, etc. The node refinement for this place is shown in fig 2.5.

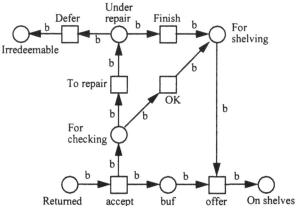

Fig 2.5: Subnet indicating the processing of returned books

Note that this has one input border place called *Returned* and one output border place called *On shelves*. The *accept* and *offer* transitions, together with the internal place *buf* constitute the basis of the canonical place refinement. Further activity is achieved by the subnet refinement which extends transitions *accept* and *offer*. This subnet retains the identity of the books, and hence it also determines the abstract marking of the subnet. Thus, the place *buf* is redundant, since its marking is equivalent to the sum of markings of places *For checking*, *Under repair*, *Irredeemable*, *For shelving*. (It is commonly the case that the place *buf* is redundant in such place refinements.) Further, this abstract marking is not modified by the various actions internal to the subnet. Clearly, a refined behaviour of the net will have a corresponding abstract behaviour, though the reverse will not necessarily be the case. For example, it may be that a book is so damaged that its return to the shelves would be indefinitely delayed, in which case further borrowing of that book would be disallowed.

For transition refinement, the canonical basis is given in fig 2.6. It has separate input and output border transitions — in this case there are two of each. Each input (output) border transition may have more than one incident input (output) arc from (to) the environment. Each input border transition has an associated place *recd*, which receives a token equal to the abstract firing mode, when the input border transition has fired with that mode. The transition *switch* can fire when all the input border transitions have fired (with the matching abstract firing mode), thereby completing the input phase. It removes the matching tokens from the *recd* places and puts corresponding tokens into all the *send* places. There is one such *send* place associated with each output border transition. Once such a token is available the output border transition can fire (with the same abstract firing mode). Initially, all the *recd* and *send* places are empty.

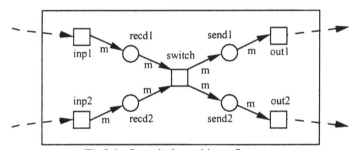

Fig 2.6: Canonical transition refinement

The abstract firing of the transition refinement commences with the firing of any of the input border transitions and is *completed* when all the matching output border transitions have fired and the *recd* and *send* places are again empty. Only such completed firing sequences will have corresponding abstract firing sequences. The canonical construction ensures that input border transitions fire before the corresponding output border transitions, ensuring the enabling of the corresponding abstract transition.

An arbitrary transition refinement will be of the form of a canonical refinement (as above) together with a subnet refinement which augments the border transitions. In our running example, we may wish to refine the *Borrow* transition to reflect the component activities of checking the student identity and processing the book. This might be achieved in the subnet of fig 2.7.

Note that the transition *validate user* may fail to fire (because the user is not acceptable). In this case, the abstract firing of this subnet will never complete, and hence such an incomplete refined activity will have no corresponding abstract activity.

Even though the above three forms of refinement can be clearly identified and analysed in isolation, they will commonly be used in concert in practical applications.

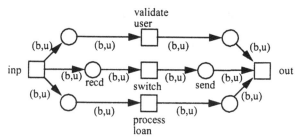

Fig 2.7: Refined Borrow transition

3 Formal Definitions

In this section we present the formal definitions of CPNs in §3.1, and the definition of CPN morphisms in §3.2. We identify the notion of system morphisms to capture the notion of refinement with behavioural compatibility, in contrast to the more traditional net morphisms which (only) specify structural constraints. The definition of our proposed refinements is given in §3.3. The definitions and their associated explanatory notes should be read together. The explanatory notes in §3.1 are minimal, since we assume that the reader has some familiarity with the definition of CPNs (e.g. [9]).

3.1 Formal Definition of Coloured Petri Nets

We define Coloured Petri Nets in the context of a **universe of non-empty colour sets** Σ with an associated **partial order** $<: \prod \Sigma \infty \Sigma$ which is derived from type compatibility in the theory of object-oriented languages [2]. $X <: Y$ means that the values of X can be used in contexts expecting values of Y, and typically it means that X has extra data components over Y. In this case, we assume the existence of a (polymorhpic) projection function $\prod_Y$ from the values of X to those of Y (which do not appear in any proper subtype). For our purposes, we only make use of the fact that values of X are also values of Y.

Given a universe of colour sets Σ, we define $\Phi\Sigma = \{ X \oslash Y \mid X, Y \in \Sigma \}$, the functions over Σ, $\mu X = \{ X \oslash \mathbb{N} \}$ the multisets over X (where $\mathbb{N}$ is the set of non-negative integers), and $\sigma X = \{ x_1 x_2 \ldots x_n \mid x_i \in X \}$ the sequences over X. Where appropriate, multisets are elements of Σ, and thus mappings between multisets will occur in $\Phi\Sigma$ as required. Sum, difference and inequality are defined on multisets in the usual way, and a common notation for multisets [9] is adopted, e.g. $m = 3\text{'}b+4\text{'}c$ is a multiset with $m(b) = 3$ and $m(c) = 4$, the only elements of m. We take the liberty of abbreviating $1\text{'}x$ as x. We usually name sequences with an asterisk suffix.

Definition 3.1: A **Coloured Petri Net** is a tuple $N = (P, T, A, C, E, \mathbb{M}, \mathbb{Y}, M_0)$ where:

(a) P = set of places

(b) T = set of transitions, with $P \leftrightarrow T = \emptyset$

(c)   A = set of arcs, with $A \prod P \infty T \approx T \infty P$

(d)   C = colours of places and (modes) of transitions, i.e. $C: P \approx T \oslash \Sigma$

(e)   E = arc inscriptions with E: $A \oslash \Phi\Sigma$ where $E(p,t), E(t,p): C(t) \oslash \mu C(p)$

(f) $\mathbb{M}$= set of markings, i.e. $\mathbb{M} = \mu\{(p,c) \mid p \in P, c \in C(p)\}$

(g) $\mathbb{Y}$= set of steps, i.e. $\mathbb{Y} = \mu\{(t,c) \mid t \in T, c \in C(t)\}$

(h) M_0 = initial marking, i.e. $M_0 \in \mathbb{M}$

Note: The above definition is, to all intents and purposes, equivalent to the common definition [9]. Note however, that there is at most one arc in each direction for any (place, transition) pair, and that we refer to the firing modes of transitions as colours and not as bindings. Thus, the effect of an arc is given by the arc inscription in conjunction with a particular mode, rather than the evaluation of an expression in a given binding. There is no guard function defined on transitions, but the same effect is achieved by limiting the associated set of firing modes. As usual, the markings of N are the multisets of (place, colour) pairs, while the steps are the multisets of (transition, colour) pairs. While these are derivative quantities, they are included in the tuple so that later it will be clear that morphisms map markings and steps to markings and steps respectively.

Definition 3.2: For CPN N, $x \in P \approx T$, $X \prod P \approx T$ we define:

(a)   **the inputs of x, •x** = $\{y \in P \approx T \mid (y,x) \in A\}$

(b)   **the outputs of x, x•** = $\{y \in P \approx T \mid (x,y) \in A\}$

(c)   **the border of X, bd(X)** = $\{x \in X \mid \exists y \in P \approx T X: y \in •x \approx x•\}$

(d)   **the environment of X, env(X)** = $\{y \in P \approx T X \mid \exists x \in X: y \in •x \approx x•\}$

Thus, the border of a set of nodes X, are those nodes connected to other nodes not in X, and those other nodes constitute the environment of X.

Definition 3.3: For CPN N, the **incremental effects** E^-, E^+: $\mathbb{Y} \oslash \mathbb{M}$ of the occurrence of a step Y are given by:

(a)
$$E^-(Y) = \sum_{(t,m)\in Y} \sum_{(p,t)\in A} \{p\} \times E(p,t)(m)$$

(b)
$$E^+(Y) = \sum_{(t,m)\in Y} \sum_{(t,p)\in A} \{p\} \times E(t,p)(m)$$

Definition 3.4: For CPN N, a step $Y \in \mathbb{Y}$ is **enabled** in marking $M \in \mathbb{M}$, written M [Y>, if $M \geq E^-(Y)$. If step $Y \in \mathbb{Y}$ of CPN N is enabled in marking $M_1 \in \mathbb{M}$, it may **fire** leading to marking $M_2 \in \mathbb{M}$, written M_1 [Y> M_2, with $M_2 = M_1 - E^-(Y) + E^+(Y)$. A step sequence $Y^* = Y_1 Y_2 ... Y_n \in \sigma\mathbb{Y}$ of CPN N is enabled in marking $M_1 \in \mathbb{M}$ and may occur, leading to marking $M_2 \in \mathbb{M}$, written M_1 [Y*> M_2 if there exists intermediate markings M_2', M_3', ... $M_n' \in \mathbb{M}$ such that M_1 [Y_1> M_2', M_i' [Y_i> M_{i+1}' for $i \in 2, ... n-1$, and M_n' [Y_n> M_2

Definition 3.5: For CPN N, step $Y \in \mathbb{Y}$ is **realisable** by $Y^* \in \sigma\mathbb{Y}$ in marking $M_1 \in \mathbb{M}$ leading to marking $M_2 \in \mathbb{M}$ if M_1 [Y*> M_2 and
$$\sum_{y\in Y^*} y = Y$$

Note: If a step Y is enabled in marking M, then it is realisable by Y.

Definition 3.6: For CPN N, the set of **reachable markings** $\mathbb{M}_R \Pi \mathbb{M}$ is given by

$\mathbb{M}_R = \{ M \in \mathbb{M} \mid \exists Y^* \in \sigma Y\!: M_0 [Y^*> M \}$. The set of **enabled step sequences** $\mathbb{Y}^*_E \Pi \sigma \mathbb{Y}$ is given by $\mathbb{Y}^*_E = \{ Y^* \in \sigma \mathbb{Y} \mid \exists M \in \mathbb{M}: M_0 [Y^*> M \}$.

3.2 Formal Definition of CPN Morphisms

This section defines morphisms between CPNs and distinguishes between net morphisms (which respect structure) and system morphisms (which respect behaviour). We allow morphisms to ignore components, but where components do have images they will be consistent, i.e. nodes will map to nodes and arcs to arcs. Further, given our interest in abstraction, we only consider morphisms which are surjective with respect to P', T' and A', i.e. refinements may add or modify components but will not delete them. Consequently, we can define the preimage of every node. (The fact that ϕ is surjective means that ϕ^{-1} is defined on P' and T'.)

Definition 3.7: Given a morphism $\phi: N \varnothing N'$ and $x' \in P' \approx T'$, we define its **preimage** by $N_{x'} = \phi^{-1}(x')$ and write $N_{x'} = (P_{x'}, T_{x'}, A_{x'}, C_{x'}, E_{x'}, \mathbb{M}_{x'}, \mathbb{Y}_{x'}, M_{0x'})$. .

Definition 3.8: A **net morphism** $\phi: N \varnothing N'$ is a mapping from N to N' which is **structure-respecting**, namely:

(a) ϕ is surjective with respect to P', T', A'
(b) $\forall\ (x,y) \in A \leftrightarrow P \infty T \implies \phi(x) = \phi(y) \lor (\phi(x),\phi(y)) \in A' \leftrightarrow (P' \infty T')$
(c) $\forall\ (x,y) \in A \leftrightarrow T \infty P \implies \phi(x) = \phi(y) \lor (\phi(x),\phi(y)) \in A' \leftrightarrow (T' \infty P')$

This is the common definition of net morphism (albeit with various additional constraints) which is primarily concerned with respecting the adjacency properties of the net [1, 3, 4, 6, 16, 17]. It does not constrain behaviour (except indirectly) — in fact, sets of markings and steps are not normally included. It also does not encompass restriction on places and transitions, i.e. where selected places and transitions (and their associated arcs) are ignored by the morphism. The compositionality of net morphisms is captured by the following proposition.

Proposition 3.9: Given two net morphisms $\phi_1: N \varnothing N'$ and $\phi_2: N' \varnothing N''$ their composition $\phi = \phi_2 \circ \phi_1: N \varnothing N''$ is a net morphism.

Proof:

The composition of two surjective morphisms is surjective.

Consider $(x,y) \in A \leftrightarrow P \infty T$ and write $x' = \phi_1(x)$, $x'' = \phi_2(x')$, etc.

By the properties of ϕ_1, $x' = y'$ (in which case $x'' = y''$) or $(x', y') \in A' \leftrightarrow P' \infty T'$

By the properties of ϕ_2, $x'' = y''$ or $(x'', y'') \in A'' \leftrightarrow P'' \infty T''$

Thus $(x,y) \in A \leftrightarrow P \infty T \implies \phi(x) = \phi(y) \lor (\phi(x),\phi(y)) \in A'' \leftrightarrow P'' \infty T''$

A similar result follows for $(x,y) \in A \leftrightarrow T \infty P$.   ◊

In order to define the behaviour-respecting properties of a system morphism, it is desirable to consider complete steps, since the firing of multiple transitions in the refinement may correspond to the firing of one transition in the abstraction.

Definition 3.10: Given a morphism ϕ: N $\varnothing$ N', a step Y of N is **complete** if $\forall$ t' $\in$ T': $\forall$ t $\in$ bd(N$_{t'}$): $\phi(\{t\} \infty Y(t)) = \{t'\} \infty \phi(Y)(t')$.

Thus, a step is complete if all the border transitions occur with matching modes (which also match the mode occurrence of the corresponding abstract transition).

Definition 3.11: A **system morphism** ϕ: N $\varnothing$ N' is a mapping from N to N' which is **behaviour-respecting**, namely:

(a) ϕ is surjective with respect to P', T', A'

(b) ϕ is linear and total over both $\mathbb{M}$ and $\mathbb{Y}$.

(c) $\forall$ M $\in$ $\mathbb{M}_R$: $\forall$ Y $\in$ $\mathbb{Y}$:

   Y is complete and realisable as $Y_1 Y_2 ... Y_n$ at marking M $\Rightarrow$

   $\phi(Y)$ is realisable as $\phi(Y_1)\phi(Y_2)...\phi(Y_n)$ at marking $\phi(M)$

(d) $\forall$ M $\in$ $\mathbb{M}_R$: $\forall$ Y $\in \mathbb{Y}$: Y is complete $\Rightarrow$

   $\phi(M + E^+(Y) - E^-(Y)) = \phi(M) + \phi(E^+)(\phi(Y)) - \phi(E^-)(\phi(Y))$

Note:

(c)   If the refined step is complete, then its realisation can be used to derive the realisation of the corresponding abstract step, by projecting or restricting each component step.

(d)   This modified rule for the effect of a refined step (cf. [18]) is used since we cannot consider the component steps (of its realisation) in isolation. Thus, part (c) guarantees the enabling of the abstract sequence, while part (d) captures its overall effect

The above definition clarifies what we mean by behaviour-respecting, also called *behavioural compatibility*. This kind of morphism is called a system morphism to emphasise that it is concerned with the behaviour of the net (following the common Petri Net distinction of a net concerning the structure, and a system including behaviour). This definition is akin to that of Winskel [18], where a Petri Net is a two-sorted algebra (the sorts being the markings and the steps) and the morphism is a homomorphism which respects the input and output operations (our E^- and E^+).

This definition captures the requirement that we identified earlier, namely that refinement constrains the behaviour of the system since refined behaviours have a corresponding abstract behaviour.

Proposition 3.12: Given two system morphisms ϕ_1: N $\varnothing$ N' and ϕ_2: N' $\varnothing$ N" their composition $\phi = \phi_2 \circ \phi_1$: N $\varnothing$ N" is a system morphism.

Proof:

The composition of surjective morphisms is surjective.

The composition of total, linear mappings is total and linear.

Now, consider an arbitrary marking M$\in$ $\mathbb{M}$, $\phi_1(M)$=M'$\in$ $\mathbb{M}'$, $\phi_2(M')$=M"$\in$ $\mathbb{M}$".

Consider an arbitrary step Y$\in \mathbb{Y}$, $\phi_1(Y)$=Y'$\in \mathbb{Y}'$, $\phi_2(Y')$=Y"$\in \mathbb{Y}$".

Suppose Y and Y' are complete with respect to ϕ_1 and ϕ_2 respectively.

It follows that Y is complete with respect to ϕ.

Then if Y is realisable as $Y_1Y_2...Y_n$, then Y' is realisable as $\phi_1(Y_1)\phi_1(Y_2)...\phi_1(Y_n)$.

Hence Y" is realisable as $\phi_2(\phi_1(Y_1))\phi_2(\phi_1(Y_2))...\phi_2(\phi_1(Y_n))$.

Also: $\phi_2(\phi_1(M + E^+(Y) - E^-(Y)))$

$\quad = \phi_2(\phi_1(M) + \phi_1(E^+)(\phi_1(Y)) - \phi_1(E^-)(\phi_1(Y)))$ by properties of ϕ_1

$\quad = \phi(M) + \phi(E^+)(\phi(Y)) - \phi(E^-)(\phi(Y))$ by properties of ϕ_2 ◊

The above proposition tells us that we can combine the different forms of refinement (considered in the subsequent section) and we still have a system morphism.

3.3 CPN Refinements

We now consider each of the proposed kinds of refinement. For each one, we give a formal definition, show that it satisfies our generic constraint, and then indicate how it relates to the underlying Place Transition Net (PTN). In each case, the terminology reflects the way the incremental change is used, i.e. from abstract to refined, but the morphism is always expressed as a mapping from refined to abstract.

3.3.1 Type Refinement

The first, and perhaps simplest, form of refinement is to retain the structure of the net without modification but to replace some (or all) of the token and mode types by subtypes. Given our formulation of CPNs, where the arc inscriptions are functions (from modes to token multisets), it may not even be necessary to change the arc inscriptions, provided that they are given by polymorphic functions. For example, if an arc inscription is the identity function, i.e. the mode determines the token to be removed or added, then a simple change to both mode and token type would give a refined behaviour. If the arc inscription functions are *not* given by polymorphic functions, then they will need to be replaced, but the new versions must be consistent with the old. The distinctive thing about type refinement is that there is a projection function from subtype to supertype so that every refined state or action has a corresponding abstract state or action.

Definition 3.13: A morphism $\phi: N \oslash N'$ **captures a type refinement** if:
(a) ϕ is an identity function on P, T, A, i.e. $\forall x \in P: \phi(p) = p$, etc.
(b) $\forall x \in P \approx T: C(x) <: \phi(C)(x)$
(c) $\forall x \in P \approx T: \forall c \in C(x): \phi(1`(x,c)) = 1`(x, {}^\bullet\phi(C)(x)^{(c)})$
(d) $\forall (p,t) \in A: \forall (t,m) \in \mathbb{Y}:$

$\qquad \phi(E^-(1`(t,m)))(p) = {}^\bullet\phi(C)(p)^{(E(p,t)(m))} = \phi(E)(p,t)({}^\bullet\phi(C)(t)^{(m)})$

$\qquad \forall (t,p) \in A: \forall (t,m) \in \mathbb{Y}:$

$\qquad \phi(E^+(1`(t,m)))(p) = {}^\bullet\phi(C)(p)^{(E(t,p)(m))} = \phi(E)(t,p)({}^\bullet\phi(C)(t)^{(m)})$

Note:

(a)   There is no change to the structure of the net — to places, transitions and arcs.

(b)   The colours associated with the places and transitions are consistently subtyped.

(c)   Given the consistent structure, the morphism is defined for all markings and steps using the appropriate projection functions (see §3.1).

(d)   The arc inscriptions are consistent, i.e. the projected effect of a mode or step is the same as the effect of the projected mode or step.

Proposition 3.14:   A morphism ϕ: N $\varnothing$ N' which captures a type refinement is a system morphism (in the sense of def 3.11).

Proof:

ϕ is surjective on P', T', A', and is linear and total on $\mathbb{M}, \mathbb{Y}$.

Given M $\in$ $\mathbb{M}_R$ and an enabled step Y $\in \mathbb{Y}$, i.e. M [Y>

Y is necessarily complete since N_t' is always a single transition.

From def 3.13(c), M' = ϕ(M) $\in$ $\mathbb{M}$' and Y' = ϕ(Y) $\in \mathbb{Y}$ ' are both defined

The enabling in N means M $\geq$ E$^-$(Y) and hence ϕ(M) $\geq$ ϕ(E$^-$(Y))

Def 3.13(c, d) implies that ϕ(M) $\geq$ ϕ(E$^-$)ϕ(Y) and hence Y' is enabled, i.e. M' [Y'>

Hence, ϕ(M – E$^-$(Y) + E$^+$(Y)) = M' – ϕ(E$^-$)(Y') + ϕ(E$^+$)(Y')

Hence, ϕ(M + E$^+$(Y) – E$^-$(Y)) = M' + ϕ(E$^+$)(Y') – ϕ(E$^-$)(Y') $\Diamond$

This result is actually stronger than that required by def 3.11 in that every enabled refined step is complete and has a corresponding enabled abstract step. In other words, given the realisation of a refined step, we can determine a realisation of the corresponding abstract step.

In §2 we gave an example of type refinement, where the book token type was extended with information about the loan type. There were corresponding changes to the relevant transition firing modes. This form of refinement can eliminate some abstract behaviour since the refined token requirements of a transition may not be satisfied, even though the abstract requirements are. Thus, the firing of the *Loan expires* transition may now be constrained by the loan type.

The underlying PTN for a CPN has one (colourless) place for each (place, colour) combination of the CPN, and one (colourless) transition for each (transition, colour) combination. If we relate this morphism to the underlying PTN (in the sense of refining the behaviour), then each value will potentially be replaced by a number of values (given the additional data components), and hence it is apparent that it corresponds to unfolding a place (transition) into a number of places (transitions). This form of net transformation (and its converse) have been classified as *unfolding* and *folding* [1].

3.3.2 Subnet Refinement or Extension

The second form of refinement is to add net components — places, transitions and arcs, or even additional token or mode values. As a morphism (from refined to abstract nets), this would be called a restriction, since net components are being discarded or ignored. Where token or mode types are extended, then in contrast to §3.3.1, abstraction does *not* project the additional refined values onto abstract values but rather ignores them. (In the equivalent unfolded PTN, this is the same as ignoring places, transitions and arcs.)   This does not satisfy the structure-respecting requirements for a net morphism (def 3.8), but it does qualify as a system morphism (def 3.11).

Definition 3.15: A morphism ϕ:N $\varnothing$ N' **captures a subnet refinement** if:

(a)   The net structure is restricted, i.e. $\forall p \in$ P: $\phi(p)$ is defined $\Rightarrow \phi(p) \in$ P, and similarly for T, A.

(b) $\forall x \in \phi(P) \approx \phi(T)$: C(x) $\phi(C)(x)$

(c) $\forall x \in P \approx T$: $\forall c \in$ C(x): $\phi(1`(x,c)) = 1`(x,c)$ if $x \in \phi(P) \approx \phi(T)$, otherwise $\varnothing$

(d) $\forall Y \in \mathbb{Y}$: $\phi(E^+(Y)) = \phi(E^+)(\phi(Y))$ and $\phi(E^-(Y)) = \phi(E^-)(\phi(Y))$

Note:

(a)   The sets of places, transitions and arcs may be restricted by ϕ.

(b)   The colours associated with retained places and transitions may be restricted.

(c)   Given the consistent colouring, the morphism is simply defined for all markings and steps.

(d)   The restricted incremental effect of the step is the same as the incremental effect of the restricted step. This implies that ignored components can refer to, but cannot permanently modify, retained components.

Proposition 3.16: A morphism ϕ: N $\varnothing$ N' which captures a subnet refinement is a system morphism (in the sense of def 3.11).

Proof:

ϕ is surjective on P', T', A', and is linear and total on $\mathbb{M}, \mathbb{Y}$.

Given $M \in \mathbb{M}_R$ and an enabled step $Y \in \mathbb{Y}$, i.e. M [Y>

From def 3.15 (c), $M' = \phi(M) \in \mathbb{M}$ and $Y' = \phi(Y) \in \mathbb{Y}$ ' are both defined

Y is necessarily complete since $N_{t'}$ is always a single transition.

The enabling in N means $M \geq E^-(Y)$ and hence $\phi(M) \geq \phi(E^-(Y))$

Def 3.15 (c, d) implies that $\phi(M) \geq \phi(E^-)\phi(Y)$ and hence Y' is enabled, i.e. M' [Y'>

Hence, $\phi(M - E^-(Y) + E^+(Y)) = M' - \phi(E^-)(Y') + \phi(E^+)(Y')$

Hence, $\phi(M + E^+(Y) - E^-(Y)) = M' + \phi(E^+)(Y') - \phi(E^-)(Y')$ $\Diamond$

As with type refinement, this result is stronger than that required by def 3.11 in that every enabled refined step is complete and has a corresponding enabled abstract step. In other words, given the realisation of a refined step, we can determine a realisation of the corresponding abstract step.

In §2 we gave an example of subnet refinement, where the subnet was augmented to cater for book reservations. The abstract behaviour of the net was restricted by the refinement — if a book can be borrowed given the possibility of reservations (the refined case), then it can also be borrowed if reservations are ignored (i.e. the abstract case).

If we relate this morphism to the underlying PTN, then it is immediately apparent that it corresponds to a net transformation (and its converse) which have been classified as *embedding* (or extension) and *restriction* [1].

3.3.3 Node Refinement

The third form of refinement is the replacement of a place (transition) by a place (transition) bordered subnet, also called a *superplace* (*supertransition*). We refer to this as *node refinement* to distinguish it from the other forms of refinement being considered, even though traditional Petri Net theory simply refers to it as refinement [1]. The desirable properties of node refinement for CPNs have been considered elsewhere [12]. For behavioural consistency, it was argued that the subnet which refines a place should have the notion of an abstract marking, and the subnet which refines a transition should have the notion of abstract firing. These constraints were captured by requiring the morphism to be structure-respecting, colour-respecting, marking-respecting and step-respecting. Here, we propose canonical place and transition refinements, and in §4 compare this approach to the previous one. In any case, the morphism for a node refinement is both a net morphism and a system morphism.

Definition 3.17: Given a morphism $\phi: N \oslash N'$ then $N_{p'}$ is a **canonical place refinement** of $p' \in P'$ provided:

(a) $\forall p \in P - P_{p'}: \forall t \in T - T_{p'}:$ $\phi(p) = p \wedge \phi(t) = t \wedge$

$(p,t) \in A \Rightarrow ((p,t) \in A' \wedge E(p,t) = E'(p,t)) \wedge$
$(t,p) \in A \Rightarrow ((t,p) \in A' \wedge E(t,p) = E'(t,p))$

(b) $\forall t \in env(N_{p'}): E'(p',t) = \sum\limits_{(p,t) \in A \cap (P_{p'} \times T)} E(p,t)$ and $E'(t,p') = \sum\limits_{(t,p) \in A \cap (T \times P_{p'})} E(t,p)$

(c) $bd(N_{p'}) = \{inp1, inp2, \ldots out1, out2, \ldots\}$

(d) $P_{p'} = bd(N_{p'}) \approx \{buf\} \approx P_{other}$

(e) $T_{p'} = \{accept1, accept2, \ldots offer1, offer2, \ldots\} \approx T_{other}$

(f) $A_{p'} = \{(inp1,accept1), (accept1,buf), \ldots (buf,offer1), (offer1,out1), \ldots\} \approx A_{other}$

(g) $\bullet inp1 \prod env(N_{p'}) \wedge inp1 \bullet = \{accept1\} \wedge \ldots$

$out1 \bullet \prod env(N_{p'}) \wedge \bullet out1 = \{offer1\} \wedge \ldots$

$\bullet buf = \{accept1, \ldots\} \wedge buf \bullet = \{offer1, \ldots\}$

(h) $\forall x \in P_{p'} - P_{other} \approx T_{p'} - T_{other}: C_{p'}(x) = C(x) = C'(p')$

(i) $\forall a \in A_{p'} - A_{other}: E_{p'}(a) = Id$

(j) $\forall p \in P: \forall c \in C(p): \phi(1`(p,c)) =$     $1`(p',c)$ if p $\in$ $P_{p'} - P_{other}$
$= \oslash$ if $p \in P_{other}$
$= 1`(p,c)$ otherwise

$\forall t \in T: \forall c \in C(t): \phi(1`(t,c)) = 1`(t,c)$ if $t \notin T_{p'}$ otherwise $\oslash$

Note:

(a) Apart from the subnet $N_{p'}$ and its incident arcs, the rest of the net is unchanged.

(b) The flow of tokens across the boundary of the place refinement is consistent.

(c) The border of the subnet consists of the places *inp1, …, out1, …*.

(d-f) The component places, transitions and arcs consist of the basis places, transitions and arcs respectively, plus others which constitute the subnet refinement of the *accept* and *offer* transitions (cf. fig 2.4).

(g)   The input (output) border places only have input (output) from (to) environment transitions, other arcs incident on basis places are exactly those of part (e).

(h)   The colour of the basis places and transitions are all the same.

(i)   The arc inscription for all the basis arcs is the identity function.

(j)   In abstracting the place refinement, the abstract marking is given by the marking of the basis places and the internal actions are ignored.

Proposition 3.18:   A morphism ϕ: N $\emptyset$ N' with a canonical place refinement constitutes a system morphism.

Proof:

ϕ is surjective on P', T', A', and is linear and total on $\mathbb{M}$,$\mathbb{Y}$.

Given M $\in \mathbb{M}_R$ and an enabled step Y $\in \mathbb{Y}$, i.e. M [Y>

From def 3.17 (j), M' = ϕ(M) $\in \mathbb{M}$' and Y' = ϕ(Y) $\in \mathbb{Y}$' are both defined

Y is necessarily complete since $N_{t'}$ is always a single transition.

Given the linearity of ϕ, M $\geq E^-$(Y) implies ϕ(M) $\geq \phi(E^-$(Y))

From def 3.17 (a, b), ϕ(M) $\geq \phi(E^-)(\phi$(Y)) and hence Y' is enabled, i.e. M' [Y'>

Hence, ϕ(M – E^-(Y) + E^+(Y)) = ϕ(M) – $\phi(E^-)$(Y') + $\phi(E^+)$(Y')

Hence, ϕ(M + E^+(Y) – E^-(Y)) = ϕ(M)+ $\phi(E^+)$(Y') – $\phi(E^-)$(Y')   ◊

As with type and subnet refinement, this result is stronger than that required by def 3.11 in that every enabled refined step is complete and has a corresponding enabled abstract step. In other words, given the realisation of a refined step, we can determine a realisation of the corresponding abstract step. This is not the case for the following transition refinement.

Definition 3.19:   Given a morphism ϕ: N $\emptyset$ N' then $N_{t'}$ is a **canonical transition refinement** of t' $\in$ T' provided:

(a) $\forall p \in P-P_{t'}$: $\forall t \in T-T_{t'}$: ϕ(p) = p $\wedge \phi$(t) = t $\wedge$

        (p,t) $\in$ A $\Rightarrow$ ((p,t) $\in$ A' $\wedge$ E(p,t) = E'(p,t)) $\wedge$
        (t,p) $\in$ A $\Rightarrow$ ((t,p) $\in$ A' $\wedge$ E(t,p) = E'(t,p))

(b) $\forall p \in$ env($N_{t'}$): E'(p,t') = $\sum_{(p,t) \in A \cap (P \times T_{t'})} E(p, t)$ and E'(t',p) = $\sum_{(t, p) \in A \cap (T_{t'} \times P)} E(t, p)$

(c)   bd($N_{t'}$) = {inp1, ..., out1, ... }

(d) $P_{t'}$ = {recd1, ..., send1, ...} $\approx P_{other}$

(e) $T_{t'}$ = bd($N_{t'}$) $\approx$ {switch} $\approx T_{other}$

(f) $A_{t'}$ = {(inp1,recd1), (recd1,switch), ... (switch,send1), (send1, out1), ...} $\approx A_{other}$

(g) $\bullet inp1 \prod env(N_{t'}) \approx P_{other} \wedge \bullet recd1 = \{inp1\} \wedge recd1\bullet = \{switch\} \wedge \ldots$

   $out1\bullet \prod env(N_{t'}) \approx P_{other} \wedge send1\bullet = \{out1\} \wedge \bullet send1 = \{switch\} \wedge \ldots$

   $\bullet switch = \{recd1,\ldots\} \wedge switch\bullet = \{send1,\ldots\}$

(h) $\forall x \in P_{t'}-P_{other} \approx T_{t'}-T_{other}: C_{t'}(x) = C(x) = C'(t')$

(i) $\forall a \in A_{t'}-A_{other}: E_{t'}(a) = Id$

(j) $\forall p \in P_{t'}-P_{other}: M_{0t'}(p) = \emptyset$

(k) $\forall p \in P: \forall c \in C(p): \phi(1`(p,c)) = 1`(p,c)$ if $p \in P-P_{t'}$ otherwise $\emptyset$

   $\forall t \in T: \forall c \in C(t): \phi(1`(t,c)) = \quad 1`(t',c) \quad$ if $\quad t \quad = \quad$ switch

   $= \emptyset$ if $t \in T_{t'} - \{switch\}$

   $= 1`(t,c)$ otherwise

Note:

(a)   Apart from the subnet $N_{t'}$ and its incident arcs, the rest of the net is unchanged.

(b)   The flow of tokens across the boundary of the transition refinement is consistent.

(c)   The border of the subnet consists of the transitions *inp1*, ..., *out1*, ...

(d-f) The component places, transitions and arcs consist of the basis places, transitions and arcs respectively, plus others which constitute the subnet refinement of the border transitions (cf. fig 2.6).

(g)   The input (output) border transitions only have input (output) from (to) environment places or the other internal places, the basis places only have the incident arcs of part (e).

(h)   The colour of the basis places and transitions are all the same.

(i)   The arc inscription for all the basis arcs is the identity function.

(j)   The initial marking of all basis places is empty.

(k)   The internal marking of the supertransition is ignored by the abstraction and the firing of the *switch* transition corresponds to the abstract firing of t'.

Proposition 3.20: A morphism $\phi: N \oslash N'$ with a canonical transition refinement constitutes a system morphism.

Proof:

ϕ is surjective on P', T', A', and is linear and total on $\mathbb{M}$, $\mathbb{Y}$.

Given $M \in \mathbb{M}_R$ and a complete step $Y \in \mathbb{Y}$.

From def 3.19 (k), $M' = \phi(M) \in \mathbb{M}'$ and $Y' = \phi(Y) \in \mathbb{Y}'$ are both defined

Suppose Y is realisable as $Y_1 Y_2 \ldots Y_n$ at M

Consider $\phi(Y_1)\phi(Y_2)\ldots\phi(Y_n)$ and, in particular, an arbitrary element $\phi(Y_i)$

Transitions external to $N_{t'}$ occur in $\phi(Y_i)$ exactly as in Y_i and are similarly enabled.

Transitions internal to $N_{t'}$ apart from *switch* occuring in Y_i are ignored in $\phi(Y_i)$

The transition *switch* in $N_{t'}$ occuring in Y_i occurs as t' in $\phi(Y_i)$

Completeness guarantees that border transitions occur with the same mode(s) as t'

The canonical construction guarantees that the input tokens are consumed first.

Thus, transition t' is enabled in $\phi(Y_i)$

Finally, defs 3.10, 3.19(b) guarantee that the total effect of Y and Y' are the same

Hence, $\phi(M + E^+(Y) - E^-(Y)) = \phi(M) + \phi(E^+)(Y') - \phi(E^-)(Y')$ ◊

Definition 3.21: A morphism ϕ: N ∅ N' **captures a canonical node refinement** if ϕ has a canonical place or transition refinement.

In §2 we gave examples of node refinement, one where a place was replaced by a subnet to indicate the more detailed processing of a book on its return from loan, and one where a transition was replaced by a subnet to indicate more of the details of borrowing a book. Again, such refined behaviour had corresponding abstract behaviour, but not necessarily vice versa.

If we relate this morphism to the underlying PTN then it corresponds to the refinement of a place or transition by an appropriately-bounded subnet. This form of net transformation (and its converse) have been classified as *refinement* and *abstraction* [1].

4 Properties of Abstractions

The previous section has defined three forms of CPN refinement which maintain behavioural compatibility. In this section we show that even though they appear to be quite constrained, they are at least as general as the earlier proposals [12].

The earlier proposals required that the flow of tokens across the boundaries of node refinements should be consistent with the corresponding abstractions:

Definition 4.1: A structure-respecting morphism ϕ: N ∅ N' is **colour-respecting** if C', E' of N' satisfy:

(a) $\forall x' \in P' \approx T'$: $\forall x \in bd(N_{x'})$: $C(x) <: C'(x')$

$$\forall (x', y') \in A': E'(x', y') = \sum_{x \in bd(N_{x'})} \sum_{y \in bd(N_{y'})} E(x, y)$$

(b)

Note:

(a)   The colour of a superplace or a supertransition must be compatible with the colours of the corresponding places or transitions in its border.

(b)   The morphism preserves the token transfer across the interfaces between superplaces and supertransitions.

The above property is satisfied by the node refinements proposed in §3.3.3 as is apparent from defs 3.17 (b) and 3.19 (b). The propositions which follow will guarantee the converse, namely that the colour-respecting morphisms we consider have corresponding canonical node refinements.

In the earlier paper [12], it was argued that behavioural compatibility required that place refinements should have an associated abstract marking which was invariant over the internal actions of the subnet. This was captured by the following property.

Definition 4.2: A colour-respecting morphism ϕ: N ∅ N' is **marking-respecting** if:

(a) ϕ is linear and total on $\mathbb{M}$

$$\phi(M)(p') = \sum_{p \in N_{p'}} \phi(M(p))$$

(b) $\forall p' \in P'$: $\forall M \in \mathbb{M}$:

(c) $\forall\, p' \in P'$: $M_1\, [(t,c) > M_2$:

$$\phi(M_2)(p') \;=\; \phi(M_1)(p') \;-\; \sum_{p \in bd(N_{p'})} E(p,t)(c) \;+\; \sum_{p \in bd(N_{p'})} E(t,p)(c) \qquad \text{if}$$

$$t \in env(N_{p'})$$

$$=\; \phi(M_1)(p') \qquad\qquad\qquad\qquad\qquad\qquad\qquad \text{otherwise}$$

Note:

(a)   For regularity, ϕ is linear and defined over all components of M, not just reachable markings, which would be the minimal requirement.

(b)   The abstract marking of a superplace is determined by the refined markings of its constituent places. This is largely implied by requirement (c) but it is clearer to state it explicitly.

(c)   The abstract marking of a superplace is only modified by environment transitions and not by other external transitions or internal transitions.

The following propositions elucidate the relationship between marking-respecting morphisms and the canonical place refinements of §3.3.3.

Proposition 4.3: A morphism ϕ: N $\varnothing$ N' with a canonical place refinement (as in def 3.17) is marking-respecting.

Proof: This follows directly from def 3.17(j).   ◊

Proposition 4.4: An arbitrary marking-respecting refinement can be transformed into an behaviourally compatible canonical place refinement where the internal place *buf* is redundant.

Proof:

Build the basis of a canonical place refinement as follows:

•   for each border place with one or more input arcs from the environment, introduce a new input border place which usurps those input arcs;

•   from each new input border place, add an arc to a corresponding *accept* transition (which is newly introduced), and add an arc from this *accept* transition to the original border place;

•   for each border place with one or more output arcs to the environment, introduce a new output border place which usurps those output arcs;

•   from each such original border place, add an arc to a corresponding *offer* transition (which is newly introduced), and add an arc from this *offer* transition to the new output border place;

•   add a new place *buf* with input arcs from all the newly introduced *accept* transitions and output arcs to all the newly introduced *offer* transitions;

•   the token and mode type of all introduced places and transitions is that of the token type for the abstract place, and the new arcs all have the same identity function as inscription.

This process is illustrated in fig 4.1, where the original components are shaded:

•   the border place *b* (which has both input and output arcs with the environment) is replaced by two border places *ib* (for input) and *ob* (for output)

•   there is an *accept* transition *ab* corresponding to the new input border place *ib*

- there is an *offer* transition *ob* corresponding to the new output border place *ob*
- the input arc from the environment incident on *b* is now incident on *ib*
- the output arc to the environment incident on *b* is now incident on *ob*
- there is a new input arc incident on *b* and a new output arc incident on *b*

Similarly for the other border places *a* and *c*.

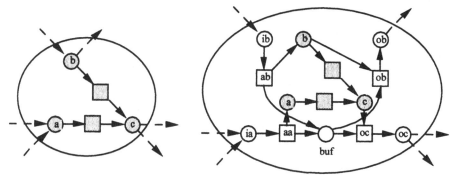

Fig 4.1: Arbitrary place refinement

The behaviour of the transformed net is compatible with the original (see def 3.11):

- as soon as a token is deposited into a new input border place, the corresponding *accept* transition can fire, depositing a copy of the token into the corresponding original border place, and a copy is also added to the place *buf*
- as soon as a token is available in one of the original border places with an output arc to the environment, then the corresponding *offer* transition can be fired, making the token available in the new border place (and thence to the environment)
- given a transition sequence of the modified net, the corresponding original sequence can be extracted by ignoring the firing of *accept* and *offer* transitions.

Note that the place *buf* is redundant (and therefore could be eliminated):

- when an environment transition of the original place refinement deposited one or more tokens into the border places, the abstract marking given by $\phi(M)(p')$ varied by exactly those tokens (see def 4.2(c)).
- with the new refinement, the same property holds (except that we now have new input border places).
- the removal of tokens from any new input border place (by firing the corresponding *accept* transition) is matched by depositing the token in the *buf* place and in the corresponding original border place.
- the firing of the original internal transitions did not affect the abstract marking, and hence it stays synchronised with the marking of place *buf* (see def 4.2(c)).
- finally, the removal of tokens from the original place refinement is now matched by the removal of tokens from the place *buf*

Finally, it is straightforward to check that the original subnet has become a subnet refinement augmenting the *accept* and *offer* transitions of the basis for a canonical place refinement.   ◊

Again, the earlier paper [12] argued that behavioural compatibility required that transition refinements should have the notion of abstract firing. This was captured by the following property.

Definition 4.5: A colour-respecting morphism ϕ: N $\varnothing$ N' is **step-respecting** if

(a) ϕ is linear on $\mathbb{Y}$

(b)   if $Y \in \mathbb{Y}$ is realisable at $M \in \mathbb{M}$, and $\phi(Y)$ is defined with $\phi(Y)=Y' \in \mathbb{Y}'$ realisable at $\phi(M) = M' \in \mathbb{M}'$, then
$$\forall\, t \in bd(N_{t'}):\ Y(t) = Y'(t')$$

(c)   if $Y \in \mathbb{Y}$ is realisable at $M_1 \in \mathbb{M}$ leading to M_2, and $\phi(Y)$ is defined with $\phi(Y)=(t',c') \in \mathbb{Y}'$ realisable at $\phi(M_1) = M_1' \in \mathbb{M}'$, then:

$$M_2(p) = \begin{array}{ll} M_1(p) - \sum\limits_{t \in bd(N_{t'})} E(p,t)(c') + \sum\limits_{t \in bd(N_{t'})} E(t,p)(c') & \text{if } p \in env(N_{t'}) \\[2mm] = M_1(p) & \text{if } p \in P - P_{t'} \end{array}$$

Note:

(a) ϕ is linear, i.e. $\phi(Y_1 + Y_2) = \phi(Y_1) + \phi(Y_2)$ provided $\phi(Y_1)$, $\phi(Y_2)$ are defined.

(b)   The occurrence of border transitions is determined by the abstract firing modes. This is largely implied by requirement (c) but it is clearer to state it explicitly.

(c)   Of the places external to the supertransition, only the environment places are modified and only by the effect of the border transitions.

Proposition 4.6:   Given a morphism ϕ: N $\varnothing$ N' with a canonical transition refinement $N_{t'}$, a realisable step Y of N has a corresponding realisable step of N' if and only if it is complete.

Proof:

Consider a complete step Y realisable as the sequence Y*:

In Y*, each abstract firing of $N_{t'}$ will include a firing of the transition *switch*.

The input border transitions of $N_{t'}$ will fire prior to the firing of *switch*.

The consumed tokens could also be consumed later — just before the *switch*.

The output border transitions of $N_{t'}$ will fire after the firing of *switch*.

The generated tokens could be generated earlier — just after the *switch*.

Hence, the complete firing of $N_{t'}$ could be replaced by the abstract firing of t'

Specifically, the firing of t' could replace the *switch* transition in the sequence.

Conversely, suppose a refined sequence is not complete:

Then the firing of some border transition is not included.

The corresponding token transfers will not occur.

Hence, the effect on the environment will not match that of t'.   ◊

The following propositions elucidate the relationship between step-respecting morphisms and the canonical transition refinements of §3.3.3.

Proposition 4.7: A morphism φ: N ∅ N' with a canonical transition refinement is step-respecting.

Proof:

Every refined step has a corresponding abstract step.

If the refined step is complete and realisable, then so is the abstract step (prop 4.6).

The construction ensures that all border transition have the same occurrence modes Hence property (b) of def 4.5 holds.

Property (c) then follows from def 3.19 (a,b).   ◊

Proposition 4.8: An arbitrary step-respecting transition refinement with distinct input and output border transitions can be transformed into a behaviourally compatible canonical transition refinement.

Proof:

An arc to a corresponding *recd* place is added from each input border transition.

An arc from a corresponding *send* place is added to each output border transition.

The subnet is augmented by a *switch* transition.

The subnet is augmented by arcs from the *recd* places to the *switch* transition.

The subnet is augmented by arcs from the *switch* transition to the *send* places.

Behavioural compatibility can be demonstrated by ignoring the firing of *switch*.

It is straightforward to check that the original subnet is a subnet refinement augmenting the *accept* and *offer* transitions of the basis for a canonical place transition. ◊

The above results demonstrate that the current proposals, while being more prescriptive, are just as general as the earlier proposals.

5 Conclusions and Further Work

This paper has significantly extended a previous one on the abstraction of CPNs [12]. The context is that of refining CPNs while maintaining behavioural compatibility. We have identified three kinds of refinement, namely *type*, *subnet* and *node* refinement. In practical applications, it is anticipated that these will be used in concert. Further, we have shown that by considering node refinement in conjunction with subnet refinement, it was possible to propose a canonical form for node refinement which could be extended by subnet refinement to be behaviourally compatibile with an arbitrary node refinement (as in [12]).

We have shown that the proposed refinements all qualify as system morphisms, and that the composition of system morphisms is also a system morphism. This notion of morphism is an extension of that proposed by Winskel [18]. By identifying the three kinds of refinement, we have proposed an answer to the question posed by Winskel, namely the appropriate kind of morphism for CPNs.

There are a number of interesting avenues for further work. Given the generality of these canonical forms, we propose to incorporate them into a Petri Net tool, thereby supporting the designer in producing well-structured specifications. By imposing behavioural constraints on abstraction, it is possible to use the abstraction to prove properties about the modelled system, which is often important for large complex systems with excessively large state spaces. We are continuing to investigate the range of applicability of these forms of refinement to practical problems [11]. We are also developing incremental reachability algorithms to take advantage of the behavioural compatibility [14]. Further down the track, we would like to identify when refinements are in some sense independent and hence can be analysed in isolation.

Acknowledgements

The author is pleased to acknowledge the helpful discussions held with Robert Esser, Glenn Lewis and Joachim Wehler. The careful comments of the reviewers have helped to improve this paper significantly.

References

[1]   B. Baumgarten *Petri-Netze: Grundlagen und Anwendungen* Wissenschaftsverlag (1990).

[2]   L. Cardelli and J.C. Mitchell *Operations on Records* Mathematical Foundations of Programming Semantics, Lecture Notes in Computer Science 442, Springer Verlag (1989).

[3]   J. Desel and A. Merceron *Vicinity Respecting Net Morphisms* Advances in Petri Nets 1990, G. Rozenberg (ed.), Lecture Notes in Computer Science 483, pp 165-185, Springer-Verlag (1990).

[4]   J. Desel and A. Merceron *Vicinity Respecting Homomorphisms for Abstracting System Requirements* Report 337, Universität Karlsruhe (1996).

[5]   A. Diller *Z: An Introduction to Formal Methods* 2nd edn., Wiley (1994).

[6]   R. Fehling *A Concept of Hierarchical Petri Nets with Building Blocks* Advances in Petri Nets 1993, G. Rozenberg (ed.), Lecture Notes in Computer Science , pp 148-168, Springer-Verlag (1993).

[7]   H.J. Genrich *A Dictionary of Some Basic Notions of Net Theory* Net Theory and Applications, W. Brauer (ed.), Lecture Notes in Computer Science 84, pp 519-535, Springer-Verlag (1980).

[8]   H.J. Genrich *Predicate/Transition Nets* Advances in Petri Nets 1986 – Part 1, W. Brauer, W. Reisig, and G. Rozenberg (eds.), Lecture Notes in Computer Science 254, Springer-Verlag (1987).

[9]   K. Jensen *Coloured Petri Nets: Basic Concepts, Analysis Methods and Practical Use – Volume 1: Basic Concepts* EATCS Monographs in Computer Science, Vol. 26, Springer-Verlag (1992).

[10]   K. Jensen, S. Christensen, P. Huber, and M. Holla *Design/CPN™: A Reference Manual* MetaSoftware Corporation (1992).

[11]   C. Lakos and G. Lewis *A Catalogue of Incremental Changes for Coloured Petri Nets* Technical Report TR99-02, Department of Computer Science, University of Adelaide (1999).

[12]   C.A. Lakos *On the Abstraction of Coloured Petri Nets* Proceedings of 18th International Conference on the Application and Theory of Petri Nets, Lecture Notes in Computer Science 1248, pp 42-61, Toulouse, France, Springer-Verlag (1997).

[13] C.A. Lakos *The Compositionality of Abstraction for Coloured Petri Nets* Technical Report TR99-01, Department of Computer Science, University of Adelaide (1999).

[14] G. Lewis and C. Lakos *Incremental Reachability Algorithms* Technical Report TR99-1, Department of Electrical Engineering and Computer Science, University of Tasmania (1999).

[15] J. Meseguer and U. Montanari *Petri Nets are Monoids* Information and Computation, **88**, 2, pp 105-155 (1990).

[16] C.A. Petri *Introduction to General Net Theory* Net Theory and Applications, W. Brauer (ed.), Lecture Notes in Computer Science 84, pp 1-19, Springer-Verlag (1980).

[17] W. Reisig *Petri Nets in Software Engineering* Advances in Petri Nets 1986 – Part 2, W. Brauer, W. Reisig, and G. Rozenberg (eds.), Lecture Notes in Computer Science 255, pp 63–96, Springer-Verlag (1987).

[18] G. Winskel *Petri Nets, Algebras, Morphisms, and Compositionality* Information and Computation, **72**, pp 197-238 (1987).

Modelling and Analysis of a DANFOSS Flowmeter System Using Coloured Petri Nets

Louise Lorentsen and Lars Michael Kristensen

Department of Computer Science, University of Aarhus,
Aabogade 34, DK-8200 Aarhus N. DENMARK,
louisel,kris @daimi.au.dk

Abstract. DANFOSS is a Danish manufacturer of refrigeration, motion, heating, and water controls. This paper describes the main results of a project on the modelling and analysis of a DANFOSS flowmeter system using Coloured Petri Nets (CP-nets or CPNs). A modern flowmeter system consists of a number of communicating processes, cooperating to make various measurements on, e.g., the flow of water through a pipe. The purpose of the project was to investigate the application of CP-nets for validation of the communication protocols used in the flowmeter system. Analysis by means of state spaces successfully identified problems in the proposed communication protocols. An alternative design was analysed using state spaces reduced by taking advantage of the inherent symmetries in the system. Exploiting the symmetries made it possible to analyse configurations of the flowmeter system approaching the size of typical flowmeter systems.

1   Introduction

The Danish company DANFOSS is one of the largest industrial groups in Denmark, and it is one the world leaders within the area of refrigeration, motion, heating, and water controls. This paper presents the main results of a project focusing on modelling and analysis of a DANFOSS flowmeter system. The project was carried out as a joint project between DANFOSS INSTRUMENTATION, which is a subgroup of DANFOSS, and the CPN group at the University of Aarhus.

Flowmeters are primarily used to make measurements on the flow of water through pipes. The concrete flowmeter system studied in the project consisted of several processes each doing measurements on the flow of water. Examples are processes measuring the amount of water flowing through the pipe, processes measuring the temperature of the water, and processes doing calculations based on measurements obtained by other processes.

DANFOSS has in recent years changed the architecture in the flowmeter systems from a centralised solution to a more flexible and distributed solution. The main advantage of the distributed architecture is that it allows the flowmeter systems to be adapted to the specific needs of customers. However, the distributed architecture also makes new problems arise. In the distributed solution

M. Nielsen, D. Simpson (Eds.): ICATPN 2000, LNCS 1825, pp. 346–366, 2000.

it is much more difficult to reason about the individual processes and their influence on each other. Practical tests at DANFOSS have shown that the process communication in the first design of the flowmeter system contained at least one deadlock, and it was therefore realised that a more thorough investigation of the proposed designs was needed.

The overall aim of the project reported on in this paper was to demonstrate the use of Coloured Petri Nets (CP-nets or CPNs) [11, 14] and its supporting Design/CPN tool [16] for investigating whether the different design alternatives have the desired properties as formulated by the producer. An example of such a property is the absence of deadlocks in the flowmeter system. CP-nets have previously been used in a number of projects in an industrial setting. Examples of this include the modelling of software architectures for mobile phones at Nokia [19], validation of communication protocols used in Bang & Olufsens's Beolink Audio/Video systems [5], and communication gateways at Australian Defence Forces [10]. It was therefore envisioned that CP-nets would also be applicable for raising the quality of the design process for the new flowmeter system at DANFOSS.

The paper is organised as follows. Section 2 gives an overview of the project organisation and the different activities. Section 3 contains an introduction to the DANFOSS flowmeter system. Section 4 presents selected parts of the CPN model of the flowmeter system. Section 5 describes how state spaces were used to investigate the correctness of the flowmeter system. Section 6 describes a new design proposal and explains how it was verified using state spaces reduced by means of symmetries. Finally, Sect. 7 contains the conclusions. The reader is assumed to be familiar with the basic ideas of High-level Petri Nets and state spaces (also called occurrence graphs or reachability graphs/trees).

2   Overview of the Project

The project was organised in three phases and involved engineers from DANFOSS and people from the CPN group, including the authors of this paper. Hence the project group consisted of persons with expertise in the application domain, i.e., the flowmeter system, as well as members with expertise in the methods and tools to be applied, i.e., CP-nets and Design/CPN. The project was divided into three main phases which altogether ran over a period of 15 months. There have however been gaps between the different phases which means that the project ran actively for a period of 7 months.

The first phase of the project focused on the construction of a CPN model of the flowmeter system. An initial CPN model captured two different design proposals for the flowmeter system. Overview information was provided at meetings with engineers at DANFOSS and complemented by internal documentation about the flowmeter system made available by DANFOSS. The primary purpose of the meetings at this stage was to obtain an overview of the flowmeter system and discuss the main issues related to the design. Throughout the first phase, focus gradually shifted from getting information about the flowmeter system and understanding the design issues to discussions of the CPN models which were constructed by the people from the CPN group. The main purpose of these discussions was to ensure that the CPN model correctly reflected the two different design proposals. Interactive simulations, i.e., single step simulations with detailed graphical feedback, were used to investigate the behaviour of the CPN model as well as for debugging purposes.

Using the Message Sequence Charts library [4], extensions were made to the CPN model allowing Message Sequence Charts (MSCs) [2] to be automatically constructed as graphical feedback from simulations. During project meetings these interactive simulations and MSCs were used to discuss and review the behaviour of the CPN model. Combined with the graphical nature of Petri Nets this mediated the identification of a number of discrepancies between the design and the CPN model. Subsequent to the meetings the CPN model was modified according to the identified discrepancies. Eventually these reviews lead to a validated CPN model in the sense that it correctly captured what was considered to be the relevant aspects of the two design proposals. The reason for choosing MSCs to visualise the behaviour of the CPN model was that diagrams very close to MSCs were already used in the design process at DANFOSS. This allowed the behaviour of the CPN model to be visualised in a way that was very familiar to the engineers from DANFOSS.

In the second phase of the project, focus changed from modelling to state space analysis [12]. The analysis was done by means of the Design/CPN Occurrence Graph Tool (OG Tool) [16]. The aim of the second phase was to investigate whether the design alternatives had the desired properties specified by DANFOSS. In this phase we encountered the *state explosion problem*, i.e., the state spaces started to grow rapidly when analysing configurations of flowmeter systems with more than three processes. However, even the analysis of small configurations of the flowmeter system identified deadlocks and problems with the consistency of data in the two design proposals.

In the third phase of the project, a modified design of the flowmeter system was analysed. The CPN model was revised to capture the modified design and by means of state space analysis it was verified that the desired properties of the flowmeter system were fulfilled for small configurations of the flowmeter system. To be able to verify larger configurations, symmetries in the flowmeter system were exploited. The symmetries in the flowmeter system made it possible to apply state spaces reduced by means of permutation symmetries [13, 12] to alleviate the state explosion problem. The analysis was done by means of the Design/CPN OE/OS Graph Tool [16]. Exploiting the symmetries made it possible to analyse configurations of the flowmeter system approaching the size of typical flowmeter systems.

3   The DANFOSS Flowmeter System

This section presents the DANFOSS flowmeter system modelled in the project. The overall architecture of the system is described first, followed by a description of the communication protocols. We describe the main issues in the design of the flowmeter system, and finally we list some of the most crucial properties which the flowmeter system is required to fulfil.

3.1   System Architecture and Communication Protocols

Figure 1 shows the overall architecture of the flowmeter system. A flowmeter system consists of one or more *modules* connected via a *Controller Area Network* (CAN) [15]. Each module consists of a number of processes called *CAN Applications* (CANAPPs) and a *driver* that interfaces the module to the CAN. Figure 1 shows an example of a flowmeter system consisting of three modules containing two, three, and four CANAPPs, respectively. Each CANAPP in the system has a small piece of local memory which holds a number of so-called *attributes.* The communication in the system consists of asynchronous message passing between the CANAPPs. This message passing allows each CANAPP to read and write the attributes of the other CANAPPs.

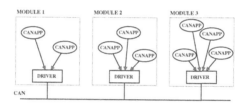

Fig. 1. Overall architecture of the flowmeter system.

The concrete location of the CANAPPs and the number of modules are flexible, e.g., in a system with four CANAPPs it is possible to put each of the CANAPPs in an individual module or to have two modules with two CANAPPs each. A typical flowmeter system consists of 3-10 CANAPPs and 1-5 modules depending on the location of the CANAPPs. It is, however, important to mention that the location of the CANAPPs is fixed during the operation of the flowmeter system, i.e., it is not possible for one CANAPP to migrate from one module to another module. In the rest of this paper we will use the notation $CANAPP_{(i\,j)}$ to denote CANAPP j on module i. Similarly, we will use $Driver_j$ to denote the driver on module j, and $Module_j$ to denote module j.

The communication between the CANAPPs is based on a protocol architecture with three layers. The layers constitute a collapsed form of the OSI seven layer architecture, mapping onto the physical, data link, and application layers of the OSI Reference Model [8].

All communication in the flowmeter system consists of asynchronous message passing between the CANAPPs. We will illustrate a representative communication pattern shortly. The different message types are listed in Table 1. Basically the messages can be divided into two groups depending on their use.

Table 1. The message types and their function.

Message	Function
ReadRequest	Request to read an attribute of another CANAPP
ReadResponse	Response to a ReadRequest containing the value of an attribute
WriteRequest	Request to write an attribute of another CANAPP
WriteResponse	Response to a WriteRequest indicating a change of an attribute
Broadcast	Distribute a value to all other CANAPPs in the system (without acknowledgement from each driver)
Event	Distribute a value to all other CANAPPs in the system (with acknowledgement from each driver)

The *Read* messages (ReadRequest and ReadResponse) and the *Write* messages (WriteRequest and WriteResponse) are used in point-to-point communication between the CANAPPs, whereas the Broadcast and Event messages are used in broadcast communication. In point-to-point Read (Write) communication, a ReadRequest (WriteRequest) is sent to read (write) the attribute of another CANAPP. A ReadResponse (WriteResponse) is then generated by the receiving CANAPP containing either the value of the attribute (in case of a read message) or just a value indicating an accept (in case of a write message). The delivery of the point-to-point messages is guaranteed by use of an acknowledgement mechanism in the communication between the drivers. The Broadcast and Event messages are used to distribute information to all CANAPPs in the flowmeter system – either as a Broadcast message without any guaranty of delivery, or as an Event message, with guaranteed delivery.

A typical point-to-point communication in a flowmeter system consisting of two CANAPPs located in different modules is shown in the MSC in Fig. 2. The MSC illustrates a write communication between two CANAPPs located in different modules. The MSC is identical to the kind of MSCs used intensively in phase one of the project.

The MSC contains a vertical line for each of the two CANAPPs, the drivers of the modules, and the CAN. The arrows between the vertical lines correspond to messages sent in the flowmeter system. The communication sequence considered corresponds to a write request/response communication to $CANAPP_{(2\ 1)}$ initiated by $CANAPP_{(1\ 1)}$, and causes the following sequence of events to occur. The numbers in the list below correspond to the numbers found below the arrows and next to the mark on the rightmost line of Fig. 2.

1. $CANAPP_{(1\ 1)}$ generates a WriteRequest message to be sent to $CANAPP_{(2\ 1)}$ to write the value of an attribute, and passes the message to $Driver_1$.
2. $Driver_1$ sends the message on the CAN.
3. $Driver_2$ receives the message from the CAN.
4. $Driver_2$ generates an acknowledgement which is sent on the CAN.
5. $Driver_1$ receives the acknowledgement from the CAN.

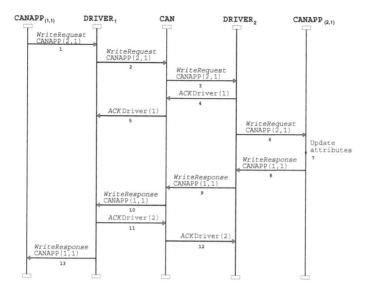

Fig. 2. Write communication based on the request/response mechanism.

6. Driver$_2$ delivers the WriteRequest message to CANAPP$_{(2\ 1)}$, which accepts to have its attribute written by CANAPP$_{(1\ 1)}$.
7. CANAPP$_{(2\ 1)}$ updates the requested attribute.
8. CANAPP$_{(2\ 1)}$ now generates a WriteResponse message indicating that the attribute has been updated. The response is handed to Driver$_2$.
9. Driver$_2$ sends the message on the CAN.
10. Driver$_1$ receives the WriteResponse message from the CAN.
11. Driver$_1$ generates an acknowledgement which is sent on the CAN.
12. Driver$_2$ receives the acknowledgement from the CAN.
13. Driver$_1$ delivers the WriteResponse message from CANAPP$_{(2\ 1)}$ to CANAPP$_{(1\ 1)}$.

3.2   CANAPP Design Patterns

When two CANAPPs communicate they do it in an asynchronous way as illustrated above. When receiving a ReadRequest (WriteRequest) a ReadResponse (WriteResponse) is generated and sent back to the sender of the request. When designing the flowmeter system DANFOSS used the OCTOPUS method [1]. This method describes two different design alternatives/patterns for such asynchronous communication between objects (processes). The two design alternatives are called *Internal Wait Point* and *Primary Wait Point*. Below we briefly describe each of the two design patterns.

Internal Wait Point (IWP) approach. When a CANAPP sends a request the execution of the CANAPP is blocked. If the CANAPP is located in the same module as other CANAPPs, then all CANAPPs in the module are blocked. The CANAPPs are not released until a response matching the request has been received.

Primary Wait Point (PWP) approach. In this approach the response message is treated as an event. The requesting CANAPP and all other CANAPPs residing in the same module are not blocked as in the IWP approach. This means that a CANAPP can receive a request from another CANAPP even if it is temporarily waiting for a response to a previously sent request. This is typically implemented by means of two threads.

As identified in practical tests at DANFOSS, the CANAPPs need to be carefully designed, e.g., to avoid deadlocks. This leads to the formulation of three crucial properties which the final design of the flowmeter system is required to posses. The properties are given here in an informal way. We will show in Sect. 5 how they can be translated into dynamic properties of the constructed CPN model.

Absence of Deadlocks. It should not be possible to bring the flowmeter system into a situation in which all the CANAPPs in the flowmeter system are blocked.

Absence of Attribute Corruption. It is important for the correct operation of the flowmeter system that when a CANAPP has initiated a request its attributes are not modified before the request has been completed.

Topology Independence. One of main advantages envisioned for the flowmeter system was that it could easily be adapted to customers' needs by providing flexibility in the choice as to which and how many CANAPPs should go into the modules. It is therefore important that the two properties above are valid independently of how the CANAPPs are distributed in the modules.

4   CPN Model of the Flowmeter System

This section presents selected parts of the CPN model of the flowmeter system. The purpose of this section is twofold. Firstly, to provide an overview of the CPN model, and secondly, to give an idea of the complexity of the CPN model and the abstraction level chosen. The CPN model has been put together in such a way that it captures both a design based on the Primary Wait Point (PWP) approach and a design based on the Internal Wait Point (IWP) approach. The configuration, i.e., the distribution of CANAPPs, is captured in the initial marking. Hence, the two design alternatives and different configurations of the flowmeter system can be analysed using only one CPN model but with different initial markings.

4.1   CPN Model Overview

Figure 3 gives an overview of the CPN model by showing how it has been hierarchically structured into 15 modules (subnets). The subnets of the model are in CPN terminology referred to as *pages*. Each node in Fig. 3 represents a page of the CPN model. An arc between two nodes indicates that the source node contains a so-called *substitution transition* whose behaviour is described on the page represented by the destination node.

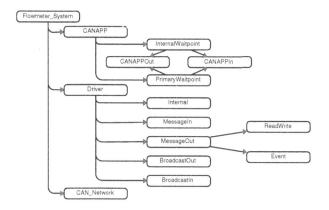

Fig. 3. The hierarchy page.

The CPN model consists of three main parts. One part modelling the CANAPPs consisting of the pages CANAPP, PrimaryWaitPoint, InternalWaitPoint, CANAPPIn, and CANAPPOut. A second part modelling the drivers consisting of the pages Driver, Internal, MessageIn, MessageOut, ReadWrite, Event, BroadcastOut, and BroadcastIn. The third part modelling the CAN consists of page CAN_Network.

Page Flowmeter_System, depicted in Fig. 4, is the top-most page of the CPN model and provides the most abstract view on the CPN model. The page consists of three substitution transitions corresponding to the three layers of the protocol architecture of the flowmeter system. Between each of the layers there are a number of places modelling the buffers between the layers. The detailed behaviour of CAN_Network, Driver, and CANAPP is modelled on subpages associated with the substitution transitions. In the following we will explain in more detail how the CANAPPs are modelled in the design based on the PWP approach. The modelling of the IWP approach is similar in complexity.

4.2   Modelling of the CANAPPs

Figure 5 depicts the page PrimaryWaitPoint which is the top-most page in the part of the model concerned with the CANAPPs in the PWP approach. The modelling of the CANAPPs has been split in two parts. The part of the CANAPP responsible for sending requests and receiving responses is modelled by the substitution transition CANAPPOut. The part of the CANAPP responsible for handling incoming requests and generating responses is modelled by the substitution

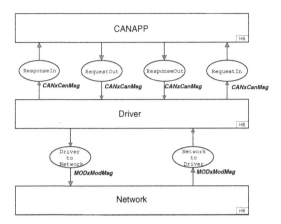

Fig. 4. The page Flowmeter_System.

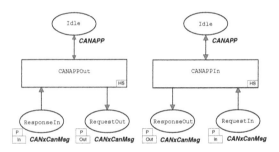

Fig. 5. The page PrimaryWaitPoint.

transition CANAPPIn. The places in the lower part of the page model the buffers between the CANAPP layer and the driver layer. The two Idle places are used to model the initial state of the two threads in the CANAPP.

Figure 6 depicts the page CANAPPOut which is the subpage of the substitution transition CANAPPOut shown in Fig. 5. This page is an example of a page at the lowest level of the CPN model. It models the control flow in the part of the CANAPP generating requests to the other CANAPPs and awaiting responses. The sending of a request is modelled by the transition Request, which causes the CANAPP to change its state from being Idle to Waiting, and pass the message to the driver by putting it into the buffer modelled by the place RequestOut. This corresponds to event 1 in Fig. 2. The driver in the module will remove the message from the place RequestOut and deliver it to the destination. When the response returns, the corresponding message is put in the buffer modelled by the place ResponseIn. This corresponds to events 2-12 in Fig. 2. The actual reception of a response is modelled by the transition Con rm. An occurrence of this transition removes the response from the place ResponseIn, updates the attributes of the CANAPP, modelled by place **Attributes**, and causes the CANAPP to change its state from **Waiting** to **Idle**. This corresponds to event 13 in Fig. 2.

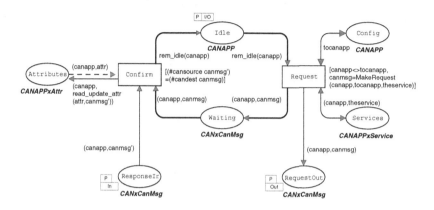

Fig. 6. The page CANAPPOut.

The places **Con g** and **Services** are used to model configuration information which can be accessed by the CANAPP. Page CANAPPIn, modelling the handling of an incoming request and the sending of a response, is similar to the CANAPPOut page. The two parts of the CANAPPs run in parallel reflecting that in the PWP approach the CANAPPs are able to make outgoing requests as well as handle incoming requests concurrently.

5   Analysis of Two Initial Design Proposals

This section describes how the two design proposals based on the PWP and the IWP approaches have been analysed by means of state spaces and the Design/CPN Occurrence Graph Tool (OG Tool) [16]. This section also presents the modifications made to the CPN model in order to make it suited for state space analysis, it states the goals of the analysis, and it presents the obtained results.

To make the CPN model suited for state space analysis some adjustments were needed. Essentially two modifications of the CPN model were made. The modelling of the attributes in the CANAPPs was simplified such that the concrete values of the attributes were no longer modelled. This simplification is justified since no parts of the CPN model make choices depending on concrete values of the attributes. When the concrete values of the attributes are not modelled, then the read and write messages become similar since their effect is no longer modelled. The **Broadcast** and **Event** messages were not considered. Handling these messages in the context of state space analysis would require more extensive modifications to the CPN model in order to obtain a CPN model with a finite state space. It was therefore decided to leave these two kinds of messages out.

5.1   Analysis Goals

The primary goal of the state space analysis was to investigate whether the two design proposals fulfilled the three requirements stated in the end of Sect. 3. The first step in order to investigate this was to translate them into dynamic properties of the CPN model. This makes it possible to formulate the requirements as *queries* in the OG Tool. The answers to the queries can then be automatically determined by the OG Tool when a state space has been generated. Below we show how to translate each of the three properties into queries which can be invoked in the OG Tool.

Absence of Deadlocks. This property can be formulated as the absence of reachable *dead markings* in the CPN model. A dead marking is a marking without enabled transitions, and it is an example of a standard dynamic property of a CPN model. The OG Tool has a built-in standard query function ListDead-Markings, which lists the reachable dead markings (if such markings exist).

Absence of Attribute Corruption. When a CANAPP has initiated a request it is not allowed to start handling an incoming request before the request has been completed, i.e., a response has been received. This property cannot easily be formulated as a standard dynamic property of the CPN model. However, it can be conveniently formulated using temporal logic [6]. The OG Tool library ASK-CTL [16] makes it possible to make queries formulated in a state and action oriented variant of CTL [3]. That the attributes cannot be corrupted for a given $CANAPP_{(i\,j)}$ can be expressed as the following action-based CTL formula. An explanation is given below.

AG((Request $canapp$ =$CANAPP_{(i\,j)}$)
A(((Indication $canapp$=$CANAPP_{(i\,j)}$))U(Con rm $canapp$=$CANAPP_{(i\,j)}$))

The formula states that whenever (denoted AG) the transition Request occurs in a binding corresponding to $CANAPP_{(i\,j)}$, then in all futures (denoted A) the transition Indication cannot occur in a binding corresponding to $CANAPP_{(i\,j)}$ until (denoted U) the transition Con rm has occurred in a binding corresponding to $CANAPP_{(i\,j)}$. An occurrence of the transition Request in a binding corresponding to $CANAPP_{(i\,j)}$ has been written as (Request $canapp$ =$CANAPP_{(i\,j)}$). The binding of Indication and Con rm is written in a similar way. An occurrence of the transition Request (see Fig. 6) models the start of a request, an occurrence of transition Indication models the start of handling an incoming request, and an occurrence of transition Con rm (see Fig. 6) models the reception of a response.

Topology Independence. The two properties above should be valid for the system independently of how the CANAPPs are placed in the modules in the system. This property can therefore be investigated by analysing different configurations of the flowmeter system. Investigating different configurations can be done by simply changing the initial marking of the CPN model. Hence, topology independence can be investigated by constructing state spaces for different initial markings.

5.2 Analysis Results

State spaces have been constructed for a number of configurations of the flowmeter system. Table 2 gives some statistical information on the state spaces for different configurations. The Con guration column depicts the configuration in question. We have used a graphical notation to indicate the configuration considered. For instance, the second row in the right-most part of Table 2 gives statistics for a configuration with two modules with one and two CANAPPs, respectively. The WP column shows which wait point approach was considered. The Nodes and Arcs columns give the number of nodes and arcs in the state space, respectively. The Time column gives the time in seconds it took to generate the state space. All state spaces were generated on a Sun Workstation with 512 MB of memory. It is worth observing that the IWP approach gives smaller state spaces than the PWP approach. The reason for this is that in the IWP approach the CANAPPs block after transmission of a message, and hence there is less concurrency between the CANAPPs compared to the PWP approach. With the available computing power and due to the state explosion problem, it was not possible to construct the full state space for large configurations in the PWP approach. Therefore, in some of the configurations only external communication (communication between CANAPPs located in different modules) is analysed. An asterisk ( ) indicates that only external communication is analysed in the given configuration. However, even the analysis of small configurations of the flowmeter system showed that neither of the two design alternatives satisfies the required properties.

Table 2. Generation statistics for full state spaces.

Configuration	WP	Nodes	Arcs	Time	Configuration	WP	Nodes	Arcs	Time
	PWP	37	73	1		PWP	133	281	1
	IWP	6	5	1		IWP	80	133	1
	PWP	1,299	4,189	7		PWP	5,581	18,707	44
	IWP	14	13	1		IWP	446	869	1
	PWP	19,770	74,941	223		PWP	62,605	276,721	1,484
	IWP	11,451	37,513	109		IWP	26	25	1
*	PWP	37,825	146,721	1,499		PWP	44,470	190,945	544
	IWP	2,266	4,841	15	*	IWP	3,866	8,473	33

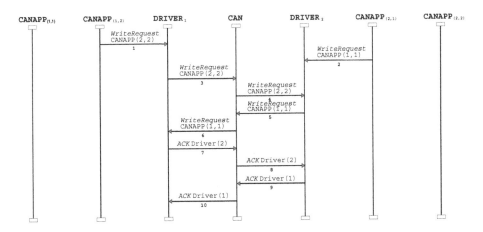

Fig. 7. Visualisation of a path leading to a deadlock.

In all the configurations considered, the analysis revealed that the design based on the IWP approach had several deadlocks. Below we will concentrate on a specific deadlock situation found in the analysis of the IWP approach. The example given is in a configuration of the flowmeter system consisting of two modules each with two CANAPPs. Figure 7 visualises a path/execution leading to a dead marking of the CPN model. This marking was identified using the query function ListDeadMarkings followed by the use of another query function which is able to find one of the shortest paths (occurrence sequences) in the state space leading from the initial marking to a specified marking. Once such an occurrence sequence has been found, it is straightforward to visualise it using MSCs. The MSC in Fig. 7 shows the following sequence of events.

1. CANAPP$_{(1\,2)}$ sends a WriteRequest to CANAPP$_{(2\,2)}$. As the model is based on the IWP approach both CANAPP$_{(1\,1)}$ and CANAPP$_{(1\,2)}$ are blocked until a WriteResponse is received.
2. CANAPP$_{(2\,1)}$ sends a WriteRequest to CANAPP$_{(1\,1)}$. Both CANAPP$_{(2\,1)}$ and CANAPP$_{(2\,2)}$ are blocked until a WriteResponse is received.
3-10. The messages are sent over the CAN. All of the CANAPPs in the flowmeter system are blocked. Thus, neither CANAPP$_{(1\,1)}$ nor CANAPP$_{(2\,2)}$ can receive the WriteRequests, and a deadlock has occurred.

All deadlocks are of course unwanted, but especially the kind of deadlock visualised in Fig. 7 is problematic in a flowmeter system. The reason is that it should be possible to distribute the CANAPPs freely in the modules in the system and the concrete location of the CANAPPs should not affect the correctness of the system. If CANAPP$_{(2\,2)}$ in the example above instead is placed in a separate module (if the flowmeter system consisted of three modules instead of two) no

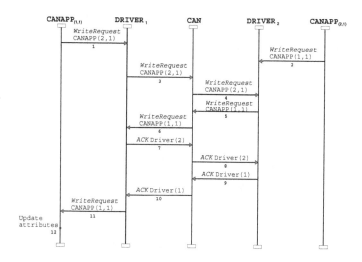

Fig. 8. Visualisation of a path leading to corruption of an attribute.

deadlock would occur for the above communication pattern. The above also illustrates topology independence as a non-trivial property of the flowmeter system. Analysis of the PWP approach showed that none of the configurations analysed had any nodes corresponding to dead markings of the CPN model. Therefore, all the configurations analysed in the PWP approach fulfil the absence of deadlocks property.

The analysis of the flowmeter system regarding the absence of attribute corruption showed that all configurations analysed for the IWP approach fulfil this property. However, the analysis in the case of the PWP approach showed that this approach does *not* fulfil the property. Figure 8 visualises a path/execution to a node in the state space representing a marking in which an attribute has been corrupted. The configuration of the flowmeter system considered consists of two modules each with one CANAPP. The MSC shows the following sequence of events.

1. $CANAPP_{(1\ 1)}$ sends a WriteRequest to write an attribute of $CANAPP_{(2\ 1)}$. $CANAPP_{(1\ 1)}$ now waits for a response but is not blocked.
2. $CANAPP_{(2\ 1)}$ sends a WriteRequest to write an attribute of $CANAPP_{(1\ 1)}$.
3-10. The messages and corresponding acknowledgements are sent over the CAN.
11. $CANAPP_{(1\ 1)}$ receives the WriteRequest message from $CANAPP_{(2\ 1)}$.
12. $CANAPP_{(1\ 1)}$ changes its attributes according to the value in the WriteRequest from $CANAPP_{(2\ 1)}$. A corruption has occurred.

To summarise the results of the analysis: the design based on the IWP approach violates the first property (absence of deadlocks) but fulfils the second (absence of attribute corruption). The design based on the PWP approach on the other hand fulfils the first property but violates the second. Thus, neither of the two design proposals are suitable for a final implementation of the flowmeter system. In the next section we will consider a variant of the PWP approach which avoids the problem of attribute corruption identified above.

6   Analysis of a Third Design Proposal

The identification of the deadlocks and attribute corruption in the two initial design proposals naturally leads to the suggestion of a third design proposal which resolves the identified problems. In this section we present such a design proposal based on the PWP approach which overcomes the problem with attribute corruption. State space analysis of the first two design proposals had shown that the state space explosion problem prohibited analysis of larger configurations. It was therefore decided to attempt to use a state space reduction method in order to alleviate the state explosion problem when analysing the third design proposal. It was decided to use the symmetry method [13,12] since the flowmeter system is made up of a number of identical components (the modules, drivers, and CANAPPs) whose behaviour are identical. Moreover, the symmetry method is supported by the Design/CPN OE/OS Graph Tool (OE/OS Tool) [16], and as we will see, the method preserves the properties which we want to verify. In this section we briefly present the idea in the modified design, and we then explain how it was analysed using state spaces reduced by taking advantage of the symmetries in the flowmeter system.

The basic idea in the new design proposal is to introduce a possibility for the CANAPPs to send a negative response to a request message if the CANAPP is currently waiting for a response to a previously sent request. Therefore, if the CANAPP is not in the process of performing a request when the incoming request arrives, it will access the attribute and send a positive response. If the CANAPP is in the process of performing a request, then it will not access the attribute but instead send a negative response. This can be reflected in the CPN model by adding a place modelling a shared variable between the two threads of the CANAPP (see Fig. 5). This variable can then be used by the thread handling the incoming requests to decide whether it is safe to access the attributes.

6.1   Symmetry Specification

For capturing the symmetries in the flowmeter system *state spaces with permutation symmetries* (OS-graphs in [12]) were applied. The OE/OS Tool supports state spaces with permutation symmetries based on user supplied symmetry specifications. This means that the user of the tool is required to provide the permutation symmetries, and the tool then uses these as a basis for the reduction. A symmetry specification consists of assigning *symmetry groups* of permutations to the *atomic colour sets* of the CPN model. An atomic colour set is a colour set defined without reference to other colour sets. The symmetry group determines how the colours of the atomic colour sets are allowed to be permuted, and in turn induces permutation symmetries on the markings and the binding elements of the CPN model. The idea of state spaces with permutation symmetries is to construct equivalence classes of markings which are symmetric in the sense that they can be obtained from each other by one of the permutation symmetries. Instead of representing all markings it suffices to store a representative for each equivalence class containing a reachable marking. A similar remark applies to binding elements. In this way a condensed state space is obtained which is typically orders of magnitude smaller than the full state space.

For the flowmeter system the symmetry specification should capture the symmetry in the CANAPPs as well as in the modules. To illustrate the symmetry specification applied for the flowmeter system consider Fig. 9. Figure 9 shows the initial part of the state space for a flowmeter system consisting of two modules each containing two CANAPPs. Communication between CANAPPs in the same module has been disabled for presentation purposes. Node 1 corresponds to the initial marking and has eight immediate successor nodes corresponding to the possible requests which can be initiated in the system (each CANAPP can initiate a request to the two CANAPPs on the other module). For node n we will denote the corresponding marking M_n.

For the nodes 2, 3, 7, and 9 we have indicated in the associated dashed box what communication has been initiated, e.g., node 9 corresponds to a state of the system in which $\text{CANAPP}_{(1\ 1)}$ has initiated a request towards $\text{CANAPP}_{(2\ 1)}$. The symmetry specification is based on the observation that M_9 is symmetric to M_7 except for a permutation which swaps the two CANAPPs in Module$_2$, In a similar way it can be observed that M_2 can be obtained from M_9 by a permutation which swaps the two modules, and M_3 can be obtained from M_9 by a permutation which swaps the two modules and which swaps the two CANAPPs in Module$_2$. Furthermore, it is possible to obtain M_4, M_5, M_6, and M_8 from M_9 by permutation of the CANAPPs and the modules. If such symmetric markings are grouped into equivalence classes, it is possible to represent this initial fragment of the full state space by the condensed state space shown in Fig. 10. Here node 1 represents the equivalence class containing the initial marking only, and node 2 represents the equivalence class containing the markings M_2 to M_9. The marking M_9 can be chosen as a representative for this equivalence class. The successors of M_9 are grouped into equivalence classes in a similar way. Altogether this means that the 65 markings which can be reached in two steps from the initial marking can be represented using only 8 nodes in the condensed state space. In summary, the symmetry specification used for the flowmeter system allows permutation of CANAPPs within the same module, and it allows permutation of modules containing the same number of CANAPPs.

6.2 Consistency Check

Since the OE/OS Tool supports user supplied symmetry specifications, a *consistency check* is needed to ensure that the symmetries supplied are symmetries which are actually present in the CPN model. This amounts to checking that the initial marking, the guards, and the arc expressions of the CPN model are symmetric in a way precisely defined in [12]. The currently released version of the OE/OS Tool does not support an automatic check for consistency, but as part of the project we have developed an extension of the tool which supports a semi-automatic check for consistency. It is based on a combination of syntactical and semantical checks. The semantic check is based on evaluating net inscriptions in all possible bindings. As a consequence the semantic check is time-consuming, but together with the syntactical check it is possible to make a fully automatic check for consistency for the CPN model of the flowmeter system.

6.3 Analysis Results

The analysis focuses on the same three properties as for the two initial design proposals. For investigating the absence of dead markings a built-in query function of the OE/OS Tool was used which lists the equivalence classes containing the reachable dead markings (if such equivalence classes exist). The query related to the absence of attribute corruption has to be modified to take into account that the transition Indication corresponding to the reception of an incoming request can now occur in two modes depending on whether the request is accepted or rejected. The modified query is shown below and is identical to the original query except that it is only required that the Indication transition does not occur in a binding where the request is accepted, i.e., the variable mode is bound to accept.

$$\mathsf{AG}((\mathsf{Request}\ \ canapp = \mathrm{CANAPP}_{(i\ j)}\ )$$
$$\mathsf{A}((\ (\mathsf{Indication}\ \ canapp = \mathrm{CANAPP}_{(i\ j)}\ mode = accept\ ))\ \mathsf{U}$$
$$(\mathsf{Con\ rm}\ \ canapp = \mathrm{CANAPP}_{(i\ j)}\ ))$$

Another problem which has to be resolved with the above query is that it refers to occurrence sequences related to a specific CANAPP. Since we allow permutations of CANAPPs this property is not a priori preserved by the symmetry reduction. However, it was shown in [7, 9] that symmetry reduction preserves the truth value of a CTL formula if the truth value of its atomic state propositions are invariant under the permutation symmetries. The absence of attribute corruption can therefore be checked from the symmetry reduced state space by checking the property for a $\mathrm{CANAPP}_{(i\ j)}$ for which permutation is not allowed. CANAPPs

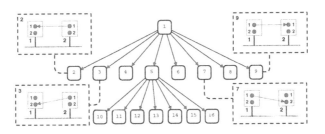

Fig. 9. Initial fragment of the full state space.

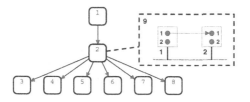

Fig. 10. Initial fragment of the condensed state space.

Configuration	Full State Space				Condensed State Space				Ratio	Sym
	Nodes	Arcs	Time	T/N	Nodes	Arcs	Time	T/N		
	19,970	74,941	223	0.011	3,410	12,478	121	0.035	5.7	6
	62,605	276,721	1,484	0.024	2,776	4,767	327	0.118	22.5	24
	295,965	1,329,113	–	–	37,156	166,316	4,123	0.111	7.9	8
	319,337	1,460,785	–	–	54,019	245,139	2,523	0.047	5.9	6
*	456,174	2,148,585	–	–	114,370	537,857	7,428	0.065	3.9	4
*	44,470	190,995	544	0.012	2,971	11,075	197	0.066	14.9	24
*	578,376	2,746,401	–	–	49,848	234,431	15,214	0.305	11.6	12
*	209,629	1,044,361	–	–	4,759	18,699	1,438	0.302	44.0	120

Table 3. Generation statistics for full and condensed state spaces.

which cannot be permuted are present in configurations of the flowmeter system containing a module with a single CANAPP. An alternative approach to this is to strengthen the symmetry specification such that the CANAPP in question cannot be permuted.

Table 3 gives some statistical information on the generation of the condensed state space for some selected representative configurations. For comparison it also gives the size of the full state space. The size of the full state space has been calculated from the condensed state space in the cases where the full state space could not be generated given the available computing power. The Time column gives the time in seconds it took to generate the state space. T/N gives the processing time (in seconds) per node in the state space. The Ratio gives the reduction obtained in the number of nodes. The column Sym gives the number of permutation symmetries for the configurations, which is an upper limit on the reduction which can be obtained. The condensed state spaces were generated on a Sun Workstation with 512 MB of memory.

As can be seen from Table 3, the use of symmetries yielded significant reductions in the size of the state spaces and allowed us to consider configurations of the flowmeter system which could not be handled with full state spaces. This is what we would have expected. It is, however, also worth observing that the generation of condensed state spaces was faster than the generation of the full state spaces. Even though we only have three observations, they indicate what seems to be a general fact: the time that is lost on a more expensive test on equivalence of markings and binding elements, is accounted for by having fewer nodes and arcs to generate; and also to compare with before a new node or arc can be inserted in the state space.

We were able to verify the absence of reachable dead markings in all configurations for which a condensed state space could be generated. Moreover, the absence of attribute corruption was verified for the configurations with a module containing a single CANAPP. In addition to these properties we were able to verify that the initial marking was a *home marking*. This means that from any reachable marking it is always possible to return to the initial marking. This is a very attractive property since it means that the flowmeter system can never be brought in a situation in which the system cannot be made to reenter its initial state. Moreover, it could be verified that the binding elements corresponding to the initiation of a request, reception of a response, handling of a request, and transmission of a response were *live*. This means that all CANAPPs in the system always have the possibility of completing requests and handling incoming requests.

7   Conclusions

In this paper we have presented the main results of a project in which CP-nets was put into practical use in an industrial setting at DANFOSS for the modelling and analysis of a flowmeter system. The project was divided into three phases. During the first phase the graphical nature of Petri Nets and the capability to visualise the behaviour of a CPN model were extremely important tools in the process of validating that the CPN model correctly reflected the intended design of the flowmeter system. This observation is consistent with observations made in other industrial projects such as [5].

The second phase clearly showed the practical limits of using full state spaces in an industrial setting, since the size of the state space started to grow rapidly once larger configurations were considered. Given the available computing power it was only possible to analyse configurations with up to 3-4 CANAPPs. However, even the analysis of small configurations identified errors in the proposed designs. This also demonstrates that errors in systems tend to manifest themselves in even small configurations of the system.

The application of the symmetry method made it possible to verify configurations of the flowmeter system containing up to six CANAPPs. Since concrete configurations may contain more CANAPPs (typically up to 10), it is relevant to ask whether anything could have been done to handle even larger configurations, i.e., whether other reduction methods could have been applied in addition to the symmetry method. Since the CANAPPs in the system are asynchronous it would be obvious to consider the stubborn set method [17]. It was proved in [18] that the stubborn set method can be combined with the symmetry method.

The stubborn set method was however not applied in the project since the tool support for stubborn sets in Design/CPN is currently not mature enough to be used on models of the size considered in this project.

In conclusion we believe that this project has demonstrated that CP-nets and the state space method may indeed be relevant and valuable methods to be used in the design process of not only the flowmeter system but also in future products at DANFOSS which was the main goal of the project.

Acknowledgements. We would like to thank DANFOSS INSTRUMENTATION, especially Arne Peters, for their contributions to this project.

References

1. M. Awad, J. Kuusela, and J. Ziegler. *Object-Oriented Technology for Real-Time Systems: A Practical Approach using OMT and Fusion.* Prentice-Hall, 1996.
2. ITU (CCITT). Recommendation Z.120: MSC. Technical report, International Telecommunication Union, 1992.
3. A. Cheng, S. Christensen, and K.H. Mortensen. Model Checking Coloured Petri Nets Exploiting Strongly Connected Components. In M.P. Spathopoulos, R. Smedinga, and P. Kozák, editors, *Proceedings WODES '96.* Institution of Electrical Engineers, Computing and Control Division, Edinburgh, UK, 1996.
4. S. Christensen. *Message Sequence Charts. User's Manual,* January 1997. Available from http://www.daimi.au.dk/designCPN/.
5. S. Christensen and J.B. Jørgensen. Analysis of Bang and Olufsen's BeoLink Audio/Video System Using Coloured Petri Nets. In P. Azéma and G. Balbo, editors, *Proceedings of ICATPN'97,* volume 1248 of *Lecture Notes in Computer Science,* pages 387–406. Springer-Verlag, 1997.
6. E.M. Clarke, E.A. Emerson, and A.P. Sistla. Automatic Verification of Finite State Concurrent Systems using Temporal Logic. *ACM Transactions on Programming Languages and Systems,* 8(2):244–263, 1986.
7. E.M. Clarke, R. Enders, T. Filkorn, and S. Jha. Exploiting Symmetries in Temporal Model Logic Model Checking. *Formal Methods in System Design,* 9, 1996.
8. J.D. Day and H. Zimmermann. The OSI Reference Model. *Proceedings of the IEEE,* 71, December 1983.
9. E.A. Emerson and A. Prasad Sistla. Symmetry and Model Checking. *Formal Methods in System Design,* 9, 1996.
10. D.J. Floreani, J. Billington, and A. Dadej. Designing and Verifying a Communications Gateway Using Coloured Petri Nets and Design/CPN. In J. Billington and W. Reisig, editors, *Proceedings of ICATPN'96,* volume 1091 of *Lecture Notes in Computer Science,* pages 153–171. Springer-Verlag, 1996.
11. K. Jensen. *Coloured Petri Nets. Basic Concepts, Analysis Methods and Practical Use. Volume 1, Basic Concepts.* Monographs in Theoretical Computer Science. Springer-Verlag, 1992.
12. K. Jensen. *Coloured Petri Nets. Basic Concepts, Analysis Methods and Practical Use. Volume 2, Analysis Methods.* Monographs in Theoretical Computer Science. Springer-Verlag, 1994.
13. K. Jensen. Condensed State Spaces for Symmetrical Coloured Petri Nets. *Formal Methods in System Design,* 9, 1996.

14. L.M. Kristensen, S. Christensen, and K. Jensen. The Practitioner's Guide to Coloured Petri Nets. *International Journal on Software Tools for Technology Transfer*, 2(2):98–132, December 1998.
15. W. Lawrenz. *CAN Contoller Area Network, Grundlagen und Praxis*. Hüttig Buch Verlag, Heidelberg, 1994.
16. Design/CPN Online. http://www.daimi.au.dk/designCPN/.
17. A. Valmari. Error Detection by Reduced Reachability Graph Generation. In *Proceedings of the 9th European Workshop on Application and Theory of Petri Nets*, pages 95–112, 1988.
18. A. Valmari. Stubborn Sets of Coloured Petri Nets. In G. Rozenberg, editor, *Proceedings of ICATPN'91*, pages 102–121, 1991.
19. J. Xu and J. Kuusela. Analyzing the Execution Architecture of Mobile Phone Software with Colored Petri Nets. *Software Tools for Technology Transfer*, 2(2):133–143, December 1998.

Automatic Code Generation Method Based on Coloured Petri Net Models Applied on an Access Control System

Kjeld H. Mortensen

Department of Computer Science, University of Aarhus,
Aabogade 34, DK-8200 Aarhus N, Denmark
k.h.mortensen@daimi.au.dk

Abstract. In this paper we describe a method for automatic implementation of systems based on models made by means of Coloured Petri
Nets (CP-nets or CPN). The Design/CPN tool has been extended in
order to support this method. We do not describe the algorithms and
data-structures used to implement the code generation tool but rather
the context such a tool is used in.

The contribution of this work origins from the fact that the code used to
simulate the CPN model and the code used to generate the final system
implementation are identical. Hence the behaviour of the model and final
system are the same, and analysis results found by means of Design/CPN
also hold for the final running system. This is different from other CPN-
based code generation methods. Furthermore, since the method is fully
automatic the traditional manual implementation phase has been eliminated. Thus the method described in this paper dramatically reduces
development time and cost compared with prevailing system development methods where system implementation is accomplished manually.

In this paper we demonstrate that the method is usable in practice for
an industrial case, namely an access control system developed by the
Danish security company Dalcotech A/S. A CPN model was made of a
realistic access control system scenario. We describe this model and how
Dalcotech applied the automatic code generation method in order to obtain a system implementation quickly and safely. In this way Dalcotech
now has the capability to reduce the resources spent on the implementation phase.

Topics: Applications of nets to embedded systems; higher-level net models; experience with using nets, case studies; automatic code generation
from Coloured Petri Nets.

1 Introduction

Models based on Coloured Petri Nets (henceforth CP-nets or CPN) [7, 10] are
currently typically executed by means of a simulation tool, which is also the case
for the Design/CPN tool [29]. In this paper we extend this approach such that a

M. Nielsen, D. Simpson (Eds.): ICATPN 2000, LNCS 1825, pp. 367 386, 2000.

system implementation can be automatically generated in Design/CPN in order to execute CPN models directly on the target hardware in question.

In an earlier project with Dalcotech [3] we made a proof-of-concept experiment with automatic code generation for a complex alarm system [18]. Standard ML code was extracted from a CPN model and burned into two PROMs that were mounted in a prototype of the final hardware. (See [13, 17] for more information on the functional computer language ML.) The experiment suggested that automatic code generation from CPN models can be used in practice, but the experiment also showed a need to generate more efficient code and to develop integrated tool support, e.g., for debugging, testing, and external libraries.

However, automatic code generation was not the focus of that project. In fact the final implementation of the system was done by hand in C++ based on the CPN model that was designed. This was very time-consuming. Therefore it was decided to follow up on the challenges in a new project dedicated to automatic code generation. The project is called AC/DC (Automatic Code Generation from Design/CPN) [23] which is the main topic of this paper. The AC/DC project was supported by the Danish National Centre for IT-Research (CIT) [25].

The AC/DC project was initiated in April 1997 and ended successfully in April 1999. The project was carried out by a consultancy group of 4 people (2 from the University of Aarhus [26], 1 from Dalcotech [27], and 1 from DELTA [28]) and a work group of 2 people from Dalcotech and 2 people from University of Aarhus. About 2 man-years was spent on the project. The total budget was 1 million Danish kroner of which CIT contributed with half of the funding.

The purpose of the AC/DC project was to develop techniques and tools for automatic generation of code from CPN models. In this way it is possible to obtain an implementation automatically of systems that are designed by means of CP-nets and the Design/CPN tool. We wish to eliminate a time-consuming and error-prone manual implementation phase.

In the AC/DC project we applied the techniques and tools in practice. Dalcotech is expanding their product line to include building control and a CPN model was built of a realistic access control system. The system implementation was then obtained automatically by means of our code generation method based on the CPN model alone.

The outline of the paper is as follows: In Sect. 2 we describe the method for automatic code generation as it is used specifically by Dalcotech, and we describe how the method is applied on the automatic implementation of access control systems in Sect. 3. We generalise the method in Sect. 4. In Sect. 5 we relate our work with similar methods, and we conclude the paper with Sect. 6 where we also make suggestions for further research.

2   Method as Applied by Dalcotech

In this section we describe the automatic code generation method as it is used in industry by Dalcotech A/S. Their goal is to obtain automatic implementation of an access control system designs. Later in this paper (Sect. 4) we show that the method not only applies to access control systems, but is sufficiently general to cover a wide range of CPN models.

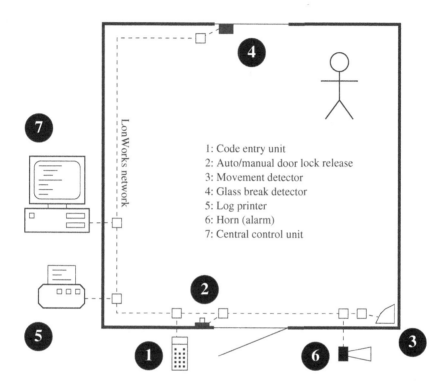

Fig. 1. Access control system scenario. White squares on the network represents the standard interface (Neurons) to various devices.

2.1   Access Control System Scenario

In order to understand the code generation method better we need to provide some background on Dalcotech's access control systems and the environment in which they are used.

Basic System Functionality: The access control system works typically as follows (see Fig. 1): A person outside the room may open the door by punching his personal code into the code entry unit (1). The control system (7) recognises the person, checks that the person is allowed in the area at this time, disables the alarm, and unlocks the door for a limited period of time (2). The system log is updated by printing status information on the log printer (5). More people may enter, each punching in a personal code, and the system updates the log. In order to leave the room a person must push the unlock door button (2) and punch in the personal code again (1). The last person leaving the room enables the alarm again. Once the room is armed the horn (6) can be triggered in case one of the entry detectors are activated (3 and 4). There are other interaction possibilities and combinations of actions, but they are left out of this paper.

LonWorks, LonTalk, and Neurons: Every device is monitored and manipulated by the central control unit in a LonWorks network.[1] LonWorks is a networking architecture typically used for process control in building automation and manufacturing environments. The network speed is 39.4 kbit/s, and the protocol used, LonTalk, is OSI based. LonTalk supports an abstract concept of *network variables*, which is a representation of the states of devices in the network. For instance, we can associate an *input* network variable with the state of a horn (actuator) or an *output* network variable with the state of a glass break detector (sensor).

Each device must be connected to the LonWorks network via a *Neuron chip* (white squares in Fig. 1). A Neuron supports the LonTalk protocol, and input/output to the device in question. A Neuron is responsible for maintaining and updating network variables. Hence if we use the LonTalk protocol to set the input network variable of the horn to "1" then the result is a sounding alarm. Network variables are configured at system startup time.

As LonTalk is rather essential in access control systems it was decided to make Design/CPN library support for this protocol such that Dalcotech can access network variables directly in the CPN model [16]. The library needs to support two situations: One is for simulation without the LonWorks network and the other is in the final running system. Hence we have built two variants of the library: A LonTalk emulation library and a run-time library. The modeller can use the emulation library to play-back a sequence of LonWorks messages, such that he or she can create message event scripts which can be used during simulation in Design/CPN. The run-time library implements the full LonTalk protocol and is used when generating the final system implementation.

Characteristics of Dalcotech's Environment: The central control unit (cf. (7) in Fig. 1) is of our primary concern in the AC/DC project as it is a standard component used in Dalcotech's products. It runs MS DOS which is something we must take into account with the code generation method. The central control unit is an embedded system which in principle runs forever. (It has a backup battery for the case of external power failure.)

LonWorks is also a standard component in Dalcotech's products, thus we must support the inclusion of a LonTalk library in the code generation method.

Another characteristic of Dalcotech's access control systems is that the system response time is allowed to be several seconds. We can live with that it

---

[1] LonWorks and LonTalk are trademarks of Echelon Corporation [30].

takes two seconds, from the point where an access code is punched in, until the door unlocks, and we allow even several seconds to pass from a window breaks until the alarm is sounding. We call these kinds of systems for "soft" real-time systems.

Finally we must be prepared for the fact that customers have individual feature requirements. One customer may request access control of five rooms while another has ten rooms, say. Some may even need connection with external security companies via a telephone line. Thus we expect that each system variant is sold in small series.

2.2   Automatic Code Generation for Access Control Systems

Figure 2 provides an overview of how Dalcotech uses the CPN methodology. The gray rectangle includes those stages involved with the method for automatic code generation. The code generation method is based on the ML computer language [13, 12], primarily because the inscription language in Design/CPN is based on ML. In the following we describe the details of the figure, which also is documented in a manual [15].

Analysis: Dalcotech models and designs the access control system by means of CP-nets and Design/CPN. In the analysis stage the model is simulated by means of the LonTalk emulation library as previously described. The result is a debugged and more reliable model of the system.

Code generation: Once we are sufficiently confident in the model, we wish to generate code. Dalcotech can virtually push a button in Design/CPN and as a result extract the ML source code necessary to support the execution of the given access control model. The source code essentially constitutes two parts, namely a generic simulation kernel and then the representation of the CPN model. The simulation kernel contains a transition scheduler and the CPN representation contains the ML implementation of the transitions and places.

Specialising to environment: We now begin a series of stages with the purpose of specialising the source code towards the final system. In this stage the generated source code is linked with the LonTalk run-time library.

Specialising of ML dialect: This stage translates between two ML dialects. The compiler, Moscow ML, we have chosen to use in the next stage can only understand a dialect of ML which does not support modules. Thus we need to unfold the code by means of a "flattener" tool called MLton [22].

Generation of system to central controller: The final specialisation stage creates the executable program which is going to run on the central control unit. The program is generated on the MS DOS platform where Moscow ML [19] is the only ML compiler available.

In the next section, where we describe the access control model, we also discuss the primary benefits of the method for Dalcotech.

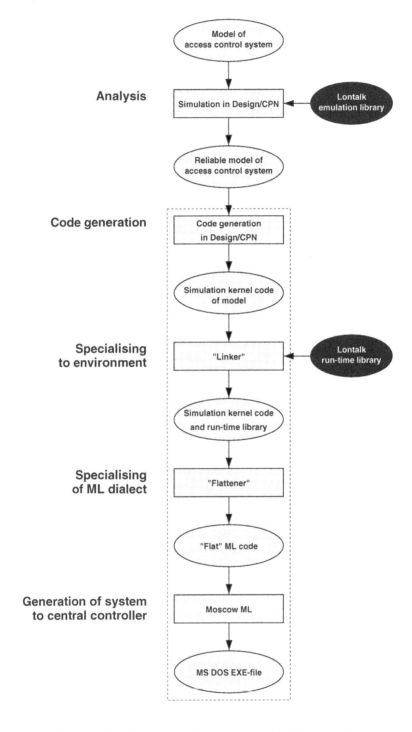

Fig. 2. Overview of method as applied by Dalcotech.

3 CPN Model of Access Control System

In this section we present the access control model which was made by Dalcotech in the AC/DC project. We demonstrate how the code generation method, described in the previous section, was successfully applied, and we show that the model architecture is flexible in case a customer requests new features.

3.1 Selected Parts of Model

The model was built mainly by two people from Dalcotech, by means of the Design/CPN tool, which took less than two man-months including specification, design/modelling, analysis, and code generation. People from the consultancy group were only involved in two model reviews. The reviews were very useful and effective means for improving the modelling strategy and architecture. The first review resulted in guidelines for continuing the modelling work while the second review was used for minor adjustments in order to finish the model.

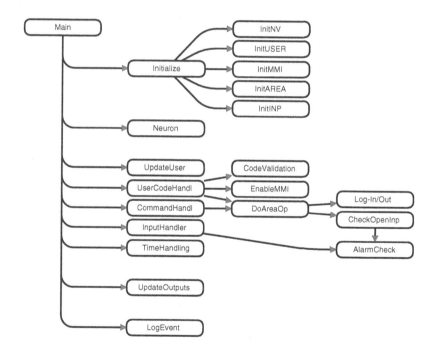

Fig. 3. Hierarchy page of CPN model.

Hierarchical Structure: The structure of the CPN model of the access control system is depicted in Fig. 3. The model is of moderate size, and consists of 21 pages and 70 transitions. There are three important parts in the model, namely initialisation (`Initialize`), Neuron (`Neuron`), and event handler (`Main`). They are described in the following sections.

Initialisation: The model is initialised in several steps in a sequential fashion. We do not show any pages from the model as each step is simply modelled with one transition (a design choice made by the Dalcotech modellers).

InitNV: All network variables are initialised in order to relate each network variable with a device in the network. The configuration is flexible within the range of the number of rooms to cover and the number of users. The CPN model structure does not need to change in case the configuration is changed.

InitUSER: The user database is initialised with names related with entry codes and where people are allowed to be and when.

InitMMI: The database containing information about human-computer interfaces, such as the code entry unit, are initialised with logic for managing enabling and disabling of alarms in case the correct code is punched.

InitAREA: Finally we initialised the status of the areas. All areas a initially armed.

InitINP: The internal representation of the states of input devices are initialised to off.

After this sequence the configuration data is extracted from the initialised databases in the model and sent to the network.

Neuron: The Neuron is a standard component in LonWorks based systems. Care should be taken to model such a component as it is expected to be reused in other models of LonWorks systems. The model made by Dalcotech is depicted in Fig. 4. Relevant colour sets are shown below:

```
colorset NV_UPDATE = product
    NV_IDX                 (* LonWorks Network variable index *)
  * NV_VAL;                (* LonWorks Network variable value *)
colorset LON_EVENT = union
    NO_EVENT               (* No events from LonWorks *)
  + UPDATE: NV_UPDATE      (* Update of network variable *)
  + ...
colorset LON_TIMED_EVENT = product
    LON_EVENT              (* LonWorks event *)
  * INT;                   (* Time passed since last event read *)
```

As with the physical Neuron, the model is divided into two parts: A device driver part (left) and a part providing an interface to the application (right). The device driver part just makes basic input/output calls to the LonTalk library which is the basic interface to the network. We could have modelled the

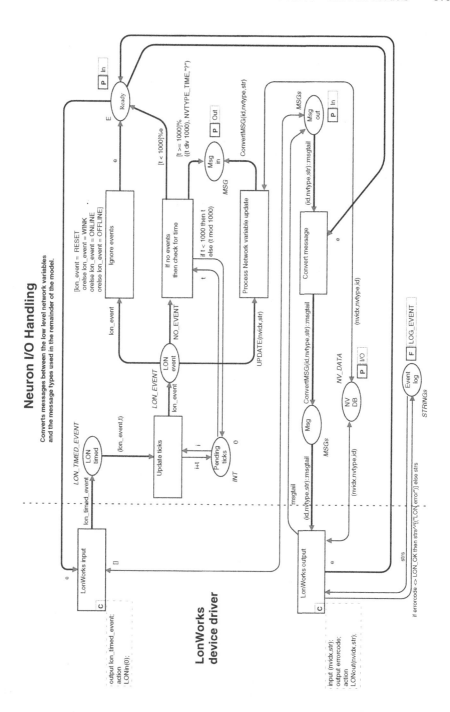

Fig. 4. CPN page modelling the Neuron.

LonTalk protocol, but time constraints did not permit to do so as the protocol is rather complicated. Instead code segments of the transitions LonWorks input and LonWorks output are used to receive and send messages respectively. Code segments are typically used to make library calls which have side-effects. (Design/CPN does not allow side effects in arc inscriptions or guards.) The LonTalk library functions LONin and LONout are used for this in the code segments. The LONin call is blocking with a timeout, i.e., if no message appears on the network within timeout, then the call returns NO_EVENT.

The application part converts network messages to higher level messages (events) appropriate for the application (upper half) and also converts higher level messages to network messages (lower half). In case no message (NO_EVENT) is received from the network then timer information is propagated. This information can be used elsewhere in the access control model for managing user access times and alarm timeouts. Notice that the model does not use timed CP-nets but instead stores information on the (clock) time passed since the last event was read (LON_TIMED_EVENT).

We could imagine that the model would benefit from using timed CP-nets. Timeouts and delays could then be expressed directly as modelling primitives and then potentially make the model simpler. The timing logic of the transition scheduler of the simulation engine would then need to be related with the system clock. Otherwise the timeouts and delays would not be real-time. Unfortunately this simulation architecture is not very well studied, and there are problems to be solved. For instance, what if the model gets significantly behind the real-time clock? This could happen if a calculation in a transition takes up too much CPU time. We leave this as a research topic for the future.

Event Handling: Event handling is a central cycle in many systems, in particular reactive systems where user input is involved. The access control model also has an event cycle which is depicted in Fig. 5. The relevant colour sets are listed below:

```
colorset MSG = product
   INT                      (* Id *)
 * NV_TYPE                  (* Network variable type *)
 * NV_VAL;                  (* Data *)
colorset MSGs = list MSG;
```

Once the system is initialised (top left substitution transition) the Neuron is activated such that new messages from the network can be read. The Neuron, Neuron Input/Output, converts a network message to an event which is put on the place Msg in. This is subsequently dispatched to the appropriate event handler which can be one of User changed, User code handling, Command handling, Input change handling, or Time handling. Once an event is fully treated then the event cycle checks if we need to generate any messages to the network (Update outputs), and then generates messages for the log printer on the network. For instance, if a user has punched in an access code, then the system registers this and then sends an unlock door message, a turn on green light

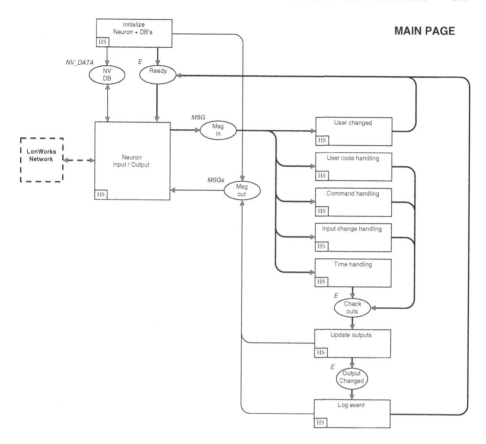

Fig. 5. CPN page modelling the main event handling cycle.

indicator message, and a log entry message to the printer. Finally the Neuron is activated again for processing all assembled messages on `Msg out`, and then the Neuron is ready to process the next input message from the network.

Note that the description just provided is purely sequential. Concurrency of message handling is primarily controlled on the Neuron model page. Handling of multiple events concurrently would make the model much more complicated because there are several data bases accessed throughout the model, and ensuring mutual data base consistency would make the first model design unnecessarily complicated. The choice of sequential message handling was made early in order to keep the first design simple, and then later extend the model with multi-threaded behaviour when the design has stabilised.

3.2   Bene ts of Model Architecture

The model we have describe above in Sect. 3.1 has a number of useful engineering properties. It has reusable components, is extensible, and adaptable. Usually a customer will need an access control system with special features and it is therefore important to avoid building the system from scratch every time. In the following we describe why the model has such properties.

Extensible: Consider the CPN module in Fig. 5 which depicts the event handler of the access control model.

The structure of the event handler is divided into cases for each kind of possible event. A new customer would typically request support for new devices in the system which consequently would induce new kinds of events. Extending the case structure is relatively easy as one just needs to add a new substitution transition to handle the new kind of event. An event may require updating some of the access control data bases in the system, but these are easily accessible as they are located on fusion places.

Reusable: Consider the CPN module in Fig. 4 which depicts the Neuron of the access control model. As described in Sect. 3.1 the Neuron is a standard component for LonWorks based systems. We therefore only need to model a Neuron once.

If we model the Neuron as one page, as in our case, we can easily use the save/load sub-page feature in Design/CPN in order to imitate a simple module facility. Should we need the Neuron module in another model, all we have to do is to make a substitution transition and load the Neuron module via load sub-page and then assign the ports to the sockets around the substitution transition.

As the Neuron is a standard component it will be used frequently and therefore also tested in many different contexts. We therefore expect such a module to become increasingly robust over time. As the module matures it will become safer to use in new contexts.

Adaptable: Consider the initialisation phase which is described in Sect 3.1. One step in the initialisation consists of sending configuration information to the LonWorks network such that network variables are configured for the attached devices. The configuration is in fact adaptable by the end-user within a limited range. For instance, a specific code generated system targeted for small buildings will typically be pre-configured to be able to handle about three rooms with 20 sensors and actuators. In this way Dalcotech does not need to code generate a new system for every new customer, but can instead have a limited number of versions each covering a range of building sizes.

3.3 Analysis of Access Control System

Dalcotech has analysed the access control system chiefly by means of simulation. Although Design/CPN could have been used to generate a (partial) state space, this was not pursued mainly due to lack of time. The state space method has, however, been applied by Dalcotech in an earlier project (cf. [3]). In the following we elaborate on the use of simulation and the LonTalk emulation library.

When using the Design/CPN simulator, we do not have access to the physical LonWorks network. This is where the LonTalk emulation library is useful, as it can be used as a simplified play-back mechanism of network messages. When we call the `LONin` function then the library reads in a message specification file in order to see if any LonWorks messages are pending. For instance a file containing:

```
3.000
UPDATE(18,"\001")
2.000
UPDATE(18,"\000")
```

specifies that after a delay of 3 seconds the network variable numbered 18 has changed value to 1 (say, movement detected in the room). After additional 2 seconds the variable changes back to zero (no movement). In this way the modeller can easily make many different simulation runs, and thus avoid step-wise interaction with the simulator. This is documented in [16].

3.4 Code Generation of Access Control System

The automatic code generation method uses two tools, namely Design/CPN (Linux version) for generating the general ML source code and a post-processing tool for specialising the code to the environment in question. In the following we describe the code generation method (cf. Fig. 2) as applied by Dalcotech on the access control system model.

Techniques and Tools: First of all the modeller needs to declare in the model that the LonTalk library is used and that it should be recognised later by the linker. In the global declarations Dalcotech only needs to include the statement "`use "_LonTalk.sml"`".

Dalcotech invokes a menu entry in Design/CPN such that source code is generated from the access control model. The code is automatically written to a file, say `/tmp/access-ctrl.sml`. For the specific model the source code generated is 4800 lines of ML code (240 kb).

We then use a post-processing tool, `cpnsim2mosml.pl`, which both includes the LonTalk run-time library and specialises the source code to be compilable with Moscow ML by means of a "flattener" tool:

```
linux% cpnsim2mosml.pl /tmp/access-ctrl.sml /tmp/access-ctrl.mml
```

The generated file, `access-ctrl.mml`, contains 29,000 lines of code (920 kb).

We move `access-ctrl.mml` to the MS DOS platform where Moscow ML is used to compile the final system executable:

```
msdos% mosmlc access-ctrl.mml
```

This command creates an MS DOS executable, `mosmlout.exe`, which now can be started on the central control unit, assuming the LonWorks network is attached to the hardware. The size of the file is 206 kb excluding a run-time ML byte-code interpreter which has the fixed size of 160 kb. Closer investigation shows that 500–2000 bytes are generated for each transition on average.

The whole procedure takes 10–20 minutes on a 200MHz Pentium PC with 64Mb RAM for the access control model.

3.5 Performance of Access Control System

Once the automatically generated implementation is executing on the MS DOS platform we can investigate the behaviour and responsiveness of the system. We have available a simplified physical demonstration hardware scenario similar to Fig. 1.

The performance of the system has been tested on a 90MHz MS DOS Pentium machine with 64Mb RAM. Such a system is certainly realistic for an installation of the access control system. In this configuration the system reaction time is measured to be in the order of 0.25–0.5 seconds, which is well within the margins of this kind of access control systems made by Dalcotech. Thus the tools used in the method are satisfactory for Dalcotech's purpose.

3.6 Main Benefits of Method for Dalcotech

There are two major advantages of taking the implementation directly from the simulator. Firstly simulation in Design/CPN and generation of the final executable system are both based on the *same* ML source code (see Fig. 6). This immediately gives a number of benefits:

Analysis results found when simulating the model can also be applied on the running system.
Always consistent documentation of the system. The model is the only documentation necessary of the system.

Fig. 6. Generated ML code is used *both* in Design/CPN simulator and final system implementation.

Secondly since the code generation method is fully automatic, we eliminate a traditionally time consuming and error-prone manual implementation phase:

Time spent in the implementation phase is dramatically reduced; from weeks to 20 minutes.

Errors do not originate from implementation but are identified already in the modelling and analysis phase.

We described earlier that Dalcotech expects to produce access control systems in small series, i.e., each system variant is only sold in a few copies. Thus time spent on development of each system variant is critical in order to be profitable. As stated above the method presented in this paper indeed reduces development time because the time spent on implementation is reduced. We are not so concerned with producing a slower and more resource demanding system with the method because hardware costs are of minor importance.

Hence we can conclude that the method for automatic code generation is applicable and useful for Dalcotech.

4   General Method Developed in AC/DC Project

In this section we show that the automatic code generation method not only applies to Dalcotech's access control systems but is a general method. This is a result of refining the method during the AC/DC project.

The generalised method for automatic code generation is summarised in Fig. 7 and should be compared with Fig. 2. The method is generalised in two areas, namely with respect to inclusion of external application specific libraries and with respect to the choice of ML compiler to make the final system executable. This ensures that in principle any CPN model can be treated by the method.

The post-processing tool does not depend on the LonTalk library as used with the access control model. The tool is in fact generic in the sense that it can be parameterised with whatever platform dependent library is needed. Thus the code generation method supports external libraries in case one needs application specific code. As indicated in Fig. 7 there needs to be an emulation and a run-time version of the library. Both versions need to be written in the ML language and must provide the same interface (cf. [15]).

Once we have generated ML source code from Design/CPN we are not forced to use Moscow ML, but can choose among several alternative ML compilers (see bottom rectangle of Fig. 7). Some compilers generate byte-code which needs to be interpreted, such as Moscow ML, others generate native machine code, such as SML/NJ [12]. Some compilers have the advantage of producing fast executable code, others have the advantage of producing compact code. The table below summarises execution time and generated code size for a number of ML compilers. The measures are given relative to SML/NJ and we note that the numbers are not exact but provides instead typical trends. (The data is taken from existing SML/NJ benchmarks and the model presented in this paper.)

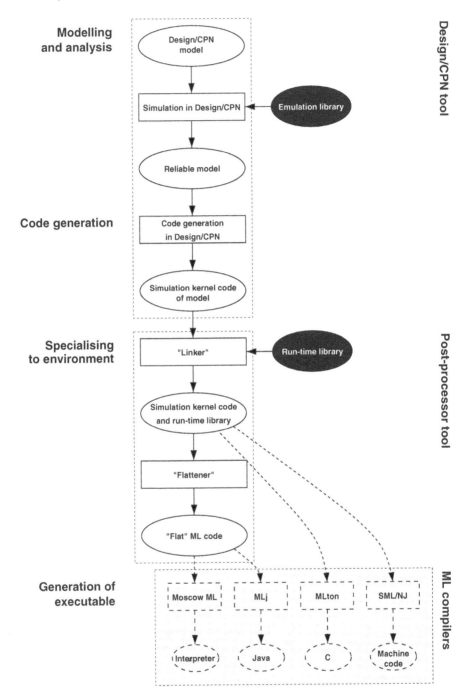

Fig. 7. Overview of general code generation method as developed in the AC/DC project.

ML compiler	Output	Execution time relative to SML/NJ	Code size relative to SML/NJ
SML/NJ	native	1:1	1:1
Moscow ML	interpreted	50:1	1:75
MLton	C	3:1	1:5

There is also an ML compiler, MLj [6], which can generate Java byte-code, but we have not used this compiler in the AC/DC project. As the table suggests there are compilers which can be used for different situations. Observe that there appears to be a trade-off between generating fast code and small executables. In the AC/DC project we are more concerned with generating small executables as we generate code for an embedded system with "soft" real-time demands, i.e., the central control unit.

5   Related Work

There are more than 50 tools for Petri nets which are actively being developed [20, 31]. Many of these tools have some sort of simulation support for Petri nets. Typically a given Petri net model is translated into a different language (C/C++, Java, SML, etc.) which subsequently is interpreted or compiled by already established and well-known tools. In this section we compare our work with other tools and approaches, however limited to tools supporting high-level Petri nets or other high-level languages such as object-oriented languages. We cannot compare the performance of generated code with other tools because the inscription languages are radically different and therefore we cannot easily transfer a CPN model from one tool to another.

CPN-AMI: This is a CASE environment based on some kind of high-level Petri nets [11]. Among other components CPN-AMI contains a simulator (CPN/-DESIR) and a code generator (H-Tagada) which generates ADA code. The code executed by the simulator and the generated ADA code are different. Thus there is not as strong relationship between simulation and generated code as with the code generation method of Design/CPN (cf. discussion of Fig. 6). However, the H-Tagada code generator is able to separate a Petri net model into processes (G-objects), thus making potential for true concurrent executing in multi-processor environments. This is not supported in the Design/CPN code generator.

Another tool called Artifex [24] is similar to CPN-AMI in that it has both a simulator and code generator, however Artifex generates C/C++ code. There are a number of other tools which have the same code generation architecture as these.

PEP: This tool has many different kinds of translators which can translate back and forth between many different kinds of languages such as P/T nets (Petri box), high-level nets (M-net), and textual language $(B(PN)^2)$ [4]. A modification in one language representation can automatically be updated in another, in a consistent fashion. This is analogous to what sometimes can be found in UML notation based environments: Updating a textual representation of a class will consistently update a diagrammatic representation of a class and vice versa. An example of such system is the Mjølner System [9].

The code generation method of Design/CPN does not support what can be called reverse engineering, i.e., translation from implementation to a CPN model. However, this is not the intention either. Once the code has been generated it should neither be investigated nor modified manually. One could then ask if we would support reverse engineering, but this is difficult in the case of ML and CP-nets because they are radically different languages. This is not the case for the UML based environments where the diagrams and textual languages are very similar in structure. Also in the case of PEP the various languages supported are mutually similar, hence if we use this idea with UML we should look at object-oriented Petri nets [1, 2] in order to close the language gap.

6   Conclusion

Although Dalcotech has used CP-nets in practice for a number of years, it is in AC/DC the first time they use CP-nets directly for automatically generating the implementation. In this way Dalcotech now has the capability to dramatically reduce the time spent in the implementation phase. Additionally they also dramatically reduce the amount of time spent on debugging because errors in the implementation in principle originate from the model only.

In the AC/DC project Dalcotech has successfully conducted a code generation case study where a CPN model has been built for the next generation of access control systems the company has planned to produce. By means of the code generation method and their model, Dalcotech has obtained an automatic implementation of the embedded controlling unit of the access control system. This non-trivial case study proves that the method can be applied in practice by engineers who are not completely familiar with the advanced technology behind automatic code generation (knowledge of CP-nets and Design/CPN assumed). Additionally, Dalcotech has during the modelling process gained better insight into access control systems in general.

The AC/DC project has for our part provided an excellent opportunity to develop and apply the techniques and tools in the context of an interesting industrial case study. Furthermore, we have been able to extend the use of CP-nets into the area of automatic code generation for embedded systems and developed a method to support this [14]. It has been a useful challenge to match the needs of Dalcotech, and we have thus become more experienced with the techniques and tools and more confident that our method works in practice [8].

Future Work

An ongoing activity is the effort to conduct technology transfer of the method to industry. A full day seminar was conducted for a number of Danish companies.

The seminar included introduction to CP-nets, the method for automatic code generation, and its application. The method is presented as something which will complement, support, and enrich existing methodologies in industry, such as UML — not a replacement for well-known technologies. In this way we help to make advanced computer science technology more easily available for a wider audience.

Our method is currently limited to "soft" real-time systems where the response time tolerance is relatively high. We allow, e.g., that an alarm is activated 1–2 seconds after a detector device has been triggered. The limitation origins primarily from the design of the ML language which always demands the presence of some sort of automatic garbage collection and memory management system. Most ML compilers do not allow control over spontaneous triggering of a time consuming garbage collection, and it is therefore nearly practically impossible to reason on execution times. We have consulted a research group in Copenhagen which is working with controlled garbage collection for ML compilers such that one can begin to reason on execution times [21]. They are interested in AC/DC because they are looking for an industrial case study. We are interested because we may be able to improve our method such that it can also cover systems with "hard" real-time demands.

Acknowledgements

The AC/DC project was partly funded by the Danish National Centre for IT-Research (CIT) [25].

The first version of the CPN code generator was developed by T.B. Haagh and T.R. Hansen and is documented in their thesis [5].

References

[1] G. Agha and F. de Cindio, editors. *Workshop on Object-oriented Programming and Models of Concurrency of the 16th International Conference on Application and Theory of Petri Nets*, Turin, Italy, 1995.

[2] G. Agha, F. de Cindio, and A. Yonezawa, editors. *Workshop on Object-oriented Programming and Models of Concurrency of the 17th International Conference on Application and Theory of Petri Nets*, Osaka, Japan, 1996.

[3] T. Andersen. SECU-DES, Improved Methodology for the Design of Communication Protocols in Security Systems. Final Report, ESSI Project 10937, Dalcotech A/S, May 1996. Online version: www.esi.es/ESSI/Reports/All/10937/.

[4] B. Grahlmann. The PEP Tool. In *Tool Presentations of ATPN'97 (Application and Theory of Petri Nets)*, June 1997.

[5] T.B. Haagh and T.R. Hansen. Optimising a Coloured Petri Net Simulator. Master's thesis, University of Aarhus, Department of Computer Science, Denmark, 1994.

[6] Persimmon IT. MLj a SML to Java Bytecode Compiler. Web site. www.dcs.ed.ac.uk/home/mlj/.

[7] K. Jensen. *Coloured Petri Nets   Basic Concepts, Analysis Methods and Practical Use. Volume 1, Basic Concepts*. Monographs in Theoretical Computer Science. An EATCS Series. Springer-Verlag, 1992.

[8] B. Jørgensen. Målte dybden før springet. *CIT NYT*, 2:8–9, June 1999. Article in newsletter, in Danish.

[9] J.L. Knudsen and B. Magnusson M.Löfgren, O.L. Madsen, editors. *Object-Oriented Environments: The Mj lner Approach*. Prentice-Hall, 1993.

[10] M. Kristensen, S. Christensen, and K. Jensen. The Practitioner's Guide to Coloured Petri Nets. *International Journal on Software Tools for Technology Transfer*, 2(2):98–132, December 1998.

[11] MARS-Team. The CPN-AMI environment. Technical report, MASI lab, Institut Blaise Pascal, Universit Pierre & Marie Curie, Paris, France, October 1993.

[12] D. McQueen. Standard ML of New Jersey. Web site. cm.bell-labs.com/cm/cs/what/smlnj/.

[13] R. Milner, R. Harper, and M. Tofte. *The De nition of Standard ML*. MIT Press, 1990.

[14] University of Aarhus, Dalcotech A/S, DELTA, and CIT. Automatisk kode-generering fra Farvede Petri Net. Final report for CIT-project number 106. In Danish, May 1999.

[15] J.-H. Paulsen. *Design/CPN Automatic Code Generation User's Guide*. University of Aarhus, Department of Computer Science, Denmark, June 1999.

[16] J.-H. Paulsen. *Design/CPN LonTalk Library User's Guide*. University of Aarhus, Department of Computer Science, Denmark, June 1999.

[17] L.C. Paulson. *ML for the Working Programmer*. Cambridge University Press, 2nd edition, 1996.

[18] J.L. Rasmussen and M. Singh. Designing a Security System by Means of Coloured Petri Nets. In J. Billington and W. Reisig, editors, *17th International Conference on Application and Theory of Petri Nets 1996*, volume LNCS 1091 of *Lecture Notes in Computer Science*, pages 400–419, Osaka, Japan, June 1996. Springer-Verlag.

[19] S. Romanenko and P. Sestoft. *Moscow ML owner's manual*, April 1998. Web site: www.dina.kvl.dk/ sestoft/mosml.html.

[20] Harald Störrle. An Evaluation of High-End Tools for Petri-Nets. Technical Report 9802, Ludwig-Maximilians-Universität München, 1997.

[21] M. Tofte, L. Birkedal, M. Elsman, N. Hallenberg, T.H. Olesen, P. Sestoft, and P. Bertelsen. Programming with Regions in the ML Kit. Technical Report 25, University of Copenhagen, Department of Computer Science, 1998.

[22] S. Weeks. MLton User's Guide. Available online: www.neci.nj.nec.com/PLS/MLton/.

[23] AC/DC Project. CIT project number 106. Web Site. www.daimi.au.dk/CPnets/ACDC.

[24] Artifex by ARTIS. Web Site. www.artis-software.com.

[25] The Danish National Centre for IT Research. Web Site. www.cit.dk.

[26] Coloured Petri Nets at the University of Aarhus. Web Site. www.daimi.au.dk/CPnets.

[27] Dalcotech A/S. Web Site. www.dalcotech.dk.

[28] DELTA Software Engineering. Web Site. www.delta.dk.

[29] Design/CPN Online. Web Site. www.daimi.au.dk/designCPN.

[30] Echelon Corporation. Web Site. www.echelon.com.

[31] World of Petri Nets. Web Site. www.daimi.au.dk/PetriNets.

Pre- and Post-agglomerations for LTL Model Checking

Denis Poitrenaud[1] and Jean-Francois Pradat-Peyre[2]

[1] SRC - LIP6 - UMR 7606, Université Paris VI, Jussieu
4, Place Jussieu, 75252 Paris cedex 05
[2] CEDRIC - CNAM Paris
292, rue St Martin, 75003 Paris

Abstract. One of the most efficient analysis technique is to reduce an original model into a simpler one such that the reduced model has the same properties than the original one. G. Berthelot defined in this thesis some reductions of Petri nets that are based on local structural conditions and that simplify significantly the net. However, the author focused only on the preservation of classical properties (such that liveness, boundedness, ...) that are not necessarily the most useful in practice. In this paper, we prove that two of these structural reductions (the pre and post transitions agglomerations) preserve also a large set of properties expressed in linear-time temporal logics under simple conditions.

1 Introduction

Checking properties of a Petri net can be performed either by structural analysis or by model checking. With structural analysis, such invariants computation, structural reductions or rank theorem, one tries to obtain behavioral properties of the model by only using the structure of the net. The construction and the exploration of the state space is not required and then, this kinds of techniques can be performed on very large systems. The drawbacks of this approach are that properties that can be check are restricted (liveness, boundedness, ...) and that results obtained need sometimes expert interpretations. On the other hand, model checking is fully automatic and can be used to check a large set of properties (often expressed in temporal logics) but is faced to the well known problem of state space complexity. Combining these two approaches is a very attractive challenge.

One of the most effective structural analysis technique consists in applying structural reductions on the net and in checking properties on the reduced one. This is very efficient since application conditions depend only on local structure of the net and since removing places or agglomerating transitions restrict the behavior of the model. The main difficulty consists in proving that the reduction preserves the equivalence between the original net and the reduced one for some general sets of properties.

In this paper, we focus on the most useful structural reductions: the pre and the post transition agglomerations. We demonstrate that these reductions

M. Nielsen, D. Simpson (Eds.): ICATPN 2000, LNCS 1825, pp. 387–408, 2000.
© Springer-Verlag Berlin Heidelberg 2000

preserve a large set of properties expressed in linear-time temporal logic under simple conditions. This result is obtained in three steps:

1. first, we prove that for particular sequences (sequences that are fireable in the reduced net and in the original one) we have the equivalence for a large set of formulas (formulas that do not take into account some places or transitions) between the reduced net and the original one; this result is stated in Theorem 1;
2. second, we demonstrate that to every sequence (infinite or blocking) of the original net corresponds a reordered sequence that is fireable both in the reduced net and in the original one and that has the same projection over a subset of transitions as the original one (Lemma 1, 2, 3 and 4);
3. at last, we prove that the pre and the post-agglomerations define an equivalence for a large set of properties (Theorems 2 and 3).

The paper is organized as follow: in section 2, we give usual definitions related to our context. In section 3, we prove the results promised above for the pre and post-agglomerations. In section 4 we give some experimentations on well known examples and we compare our approach with other king of techniques. At last, in section 5 we conclude and discuss future works.

2   Common Definitions and Properties

In this section we recall the definitions of Petri nets, the pre and the post transition agglomerations and the linear-time temporal logics.

2.1   Definition of Petri Nets

Definition 1. *A Petri net [4] is 4-tuple $\langle P, T, W^-, W^+ \rangle$ where: P is the set of places, T is the set of transitions ($P \cap T = \emptyset, P \cup T \neq \emptyset$), W^- (resp. W^+) is the backward (resp. forward) incidence application from $P \times T$ to $\mathbb{N}$.*

A marked Petri nets is a couple (N, m_0) where N is a Petri net and m_0 is the initial marking.

Definition 2. *Let $N = \langle P, T, W^-, W^+ \rangle$ be a Petri net and m_0 its initial marking. A vector of $\mathbb{N}^P$ is called a marking of N; $m(p)$ denotes the number of tokens contained in place p. A transition $t \in T$ is fireable at a marking m if and only if: $\forall p \in P, m(p) \geq W^-(p, t)$. The marking m' obtained by the firing of t is defined by: $\forall p \in P, m'(p) = m(p) - W^-(p, t) + W^+(p, t)$. One notes, $m[t > m'$ that means that t is fireable at m and reaches m'. By extension, a marking m' is reachable from a marking m if there exist a sequence $s = t_0.t_1.\ldots.t_n \in T^*$ and a set of markings $m_1, \ldots, m_n$ such that $m[t_0 > m_1, m_1[t_1 > m_2, \ldots, m_n[t_n > m'$. The set of reachable markings from m_0 is denoted by $Acc(N, m_0)$. A maximal firing sequence is either an infinite fireable sequence or a finite fireable sequence leading to a terminal marking (i.e. a marking from which none transition is fireable).*

Notation. We denote by T^* the set of finite sequences on T, by T^∞ the set of infinite sequences and $T = T^* \cup T^\infty$. λ denotes the empty sequence. $L^*(N, m_0)$ denotes the set of finite firing sequences from m_0, $L^\infty(N, m_0)$ the set of infinite firing sequences from m_0 and $L(N, m_0) = L^*(N, m_0) \cup L^\infty(N, m_0)$. $L^{max}(N, m_0)$ denotes the set of maximal firing sequences.

Notation. For a sequence $s \in T$, a transition $t \in T$ and a set $T' \subset T$,

- $t.s$ denotes the concatenation of t and s.
- $|s|$ denotes the length of s, $|s|_t$ denotes the number of occurrences of t in s, and $|s|_T$ is defined by $\sum_{r \in T} |s|_r$,
- $\Pi_T(s)$ is the sequence obtained from s by erasing the occurrences of the transitions not in T'.
- $s(i)$ denotes the ith transition of s $(i < |s|)$
- $Pref(s)$ is the set of finite prefixes of s.

2.2 The Pre- and the Post-agglomerations

Given two disjoint sets of transitions H and F, the main idea of the transitions agglomerations is that any fireable sequence can be reordered in a fireable sequence where an occurrence of a transition in H is immediately followed by an occurrence of a transition in F. This behaviors are captured by structural conditions on the net. For the pre-agglomeration, these structural conditions imply that one can delay the firing of occurrences of H while for the post-agglomeration one can anticipate the firing of occurrences of F.

Both of these reductions are based on a "good" structure of the net: there is a place p modeling an intermediate state reached by the firing of a transition of H and left by the firing of a transition of F.

De nition 3 (Agglomeration scheme). *Let (N, m_0) be a marked Petri net. Two disjoint subsets of transitions H and F and a place p satisfy a structural scheme of agglomeration if and only if:*

- $^\bullet p = H$ *and* $p^\bullet = F$
- $m_0(p) = 0$
- $\forall h \in H, \forall f \in F : W^-(p, f) = W^+(p, h) = 1$ [1]

The pre and the post-agglomerations differ principally by the application conditions of these two reductions. For the pre-agglomeration, one imposes that H is a singleton and also certain structural conditions around this transition while for the post-agglomeration, one imposes conditions around transitions F. We recall these structural conditions.

---

[1] We restrict the valuation to one only for sake of simplicity: extension to homogeneous valuations around place p is not a theoretical difficulty

The Pre-agglomeration Conditions Here, we consider that H is a singleton ($H = \{h\}$). The principle of the pre-agglomeration is the following: in every firing sequence with an occurrence of h, followed later by an occurrence of a transition f of F, one can postpone the firing of h and merge it with the firing of f. The transition h can be viewed as a local action while the transitions f of F model global actions.

Definition 4 (Pre-agglomeration [1]).

  Let (N, m_0) be a marked Petri net. A subset of transitions F is pre-agglomerateable if and only if there exist a place p and a transition $h \notin F$ such that:

1. *$\{h\}$, F and p satisfy the agglomeration scheme*
2. *$h^\bullet = \{p\}$*
3. *$^\bullet h \neq \emptyset$ and $\forall q \in {}^\bullet h, q^\bullet = \{h\}$*

  The conditions imposed by this definition can be interpreted as it follows: the point 1 ensures that the place p models an intermediate state between the firings of h and a transition of F; the point 2 ensures that it is necessary to fire h only in order to fire a transition of F; the first part of point 3 is necessary to preserve the boundedness of the net and the second part ensures that h is not in conflict with other transitions (which is the key point of this reduction).

The Post-agglomeration Conditions The principle of the post-agglomeration is the following: in every firing sequence with an occurrence of a transition h of H followed later by an occurrence of a transition f of F, one can advance the firing of f and merge it with the firing of h. Transitions of F can be viewed as local actions while transitions of H models synchronizations.

Definition 5 (Post-agglomeration [1]).

  Let (N, m_0) be a marked Petri net. A subset of transitions F is post-agglomerateable if and only if there exist a place p and a set of transitions H, $H \cap F = \emptyset$ such that:

1. *H, F and p satisfy the agglomeration scheme*
2. *$^\bullet F = \{p\}$*
3. *$h^\bullet \neq \emptyset$*

  Conditions imposed by this definition can be interpreted as it follows: point 1 ensures that in any fireable sequence, all occurrences of transitions of F are preceded by an occurrence of a transition of H: point 2, which is the key point of this reduction, ensures that transitions of F are immediately fireable after the firing of any transition of H; point 3 guarantees the boundedness equivalence.

The Reduced Net The reduced net is obtained from the original one by merging transitions of H with transitions of F. The following definitions explicit this transformation.

Definition 6 (Net reduced by a pre or a post-agglomeration).
  The reduced net (N_r, m_{0r}) *obtained by the application of a pre-agglomeration (resp. of a post-agglomeration) on H and F on the net* (N, m_0) *is defined by:*

$P_r = P$, $T_r = T_0 \uplus (H \times F)$, *we note hf the transition* $(h, f) \in (H \times F)$
$m_{0r} = m_0$
$\forall t \in T_r,\ W_r^+(p, t) = W_r^-(p, t) = 0$
$\forall t \in T_r \setminus (H \times F), \forall q \in P_r \setminus \{p\},\ W_r^+(q, t) = W^+(q, t)$ *and* $W_r^-(q, t) = W^-(q, t)$
$\forall hf \in (H \times F), \forall q \in P_r \setminus \{p\},$

- $W_r^+(q, hf) = W^+(q, f)$ *(resp. $W_r^+(q, hf) = W^+(q, h) + W^+(q, f)$) and*
- $W_r^-(q, hf) = W^-(q, f) + W^-(q, h)$ *(resp. $W_r^-(q, hf) = W^-(q, h)$).*

Remark 1. Our definition slightly differs from the original one. Effectively, the place p is not removed in the reduced net in order to preserve the marking definition. However, taking into account the incidence matrix this place will be isolated and will be never marked and then our definition is equivalent to the original one.

Fundamental Properties Equivalence It is a well known result ([1]) that the liveness, boundedness, home state existence and other global properties are preserved by both pre and a post-agglomerations.

Proposition 1 (Berthelot 85 [1]). *Let $\mathcal{P}$ be a main property (liveness, boundedness, home state existence, ...). Then :*

$$(N, m_0) \text{ satisfies } \mathcal{P} \text{ iff } (N_r, m_{0r}) \text{ satisfies } \mathcal{P}.$$

2.3   Linear-Time Temporal Logic

Linear-time temporal logic (*LTL*) is a specification language used to specify properties of a model. The syntax and the semantic of this logic is based on an alphabet A, a set of atomic propositions AP and a mapping *Verif* defined from $A \times AP$ to $\{true, false\}$. The formal syntax for *LTL* is given below.

1. Every atomic proposition $p \in AP$ is a *LTL* formula.
2. If f and g are *LTL* formulas, then so are $\neg f$, $f \wedge g$, Xf, fUg.

An interpretation for a *LTL* formula is an infinite word $\sigma = x_0.x_1.x_2 \ldots$ over A. We use the standard notation $\sigma \models f$ to indicate the truth of a *LTL* formula f for an infinite word σ. We write σ_i, for the suffix of σ starting at x_i. The relation $\models$ is defined inductively as follows: Let $p \in AP$ and f and g two *LTL* formulas.

1. $\sigma \models p$ if $Verif(x_0, p)$ for $p \in AP$;
2. $\sigma \models \neg f$ if $\neg(\sigma \models f)$;
3. $\sigma \models f \wedge g$ if $\sigma \models f$ and $\sigma \models g$;
4. $\sigma \models Xf$ if $\sigma_1 \models f$;
5. $\sigma \models fUg$ if $\exists i \geq 0 : \sigma_i \models g \wedge (\forall j < i : \sigma_j \models f)$.

Let $S \subseteq A^{\infty}$, let f be a LTL formula,

$$S \models f \Leftrightarrow \forall \sigma \in S, \sigma \models f$$

.

We are now in position to define a state-based and an action-based temporal logics for Petri nets. For each of them, we specify the set AP of atomic propositions and the alphabet A as well as the mapping $Verif$. Moreover, we define the set of words $Trace(N, m_0)$ which must satisfy the formula.

State-Based Linear Temporal Logic In this logic, the set AP refers to atomic propositions operating on the reachable marking of the considered net (i.e. $\forall prop \in AP, \forall m \in Acc(N, m_0), prop(m) \in \{true, false\}$).

Let $Truth$ be a mapping from $Acc(N, m_0)$ to 2^{AP} such that $\forall m \in Acc(N, m_0)$, $Truth(m) = \{prop \in AP \mid prop(m)\}$.

We define Sat as the extension of $Truth$ to sequences. Let $s = m_0[t_0 > m_1[t_1 > \ldots$ be a firing sequence, $Sat(s) = Truth(m_0).Truth(m_1)\ldots$

Moreover, we define the mapping Sat^{∞} as it follows. Let $s = m_0[t_0 > m_1[t_1 > \ldots$ be a firing sequence.

if $\mid s \mid < \infty$ then $Sat^{\infty}(s) = Sat(s).Sat(m_{|s|})^{\infty}$,
if $\mid s \mid = \infty$ then $Sat^{\infty}(s) = Sat(s)$

We are now in position to define our state-based temporal logic.

AP is a set of atomic propositions operating on the markings,
$A = 2^{AP}$.
$\forall Props \in A, \forall prop \in AP, Verif(Props, prop) \Leftrightarrow prop \in Props$.
$Trace(N, m_0) = \{\sigma \in A^{\infty} \mid \exists s \in L^{max}(N, m_0) : Sat^{\infty}(s) = \sigma\}$.

The logic constructed on the formulas which do not use the operator X is noted $LTL \setminus X$. Formulas in this restricted logic fulfill the stuttering property formalized in the following proposition.

Proposition 2 (Stuttering property [3]).
Let f be a state-based $LTL \setminus X$ formula. Let $\sigma = x_0 \cdot x_1 \cdot x_2 \ldots$ be an in nite word over $AP \rightarrow \{true, false\}$. Let $\sigma' = x_{i_0} \cdot x_{i_1} \cdot x_{i_2} \ldots$ the in nite extracted word of σ de ned by:

$\forall k : i_k < i_{k+1}$
$\forall k, \forall i : i_k \leq i < i_{k+1} \Rightarrow x_i = x_{i_k}$
$\forall k : x_{i_k} = x_{i_{k+1}} \Rightarrow \forall i \geq i_k : x_i = x_{i_k}$

Then $\sigma \models f \Leftrightarrow \sigma' \models f$.

In the remainder of the paper, we use a set of observed transitions associated to a set AP denoted by $Obs(AP)$. We impose that this set satisfies $Obs(AP) \supseteq \{t \in T \mid \exists m, m' \in Acc(N, m_0), m[t > m' \wedge (\exists prop \in AP, prop(m) \neq prop(m'))\}$. Using the stuttering property, one obtains the following corollary.

Corollary 1. *Let f be a state-based LTL \ X formula.*
Let $s, s' \in L$ (N, m_0) such that $\Pi_{Obs(AP)}(s) = \Pi_{Obs(AP)}(s')$ then

$$Sat^\infty(s) \models f \Leftrightarrow Sat^\infty(s') \models f$$

Computing the minimal set of transitions $Obs(AP)$ satisfying the previous requirement can be a complex task in the general case. However, in practice, it is sufficient to compute a super-set by considering all the transitions that are adjacent to places referenced in atomic propositions used in the formula.

Action-Based Linear Temporal Logic For this kind of logic, we consider labelled Petri nets. Then, we suppose the existence of a labelling mapping *Lab* from T to a finite alphabet AP augmented by the empty sequence λ. The transitions which are not observed are considered labelled by λ. *Lab* is extended to sequences as usual.

We define the mapping Lab^∞ as it follows. Let $s = m_0[t_0 > m_1[t_1 > \ldots$ be a firing sequence.

if $\mid Lab(s) \mid < \infty$ then $Lab^\infty(s) = Lab(s).(\tau)^\infty$ (with $\tau \notin AP$),
if $\mid Lab(s) \mid = \infty$ then $Lab^\infty(s) = Lab(s)$

We are now in position to define our action-based temporal logic:

the alphabet AP (such that $\tau \notin AP$),
$A = AP \cup \{\tau\}$,
$\forall a \in A, \forall a' \in AP, Verif(a, a') \Leftrightarrow a = a'$.
$Trace(N, m_0) = \{\sigma \in 2^{AP} \mid \exists s \in L^{max}(N, m_0) : Lab^\infty(s) = \sigma\}$

We define the set of observed transitions via the labelling mapping:
$Obs(AP) = \{t \in T \mid Lab(t) \in AP\}$.

Proposition 3. *Let f be a action-based LTL formula. Let $s, s' \in L$ (N, m_0) such that $\Pi_{Obs(AP)}(s) = \Pi_{Obs(AP)}(s')$ then*

$$Lab^\infty(s) \models f \Leftrightarrow Lab^\infty(s') \models f$$

Proof. Obvious since $\lambda.Lab(t) = Lab(t)$ and then $Lab^\infty(s) = Lab^\infty(s')$.   □

3   Agglomerations and *LTL* Model Checking

In this section, we consider a marked Petri net (N, m_0). We consider also two non empty subsets of transitions H and F such that (N, m_0) can be reduced by a pre or a post-agglomeration with respect to H and F. We denote $T_0 = T \setminus \{F \cup H\}$.

Our aim is to demonstrate that when no occurrence of transition in H (resp. F) is observed for a pre-agglomeration (resp. post-agglomeration), it is equivalent to check a formula on the reduced net or the original one. It is easy to prove that the existence of a counter example in the reduced net implies the non-satisfaction of the formula by the original one (sect. 3.1). Indeed, in the sequences simulated by the reduced net, any occurrence of a transition in H is followed by an occurrence of a transition in F.

For the general case, we have to take into account the difference between the number of occurrences of transitions in H and the one in F (sect. 3.2). Then, for both agglomerations (sect. 3.3 and 3.4), we demonstrate that it is always possible to reorder the sequences of the original net in equivalent sequences (i.e preserving the satisfaction of the formula) of the reduced one.

3.1   Restricted Equivalence

We define now an application translating sequences of the reduced net into sequences of the original one. Using this definition, we state on the reachability equivalence between the reduced net and the original one for particular sequences. It is important to remark that this proposition does not state this equivalence for sequences of general form.

Definition 7. ϕ is the homomorphism from T_r^ω to T^ω defined by:

- $\forall t \in T_0, \forall s \in T_r^\omega : \phi(t.s) = t.\phi(s)$
- $\forall h \in H, \forall f \in F, \forall s \in T_r^\omega : \phi(hf.s) = h.f.\phi(s)$

Proposition 4. $\forall s_r \in T_r^*, m[s_r >_r m' \iff m[\phi(s_r) > m'$

Proof (sketch of). We prove this proposition by recurrence on the length of s_r.

1. $|s_r| = 0$: the proposition is obviously true.
2. $|s_r| \geq 1$: $s_r = s'_r.t_r$, $m[s'_r >_r m_1[t_r >_r m'$.
   By recurrence hypothesis,

$$m[s'_r >_r m_1 \iff m[\phi(s'_r) > m_1$$

As $\phi(s'_r).\phi(t_r) = \phi(s'_r.t_r)$ it is sufficient to prove that

$$m_1[t_r >_r m' \iff m_1[\phi(t_r) > m'$$

If $t_r \in T_0$, then $\phi(t_r) = t_r$ and then the proposition is demonstrated. Otherwise, $\phi(t_r) = h.f$. As $h.f$ is enabled at m_1 if and only if h is enabled at m_1 and f is enabled when h has been fired, as $h^\bullet = \{p\}$ in the case of the pre-agglomeration and $^\bullet f = \{p\}$ in the case of the post-agglomeration it comes that

$$m_1[t_r >_r m'^1 \iff m_1[\phi(t_r) > m'^2$$

Taking into account the definition of the reduced net, it comes that $m'^1 = m'^2$. □

We can now define an equivalence on the *LTL* formula satisfaction between the original and the reduced net for very particular sequences. We first define the mappings Lab_r and Sat_r associated to the reduced net.

Definition 8. *Given a logic (AP, A and Lab or Truth) and a marked Petri net (N, m_0) the mappings Lab_r and Sat_r associated to the reduced net (N_r, m_{0_r}) are defined by:*

If (N_r, m_{0_r}) is reduced by a pre-agglomeration then Lab_r is the mapping defined from T_r to $AP \cup \{\lambda\}$ by: $Lab_r(t) = Lab(t)$ if $t \in T_0$ and $Lab_r(hf) = Lab(f)$ if $hf \in \{h\} \times F$.
If (N_r, m_{0_r}) is reduced by a post-agglomeration then Lab_r is the mapping defined from T_r to $AP \cup \{\lambda\}$ by: $Lab_r(t) = Lab(t)$ if $t \in T_0$ and $Lab_r(hf) = Lab(h)$ if $hf \in H \times F$.
For both pre and post-agglomeration, $Truth_r$ is the mapping defined from $Acc(N_r,$
$m_0)$ to (2^{AP}) by: $\forall m \in Acc(N_r, m_0)$, $Truth_r(m) = Truth(m)$. The mapping Sat_r is defined from $Truth_r$ as in section 2.3.

The set $Trace_r$ is defined by extension of Lab_r or Sat_r.

Theorem 1 (Restricted equivalence).
Let f be state-based $LTL \setminus X$ formula (resp. an action-based LTL formula). Let s_r be a maximal firing sequence of the reduced net obtained by a pre or a post-agglomeration.
If $h \notin Obs(AP)$ in the case of the pre-agglomeration or $F \cap Obs(AP) = \emptyset$ in the case of the post-agglomeration, then

$$Sat_r^\infty(s_r) \models f \Leftrightarrow Sat^\infty(\phi(s_r)) \models f$$
$$(resp. \ Lab_r^\infty(s_r) \models f \Leftrightarrow Lab^\infty(\phi(s_r)) \models f)$$

Proof. We distinguish the two kinds of logics:

(state-based formula) As the mapping ϕ only renames transitions in $H \times F$ it is sufficient to observe the occurrence of these transitions. Let $s.hf$ be a prefix of s_r. Suppose that $m_0[s >_r m_1[hf >_r m_2$. From Prop 4, we have $m_0[s > m_1[h > m_1' > f > m_2$. Because $h \notin Obs(AP)$, we have $Truth(m_1) = Truth(m_1')$ and by definition $Truth_r(m_1) = Truth(m_1)$.
Then $Sat_r(s_r)$ and $Sat(\phi(s_r))$ differ only by stuttering. From Prop. 2, we can deduce that $Sat_r^\infty(s_r) \models f \Leftrightarrow Sat^\infty(\phi(s_r))$.
(action-based formula) By hypothesis $Lab(h.f) = \lambda.Lab(f) = Lab(f)$. So by definition of Lab_r, $Lab_r^\infty(s_r) = Lab^\infty(\phi(s_r))$.   □

This first result tells us that if there exists a counter example of a formula in the reduced net then this counter example is also valid in the original net. Unfortunately, from up to now, nothing ensures that the controversy is correct: it may exist a counter example of a formula in the original net that is not observed in the reduced one. This is due to the fact that every sequence in $L^\infty(N, m_0)$ is not necessarily of the form $\phi(s_r)$. In particular, nothing imposes that all occurrences of transitions of H are immediately followed by an occurrence of transitions of F. As we have to extend our results to any maximal sequence of the net, we first propose characterizations of sequences before demonstrate that any sequence can be reordered in a sequence of the form $\phi(s_r)$.

3.2   Sequence Characterizations

Sequences of the original net that can be written $\phi(s_r)$ with s_r a sequence of the reduced net are of particular interest. We call them *simulateable* sequences.

Definition 9 ((H, F)-simulateable sequence).
   *A sequence $s \in T^*$ is said to be (H, F)-**simulateable** (or simulateable for short) if $s(|s|) \notin H$ and $\forall k \in [1..(|s| - 1)]$*

$$s(k) \in H \Leftrightarrow s(k + 1) \in F$$

Corollary 2 (of proposition 4). *If s is a simulateable sequence, then ϕ^{-1} is defined by $\phi(\phi^{-1}(s)) = s$ and we have that $m_0[s > m'$ iff $m_{0r}[\phi^{-1}(s) >_r m'$.*

As mentioned before, simulateable sequences can be written $\phi(s_r)$. So, if all maximal sequences of (N, m_0) are simulateable then the theorem of the restricted equivalence (th. 1) ensures that any formula can be verified either in the original net or in the reduced one. However, we have to deal with the non simulateable sequences and two situations have to be taken into account: first, in a sequence, each occurrence of transition of H is not necessarily followed immediately by an occurrence of a transition of F; second, it is possible that there exists a difference between the number of occurrences of H and the number of occurrences of F. In order to characterize this difference, we introduce a mapping, called Γ, and we call *balanced* the sequence for which this difference is null.

Definition 10. *For a sequence $s \in T^*$, $\Gamma(s) = |s|_H - |s|_F$ and then, by construction, $\forall t \in T, \Gamma(s.t) - \Gamma(s) \in \{-1, 0, 1\}$.*

Definition 11 ((H, F)-balanced sequence).
   *A sequence $s \in T^*$ is said to be (H, F)-**balanced** (or balanced for short) if $\Gamma(s) = 0$. In the other case, s is said to be unbalanced.*

Remark 2. If a sequence s is simulateable then any balanced prefix of s is also simulateable. Furthermore, if s_1 and s_2 are two simulateable sequences, then the sequence $s_1.s_2$ is also simulateable.

Using application conditions of both pre and post-agglomerations, we can remark that the mapping Γ takes only natural value ($\in \mathbb{N}$) even for unbalanced sequences.

Proposition 5 (h-precedence).
   If (N, m_0) is a pre or post-agglomerateable net, then $\forall s \in L^*(N, m_0), \Gamma(s) \geq 0$.

Proof. It sufficient to note that for all reachable marking m from m_0 ($m_0[s > m$) we have $m(p) = \Gamma(s)$ □

Nevertheless, these definitions are not sufficient to classify any sequences. Indeed, we have to distinguish two kinds of sequences: the sequences for which we can isolated an unbalanced prefix and such that the following of the sequence can be decomposed into balanced sub-sequences, and the others. The following definition highlights this possible decomposition.

De nition 12 (Degree of a sequence).
   The degree of a sequence $s \in T$ (denoted by $d^\circ(s)$) is de ned by:

$d^\circ(s) = n \in \mathbb{N}$ *if $\exists s' \in Pref(s)$ such that*
  - $\Gamma(s') = n$
  - $\forall s'' \in Pref(s)$ *with $|s''| > |s'|$ then*
    1. $\Gamma(s'') \geq n$
    2. $\Gamma(s'') > n \Rightarrow \exists s''' \in Pref(s)$ *with $|s'''| > |s''|$ and $\Gamma(s''') = n$*
  We note $\theta(s)$ the smallest pre x s' satisfying the de nition.
  $d^\circ(s) = \infty$ if such a pre x $\theta(s)$ does not exist.

We establish now two technical results characterizing infinite sequences with a finite degree.

Proposition 6. *Let s be an in nite sequence. We have*

$$d^\circ(s) = n \Rightarrow |\{s' \in Pref(s) \mid \Gamma(s') = n\}| = \infty$$

Proof. Since s is infinite, using definition of the degree it is easy to construct an infinite series of prefixes s'' longer than $\theta(s)$ and with $\Gamma(s'') = n$. □

Corollary 3. *Let $s \in T$. If $\forall s' \in Pref(s), \Gamma(s') < n$ then $d^\circ(s) < n$*

We can now characterize sequences with an infinite degree. For such a sequence s, if we consider the series defined by the growing prefixes of s, then for any natural value n, there exists an unique item of this series from which the mapping Γ takes only values greater than n.

Proposition 7. *Let $s \in T^\infty$. If $d^\circ(s) = \infty$ then $\forall n \in \mathbb{N}, \exists! s_n \in Pref(s)$ such that $\Gamma(s_n) = n$ and such that $\forall s' \in Pref(s)$ with $|s'| > |s_n|$, we have $\Gamma(s') > n$.*

Proof. Using the definition of Γ and Corollary 3 the existence of s_n is ensured (otherwise, either Γ will not increase or decrease only by one or $d^\circ(s)$ will be bounded). Suppose that the uniqueness is not satisfied. Let then note s_n^1 and s_n^2 be two of these prefixes. Necessarily, one of both is the prefix of the other and so by definition either $\Gamma(s_n^1) > n$ or $\Gamma(s_n^2) > n$. □

3.3   The Pre-agglomeration and *LTL* Model Checking

Every maximal sequence in $L^\infty(N, m_0)$ is not necessarily simulateable and then we have to extend the theorem 1 to any sequence of the net. Using the conditions of the pre-agglomeration, we will prove that any sequence can be reordered in a fireable sequence exhibiting simulateable sub-sequences. Before doing that, we first establish the independence of the occurrences of the transition h.

Proposition 8. *Let m be a reachable marking and $s \in T^*$.*
If $\forall s' \in Pref(s)$, $\Gamma(s') \geq 0$ then $m[h.s > m' \implies m[s.h > m'$.

Proof. We make a proof by induction on the length of s.

- $|s| = 0$: Obvious
- $|s| > 0$: We consider two cases:
  1. $|s|_F = 0$: We note $s = t.s'$ $(m[h.t.s' > m')$. Suppose that $m[t \not>$. This implies that $\exists q \in {}^\bullet t$ such that $q \in h^\bullet$ and this is forbidden by the point 2 of the definition 4.
     Suppose now that $m[t.h \not>$. This implies that t takes a token needed for the firing of h $({}^\bullet t \cap {}^\bullet h \neq \emptyset)$ and this is forbidden by the point 3 of the definition 4.
     Let m'' defined by $m[t > m''$. We have just proven that $m[t > m''[h.s' > m'$. Using the inductive hypothesis on s' from m'' it comes that $m[t.s'.h > m'$.
  2. $|s|_F > 0$: As $\forall s' \in Pref(s)$, $\Gamma(s') \geq 0$, the sequence s can be written $s = s_0.f.s_2$ with $s_2 \in (T_0)^*$, $f \in F$, $\Gamma(s_0) > 0$ and $\forall s_0' \in Pref(s_0), \Gamma(s_0') \geq 0$ $(m[h.(s_0.f.s_2) > m')$.
     Using the inductive hypothesis on the sequence s_0 it comes that $m[h.s_0.f.s_2 > m' \implies m[s_0.h.f.s_2 > m'$. Let us note m'' the marking reached after the firing of m_0 $(m[s_0 > m''$.
     As $\Gamma(s_0) > 0$ then $m''(p) \geq m(p) + 1 > 0$.
     Suppose that $m''[f \not>$. As p is marked, this implies that $\exists q \neq p \in {}^\bullet f$ such that $q \in h^\bullet$ and this is forbidden by the point 2 of the definition 4.
     Suppose that $m''[f.h \not>$. This implies that f takes a token needed for the firing of h $(f^\bullet \cap {}^\bullet h \neq \emptyset)$ and this is forbidden by the point 3 of the definition 4. So $m[s_0.f.h.s_2 > m'$. Using again the inductive hypothesis on s_2 $(s_2 \in (T_0)^*)$ it comes that $m[s_0.f.s_2.h > m'$ and then $m[s.h > m'$.
     □

Given a finite sequence s we can now prove that this sequence can be well reordered; i.e. there exists a permutation of s exhibiting a simulateable sequence followed by a series of h. Furthermore, this permutation preserves the projection of the sequence on $T_0 \cup F$.

Proposition 9. *Let m be a reachable marking from m_0 and $s \in T^*$ be a sequence from m to a marking m' $(m[s > m')$ such that $\forall s' \in Pref(s), \Gamma(s') \geq 0$.*

There exists a permutation of s, $s = s_{\bowtie}.s_{\lhd}$, such that:

1. $m[s > m'$
2. $\Pi_{T_0 \cup F}(s_{\lhd}) = \lambda$ *and* $\Pi_{T_0 \cup F}(s_{\bowtie}) = \Pi_{T_0 \cup F}(s)$
3. $s_{\bowtie}$ *is simulateable*

Any sequence $s = s_{\bowtie}.s_{\lhd}$ fulfilling the above requirements is called an ordered representative *of s.*

Proof. We prove by induction on the length of s that there exists at least one ordered representative of s.

$|s| = 0$: The sequence $s = s$ fulfills the conditions of the proposition.
$|s| > 0$: If $s \in T_0^*$ then the sequence $s = s_{\bowtie} = s$ is the unique ordered representative of s and fulfills the conditions of the proposition. Otherwise, the sequence s can be written $s = s'.t.s''$ with $s'' \in T_0^*$ and $t \in H \cup F$. There are two cases:

- $t = h$ (i.e. $s = s'.h.s''$)
  As $s'' \in T_0^*$, proposition 8 implies that $m[s'.s''.h > m'$. By inductive hypothesis, there exists an ordered representative of $s'.s''$ and by construction $s'.s''.h$ is an ordered representative of s and verifies $m[s'.s''.h > m'$.
- $t \in F$ (i.e. $s = s'.f.s''$ with $s'' \in T_0^*$).
  As $\Gamma(s'.f) \geq 0$, then $\Gamma(s') > 0$.
  Let s_m be the longest prefix of s' such that $\forall s^2 \in Pref(s')$, $\Gamma(s^2) > 0$. This prefix exists since $\forall s^2 \in Pref(s')$, $\Gamma(s^2) \geq 0$ and since $\Gamma(s') > 0$. Necessarily, the sequence s' can be written $s' = s_m.h.s_s$ with $\forall s^2 \in Pref(s_s)$, $\Gamma(s^2) \geq 0$ ($s = s_m.h.s_s.f.s''$).
  Using proposition 8 on $h.s_s$ it comes that $m[s_m.s_s.h.f.s'' > m'$.
  By inductive hypothesis, $s_m.s_s = s_1.h^k$ and $m[s_1.h^k.h.f.s'' > m'$.
  As $s'' \in T_0^*$ and as $\Gamma(h.f) = 0$, we can apply k times proposition 8 on the sequence $h.f.s''$ and it comes that $m[s_1.h.f.s''.h^k > m'$.
  By construction, $s_1.h.f.s''.h^k$ is an ordered representative of s.   □

We extend now the previous result to infinite sequences : any infinite sequence can also be reordered while preserving the projection of the sequence on $T_0 \cup F$.

Proposition 10. *Let $s \in L^\infty(N, m_0)$. There exists a permutation of s, s^∞ such that*

1. $\Pi_{T_0 \cup F}(s^\infty) = \Pi_{T_0 \cup F}(s)$
2. $\exists(s_{\bowtie}^i)_{i \geq 0}$ *an infinite series of simulateable sequences such that*
   (a) $d^\circ(s) = n \Rightarrow s^\infty = (s_{\bowtie}^0).h^n.(s_{\bowtie}^1).(s_{\bowtie}^2)\ldots(s_{\bowtie}^k)\ldots$
   (b) $d^\circ(s) = \infty \Rightarrow s^\infty = (s_{\bowtie}^1.h).(s_{\bowtie}^2.h).(s_{\bowtie}^3.h)\ldots(s_{\bowtie}^k.h)\ldots$

Proof. We distinguish the two cases:

1. Suppose that $d°(s) = n$.
   From Definition 12 of the degree, $s = \theta(s).s'$ with $\Gamma(\theta(s)) = n$ and $\forall s'' \in Pref(s'), \Gamma(s'') \geq 0$.
   Let $(s'_i)_{i \geq 1}$ the set of prefixes of s' such that $\forall i \geq 1, \Gamma(s'_i) = 0$. From Definition 12 and Proposition 6 this series is infinite.
   Let then the series $(s^i)_{i \geq 0}$ defined by $s^0 = \theta(s), s^1 = s'_1$ and $\forall i > 1, s'_i = s'_{i-1}.s^i$. The sequence s can be written $s = s^0.s^1 \ldots s^k \ldots$
   By construction, $\Gamma(s^0) = n$. Applying Proposition 9, it comes that $s^0 = s^0_{\bowtie}.h^n$ and if $m_0[s^0 > m$ then $m_0[s^0_{\bowtie}.h^n > m$. Moreover, $\Pi_{T_0 \cup F}(s^0.h^n) = \Pi_{T_0 \cup F}(s^0)$
   As $\forall i \geq 1, \Gamma(s^i) = 0$ and as each prefix σ of s^i verifies $\Gamma(\sigma) \geq 0$, we can apply Proposition 9 on each s^i and it comes that $s^i = s^i_{\bowtie}$ (s^i is simulateable) and that if $m_i[s^i > m_{i+1}$ then $m_i[s^i_{\bowtie} > m_{i+1}$. Moreover, $\Pi_{T_0 \cup F}(s^i) = \Pi_{T_0 \cup F}(s^i)$
2. Suppose that $d°(s) = \infty$.
   Let $(s_i)_{i \geq 1}$ the set of prefixes of s such that $\forall i \geq 1, \Gamma(s_i) = i$ and $\forall s'_i \in Pref(s)$ with $|s'_i| > |s_i|$ we have $\Gamma(s'_i) > i$. For a given $i \geq 1$ and from the proposition 7, the prefix s_i is unique.
   Let then the series $(s^i)_{i \geq 1}$ defined by $s^1 = s_1$ and $\forall i > 1, s_i = s_{i-1}.s^i$. The sequence s can be written $s = s^1.s^2 \ldots s^k \ldots$
   By construction, $\forall i \geq 1, \Gamma(s^i) = 1$. Applying Proposition 9, it comes that $s^i = s^i_{\bowtie}.h$ and if $m_i[s^i > m_{i+1}$ then $m_i[s^i_{\bowtie}.h > m_{i+1}$. Moreover, $\Pi_{T_0 \cup F}(s^i.h) = \Pi_{T_0 \cup F}(s^i)$ □

We are now in position to claim that to any infinite sequence corresponds a simulateable sequence having the same projection on $T_0 \cup F$ (which will be the potentially observed transitions).

Lemma 1. *Let $s \in L^\infty(N, m_0)$. There exists an infinite simulateable sequence $s_{\bowtie}$ such that*

1. $s_{\bowtie} \in L^\infty(N, m_0)$
2. $\Pi_{T_0 \cup F}(s_{\bowtie}) = \Pi_{T_0 \cup F}(s)$

Proof. We distinguish the two cases:

1. Suppose that $d°(s) = n$.
   From Proposition 10, we know that the sequence $s^\infty = (s^0_{\bowtie}).h^n.(s^1_{\bowtie}).(s^2_{\bowtie}) \ldots (s^k_{\bowtie}) \ldots$ has the same projection on $T_0 \cup F$ than s and belongs to $L^\infty(N, m_0)$.
   It is clear that the sequence $s_{\bowtie} = (s^0_{\bowtie}).(s^1_{\bowtie}).(s^2_{\bowtie}) \ldots (s^k_{\bowtie}) \ldots$ is infinite and has the same projection on $T_0 \cup F$ than s.
   Suppose that $s_{\bowtie} \notin L^\infty(N, m_0)$. Then $\exists i > 0$ such that $m_0[(s^0_{\bowtie}).(s^1_{\bowtie}) \ldots (s^i_{\bowtie}) \nrightarrow$.
   Let $s_i = (s^0_{\bowtie}).h^n.(s^1_{\bowtie}) \ldots (s^i_{\bowtie})$. We know that $m_0[s_i >$ and that $s_i = s'.h^n$ (with $\Pi_{T_0 \cup F}(s_i) = \Pi_{T_0 \cup F}(s')$ and s' simulateable and fireable from m_0).
   Because $\Pi_{T_0 \cup F}(s_i) = \Pi_{T_0 \cup F}((s^0_{\bowtie}).(s^1_{\bowtie}) \ldots (s^i_{\bowtie})) = \Pi_{T_0 \cup F}(s')$ and because s' and $(s^0_{\bowtie}).(s^1_{\bowtie}) \ldots (s^i_{\bowtie})$ are two simulateable sequences, then necessarily, $s' = (s^0_{\bowtie}).(s^1_{\bowtie}) \ldots (s^i_{\bowtie})$.

2. Suppose that $d°(s) = \infty$.

   Let $s^\infty = (s_{\bowtie}^1.h).(s_{\bowtie}^2.h)\ldots(s_{\bowtie}^k.h)\ldots$. Using same reasoning, we know that $s_{\bowtie} = s_{\bowtie}^1.s_{\bowtie}^2\ldots s_{\bowtie}^k\ldots$ has the same projection on $T_0 \cup F$ than s and belongs to L (N, m_0).

   We have to demonstrate that $s_{\bowtie}$ belongs to $L^\infty(N, m_0)$. Suppose $s_{\bowtie}$ is finite. Then $\exists i \geq 1$ such that $\forall j > i, |s_{\bowtie}^j| = 0$. Then $s^\infty = (s_{\bowtie}^1.h)\ldots(s_{\bowtie}^i.h).h^\infty$. Due to the structural condition on h imposed by points 3 and 2 of Definition 4 ($^\bullet h \neq \emptyset$ and $^\bullet h \cap h^\bullet = \emptyset$), and due to the finiteness of each $s_{\bowtie}^i$, it must exist a place $q \in {}^\bullet h$ such that $m_0(q) = \infty$ which is impossible.   □

At last, before to extend Theorem 1, we have to deal with sequences leading to terminal state. For such sequences, we prove that there exists a terminal sequence of the reduced net having the same projection on $T_0 \cup F$.

Lemma 2. *Let $s \in L^*(N, m_0)$ be a sequence leading to a terminal state. We note $s = s_{\bowtie}.h^k$ a reordering of s. Then $\phi^{-1}(s_{\bowtie})$ is a sequence of the reduced net leading to a terminal state (and by construction, has the same projection over $T_0 \cup F$ than s).*

Proof. We note m the marking defined by $m_0[s_{\bowtie} > m$ (by definition $m_0[\phi^{-1}(s_{\bowtie}) >_r m$). Suppose that the lemma is false then there exists $t \in T_r$ such that $m[t >_r$. We consider two cases:

$k = 0$: by definition $m[\phi(t) >$; it is a contradiction.

$k > 0$: we have to distinguish two sub-cases:

- $t \neq hf$: $\phi(t) = t$. Since $m[t >$ and $m[h^k >$ and since $^\bullet h \cap {}^\bullet t = \emptyset$ necessarily $m[h^k.t >$ which leads to a contradiction.
- $t = hf$: because $m[h.f >$ and $m[h^k >$ and because $^\bullet h \cap {}^\bullet f = \emptyset$ necessarily $m[h^k.f >$ which leads also to a contradiction.   □

We can now prove that the pre-agglomeration defines an equivalence for any property expressed in *LTL* when the transition h is not observed (and when the operator X is not used in case of state-based logic).

Theorem 2 (Pre-agglomeration and model checking).
Let f be a state-based $LTL \setminus X$ (resp. action-based LTL) formula. Let (N, m_0) be a net and (N_r, m_{0_r}) be the corresponding reduced net by a pre-agglomeration w.r.t. h and F. If $h \notin Obs(AP)$ then

$$Trace(N, m_0) \models f \Leftrightarrow Trace_r(N_r, m_{0_r}) \models f$$

Proof. We prove that there exists a counter example of f in the original net iff there exists a counter example in the reduced net.

($\Leftarrow$): It is a direct application of Theorem 1.

($\Rightarrow$): Let $s \in L^{max}(N, m_0)$ such that $Sat^\infty(s) \not\models f$ (resp. $Lab^\infty(s) \not\models f$). We distinguish two cases:

- $s \in L^\infty(N, m_0)$: Using Lemma 1, there exists an infinite simulateable sequence $s_\bowtie$ having the same projection over $T_0 \cup F$ than s; using Corollary 1 (resp. Prop. 3), we know that $Sat^\infty(s) \not\models f \Leftrightarrow Sat^\infty(s_\bowtie) \not\models f$ (resp. $Lab^\infty(s) \not\models f \Leftrightarrow Lab^\infty(s_\bowtie) \not\models f$).
  Because $s_\bowtie$ is simulateable then $\phi(\phi^{-1}(s_\bowtie)) = s_\bowtie$ and we can apply Theorem 1 on $\phi^{-1}(s_\bowtie)$.
- $s \notin L^\infty(N, m_0)$ (s leads to a terminal state): It is a direct application of Lemma 2 and Corollary 1 (resp. Prop. 3).   □

3.4   The Post-agglomeration and *LTL* Model Checking

As for the pre-agglomeration, we will see that any sequence generated by a post-agglomerateable net can be reordered in a fireable sequence exhibiting simulateable sub-sequences.

The main difference between the post and the pre-agglomeration resides in the way of the reordering is performed: for the post-agglomeration, structural conditions imposed on the net ensure that any transition of F can be fired as soon as a transition of H has been fired. Each unbalanced sequence can be then completed by the adequate number of occurrences of F and reordered into a simulateable sequence.

Proposition 11. *Let m be a reachable marking and let s be a sequence of T^* such that $\forall s' \in Pref(s), \Gamma(s') \geq 0$. Then $\forall h \in H$,*

$$m[h.s > m' \Longrightarrow \forall f \in F, m[h.f.s > m'' \text{ and } m'' \geq m'$$

Proof. Because $^\bullet F = \{p\}$, necessarily $\forall f \in F, m[h.f >$.
Let now suppose that $m[h.f.s \not> $ (for a given $f \in F$). Necessarily, there exists t, s', s'' such that $s = s'.t.s''$ and $m[h.f.s' > m_1$ but $m_1[t \not>$. There are two cases:

- $t \in F$ (i.e. $s = s'.f'.s''$): Because $^\bullet f' = \{p\}$ then $m_1[f' \not>$ implies $m_1(p) = 0$. If we note s_0 the sequence such that $m_0[s_0 > m$ then $m_1(p) = 0$ implies that $\Gamma(s_0.h.f.s') = 0$. Due to the agglomeration scheme, $\Gamma(s_0) \geq 0$. So, $\Gamma(h.f.s') \leq 0$; then $\Gamma(s') \leq 0$ and then $\Gamma(s'.f') < 0$. As $s'.f' \in Pref(s)$, we obtain a contradiction.
- $t \in T_0 \cup H$: $m[h.s'.t >, m[h.f.s' >$ and $m[h.f.s'.t \not>$ implies that $^\bullet f \cap {}^\bullet t \neq \emptyset$ which contradicts structural conditions of the post-agglomeration.   □

Corollary 4. *Let m be a reachable marking and let s be a sequence of T^ω such that $\forall s' \in Pref(s), \Gamma(s') \geq 0$. Then $\forall h \in H, \forall f \in F$,*

$$m[h.s.f > m' \Longrightarrow m[h.f.s > m'$$

We prove now that any finite sequence can be well reordered.

Proposition 12. *Let m be a reachable marking from m_0. Let $s \in T^*$ be a sequence from m to a marking m' ($m[s > m'$) such that $\forall s' \in Pref(s), \Gamma(s') \geq 0$.*

There exists a permutation of s, $s = s_{\bowtie}.s_{\lhd}$, such that:

1. $m[s > m'$
2. $\Pi_F(s_{\lhd}) = \lambda$ and $\Pi_{T_0 \cup H}(s_{\bowtie}) = \Pi_{T_0 \cup H}(s)$
3. $s_{\bowtie}$ *is simulateable*

   Any sequence $s = s_{\bowtie}.s_{\lhd}$ fulfilling the above requirements is called an ordered representative *of s.*

Proof. We prove by induction on the length of s that there exists at least one ordered representative of s.

$|s| = 0$: The sequence $s = s = \lambda$ fulfills the conditions of the proposition.
$|s| > 0$: If $s \in T_0^*$ then the sequence $s = s_{\bowtie} = s$ is the unique ordered representative of s. Otherwise, the sequence s can be written $s = s'.t.s''$ with $s'' \in T_0^*$ and $t \in H \cup F$.
By inductive hypothesis on s', we have $s' = s'_{\bowtie}.s'_{\lhd}$ and $m[s'_{\bowtie}.s'_{\lhd}.t.s'' > m'$.
We consider two cases:

- $t \in H$ (i.e. $s = s'.h.s''$). By construction, $s'_{\bowtie}.(s'_{\lhd}.h.s'')$ is an ordered representative of s.
- $t \in F$ (i.e. $s = s'.f.s''$). Because $\Gamma(s'.f) < \Gamma(s') = \Gamma(s'_{\lhd})$, then, $|s'_{\lhd}|_H > 0$. So, $\exists h \in H$, $\exists s_0 \in T_0^*$, $\exists s_1 \in (T_0 \cup H)^*$ such that $s'_{\lhd} = s_0.h.s_1$ and so we have $m[s'_{\bowtie}.s_0.h.s_1.f.s'' > m'$.
  As $|s_1|_F = 0$, Corollary 4 implies that $m[s'_{\bowtie}.s_0.h.f.s_1.s'' > m'$.
  Because of its construction ($s'_{\bowtie}.s_0.h.f$ is a simulateable sequence, $s_1.s'' \in (T_0 \cup H)^*$) and because we have only permuted transitions of F, the sequence $s'_{\bowtie}.s_0.h.f.s_1.s''$ is an ordered representative of s. □

We can now demonstrate that any infinite sequence can also be well re-ordered.

Proposition 13. *Let $s \in L^\infty(N, m_0)$. There exists a permutation of s, s^∞ such that*

1. $\Pi_{T_0 \cup H}(s^\infty) = \Pi_{T_0 \cup H}(s)$
2. $\exists (s_{\bowtie}^i)_{i \geq 0}$ *an infinite series of simulateable sequences such that*
   (a) $d^\circ(s) = n \Rightarrow s^\infty = (s_{\bowtie}^0).s_{h^n}.(s_{\bowtie}^1).(s_{\bowtie}^2)\ldots(s_{\bowtie}^k)\ldots$ *with $s_{h^n} \in (T_0 \cup H)^*$*
   (b) $d^\circ(s) = \infty \Rightarrow s^\infty = (s_{\bowtie}^1.h^1).(s_{\bowtie}^2.h^2).(s_{\bowtie}^3.h^3)\ldots(s_{\bowtie}^k.h^k)\ldots$ *with $h^i \in H$*

Proof. The proof is similar to the one used for the pre-agglomeration excepted that for s with a finite degree $\theta(s) = s_{\bowtie}^0.s_{h^n}$ with $s_{h^n} \in (T_0 \cup H)^*$ (not only in H^*). □

Before conclude this section with the Theorem 3, we prove that to any maximal sequence of a post-agglomerateable net corresponds a sequence with the same projection on $T_0 \cup H$ and that is simulateable.

Lemma 3. *Let $s \in L^\infty(N, m_0)$. There exists an infinite simulateable sequence $s_{\bowtie}$ such that*

1. $s_{\bowtie} \in L^{\infty}(N, m_0)$
2. $\Pi_{T_0 \cup H}(s_{\bowtie}) = \Pi_{T_0 \cup H}(s)$

Proof. We distinguish the two cases:

1. $d^{\circ}(s) = n$: Let $s_{\bowtie} = (s_{\bowtie}^0).s_{h.f}.(s_{\bowtie}^1).(s_{\bowtie}^2)\ldots(s_{\bowtie}^k)\ldots$ with $s_{h.f}$ the sequence obtained from s_{h^n} by adding an occurrence of F (chosen arbitrarily) after any occurrence of H. Applying n times Prop. 11, it comes that if we note m_1 the marking reached after the firing of $s_{\bowtie}^0.s_{h^n}$ then $m_0[s_{\bowtie}^0.s_{h.f} > m_1'$ with $m_1' \geq m_1$. So $s_{\bowtie} \in L^{\infty}(N, m_0)$ and has the same projection than s over $T_0 \cup H$.
2. $d^{\circ}(s) = \infty$: Let $s_{\bowtie} = (s_{\bowtie}^1.h^1.f).(s_{\bowtie}^2.h^2.f).(s_{\bowtie}^3.h^3.f)\ldots(s_{\bowtie}^k.h^k.f)\ldots$ with f a transition of F (chosen arbitrarily). Using Prop. 11 on each $s_{\bowtie}^i.h^i$, $s_{\bowtie}$ fulfills the conditions of the proposition. $\qquad\square$

Lemma 4. *Let $s \in L^*(N, m_0)$ be a sequence leading to a terminal state. Then s is balanced and $\phi^{-1}(\hat{s})$ is a sequence of the reduced net leading to a terminal state (and by construction, has the same projection over $T_0 \cup H$ than s).*

Proof. If s is unbalanced then $\hat{s} = s_{\bowtie}.s'.h.s''$ with $h \in H$, $s' \in T_0^*$ and $s'' \in (T_0 \cup H)^*$. As $p \in h^{\bullet}$, and because ${}^{\bullet}(F) = \{p\}$ and $\Pi_F(h.s'') = 0$ then necessarily $\exists f \in F$ such that $m_0[\hat{s}.f >$. Because $m_0[s > m$ iff $m_0[\hat{s} > m$, we obtain a contradiction.

Because s is balanced then $\phi^{-1}(\hat{s})$ is a sequence of the reduced net and leads to a terminal state due to Proposition 4. $\qquad\square$

We can now prove that the post-agglomeration defines an equivalence for any property expressed in LTL when transitions F are not observed (and when the operator X is not used in case of state-based logic).

Theorem 3 (Post-agglomeration and model checking).
    Let f be a state-based $LTL\backslash X$ (resp. action-based LTL) formula. Let (N, m_0) be a system and (N_r, m_{0_r}) be the corresponding reduced net by a post-agglomeration w.r.t. H and F. If $F \cap Obs(AP) = \emptyset$ then

$$Trace(N, m_0) \models f \Leftrightarrow Trace(N_r, m_{0_r}) \models f$$

Proof. The proof is similar to the one used for the pre-agglomeration (Theorem 2): We prove that there exists a counter example of f in the original net if and only if there exists a counter example in the reduced net. $\qquad\square$

4   Experimentations

In this section, we experiment the agglomeration technique on two simple and classical examples: a model of the dining philosophers (Fig. 1) in which the terminal states are controlled by a deadlock (implying the place $Tickets$) and a model of the slotted ring protocol (Fig. 2). Both examples are scalable in such way that the number of reachable markings is exponential w.r.t. a given parameter. It is important to note that different abstractions have been already done in the models. As an example, the philosophers take a second fork and release a ticket in a same atomic transition. In a programming point of view, these action are usually developed in a more complex scheme. We use colored Petri nets for the presentation of the models but all experimentations are done on the corresponding unfolded ordinary nets for a given value of the parameter.

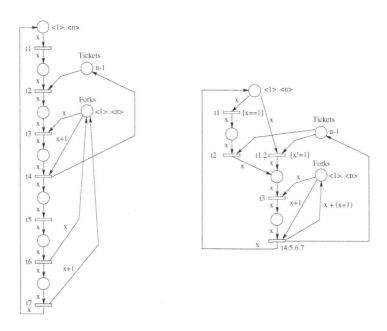

Fig. 1. a model of dinner philosophers and its reduction

An usual verification scheme consists firstly to check the absence of terminal marking. When this property is ensured, one analyzes more specific properties. We present our experimentations following this methodology by comparing the sizes of the reachability graphs generated for both original nets and agglomerated ones. Moreover, we also compare the results obtained using or not the stubborn set technic [5]. It is important to note that this technic preserves all terminal markings and allows the verification of *LTL* formulas. All the presented results have been obtained using the verification tool PROD [6].

In [1], Berthelot has shown that the presence of terminal markings is preserved by the pre and post-agglomerations. Then, we have applied all the possible agglomerations on both nets. Table 1 presents the size of the reachability graph for the two models. As expected, the graph sizes of agglomerated nets are always smaller than the ones of the original nets. The stubborn set reduction technique obtains good results on these examples (the size can vary by a factor of 10 to 40).

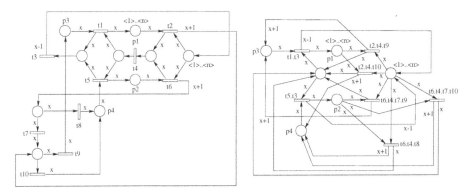

Fig. 2. a model of the slotted ring protocol and its reduction

However, one can remark that the sizes of the complete reachability graphs of the agglomerated nets are even smaller than those obtained using stubborn sets on the original nets. Furthermore, combining the two approaches (by applying stubborn sets on the agglomerated nets) allows to obtain better results.

We can remark that the effect of the stubborn set reduction on the agglomerated nets is not important. The applied agglomerations limit the concurrency of the model and then the reductions which can be done by the stubborn sets.

Model	Original Net				Agglomerated Net			
	Complete Gph.		Reduced Gph.		Complete Gph.		Reduced Gph.	
	Nodes	Edges	Nodes	Edges	Nodes	Edges	Nodes	Edges
Philo4	1158	4036	276	387	65	208	65	188
Philo6	40185	211164	3145	4685	665	3408	620	2286
Philo8	*	*	29079	44402	6305	44096	5165	21640
Philo10	*	*	247244	383189	58025	512080	40875	186690
Slotted2	208	560	60	78	24	56	22	43
Slotted3	4032	15840	380	510	160	504	116	252
Slotted4	82176	425472	2292	3095	1120	4480	632	1455
Slotted5	*	*	13073	17847	8064	39200	3525	8358

Table 1. terminal marking detection

As there is no terminal marking in these models, we now verify some *LTL* formulas. For the dining philosophers, we observe the transitions $t1$, $t2$ and $t4$ related to the philosopher 1 (we use the identity for the labelling of these transitions when the others are associated to λ) and we verify the two following action-based *LTL* formulas:

$$G(t1_1 \Rightarrow F(t4_1))$$
$$G(t1_1 \Rightarrow F(t4_1)) \Rightarrow (F(t2_1) \vee G(\neg t1_1))$$

Figure 1 presents, at its right, the net obtained after reduction. The transitions $t1_i$ and $t2_i$, $\forall i \neq 1$ have been pre-agglomerated and the transitions $t4_i$, $t5_i$, $t6_i$ and $t7_i$, $\forall i$ have been post-agglomerated successively. Notice that these agglomerations respect the conditions imposed by Theorems 2 and 3 w.r.t. the considered formulas. Notice also that one other pre-agglomeration can be applied (i.e $t1_1$ with $t2_1$) and has been realized for the experimentations of the terminal marking detection.

For the slotted ring protocol, we define a state-based logic only implying the token of the places $p1$, $p2$, $p3$ and $p4$ related to the process 1 and verify the two following formulas:

$$F(p1_1 \wedge p2_1)$$
$$G((p1_1 \wedge p2_1) \Rightarrow (p3_1 \wedge p4_1))$$

Figure 2 presents, at its right, the net obtained after reduction. The transition sets $\{t1_i, t5_i\}$ have been post-agglomerated with $\{t3_i\}$, $\forall i$ (leading to the creation of $t1t3_i$ and $t5t3_i$). Then, the following post-agglomerations have been successively done: $\{t2_i, t6_i\}$ with $\{t4_i\}$; $\{t6t4_i\}$ with $\{t7_i, t8_i\}$; $\{t6t4t7_i, t2t4_i\}$ with $\{t9_i, t10_i\}$. No other agglomeration can be done and then this net has been also used for the terminal marking detection. Notice that once again these agglomerations respect the conditions imposed by Theorems 2 and 3 w.r.t. the considered formulas.

Table 2 gives the size of the visited part of the synchronized product between the state graphs and the Büchi automata represented the negation of the formulas. Notice that the first formulas of both models are not satisfied while the examples fulfill the second ones. As the `Prod` tool implements an on-the-fly algorithm, the verification process is stopped as soon as a counter example is found. One can remark that for the agglomerated net, the counter examples are found after a longer search. However, when the formulas are satisfied, all the synchronized product must be visited. As expected, its size is smaller when the net has been agglomerated.

Applying the stubborn set technique on the agglomerated nets during the verification process has no effect and then the sizes of synchronized products are the same. We can remark that the new dependencies induced by the synchronization with the Büchi automaton limits the reduction power of the stubborn set method.

5   Conclusion

In this paper we have stated a new theoretical result: the pre and the post-agglomerations preserve a large set of properties expressed in linear temporal logics under simple conditions. Furthermore, as the corresponding colored reductions (defined by S.Haddad in [2]) are only based on the definition of the ordinary ones, results given in this paper can immediately be apply to colored Petri nets.

The complexity of applications of these reductions are linear with respect of the cardinal of $P \times T$ and reductions are made before any analysis. They offer

Model	Original Net				Agglomerated Net	
	Complete Graph		Reduced Graph			
	Nodes	Edges	Nodes	Edges	Nodes	Edges
Philo5	129	131	122	133	62	70
Philo10	2599	2914	347	358	299	350
Philo15	424368	516579	672	683	589	675
Philo5	8262	31186	8262	31182	412	1596
Philo8	*	*	*	*	10945	71514
Philo10	*	*	*	*	98427	818048
Slotted5	4801	5993	73	73	288	355
Slotted8	63773	81237	936	939	1281	1577
Slotted10	*	*	1532	1535	4733	5968
Slotted3	34942	242635	9772	33265	1292	7428
Slotted4	*	*	114100	436264	9092	65901
Slotted5	*	*	*	*	65572	576094

Table 2. *LTL* formula verification

so a very efficient solution for combating the state space explosion especially when the net is large. Furthermore, the reduced net can be analyzed with any other method and then, one can combine the use of these reductions with the use of construction/representation optimizations techniques of the state space such that partial orders, symmetries or BDD.

We plan to explore similar results for other reductions (such that implicit place simplification, equivalence place/transition simplification, ...), to develop new reductions based only on preservation of particular properties and to study the use of these techniques in other formalisms (for instance synchronized automata). The first point seems to us very easy to obtain while the others need more works.

References

[1] G. Berthelot. Checking properties of nets using transformations. In G. Rozenberg, editor, *Advances in Petri nets*, volume No. 222 of *LNCS*. Springer-Verlag, 1985.

[2] S. Haddad. A reduction theory for colored nets. In Jensen and Rozenberg, editors, *High-level Petri Nets, Theory and Application*, LNCS, pages 399–425. Springer-Verlag, 1991.

[3] Doron Peled and Thomas Wilke. Stutter-invariant temporal properties are expressible without the nexttime operator. *Information Processing Letters*, 63:243–246, 1997.

[4] W. Reisig. *EATCS-An Introduction to Petri Nets*. Springer-Verlag, 1983.

[5] A. Valmari. On-the-fly verification with stubborn sets representatives. In *Proceedings of the 5th International Conference on Computer Aided Verification, Greece*, volume 697 of *Lecture Notes in Computer Science*, pages 397–408. Springer Verlag, 1993.

[6] K. Varpaaniemi and M. Rauhamaa. The stubborn set method in practice. In *Advances in Petri Nets*, volume 616 of *Lecture Notes in Computer Science*, pages 389–393. Springer Verlag, 1992.

Bisimulation and the Reduction of Petri Nets

Philippe Schnoebelen[1] and Natalia Sidorova[2]

[1] Lab. Spécification & Vérification, ENS de Cachan & CNRS UMR 8643
61, av. Pdt. Wilson, 94235 Cachan Cedex, France
`phs@lsv.ens-cachan.fr`
[2] Eindhoven Univ. of Technology, Faculty of Electrical Engineering
P.O. Box 513, 5600 MB Eindhoven, The Netherlands
`natalia@ics.ele.tue.nl`

Abstract. We investigate structural equivalences on places of P/T nets that allow reductions compatible with bisimilarity. This comes with a study of two kinds of reductions: fusion of equivalent places, and replacement of some places by other ones. When effectivity issues are considered, we are lead to a variant of place bisimulation that takes into account a set of "relevant" markings.

1 Introduction

Place bisimulation. *Place bisimulation* is a formal notion of equivalence between places of P/T nets that combines two distinctive features: it agrees with a notion of *bisimilarity*, and it relates equivalent *places* rather than equivalent markings.

While several early papers adapted Milner's concept of bisimilarity to the transition systems generated by Petri nets [NT84, Pom86], Olderog was the first to propose a notion of bisimulation that comes from lifting a bisimulation between places [Old89, Old91]. The main advantage of this approach is that, as with many structural methods on Petri nets, it considers the elements (places and transitions) of the net rather than the reachability graph (which may be infinite). Using Olderog's method, it was possible to prove that two nets had equivalent behaviours just by exhibiting a relation between their places and checking a finite number of local constraints.

The name "place bisimulation" comes from [ABS91] where Olderog's proposal is improved and where it was proven that place bisimulation could be used to *reduce* P/T nets (rather than just state two nets have equivalent behaviours).

Reductions for Petri nets. *Reductions* are transformations of Petri nets that decrease the size. Their main use is as an optimization technique. For such applications the reduction must return a net behaviorally equivalent to the original one, but smaller in size. This smaller net can be used in place of the original net when it comes to implementing the net, or to analyzing it (e.g., for verification purposes).

Place bisimulations yield reductions that preserve the branching-time behaviour. These reductions are *structural* (one removes places and transitions,

M. Nielsen, D. Simpson (Eds.): ICATPN 2000, LNCS 1825, pp. 409–423, 2000.

and this is what may remove markings) and do not depend on the initial marking (or the set of reachable markings).

The fact that place bisimulations do not depend on the set of reachable markings has its pros and cons. On the positive side, this allows polynomial-time algorithms, and leads to reductions that are correct whatever the initial marking. On the negative side, some reductions that seem obviously correct are not given by place bisimulation simply because there exists an (irrelevant) alternative initial marking for which they are incorrect. This is the drawback we try to alleviate in this paper.

Our contribution. In this paper we investigate structural reductions of P/T nets that preserve bisimilarity and that take into account a notion of which markings are relevant. Our starting point is an investigation of what are the correct ways of de ning an equivalence on places. This departs from earlier works where behavioural equivalence is the starting point, and reductions only a byproduct.

We rst identify two di erent, equally natural, reductions based on a notion of equivalent places. They are fusion and replacement . There is a strong connection between them but correct fusions only coincide with correct replacements when we impose (rather strong) structural conditions on the set of relevant markings. These theoretical results and the accompanying examples are enlightening and they help understand why it is quite delicate to take into account a set of markings in these structural reduction problems.

Then, decidability issues impose further restrictions. We end up with a variant of place bisimulation that is parameterized by a set of markings, a partial answer to what was felt as the main inconvenience of place bisimulation. We say the answer is only partial because the set of reachable markings has to satisfy some strong closure restrictions.

Why bisimulation ? We consider bisimilarity as our basic correctness notion because it has proven to be a fundamental semantic equivalence in the algebraic theory of concurrent systems [Mil89, Mil90, Gla90], because it is more local than more classical language-theoretical notions and then often behaves better algorithmically, because there exists a very successful veri cation technology for transition systems based on bisimilarity (e.g., [CPS93]), because it has not been much studied in the context of Petri nets.

Related works. The work in [ABS91] was continued in [AS92] (studying how true concurrency is preserved, and where an algorithm for place bisimulation is given) and [APS94] (investigating place bisimulations abstracting from silent moves). These works did not start from reduction issues and we nd our new approach clearer and more natural (also, these works did not try to take into account a set of relevant markings).

Independently, Voorhoeve studied a concept very similar to place bisimulations [Voo96]. His starting point is also bisimulation + structural equivalences , but he emphasizes equivalence of behaviours and does not mention applications to reductions of Petri nets.

For optimization purposes there also exists structural reductions that do not preserve bisimilarity (e.g., [NKML95, STMD96]). For more specific purposes, reductions have also been used as a computational device [Ber86]. E.g., telling whether a net is bounded can be done by reducing it to a normal form where boundedness is obvious. Here it is not necessary that the reductions preserve all of the behaviour of the net, only the relevant aspect (boundedness in our example) is enough.

Plan of the paper. We first recall basic definitions and notations about P/T nets and bisimilarity (§ 2). Then we define reductions by fusion (§ 3) and by replacement (§ 4). It is then possible to define and study when a correct replacement gives rise to a correct fusion and vice versa (§ 5). We investigate how correct reductions combine (§ 6) and show there exists a largest correct reduction, unfortunately not computable in general (§ 7). Place bisimulation is then introduced as a computable approximation (§ 8) and exemplified (§ 9).

2   Basic Definitions

$\mathbb{N}$ denotes the set of natural numbers. Let $Act = \{a, b, c, \ldots\}$ be a finite alphabet of *action names*.

2.1   Labeled Nets and Their Behaviour

A *labelled Petri net* is a tuple $N = \langle P_N, T_N, F_N, l_N \rangle$, where:

- P_N and T_N are two disjoint non-empty finite sets of *places* and *transitions* respectively;
- $F_N : (P_N \times T_N) \cup (T_N \times P_N) \to \mathbb{N}$ is a *weight function* (also called *flow relation*);
- $l_N : T_N \to Act$ labels each transition $t \in T_N$ with some action $l_N(t)$ from *Act*.

We drop the N subscript whenever no ambiguity can arise and present nets with the usual boxes and circles graphical presentation.

Markings are configurations of a net. A *marking* M of N can be seen as a multiset over P, i.e. a mapping from P into $\mathbb{N}$. This is the viewpoint we adopt here. Then $M(p)$ is the number of appearances of place p in multiset M: this represents the number of tokens on p in marking M. The set of all possible markings is $\mathbb{N}^P$.

We write $|M|$ for the size of a marking M (its number of tokens) and, given two markings M, M' we let $\Delta(M, M')$ denote the *distance* between M and M', i.e. the number of tokens one has to move to transform M into M'.

Given a transition $t \in T$, the *preset* $^\bullet t$ and the *postset* $t^\bullet$ of t are the multisets of places given by $^\bullet t(p) \stackrel{\text{def}}{=} F(p, t)$ and $t^\bullet(p) \stackrel{\text{def}}{=} F(t, p)$ for any $p \in P$.

A transition $t \in T$ is *enabled* in marking M iff $\forall p \in P, M(p) \geq F_N(p, t)$. An enabled transition t may fire, thus performing action $l(t)$. This results in a new marking M' defined by $M' \stackrel{\text{def}}{=} M - {}^\bullet t + t^\bullet$ (that is, $M'(p) \stackrel{\text{def}}{=} M(p) - F(p, t) + F(t, p)$ for any $p \in P$).

We write $N : M \stackrel{t}{\to} M'$ to denote that $M \stackrel{t}{\to} M'$ is a step in net N. Derived notations are "$M \stackrel{t}{\to} M'$" when N is implicit, "$M \stackrel{a}{\to} M'$" when a is $l(t)$, "$M \stackrel{a}{\to}$" when M' is not relevant, etc.

2.2   Bisimulation

Let $N_1 = (P_1, T_1, F_1, l_1)$ and $N_2 = (P_2, T_2, F_2, l_2)$ be two nets and $R \subseteq \mathbb{N}^{P_1} \times \mathbb{N}^{P_2}$ be a relation between their markings.

We say that R has the *transfer property* iff for all M_1, M_2 s.t. $M_1 R M_2$, and for all steps $N_1 : M_1 \stackrel{t_1}{\to} M'_1$, there exists a step $N_2 : M_2 \stackrel{t_2}{\to} M'_2$ s.t. $l(t_2) = l(t_1)$ and $M'_1 R M'_2$.

If R and R^{-1} have the transfer property (we sometimes say that R has the transfer property in both directions) then R is a *bisimulation* between N_1 and N_2 [Par81, Mil89]. When $M_1 R M_2$ for a bisimulation R we say that (N_1, M_1) and (N_2, M_2) are bisimilar, written $N_1, M_1 \sim N_2, M_2$. When N_1 and N_2 are the same net, we simply write $M_1 \sim M_2$.

It is well-known that, given any two nets, there exists a largest bisimulation between them, and we use $\sim_{[N_1, N_2]}$ to denote it. When $N_1 = N_2$, the largest bisimulation is an equivalence. Our proofs that relations are bisimulation often use basic properties (like $\sim_{[N_2, N_3]} \circ \sim_{[N_1, N_2]} \subseteq \sim_{[N_1, N_3]}$) and Milner's "up to" technique [Mil89]: we say R has the transfer property *up to bisimulation* if $M_1 R M_2$ and $M_1 \stackrel{t_1}{\to} M'_1$ imply that there exists a $M_2 \stackrel{t_2}{\to} M'_2$ with $M_1 R' M_2$ for $R' \stackrel{\text{def}}{=} \sim_{[N_2, N_2]} \circ R \circ \sim_{[N_1, N_1]}$ (and $l(t_2) = l(t_1)$). Then R' has the transfer property.

3   Reduction by Fusion

Fusion is a natural way of reducing nets, based on the general idea of quotients of structures.

Consider a net N and assume $B \subseteq P \times P$ is an equivalence relation between places. Then N can be reduced by fusing B-equivalent places into a single place. This yields a new net denoted N/B. We write p/B for the equivalence class of p (and then for the fused place in N/B). The transitions are untouched but the arcs between places and transitions follow the fusion process: formally, if $^\bullet t$ (resp. $t^\bullet$) is $\{p_1, \ldots, p_k\}$ in N, then in N/B, $^\bullet t$ (resp. $t^\bullet$) is $\{p_1/B, \ldots, p_k/B\}$. A marking M in N is fused into a marking $m \stackrel{\text{def}}{=} M/B$ in N/B. m and M have the same number of tokens. Fig. 1 gives an example where B relates p_2 and p_4.

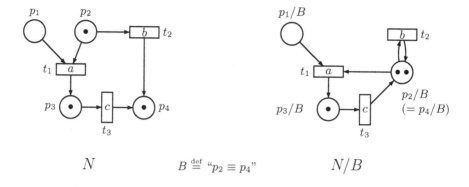

Fig. 1. Fusion of places as a reduction method

Fusing places modifies the behaviour, but only in one direction: it adds new behaviours. This is captured as

Fact 3.1. *If $M \xrightarrow{t} M'$ in N, then $M/B \xrightarrow{t} M'/B$ in N/B.*

This implies that $_-/B$, defined as $\{(M, M/B) \mid M \in \mathbb{N}^P\}$, has the transfer property from N to N/B.

A fundamental property is the following weak converse:

Proposition 3.2. *If $m \xrightarrow{t} m'$ in N/B, then there exist two markings M and M' in N s.t. $m = M/B$, $m' = M'/B$ and $M \xrightarrow{t} M'$.*

Proof. Assume $m \xrightarrow{t} m'$. Write M_1 for $^\bullet t$ in N and pick any M_2 s.t. M_2/B is $m - (M_1/B)$. Then, in N, $M \xrightarrow{t} M'$ for $M = M_1 + M_2$ and $M' = t^\bullet + M_2$. Clearly, $M'/B = m'$. $\square$

An equivalence B between places can be lifted to an equivalence, written $\overline{B}$, between markings. Formally, if $p_1 B p'_1, \ldots, p_n B p'_n$ then $\{p_1, \ldots, p_n\}\overline{B}\{p'_1, \ldots, p'_n\}$. Note that $M\overline{B}M'$ iff $M/B = M'/B$, i.e.

$$\overline{B} = (_-/B)^{-1} \circ (_-/B). \tag{1}$$

Now, a direct reading of Prop. 3.2 gives the following

Fact 3.3. *When $\overline{B}$ relates bisimilar markings only, i.e. $\overline{B} \subseteq \sim_{[N,N]}$, then $(_-/B)^{-1}$ has the transfer property up to bisimulation.*

4   Reduction by Replacement

Replacement is a second natural way of reducing nets, this time based on the general idea of homomorphic images of structures.

Consider a net N and assume $h : P \to P$ is a projection, i.e. for all $p \in P$ we have $h(h(p)) = h(p)$. In N a place p can be replaced by $h(p)$ whenever $h(p) \neq p$. This yields a new net, denoted $h(N)$.

More precisely, $h(N)$ is obtained by redirecting all output edges of transitions by their projections: if $t^\bullet = M = \{p_1 \quad p_k\}$ in N, it becomes $h(M) \overset{\text{def}}{=} \{h(p_1) \quad h(p_k)\}$ in $h(N)$. All places not in $h(P)$ are removed, all transitions that have some input place removed are removed as well. A marking M in N yields a marking $m = h(M)$ in $h(N)$. m and M have the same number of tokens. Fig. 2 gives an example where p_3, and then t_3, are removed.

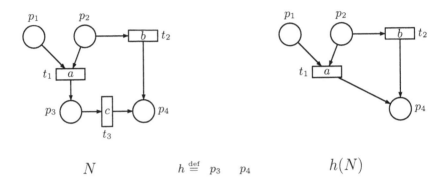

$$N \qquad\qquad h \overset{\text{def}}{=} \quad p_3 \quad p_4 \qquad\qquad h(N)$$

Fig. 2. Replacement of places as a reduction method

Replacing places modify the behaviour in the following way:

Fact 4.1. *If* $m \overset{t}{\to} m'$ *in* $h(N)$, *then* $m \overset{t}{\to} M'$ *in* N *for some* M' *s.t.* $h(M') = m'$.

There is a weak converse:

Fact 4.2. *If* $M \overset{t}{\to} M'$ *in* N *and* $h(M) = M$, *then* $M \overset{t}{\to} h(M')$ *in* $h(N)$.

5   Correct Reductions

We are only interested in *correct reductions*. Informally, a correct reduction is a reduction such that the reduced net N' is a correct variant of the original net N.

In this paper, we investigate formal notions of correctness

1. grounded into bisimilarity, and
2. only applying to a given set of markings (called *relevant markings*).

5.1   Relevant Markings

In all the following, we assume $\mathcal{A} \subseteq \mathbb{N}^P$ is a set of relevant markings. Typical examples are "the set of all markings reachable from some initial M_0", or "all markings with three tokens".

Compared to earlier works on place bisimulation, we remove the basic (implicit) assumption that all and every markings of the nets under study are meaningful. It was a very strong assumption, resulting into a very conservative view of when two places are bisimilar.

This is why our new assumption is that we only aim at correction relative to a given set of relevant markings. In a sense, this work can be seen as extending the earlier place bisimulation theory to a framework with relevant markings. But, in addition, the paper also brings a new viewpoint: here we focus on reductions while the earlier works focused on semantic equivalences. We think the new viewpoint is clearer and more natural.

5.2   Correctness for N with $\mathcal{A}$

Assume N comes with a set $\mathcal{A}$ of relevant markings. A reduction is correct if all relevant markings are reduced into bisimilar markings in the reduced net. The other (= irrelevant) markings are not considered for the correctness issues.

With this in mind, the following definitions are natural:

Definition 5.1. *An equivalence $B \subseteq P \times P$ is a* correct fusion *for net N with relevant markings $\mathcal{A}$ iff $N, M \sim N/B, M/B$ for all $M \in \mathcal{A}$.*

Definition 5.2. *A projection $h : P \to P$ is a* correct projection *for net N with relevant markings $\mathcal{A}$ iff $N, M \sim h(N), h(M)$ for all $M \in \mathcal{A}$.*

We are interested into the connections between these two definitions. Indeed, any projection $h : P_N \to P_N$ naturally induces an equivalence B_h given by $pB_hp' \stackrel{\text{def}}{\Leftrightarrow} h(p) = h(p')$. Can we say that B_h is a correct fusion iff h is a correct projection ? It turns out this is not the case in general.

Fact 5.3. *There exist situations where h is a correct projection for $N, \mathcal{A}$ and B_h is not a correct fusion.*

Example 5.4. Figure 3 gives an example where $h \stackrel{\text{def}}{=} \{p_3 \mapsto p_1, p_4 \mapsto p_1\}$ is correct (for $\mathcal{A}$ the set of reachable markings) but B_h, "$p_1 \equiv p_3 \equiv p_4$", is not.   □

Example 5.5. Another example is provided in Figure 4 where $h \stackrel{\text{def}}{=} \{p_3 \mapsto p_2, p_4 \mapsto p_2\}$ and $\mathcal{A} \stackrel{\text{def}}{=} \{\{p_1\}\}$.   □

Fact 5.6. *There exist situations where B is a correct fusion while there is no correct projection h s.t. $B = B_h$.*

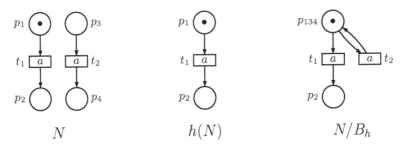

Fig. 3. $h = \{p_3 \mapsto p_1, p_4 \mapsto p_1\}$ is correct and B_h, "$p_1 \equiv p_3 \equiv p_4$", is not

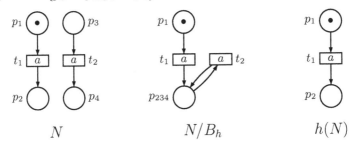

Fig. 4. $h = \{p_3 \mapsto p_2, p_4 \mapsto p_2\}$ is correct and B_h, "$p_2 \equiv p_3 \equiv p_4$", is not

Example 5.7. Figure 5 gives an example where B, "$p_1 \equiv p_2$", is a correct fusion (for $\mathcal{A}$ the set of reachable markings) but where no h is correct (the figure illustrates $h \stackrel{\text{def}}{=} \{p_1 \mapsto p_2\}$). □

Example 5.8. Another example is provided in Figure 6 with B given by "$p_2 \equiv p_3$" and $h \stackrel{\text{def}}{=} \{p_2 \mapsto p_3\}$. Here $\mathcal{A} \stackrel{\text{def}}{=} \{\{p_1\}\}$. □

It turns out that all these difficulties come from the sets $\mathcal{A}$ we picked in these examples. They disappear when we impose restrictions on possible choices for $\mathcal{A}$.

Given N and a set of markings $\mathcal{A}$, we say that $\mathcal{A}$ is *succ-closed*, if $M \in \mathcal{A}$ and $M \rightarrow M'$ entail $M' \in \mathcal{A}$. The sets $\mathcal{A}$ in examples 5.4 and 5.7 are succ-closed by construction.

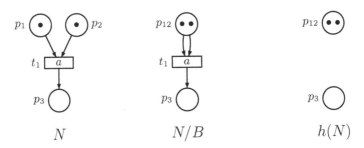

Fig. 5. B, "$p_1 \equiv p_2$", is correct and $h = \{p_1 \mapsto p_2\}$ is not

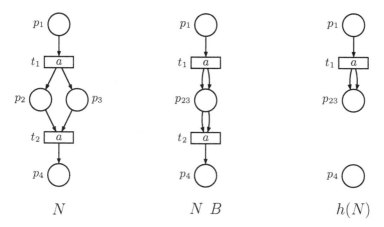

N $N\,B$ $h(N)$

Fig. 6. B, $p_2 \equiv p_3$, is correct and $h = \{p_2 \mapsto p_3\}$ is not

Secondly, given an equivalence B, we say that $\mathcal{A}$ is B-closed if $M \in \mathcal{A}$ and $M\overline{B}M'$ entail $M' \in \mathcal{A}$. The sets $\mathcal{A}$ in examples 5.5 and 5.8 are B-closed.

We can now state the following crucial property, explaining how when $\mathcal{A}$ is succ-closed and B-closed, checking whether B is correct can be done by inspecting N only. This lemma will be of great help because, in general, it is difficult to tell whether a given B or h are correct: indeed, for this we have to compare behaviours in N and in $N\,B$, or $h(N)$.

Lemma 5.9. *Assume $\mathcal{A}$ is succ-closed and B-closed. Then B is a correct fusion iff $\overline{B} \cap (\mathcal{A} \times \mathcal{A})$ is included in $\sim$, the largest bisimulation between markings of N.*

Proof. ($\Rightarrow$) This is obvious when $\mathcal{A} = \mathbb{N}^{P_N}$, using (1) and the assumption that _ B is a bisimulation. In the general case, B-closure of $\mathcal{A}$ gives that $\overline{B} \cap (\mathcal{A} \times \mathcal{A}) = R^{-1} \circ R$ where $R \overset{\text{def}}{=} \{(M\,M\,B) \mid M \in \mathcal{A}\}$ is included into $\sim_{[N\,N\,B]}$ (by assumption).

($\Leftarrow$) Now assume $\overline{B} \cap (\mathcal{A} \times \mathcal{A})$ is included in $\sim$, and define $R \overset{\text{def}}{=} \{(M\,M\,B) \mid M \in \mathcal{A}\}$. Fact 3.1 and succ-closure of $\mathcal{A}$ implies that R has the transfer property.

For the other direction, the reasoning underlying Fact 3.3 applies even when $\mathcal{A}$ is taken into account. As a consequence, R is included into $\sim_{[N\,N\,B]}$, so that B is a correct fusion. □

Theorem 5.10. *Assume h is a projection and $\mathcal{A}$ is B_h-closed and succ-closed. Then h is a correct projection for $N\,\mathcal{A}$ iff B_h is a correct fusion for $N\,\mathcal{A}$.*

Proof. ($\Rightarrow$) Assume h is a correct projection and consider $M_1 \in \mathcal{A}$. Let M_2 s.t. $h(M_1) = h(M_2)$. Correctness of h implies $N\,M_1 \sim h(N)\,h(M_1) \sim N\,M_2$ (since $M_2 \in \mathcal{A}$), hence $M_1 \sim M_2$. Thus $\overline{B_h} \subseteq \sim_{[N\,N]} \cap (\mathcal{A} \times \mathcal{A})$ and Lemma 5.9 concludes.

($\Leftarrow$) Now assume that B_h is a correct fusion. Lemma 5.9 and the fact that $M\overline{B_h}h(M)$ give that $N, M \sim h(N), h(M)$. $\square$

Our earlier examples show that succ-closure or B-closure alone is not sufficient for Theorem 5.10.

When $\mathcal{A}$ is the set of all markings, closure is guaranteed and fusions or replacements coincide. In the following, whenever $\mathcal{A}$ is succ-closed and B-closed, we speak of correct equivalences without more precision on whether we mean fusions or replacements.

6   Combining Correct Reductions

Correct reductions can be composed, but the natural ways for composing projections and for composing fusions are not the same.

Lemma 6.1. *If h_1 is a correct projection for $N, \mathcal{A}$ and h_2 is a correct projection for $h_1(N), h_1(\mathcal{A})$, where $h_1(\mathcal{A}) \stackrel{def}{=} \{h_1(M) \mid M \in \mathcal{A}\}$), then $h_2 \circ h_1$ is a correct projection for $N, \mathcal{A}$.*

Proof. Direct from Def. 5.2 and transitivity of bisimilarity. $\square$

Furthermore, if $\mathcal{A}$ is succ-closed and B_{h_1}-closed, and $h_1(\mathcal{A})$ is B_{h_2}-closed, then $\mathcal{A}$ is succ-closed and $B_{h_2 \circ h_1}$-closed.

Now let B_1 and B_2 be two equivalences and assume $\mathcal{A}$ is B_1-closed, B_2-closed and succ-closed. Then

Lemma 6.2. *If B_1 and B_2 are correct fusions for $N, \mathcal{A}$, then $B \stackrel{def}{=} (B_1 \cup B_2)^*$, the smallest equivalence larger than B_1 and B_2, is a correct fusion for $N, \mathcal{A}$.*

Proof. Once we notice B-closure of $\mathcal{A}$ and $\overline{B} = (\overline{B_1} \cup \overline{B_2})^*$, Lemma 5.9 does all the work. $\square$

A related question is whether, assuming B is a correct fusion, all more conservative reductions are correct too? Here, the intuition underlying the question is that an equivalence relation can only be incorrect by making too many identifications, not by making too few.

Let B be a correct fusion for some $N, \mathcal{A}$ (assuming $\mathcal{A}$ is B-closed and succ-closed), then

Lemma 6.3. *If B' is an equivalence s.t. $B' \subseteq B$ then B' is a correct fusion.*

Proof. Use Lemma 5.9 and $\overline{B'} \subseteq \overline{B}$. $\square$

This question can also be investigated in terms of projections: let h be a correct projection and $h = h_2 \circ h_1$. Is h_1 correct?

Lemma 6.4. *If $h = h_2 \circ h_1$ is a correct projection for $N, \mathcal{A}$, where $\mathcal{A}$ is B_h-closed and succ-closed, then h_1 is a correct projection for $N, \mathcal{A}$.*

Proof. First, note that $B_{h_1} \subseteq B_h$ and that $\mathcal{A}$ is B_{h_1}-closed. Now, if h is a correct projection then (Theorem 5.10) B_h is a correct fusion, hence (Lemma 6.3) B_{h_1} too, hence (Theorem 5.10) h_1 is a correct projection. $\square$

7   The Largest Correct Fusion

An immediate corollary of Lemma 6.2 is the following:

Theorem 7.1. *If $\mathcal{A}$ is succ-closed, there exists a largest correct fusion for N $\mathcal{A}$ (only equivalences under which $\mathcal{A}$ is closed are considered).*

We write $R(N \mathcal{A})$ for this largest correct fusion. All the equivalences that are subsets of $R(N \mathcal{A})$ are correct fusions (Lemma 6.3).

Now let N be a net with a succ-closed set of relevant markings $\mathcal{A}$. We would like to compute $R(N \mathcal{A})$ and use it for reducing N in the most e cient way (w.r.t. $\mathcal{A}$). Unfortunately this is impossible because of:

Theorem 7.2 (Undecidability). *The largest correct fusion $R(N \mathcal{A})$ is not computable in general, even when $\mathcal{A} = \mathbb{N}^P$.*

The proof is long, technical and is omitted. Basically, it shows that the problem of telling whether two places p and q are such that $Id \cup \{(p\ q)\ (q\ p)\}$ is a correct fusion for N $\mathbb{N}^P$ is undecidable. This reuses ideas from [Qui95], extending Jancar's technique [Jan95] to a setting where all possible markings may be considered.

8   Place Bisimulation

Since $R(N \mathcal{A})$ cannot be computed in general, we settle for a computable approximation: *place bisimulation*.

In the following we assume a net N and a succ-closed set $\mathcal{A}$ of relevant markings are xed. The following de nition extends earlier proposals from [AS92, APS94]:

De nition 8.1. *A reflexive symmetric relation $B \subseteq P \times P$ is a place bisimulation over N $\mathcal{A}$ if $\mathcal{A}$ is B-closed and for all $M_1\ M_2 \in \mathcal{A}$ s.t. $M_1 \overline{B} M_2$, for all steps $M_1 \xrightarrow{t} M_1'$ there is a step $M_2 \xrightarrow{u} M_2'$ with $M_1' \overline{B} M_2'$ and $l(t) = l(u)$.*

Hence a place bisimulation is a B s.t. $\overline{B} \cap (\mathcal{A} \times \mathcal{A})$ has the transfer property.

Fact 8.2. *Any transitive place bisimulation over N $\mathcal{A}$ is a correct fusion.*

Proposition 8.3. *There exists a largest place bisimulation over N $\mathcal{A}$, denoted $B(N \mathcal{A})$.*

Proof (Idea). Similar to the proof of Theo. 7.1.   □

Clearly $B(N \mathcal{A}) \subseteq R(N \mathcal{A})$.

The de nition we gave is not very e ective, and it does not give a direct way to see whether $B(N \mathcal{A})$ is computable. Indeed, given a potential B, checking whether $\overline{B} \cap (\mathcal{A} \times \mathcal{A})$ has the transfer property involves checking an in nite number of situations (if $\mathcal{A}$ is in nite).

We now explain how to reduce this to a nite number of checks.

Definition 8.4. *The set of all* minimal covers *for a transition t, denoted C(t), is the set of all minimal markings $M \in \mathcal{A}$ s.t. $^\bullet t \subseteq M$.*

Here "M minimal" means that $M' \in \mathcal{A}$ and $^\bullet t \subseteq M' \subseteq M$ imply $M = M'$.

Note that Dickson's lemma implies the finiteness of $C(t)$ for any t. $C(t)$ can be empty if there is no way to enlarge $^\bullet t$ into some element of $\mathcal{A}$.

Definition 8.5. *We say that relation $B \subseteq P \times P$ has the* weak transfer property *if $\mathcal{A}$ is B-closed, and for all $t \in T$, for all $p \in {}^\bullet t$, for all $q \in P$ s.t. pBq, for all $M_1 \in C(t)$ there is a $u \in T$ s.t. $l(u) = l(t)$ and, writing M_2 for $M_1 - p + q$, we have $M_2 \overset{u}{\to} M_2'$ with $M_1' \overline{B} M_2'$ (assuming $M_1 \overset{t}{\to} M_1'$).*

Hence the weak transfer property is the transfer property restricted to pairs M_1, M_2 and steps $M_1 \overset{t}{\to} M_1'$ where M_1 is in the minimal cover $C(t)$ and where M_2 differs from M_1 by only one token. Since this is a weaker property, we have

Fact 8.6. $B(N, \mathcal{A})$ *has the weak transfer property.*

Lemma 8.7. *Let B be an equivalence s.t. $\mathcal{A}$ is B-closed. If B has the weak transfer property, then it is a place bisimulation.*

Proof. We show that $\overline{B} \cap (\mathcal{A} \times \mathcal{A})$ has the transfer property, i.e. "for all $M_1 \overline{B} M_2$, for all $M_1 \overset{t_1}{\to} M_1'$, there is a $M_2 \overset{t_2}{\to} M_2'$ s.t. ...", by induction over the pair $(|M_1|, \Delta(M_1, M_2))$ ordered lexicographically.

So let us assume $M_1 \overset{t_1}{\to} M_1'$ and $M_1 \overline{B} M_2$. Write n for $\Delta(M_1, M_2)$. We distinguish four cases:
1. $n = 0$: then $M_2 = M_1$ and we are done.
2. $n = 1$ and $M_1 \in C(t_1)$: then we are exactly in the situation where the weak transfer property applies.
3. $n > 1$: then there is a $M_3 \in \mathcal{A}$ s.t. $M_1 \overline{B} M_3 \overline{B} M_2$ and $\Delta(M_i, M_3) < n$ for $i = 1, 2$. By ind. hyp., we can transfer $M_1 \overset{t_1}{\to} M_1'$ to some $M_3 \overset{t_3}{\to} M_3'$ that we can then transfer to some $M_2 \overset{t_2}{\to} M_2'$. We have $M_1' \overline{B} M_3' \overline{B} M_2'$, hence $M_1' \overline{B} M_2'$.
4. $M_1 \notin C(t_1)$: then there is a $M_{1,1} \in C(t_1)$ s.t. $M_{1,1} \subset M_1$. We decompose M_1 as $M_{1,1} + M_{1,2}$ and M_2 as $M_{2,1} + M_{2,2}$ s.t. $M_{1,i} \overline{B} M_{2,i}$ for $i = 1, 2$. $M_{1,1} \overset{t_1}{\to} M_{1,1}'$ and the ind. hyp. transfers this into a $M_{2,1} \overset{t_2}{\to} M_{2,1}'$ with $M_{1,1}' \overline{B} M_{2,1}'$. We complete into $M_2 \overset{t_2}{\to} M_2' \overset{\text{def}}{=} M_{2,1}' + M_{2,2}$ and have $M_1' \overline{B} M_2'$.   □

We have the immediate corollary:

Theorem 8.8. *The largest reflexive and symmetric B having the weak transfer property exists and is $B(N, \mathcal{A})$.*

Finiteness of $C(t)$ implies that checking whether a given B has the weak transfer property only involves a finite number of checks. Therefore the following abstract algorithm computes $B(N, \mathcal{A})$:

Algorithm 8.9. *input* a labeled Petri net N, and a succ-closed set $\mathcal{A}$ of relevant markings.

output $B(N, \mathcal{A})$.

step1 Let $B = E(N, \mathcal{A})$, the largest place equivalence under which $\mathcal{A}$ is closed .

step2 Check if B has the weak transfer property:

If it has, then B is $B(N, \mathcal{A})$.

Otherwise, there is a $t \in T$, a $M \in C(t)$, a $p \in M$, a $q \in P$ with pBq s.t. $M \xrightarrow{t} M'$ can not be imitated from $M - p + q$. Then remove the pairs (p, q) and (q, p) from B and return to step 2.

$\square$

This algorithm is obviously correct and stops after a nite number of steps. Its complexity depends on the size of $C(t)$, and the cost of computing $C(t)$ and $E(N, \mathcal{A})$.

Note that it is possible to run the algorithme starting with any equivalence E included in $E(N, \mathcal{A})$ (i.e. any E such that $\mathcal{A}$ is E-closed), and obtain the largest place bisimulation included in E, a safe approximation of $B(N, \mathcal{A})$.

9   Place Bisimulation in Practice

Algorithmical aspects. When we want to use Algorithm 8.9 in practice, we face two problems: we need some e ective way of computing $C(t)$ for $t \in T$, and we need to compute $E(N, \mathcal{A})$ that will be the rst upper approximation of $B(N, \mathcal{A})$.

Whether this can be done depends on how $\mathcal{A}$ is given. In the simple case where $\mathcal{A}$ is nite, then $C(t)$ and $E(N, \mathcal{A})$ are computable in a direct way. More interestingly, when $\mathcal{A}$ is given by a Presburger formula (i.e., $\mathcal{A}$ is a semilinear set) then $C(t)$ and $E(N, \mathcal{A})$ can then be de ned by Presburger formulas, and can be e ectively computed: indeed assume $P = \{p_1, \ldots, p_k\}$ and $\mathcal{A}$ is given by the Presburger formula $\phi(n_1, \ldots, n_k)$. Then $(p_i, p_j) \in E(N, \mathcal{A})$ i $\phi(n_1, \ldots, n_k) \wedge n_i > 0$ entails $\phi(n_1, \ldots, n_i - 1, \ldots, n_j + 1, \ldots, n_k)$.

Does it work ? As far as we could judge by dealing with a few examples, the equivalences we compute with Algorithm 8.9 are not much of an improvement over the classical notion of place bisimulation where all markings are considered as relevant.

The reason seems to be that, when using Algorithm 8.9, we rst have to give a succ-closed set $\mathcal{A}$. If this is a large set, then it contains many irrelevant markings and this defeats the purpose of our study. If $\mathcal{A}$ is as small as possible (e.g., it contains only the reachable markings) then this may impact the equivalence $E(N, \mathcal{A})$ computed in *step1* and make it quite drastically limited even before it is re ned into a place bisimulation.

In the end, we can only argue the usefulness of our de nition with ad-hoc examples, so that it seems like it is mostly a theoretical contribution to place bisimulation. Our hope is that this contribution can be combined smoothly with the ideas from [APS94] where silent moves are abstracted from, and that this combination does work well in practice.

10   Conclusion

This paper proposed a new point of view on Petri net reductions based on the bisimilarity notion, and that do not aim at correctness for all markings.

We started with a net N and looked at reductions that lead to bisimilar configurations for any of the markings in some set $\mathcal{A}$. It turns out restrictions must be imposed on $\mathcal{A}$ in order to get a smooth theory.

Since the largest correct reduction is in general not computable, place bisimulations is an elegant close approximation. We gave an abstract algorithm computing the largest place bisimulation when $\mathcal{A}$ is given in a sufficiently effective way.

These results shed a new light on earlier works on place bisimulation. Having reductions in mind from the start help understand why place bisimulation cannot cope easily with arbitrary set of markings. Indeed, we end up with quite strong restrictions on the set $\mathcal{A}$ which, ideally, we would like to consist of the reachable markings only.

Acknowledgments

This paper greatly benefited from the advices of anonymous referees who suggested several simplications of our earlier proofs, and who pinpointed some of the reasons why the method is hard to use in practice.

References

[ABS91]   C. Autant, Z. Belmesk, and Ph. Schnoebelen. Strong bisimilarity on nets revisited. In *Proc. Parallel Architectures and Languages Europe (PARLE'91), Eindhoven, NL, June 1991, vol. II: Parallel Languages*, volume 506 of *Lecture Notes in Computer Science*, pages 295–312. Springer, 1991.

[APS94]   C. Autant, W. Pfister, and Ph. Schnoebelen. Place bisimulations for the reduction of labeled Petri nets with silent moves. In *Proc. 6th Int. Conf. on Computing and Information (ICCI'94), Trent University, Canada, May 1994*, pages 230–246, 1994. Available at http://www.lsv.ens-cachan.fr/Publis/PAPERS/APS-icci94.ps.

[AS92]   C. Autant and Ph. Schnoebelen. Place bisimulations in Petri nets. In *Proc. 13th Int. Conf. Application and Theory of Petri Nets (ICATPN'92), Sheffield, UK, June 1992*, volume 616 of *Lecture Notes in Computer Science*, pages 45–61. Springer, 1992.

[Ber86]   G. Berthelot. Checking properties of nets using transformation. In *Advances in Petri Nets 1985*, volume 222 of *Lecture Notes in Computer Science*, pages 19–40. Springer, 1986.

[CPS93]   R. Cleaveland, J. Parrow, and B. Steffen. The Concurrency Workbench: A semantics-based tool for the verification of concurrent systems. *ACM Transactions on Programming Languages and Systems*, 15(1):36–72, 1993.

[Gla90]   R. J. van Glabbeek. The linear time – branching time spectrum. In *Proc. Theories of Concurrency (CONCUR'90), Amsterdam, NL, Aug. 1990*, volume 458 of *Lecture Notes in Computer Science*, pages 278–297. Springer, 1990.

[Jan95]   P. Jančar. Undecidability of bisimilarity for Petri nets and some related problems. *Theoretical Computer Science*, 148(2):281–301, 1995.

[Mil89]   R. Milner. *Communication and Concurrency*. Prentice Hall Int., 1989.

[Mil90]   R. Milner. Operational and algebraic semantics of concurrent processes. In J. van Leeuwen, editor, *Handbook of Theoretical Computer Science, vol. B*, chapter 19, pages 1201–1242. Elsevier Science, 1990.

[NKML95]   M. Nakagawa, S. Kumagai, T. Miyamoto, and D.-I. Lee. Equivalent net reduction for firing sequence preservation. *IEICE Trans. on Fundamentals E.*, 78-A(11):1447–1457, 1995.

[NT84]   M. Nielsen and P. S. Thiagarajan. Degrees of non-determinism and concurrency: A Petri net view. In *Proc. 4th Conf. Found. of Software Technology and Theor. Comp. Sci. (FST&TCS'84), Bangalore, India, Dec. 1984*, volume 181 of *Lecture Notes in Computer Science*, pages 89–117. Springer, 1984.

[Old89]   E.-R. Olderog. Strong bisimilarity on nets: a new concept for comparing net semantics. In *Linear Time, Branching Time and Partial Order in Logics and Models for Concurrency, Proc. REX School/Workshop, Noordwijkerhout, NL, May-June 1988*, volume 354 of *Lecture Notes in Computer Science*, pages 549–573. Springer, 1989.

[Old91]   E.-R. Olderog. *Nets, Terms and Formulas*, volume 23 of *Cambridge Tracts in Theoretical Computer Science*. Cambridge Univ. Press, 1991.

[Par81]   D. Park. Concurrency and automata on infinite sequences. In *Proc. 5th GI Conf. on Theor. Comp. Sci., Karlsruhe, FRG, Mar. 1981*, volume 104 of *Lecture Notes in Computer Science*, pages 167–183. Springer, 1981.

[Pom86]   L. Pomello. Some equivalence notions for concurrent systems. An overview. In *Advances in Petri Nets 1985*, volume 222 of *Lecture Notes in Computer Science*, pages 381–400. Springer, 1986.

[Qui95]   W. Quivrin-Pfister. *Des bisimulations de places pour la réduction des réseaux de Petri*. Thèse de Doctorat, I.N.P. de Grenoble, France, November 1995.

[STMD96]   S. M. Shatz, S. Tu, T. Murata, and S. Duri. An application of Petri net reduction for Ada tasking deadlock analysis. *IEEE Trans. Parallel and Distributed Systems*, 7(12):1309–1324, 1996.

[Voo96]   M. Voorhoeve. Structural Petri net equivalence. Tech. report, Eindhoven Technical University, Dept. Computer Science, 1996. Available at `ftp://ftp.win.tue.nl/pub/techreports/wsinmar/`.

Efficiency of Asynchronous Systems That Communicate Asynchronously

Walter Vogler

Institut für Informatik, Universität Augsburg
D-86135 Augsburg, Germany
vogler@informatik.uni-augsburg.de

Abstract. A parallel composition is introduced that combines nets (regarded as system components) by merging so-called interface places; the novel feature is a flexible typing of these places, which formulates assumptions a component makes about its environment. Based on a testing scenario, a faster-than relation is defined and shown to support modular construction, since it is a precongruence for parallel composition, hiding and renaming. The faster-than relation is characterized without reference to tests, and this characterization is used to compare the temporal efficiency of some examples.

1 Introduction

In the testing scenario of De Nicola and Hennessy [DNH84], systems are compared by the service they provide for a possible user or environment; thus, systems are regarded as components that communicate with other components, and this *communication* is *synchronous*. Service is understood as functional behaviour, i.e. which actions are performed, and since there is no consideration of time, this *behaviour* is *asynchronous*. This classical approach has been developed further in [Vog95, Vog97, BV98] in order to compare also the temporal efficiency of asynchronous systems – using Petri nets as system models; naturally, synchronous communication corresponds to combining nets by merging transitions.

From the dual nature of Petri nets, it is particularly natural to consider also the composition by merging places, which corresponds to *asynchronous communication*. This form of composition has found maybe less attention, but see e.g. [Che91, Vog92, BDE93, Val94, Gom96, Kin97]; compare e.g. [dBKPR91, dBZ99] for the mostly recent interest in asynchronous communication in the process algebra world, also for further references. In this world, one can also find some ideas how to adapt the testing approach to compare timed behaviour in a setting with synchronous communication, see e.g. [CZ91, NC96]; these ideas are unrelated to [BV98] etc. In this paper, we will develop a form of efficiency testing as in [BV98] etc., but based on asynchronous communication, i.e. merging of places. This composition allows to model not only the exchange of messages,

---

This work was partially supported by the DFG-project 'Halbordnungstesten'.

M. Nielsen, D. Simpson (Eds.): ICATPN 2000, LNCS 1825, pp. 424–444, 2000.

but also the flow of materials in manufacturing processes and the reallocation of resources.

On the one hand, to give maximal freedom in the modular construction of nets, it is desirable to allow the merge of arbitrary nets, see e.g. [Che91] or [Vog92, Section 4.3]; on the other hand, in many applications it is sensible to designate in a component N the interface places to be merged as input or output places. The latter usually means a restriction for N *and* the environment that we want to compose with N: if p is e.g. an input place of N, then often N is not allowed to have arcs to p and the environment is not allowed to have arcs from p. (E.g. [Vog92, Section 4.4] only requires the first, while [Kin97] requires both.)

In this paper, a (to my best knowledge) novel flexible typing of the interface is suggested: we will designate an interface place p of a net N as input or output place *or both*. These types correspond to assumptions about the environment, which is allowed to have arcs to p or from p or both; hence, the environment may produce some input for N or consume some output from N or both. Intuitively, N only guarantees proper behaviour if the environment or user satisfies the respective assumption. There is no assumption on N, but p is of course also an interface place of the environment, which will have its own assumptions.

This way, we cover the liberal approach with maximal freedom as the case where each interface place under consideration is an IO-place, i.e. an in- and output place; and we cover the so often useful restricted approach essentially as the case where each interface place is either an input or an output place. Our first example in the next section will demonstrate that it can also be sensible to merge e.g. an output place with an IO-place.

In the classical testing approach, a system is an implementation if it performs in all environments, i.e. for all users, functionally just as successful as the specification. Here, we also take into account the efficiency of implementation and specification: success (indicated by marking an output place ω) has to be reached within a given time. Thus, we are interested in worst case behaviour (so-called must-testing). By definition, components of asynchronous systems work with indeterminate speeds, (i.e. time cannot be used to coordinate components); most often, this is interpreted as 'each component may work arbitrarily slow'. Under this interpretation, the worst case is simply that nothing is done for a long time, hence every test is failed and we do not have a sensible theory of testing.

As a way out, [BV98] assumes that each action is performed within some given time bound (or is disabled within this time). Such an upper time bound is a reasonable basis for judging the efficiency, also see e.g. [PF77, Lyn96]; since actions can also be performed arbitrarily fast, the components work with indeterminate relative speeds also under this assumption, and we have a valid theory for asynchronous systems. While [BV98] defines a firing rule where each transition has to fire within time 1, we will also have transitions that have to fire immediately, i.e. within time 0; this demonstrates generality of our results (keeping things simple at the same time) and enhances expressivity as discussed in Sections 2 and 6.

Nets and parallel composition $|||$ with place typing as described above are introduced in Section 2. In Section 3, we give our timed firing rule using discrete time and define efficiency testing; we call a net N_1 faster than a net N_2, if it satisfies all tests that N_2 satisfies, i.e. if it exhibits at least the same functionality with at least the same efficiency as N_2 specifies. We show that $|||$ is 'almost associative', which is used for our first main result that faster-than is a precongruence for $|||$ and thus allows compositional reasoning. The analogous result for renaming and hiding is given in Section 4, where – as second main result – we characterize faster-than without referring to tests. We point out how this result can be adapted to settings where there are no 0-transitions or where use of interface places is restricted as discussed above; then we use the characterization in Section 5 to compare some example nets. In the conclusions, we discuss work in progress regarding the construction of safe nets using $|||$ or variations thereof.

I thank Elmar Bihler and Laurentiu Tiplea for our discussions.

2   Basic Notions of Petri Nets with Time Bounds and Interface

In this section, we introduce Petri nets (S/T-nets) which are extended with a function that assigns to each transition a time bound 0 or 1, and with an interface of input and output places as explained in the introduction; we define the usual basic firing rule and the parallel composition for such nets.

Thus, a *Petri net with time bounds and interface* $N = (S, T, W, \beta, I, O, M_N)$ (or just a *net* for short) consists of finite disjoint sets S of *places* and T of *transitions*, the function $W : S \times T \cup T \times S \to \{0, 1\}$ describing the *arcs*, the *time bound* $\beta : T \to \{0, 1\}$, the *input places* $I \subseteq S$, the *output places* $O \subseteq S$, and the *initial marking* $M_N : S \to I\!N_0$. We associate with N its set of *interface places* $P = I \cup O$; note that an interface place p might be in I and O at the same time. We assume that M_N is 0 on P. When we introduce a net N or N_1 etc., then we assume that implicitly this introduces its components, e.g. S, I and P or S_1, I_1 and P_1 etc.

For each $x \in S \cup T$, the *preset* of x is ${}^\bullet x = \{y \mid W(y, x) = 1\}$, the *postset* of x is $x^\bullet = \{y \mid W(x, y) = 1\}$. These notions are extended to sets as usual, e.g. ${}^\bullet X$ is the union of all ${}^\bullet x$ with $x \in X$. If $x \in {}^\bullet y \cap y^\bullet$, then x and y form a *loop*. A *marking* is a function $S \to I\!N_0$. We sometimes regard sets as characteristic functions, which map the elements of the sets to 1 and are 0 everywhere else; hence, we can e.g. add a marking and a postset of a transition or compare them componentwise.

Assumption: In all nets considered in this paper, places are not isolated (i.e. ${}^\bullet s \neq \emptyset$ or $s^\bullet \neq \emptyset$ for each $s \in S$).

As usual, we draw transitions as boxes, places as circles and arcs (i.e. pairs (x, y) with $W(x, y) = 1$) as arrows. If a place s is on a loop with a transition t with $\beta(t) = 1$ and has no arc to or from any other transition, we will not draw s but place a dot into the box of t instead; if the box representing some $t \in T$ does not have a dot, but $\beta(t) = 1$, then we give the box a double line on its right side; in both cases, t is called a *1-transition*. If there is no dot and no double line, then $\beta(t) = 0$, and t is a *0-transition*. An interface place is indicated by writing its type(s) I, O or IO next to it; we speak of an *IO-place* in the latter case. For an example, see Figure 1, which is discussed below.

We now define the usual basic firing rule, which ignores timing and interface.

- A transition t is *enabled* under a marking M, denoted by $M[t\rangle$, if $^\bullet t \leq M$. If $M[t\rangle$ and $M' = M + t^\bullet - {}^\bullet t$, then we denote this by $M[t\rangle M'$ and say that t can *occur* or *fire* under M yielding the marking M'.
- This definition of enabling and occurrence can be extended to sequences as usual: a sequence w of transitions is *enabled* under a marking M, denoted by $M[w\rangle$, and yields the follower marking M' when *occurring*, denoted by $M[w\rangle M'$, if $w = \lambda$ and $M = M'$ or $w = w't$, $M[w'\rangle M''$ and $M''[t\rangle M'$ for some marking M'' and transition t. If w is enabled under the initial marking, then it is called a *firing sequence*; the set of those is denoted by $FS(N)$.
- A marking M is called *reachable* if $M_N[w\rangle M$ for some $w \in T^*$. The net is *safe* if $M(s) \leq 1$ for all places s and reachable markings M.

Before we introduce the parallel composition $|||$, we define nets to be *isomorphic* if one can be obtained from the other by bijectively renaming the transitions and the places in $S \setminus P$, while preserving the structure (arcs, time bounds and initial marking). We regard isomorphic nets as equal. Hence, only the identity of the interface places is important – and whenever we have two nets N_1 and N_2, we can assume that they are disjoint outside their interfaces, i.e. $(S_1 \cup T_1) \cap (S_2 \cup T_2) = P_1 \cap P_2$.

Now, if we combine nets N_1 and N_2 with the *parallel composition* $|||$, then they run in parallel and communicate asynchronously over their common interface. To construct the composed net, we will merge the common interface places of N_1 and N_2. But first of all, we assume that N_1 and N_2 satisfy the above disjointness condition – otherwise, a suitable isomorphic renaming is performed automatically. Additionally, the typing of the interface places as input or output places formulates an assumption about the respective other component: e.g. N_1 may only put a token onto an interface place p of N_2, if p is an input place of N_2. Thus, $N_1 ||| N_2$ is defined if and only if, for $\{i,j\} = \{1,2\}$, $p \in P_i \cap P_j$ and $t \in T_j$ we have that $W_j(t,p) = 1$ implies $p \in I_i$ and that $W_j(p,t) = 1$ implies $p \in O_i$.

If these conditions are satisfied, $N = N_1 ||| N_2$ is obtained by componentwise union except for the interface. (In particular, we can take the union of the two time bounds since they are defined on disjoint sets; the same is true for the initial markings, since whenever they are both defined on some p, then this is an interface place where both markings are 0.) Additionally, $I = (I_1 \setminus P_2) \cup (I_2 \setminus P_1) \cup (I_1 \cap I_2)$ and $O = (O_1 \setminus P_2) \cup (O_2 \setminus P_1) \cup (O_1 \cap O_2)$.

Let us explain the treatment of the interface. The interface of N is more or less (see next paragraph) the union of the interfaces of N_1 and N_2, and the typing of an interface place p is correspondingly determined by N_1 or/and N_2.

If p is e.g. in I_1 but not in O_1, then N_1 assumes that its user will never take tokens from p; the same assumption must be made by N. Otherwise, a user U of N that takes tokens from p would indirectly use N_1 and violate its assumption.

If p is e.g. only an input for N_1 and only an output for N_2, then there is no arc to p in N_1 and no arc from p in N_2; this corresponds exactly to the restricted approach to composition by merging places that we discussed in the introduction. Note that such a p will not be in the interface of N, which is sensible, since no user U could satisfy both assumptions of N_1 and N_2; p allows point-to-point messages from N_2 to N_1 – a very useful application of our approach.

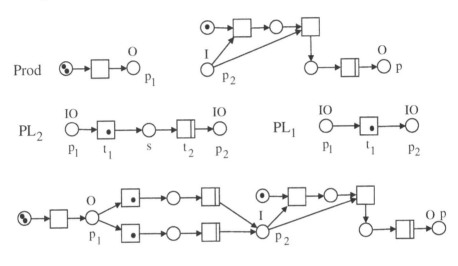

Figure 1

As an example for our nets and the application of $|||$, consider Figure 1. *Prod* is a typical user; it will quickly (i.e. with a 0-transition) put two 'orders' onto its output place p_1, and then it is willing to detect two 'products' on its input place p_2; after this, it would fire its 1-transition, i.e. 'use the products' for at most time 1, and then put a token onto p in order to signal satisfaction. Similarly, the special place ω is used in tests to signal success, i.e. satisfaction of the user.

The production line *PL2* can execute an 'order' placed on p_1 and put a 'product' onto p_2. Production has two stages (1-transitions) that each take up to time 1. The dot indicates a loop place, with the effect that in the first stage one order has to be processed after the other; one can run several orders through the second stage in parallel, taking up time 1 altogether; compare Definition 1. *PL1* is an improved version of *PL2*, needing only one stage; we certainly expect it to be faster.

In the introduction, it is already mentioned that our interface typing allows to cover a liberal and a restricted approach to place merging often found in the literature. The present example also demonstrates another use of our flexible interface typing. It is realistic to serve *Prod* with two production lines, i.e. to form the net *Prod* $|||$ (*PL2* $|||$ *PL2*) also shown in Figure 1; recall that the two copies of *PL2* are automatically made disjoint except for their interfaces. Intuitively, one would expect *PL2* ||| *PL2* to be faster than *PL2*. Note that both copies may take tokens from p_1; one copy of *PL2* would forbid the other one to do this, if p_1 would only be an input place of *PL2*; technically, *Prod* ||| (*PL2* ||| *PL2*) would not be defined. Thus, it is very reasonable to have IO-places as p_1 or similarly p_2.

Clearly, if we only consider nets where all interface places are IO-places, then parallel composition is always defined and we are in the liberal approach to place merging mentioned in the introduction. The restricted approach corresponds to the case where we only consider nets whose interface places are *either* input places with only outgoing arcs *or* output places with only ingoing arcs: consider e.g. an input place p; due to its outgoing arc, p can only be merged with a place that allows an outgoing arc in its environment, i.e. with an output place.

We mention in passing two other operations that can be useful in the construction of nets: *I- or O-hiding* an interface place p in a net N results in $N/^I p$ or $N/^O p$, which is obtained from N by removing p from I or O; *hiding* p is defined by $N/p = N/^I p/^O p$. *Renaming* an interface place p in a net N to $p' \notin P$ results in $N[p \to p']$, which is obtained from N by replacing p by p', which inherits p's arcs, marking and typing; if $p' \in S \setminus P$, it first has to be replaced isomorphically by some fresh place.

If the combined production line *PL2* ||| *PL2* is complete for the processing of orders, we should write it as (*PL2* ||| *PL2*)$/^O p_1/^I p_2$. A producer that also uses two copies of *PL2*, but decides for each order which copy to use, could e.g. use *PL2* and *PL2*$[p_1 \to p_1']$; products would still arrive at the unique place p_2.

3   Timed Behaviour of Asynchronous Systems and Testing

We will describe the asynchronous behaviour of a parallel system, taking into account the times at which things happen. The components of an asynchronous system vary in speed – but we assume that they are guaranteed to perform each enabled action within some finite given time; this upper time bound allows the relative speeds of the components to vary arbitrarily, since we have no positive lower time bound. Thus, the behaviour we define is truly asynchronous. In contrast to [BV98] etc., we do not assume that the upper time bound is 1 for all transitions; to demonstrate generality of the approach but keeping things simple at the same time, we allow 0 or 1 as bound. This will also allow to model transitions with arbitrary durations as discussed below.

The upper time bounds of transitions really correspond to firing intervals $[0, 1]$ and $[0, 0]$ attached to the transitions; instead, one could more generally attach arbitrary time intervals to the arcs from places to transitions as suggested in [SDdSS94]. Efficiency testing with synchronous communication is studied for safe nets with such arc intervals in [Bih98], and it is shown that the faster-than relation based on testing is the same, no matter whether we work with continuous or discrete time (see also [Pop91]). The basic lemma for this result almost carries over to our setting (except that we use general S/T-nets); thus, for the time being, we will only use discrete time, which is easier to handle.

We will now define what we regard as timed behaviour; it will be convenient to use a formalism different from the ones used in the papers cited above. If a transition is enabled, it can always fire since transitions – only having an upper time bound – may fire very quickly; firing itself is instantaneous. If a 0-transition is enabled, it has to fire immediately, i.e. no time will pass before it has fired or has been disabled. If a 1-transition gets enabled, it may wait for one unit of time, and we write σ for such a *time step*; to keep track how long transitions have been enabled, we record – additionally to the marking – how many tokens have an age of at least 1; then, a 1-transition has to fire immediately, if each place in its preset is marked with an 'old' token. This intuition is formalized as follows, where we assume that initially all tokens are old; if this is not adequate in some case, one can add a transition t that can fire once and produce the intended initial marking of fresh tokens.

Definition 1. A *timed marking* TM of a net is a pair (M, M^{old}) consisting of two markings of N with $M \geq M^{old}$. Again, introducing e.g. a timed marking TM_1 implicitly introduces M_1 and M^{old}_1 etc. The initial TM is $TM_N = (M_N, M_N)$.

We write $(M, M^{old})[\varepsilon\rangle(M', M^{old'})$ if one of the following cases applies:

1. $\varepsilon = t \in T$, $M[t\rangle M'$, $M^{old'} = M^{old} \ominus {}^\bullet t$, where $n \ominus m = \max(n - m, 0)$
2. $\varepsilon = \sigma$, $\forall t \in T : M[t\rangle \Rightarrow (\beta(t) = 1 \wedge \neg M^{old}[t\rangle)$, $M' = M^{old'} = M$

Generalizing this timed firing rule to sequences as above – together with a notion of reachable timed marking –, we define the set $TFS(N) = \{w \mid TM_N[w\rangle\}$ of *timed firing sequences* of N. For a timed firing sequence w, $\zeta(w)$ is the *duration*, i.e. the number of σ's in w. The behaviour in between two σ's is called a *round*.
□

Part 2 of this firing rule ensures that every 1-transition that is enabled for one unit of time and every enabled 0-transition fires before time goes on, but according to Part 1 a 1-transition may also act faster. Consider e.g. *PL2* in Figure 1 with two additional tokens on p_1 – which is not an interface place in this consideration. Since t_1 is enabled by old tokens, we cannot apply Part 2, but must fire t_1 according to 1; this leads to a timed marking with one token on each, p_1, s and the loop place of t_1, where only the token on p_1 is old. Neither t_2 nor t_1 is enabled by old tokens alone, hence as one possibility, σ can occur; then, all tokens are old, and we must fire t_1 and t_2 before the next σ. As another possibility, we can immediately fire t_1 or t_2 after the first t_1 according to Part 1.

In fact, by only applying 1, we get $FS(N) \subseteq TFS(N)$; additionally, the occurrence of σ's only changes M^{old} while the firing of transitions only depends on M as usual. Hence, deleting the time steps from all sequences in $TFS(N)$ we get exactly $FS(N)$. This shows that, despite the upper time bounds, we still deal with the full complexity of asynchronous systems; we have simply enriched the asynchronous behaviour by some timing information in an orthogonal way.

Observe the following subtle point, where our consideration of asynchronous systems simplifies things a little: if a transition is given a choice, it always takes old tokens. One could also define a timed firing rule, where it might take a fresh one, i.e.: if $M(s) > M^{old}(s)$ – indicating the presence of a fresh token on s – and $t \in s^\bullet$ fires under TM, then we could allow that this firing leaves $M^{old}(s)$ unchanged. Such a rule would give sequences where the M^{old}-components are larger; the only consequence would be that we could apply Part 2 of this rule less often. But since we are never forced to apply 2 anyway, such an alternative firing rule would not give additional timed firing sequences. Thus, we can work with our simpler rule.

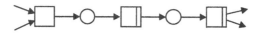

Figure 2

Quite generally, we can also model transitions with an arbitrary upper bound on their *duration*. E.g. Figure 2 shows a subnet where the first transition takes two tokens immediately once they are available, while the last transition produces two tokens after at most two time steps; thus, the subnet corresponds to a transition with maximal duration 2 with the described consumption and production. It is not clear whether we can model transitions that stay enabled up to time $n > 1$ and then fire instantaneously. See also Section 6 for expressivity.

We now define our testing scenario, where we compare nets by embedding them as components into testing environments and considering the behaviour of the resulting complete systems. Such a testing environment is something like a user that communicates with the embedded component; if after a while the user is satisfied with the service of the component, (s)he declares success by marking a special place ω – and this should happen within a prescribed period D of time even in the worst case. To be sure that we have seen everything that occurs up to time D, we only look at runs w with $\zeta(w) > D$.

Definition 2. A net is *testable* if none of its interface places is ω. A net is *a test net* if it has ω as output place with $\omega^\bullet = \emptyset$. A *timed test* is a pair (U, D), where U is a test net and $D \in I\!N_0$ (the *test duration*).

A testable net N *satisfies* a timed test (U, D) (N *must* (U, D)), if $N \,|||\, U$ is defined and each $w \in TFS(N \,|||\, U)$ with $\zeta(w) > D$ reaches a timed marking where ω is marked. For testable nets N_1 and N_2, we call N_1 a *faster implementation* of N_2 (or simply *faster than* N_2), $N_1 \sqsupseteq N_2$, if $N_1 \,|||\, U$ and $N_2 \,|||\, U$ are defined for the same test nets U and N_1 satisfies all timed tests that N_2 satisfies. $\sqsupseteq$ is also called the *efficiency preorder*.   □

The efficiency preorder has two aspects: if $N_1 \sqsupseteq N_2$, then intuitively N_1 satisfies all U that N_2 specifies, i.e. it *functionally* implements N_2; and we speak of a *faster* implementation, since N_1 might also satisfy some specified test net within a shorter time. Note that N *must* (U, D) implies N *must* (U, D') for all $D' > D$; hence, if N_1 satisfies the same U as N_2 but with a different time D, this must be a shorter time.

Our timed firing rule allows some sort of Zeno-effect, i.e. we could have arbitrary long sequences with a fixed finite duration, in particular 0. This is of course not realistic, but in a way this effect is suppressed by the testing definition: a test fails only with a w that takes enough time; note that, since $\omega^\bullet = \emptyset$ in a test net, we can restrict attention to w that stop after the last σ. The suppression may have a misleading effect if some w with $\zeta(w) \leq D$ could not be extended to a ww' with $\zeta(ww') > D$; a system with such a time stop is ill-designed and should not be considered anyway. Note that time stops can only occur due to 0-transitions (or 1-transitions with empty preset, which should be avoided anyway since they act like 0-transitions).

From the above intuitive explanation, it should be clear that we can safely use N_1 instead of N_2, only if it is faster as just defined. We will validate our preorder by checking out examples below. But to be really useful, the faster-than relation should also support modular construction, i.e. it should be a precongruence: if $N_1 \sqsupseteq N_2$ and N_2 is a component of a system $N_2 \,|||\, N$, then replacing N_2 by N_1 should give a better system $N_1 \,|||\, N$, i.e. we should have $N_1 \,|||\, N \sqsupseteq N_2 \,|||\, N$.

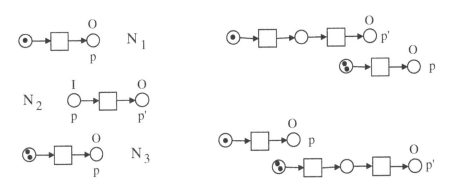

Figure 3

Testing-based relations like the above faster-than relation usually give such a precongruence for the composition operator used in the definition of testing. [Vog92] discusses (actually for a slightly different setting, in particular considering synchronous communication and congruences instead of precongruences) that this is not necessarily so, but is always the case if the composition operator is commutative and associative (and the latter argument would also work here). Clearly, $|||$ is commutative, but it may violate associativity in somewhat pathological cases: Figure 3 shows three nets N_1, N_2 and N_3; forming $N_1 \,|||\, N_2$, the place p is removed from the interface, and then it is automatically replaced isomorphically when we construct $(N_1 ||| N_2) ||| N_3$ (top right net) as explained above. $N_1 |||(N_2 ||| N_3)$ (bottom right net) is clearly different – also in behaviour.

If we obtain N_3' from N_3 by adding p to I_3, then $(N_1 ||| N_2) ||| N_3'$ looks quite the same as $(N_1 ||| N_2) ||| N_3$; but $N_1 |||(N_2 ||| N_3')$ is undefined, since there is an arc to $p \in I$ in $N = N_2 ||| N_3'$, but $p \notin I_1$.

We will now give a sufficient criterion for associativity, and we will show how associativity can be failed. Together with a first result about nets that are comparable under our efficiency preorder, this will imply that $|||$ is 'associative enough' to prove that $\sqsupseteq$ is indeed a precongruence.

Theorem 3. *Let N_1, N_2 and N_3 be nets.*

i) *If $\bigcap_{i=1}^3 P_i = (\bigcap_{i=1}^3 I_i) \cup (\bigcap_{i=1}^3 O_i)$, then $(N_1 ||| N_2) ||| N_3 = N_1 |||(N_2 ||| N_3)$, which includes that one side is defined if and only if the other is.*

ii) *If $(N_1 ||| N_2) ||| N_3$ is defined and either $N_1 |||(N_2 ||| N_3)$ is not defined or it is different from $(N_1 ||| N_2) ||| N_3$, then N_1, N_2 and N_3 have a common interface place that is only an output place of N_1 and only an input place of N_2 or vice versa; compare Figure 3.*

iii) *If N_1 is faster than N_2 (both testable), then N_1 and N_2 have the same input and the same output places, and for each common interface place p, we have that $p^\bullet \neq \emptyset$ in N_1 if and only if $p^\bullet \neq \emptyset$ in N_2, and similarly for the presets.*

iv) *Faster-than is a precongruence for testable nets.*

Proof. i) If an interface place p belongs to at most two of the nets, it is easy to see that, with respect to p, definedness of the composition and classification as input and/or output place does not depend on the bracketing. If e.g. p does not belong to N_1, then composition with N_1 is certainly defined w.r.t. p and inherits arcs and classification from the other operand; hence, definedness, classification and arcs of p depend only on $N_2 ||| N_3$.

Now let p be, say, an input place for all three nets; if it is also an output place for all three nets, then all intermediate and final results are defined and have p has in- and output place. If p is not an output place of e.g. N_1 (the other cases being similar), then definedness of the left-hand-side requires that there is no arc from p in N_2 and subsequently neither in N_3, whereas definedness of the right-hand-side requires that there is no arc from p in $N_2 ||| N_3$, which also means that there is none in N_2 and none in N_3; thus, one side is defined w.r.t. p iff the other is, and the result will be the same w.r.t. p since the three nets merge on p; p will only be an input place of the final result.

ii) Assume the hypothesis holds; then, by i), there must be a common interface place p of all three nets that is not an input of all of them and not an output of all of them. If the claim fails, then p is without loss of generality an input place for N_1 and N_2, and only an output place for N_3. By the definedness of $(N_1 ||| N_2) ||| N_3$, the latter means that neither in N_1 nor in N_2 there are arcs to p. Furthermore, p is not an output place for one of N_1 or N_2, hence in the other there are no arcs from p. Thus, p is isolated in this other net, which is a contradiction to our general assumption.

iii) If p is only an input place of N_1, consider a test net U with interface consisting of ω and the IO-place p and with an arc from p; since $N_1 \,|||\, U$ is not defined, neither is $N_2 \,|||\, U$, and the only possible reason is that p is an interface place but not an output place of N_2. With a dual argument, we conclude that the interfaces of N_1 and N_2 (including the typing) coincide – with the possible exception that there might be some IO-place p of N_1 with $p \notin P_2$ (or vice versa).

So let p be such a place, and w.l.o.g. $p^\bullet \neq \emptyset$ by our assumption on nets. Consider a test net U with interface ω plus input place p. Again, composition with U is undefined for N_1, hence for N_2; we conclude that p must be an interface place of N_2, and thus the exception is in fact not possible.

The last consideration showed that also $p^\bullet \neq \emptyset$ in N_2. If we take U in this consideration without arcs from p, we could draw the same conclusion for $p \in P_1 \setminus O_1$. Analogous arguments conclude the proof.

iv) Let N_1 be faster than N_2, and let N be another testable net. From iii) we conclude that $N_1 \,|||\, N$ is defined iff $N_2 \,|||\, N$ is (which we assume now), and that composition with $N_1 \,|||\, N$ and $N_2 \,|||\, N$ is defined for the same test nets. Now let (U, D) be a test that $N_2 \,|||\, N$ satisfies; we have to show that $N_1 \,|||\, N$ satisfies the test as well.

Satisfaction depends on the behaviour of $(N_2 \,|||\, N) \,|||\, U$, $(N_1 \,|||\, N) \,|||\, U$ resp. If we can apply associativity in both cases, then N_2 satisfies $(N \,|||\, U, D)$ because satisfaction depends on the same net $N_2 \,|||| (N \,|||\, U)$; thus; N_1 also satisfies $(N \,|||\, U, D)$, and we are done.

If associativity fails, we apply ii) to get a common interface place of N_1, N_2, N and U that w.l.o.g. is only an input place for N_1 and N_2 (apply iii)) and only an output place of N. Take a fresh place p' and $N_1' = N_1[p \to p']$, $N_2' = N_2[p \to p']$ and $N' = N[p \to p']$. Clearly, N_1' is faster than N_2' and composition with N' is defined for both these nets. Now $N_1 \,|||\, N$ and $N_1' \,|||\, N'$ are isomorphic, since p and p' are not in their interfaces. Thus, it suffices to consider N_1', N_2', N' and U; repeating this argument, we arrive at nets where associativity can be applied and are done. $\square$

Note that by ii) $|||$ is almost associative, and the proof of iv) shows that with a suitable renaming we can always achieve associativity.

4   Characterization of the Efficiency Preorder and a Proof Method

The efficiency preorder $\sqsupseteq$ formalizes observable difference in efficiency; referring to all possible tests, it is not easy to work with directly. Therefore, our aim is now to characterize $\sqsupseteq$ by only looking at the nets themselves that are compared. In classical testing [DNH84], such a characterization is based on failure pairs, which have just one so-called refusal set giving information on the final state of a run, while [BV98] uses refusal traces where refusal sets repeatedly give information on intermediate states of a run.

In a setting comparable to classical testing but using asynchronous communication, [Vog92, Section 4.3] uses a standard environment and then arrives at a refusal-type semantics that is much more involved than failure semantics. We will also use this standard environment, but pleasingly our characterization is a fairly simple refusal-type semantics similar to refusal traces (explained in more detail after Definition 4). This underlines the usefulness of the failure/refusal paradigm also for treating asynchronous communication, in contrast to [dBKPR91].

We will use our characterization to show that $\sqsupseteq$ is a precongruence w.r.t. relabelling and hiding; based on the characterization, we will define an appropriate kind of simulation, which in turn will be used in the next section to show e.g. that in the example of Figure 1 two production lines are indeed faster than one.

Definition 4. For a net N, N^{env} is obtained by adding to N the following *interface transitions*:

- for each $p \in I$, a fresh 1-transition p^+ and an arc from p^+ to p; and
- for each $p \in O$, a fresh 1-transition p^1, a fresh 0-transition p^0 and arcs from p to p^1 and p^0.

For timed markings (M, M^{old}) and $(M', M^{old'})$ of N^{env}, we define the *pr-firing* (pr stands for place refusal) $(M, M^{old})[\varepsilon\rangle_{pr}(M', M^{old'})$ if one of the following cases applies:

1. $\varepsilon = t \in T^{env}$, $M[t\rangle M'$, $M^{old'} = M^{old} \ominus {}^\bullet t$
   (if $t \in \{p^0, p^1\}$ for some $p \in O$, we will actually write p^- instead of t in this case, since p^0 and p^1 are the same w.r.t. firing);
2. $\varepsilon = X \subseteq \{p^0, p^1 \,|\, p \in O\}$, $M' = M^{old'} = M$ and, for all $t \in T^{env}$, $M[t\rangle$ implies either $\beta(t) = 1 \wedge (\neg\, M^{old}[t\rangle \vee \exists p \in O : t = p^1 \notin X)$ or $\exists p \in O : t = p^0 \notin X$ or $\exists p \in I : t = p^+$; X is called a *refusal set*.

Generalizing this pr-firing rule to sequences as above, we define the set $PRS(N) = \{w \,|\, TM_N^{env}[w\rangle_{pr}\}$ of *pr-sequences* of N – where TM_N^{env} is the initial timed marking of N^{env}. For timed markings TM and TM' of N^{env} and some w that ends with a set and satisfies $TM[w\rangle_{pr} TM'$, we write $TM[v\rangle\rangle_{pr} TM'$, if v is obtained from w by deleting all $t \in T$; w is called the pr-sequence *underlying* the pr-trace v, and $PR(N)$ is the set of these pr-traces (which consist of transitions p^+ and p^- and refusal sets). The behaviour in between two refusal sets is called a *round*. □

In N^{env}, N is wrapped into a standard environment that might put/take tokens to/from the interface of N, depending on the typing. The pr-firing rule might look very complicated at first sight, but in fact it is not: it is very similar to the timed firing rule above, except that time steps are indicated by refusal sets instead of σ's. Occurrence of such a set is a 'partial σ'; it requires from the transitions of N the same as σ, but ignores the additional transitions p^1, p^0 not listed in the set and the additional p^+. That pr-traces always end with a set (in contrast to [BV98]) is related to the fact that (w.l.o.g.) the decisive behaviour for failing a test ends with a σ (see above), and it is required due to the possibility of time stops discussed in the previous section.

Observe that we only see interface transitions in the pr-traces, and that their existence depends on the interface typing; compare the second example in Section 5. We will show that our faster-than relation coincides with PR-inclusion.

Usually, the next step in an approach as ours would be to show that the PR-semantics of a composition can be constructed from the PR-semantics of the components; this would be used in the characterization proof and imply precongruence. Unfortunately, such a denotational construction has not been found as yet. But luckily, we already have a precongruence result, and the following proposition will be good enough to verify the characterization afterwards; it shows that we can find out enough about a composition from knowing one component and the PR-semantics of the other. The proof of this proposition needs two lemmas, whose proofs we omit.

Lemma 5. *Let N be a net, $t \in T$, $p \in I$ and $q \in O$. In N^{env}, $TM[tp^+\rangle_{pr} TM'$ implies $TM[p^+t\rangle_{pr} TM'$ and $TM[p^-t\rangle_{pr} TM'$ implies $TM[tp^-\rangle_{pr} TM'$.*

Lemma 6. *Let N and U be nets such that $N_0 = N \,|||\, U$ is defined; let U_P be obtained from U by deleting P. Let TM_0 and TM'_0 be timed markings of N_0, let TM_P and TM'_P be their restrictions to the places of U_P, and TM and TM' their restrictions to the places of N^{env}. Let $t \in T_0$, such that $^\bullet t \cap P = \{p_1, \ldots, p_n\}$ and $t^\bullet \cap P = \{q_1, \ldots, q_m\}$ in case that $t \in T_U$.*

1. *If $t \in T$, then $TM_0[t\rangle TM'_0$ if and only if $TM[t\rangle_{pr} TM'$ and $TM_P = TM'_P$.*
2. *If $t \in T_U$, then $TM_0[t\rangle TM'_0$ if and only if $TM[p_1^- \ldots p_n^- q_1^+ \ldots q_m^+\rangle_{pr} TM'$ and $TM_P[t\rangle_{pr} TM'_P$.*
3. *$TM_0[\sigma\rangle TM'_0$ if and only if $M'_P = M^{old'}{}_P = M_P$ and there exists some X such that $TM[X\rangle_{pr} TM'$ and, for all $t \in T_U$, $M_P[t\rangle$ implies either $\beta_U(t) = 1 \wedge (\neg M^{old}{}_P[t\rangle \vee \exists p \in {}^\bullet t \cap P : p^1 \in X)$ or $\beta_U(t) = 0 \wedge \exists p \in {}^\bullet t \cap P : p^0 \in X$.*

Proposition 7. *Let N_1, N_2 and U be nets such that $I_1 = I_2$, $O_1 = O_2$, $PR(N_1) \subseteq PR(N_2)$ and $N_1 \,|||\, U$ and $N_2 \,|||\, U$ are defined. Then, for each $w \in TFS(N_1 \,|||\, U)$ with $\zeta(w) = n$ reaching the timed marking TM_1 and ending with a σ, there exists a $v \in TFS(N_2 \,|||\, U)$ with $\zeta(v) = n$ reaching the timed marking TM_2 and ending with a σ such that TM_1 and TM_2 coincide on $S_U \setminus P_1$.*

Proof. Applying Lemma 6, w 'splits off' a $w_1 \in PRS(N_1)$ ending with a refusal set and some behaviour on the side of U; to the pr-trace $u \in PR(N_1) \subseteq PR(N_2)$ corresponding to w_1, we find an underlying $v_2 \in PRS(N_2)$ ending with a refusal set from which we can construct v stepwise with the above behaviour on the side of U, using again Lemma 6. Note that the timed marking on $S_U \setminus P_1$ changes the same way along w and v.

The only problem we could meet when constructing v is that application of 6.2 requires a suitable 'block' $p_1^- \ldots p_n^- q_1^+ \ldots q_m^+$ in v_2, but this sequence might actually be interspersed with transitions from T_2. In this case, we transform v_2 using Lemma 5 such that v_2 has the necessary 'blocks'. □

Theorem 8. *Let N_1 and N_2 be testable nets. Then $N_1 \sqsupseteq N_2$ if and only if the syntactic and the semantic property as follows hold:*

- *syntactic property: $I_1 = I_2$, $O_1 = O_2$, and for each interface place p, we have that $p^\bullet \neq \emptyset$ in N_1 if and only if $p^\bullet \neq \emptyset$ in N_2, and similarly for the presets;*
- *semantic property: $PR(N_1) \subseteq PR(N_2)$.*

Proof. "if": Clearly, composition with N_1, N_2 resp., is defined for the same nets. Let (U, D) be a timed test. If N_1 fails the test, then due to a $w \in TFS(N_1 \mathbin{|||} U)$ with $\zeta(w) > D$ ending with a σ such that ω is not marked afterwards. Proposition 7 gives some $v \in TFS(N_2 \mathbin{|||} U)$ with which N_2 fails the test as well.

"only if": Theorem 3 iii) shows the first part. To prove the other part, we construct for each $w \in PR(N_1)$ a test that is failed by a testable net N if and only if $w \in PR(N)$; then N_1 fails this test, hence N_2 does, so $w \in PR(N_2)$ and we are done.

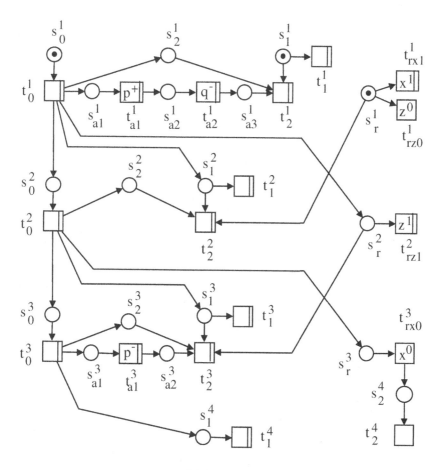

Figure 4

We sketch this construction with an example; the construction is akin to constructions in the case of synchronous communication except for the 0-transitions and the treatment of the last refusal set. For $w = p^+q^-\{x^1, z^0\}\{z^1\}p^-\{x^0\}$, we choose $D = 3$ and the test net U shown in Figure 4. The interface of U consists of the omitted IO-places p, q, x and z, and the output place w. A transition labelled e.g. p^+ (or w) has an arc to p (or w), a transition labelled e.g. p^-, p^0 or p^1 has an arc from p.

We will consider a possible timed firing sequence v that makes a net N fail the test $(U, 3)$ from the perspective of Lemma 6, i.e. we will consider what pr-trace N^{env} sees and what U has to do.

The subnet consisting of the places and transitions with lower index 0 act as a clock: since the token on s_0^1 is old initially, t_0^1 has to fire in the first round, t_0^2 at the latest in the second and t_0^3 at the latest in the third; actually, they must fire exactly in these rounds because otherwise, t_1^4 would fire at the latest in the third round and the test would not be failed. Hence, when v occurs, we have an old token on s_1^i and on s_r^i in the ith round, $i = 1, 2, 3$. Since we must not fire t_1^i in the ith round, we must fire t_2^i instead.

The first consequence is that, before t_1^2 fires, N sees exactly a token being put onto p and then a token being taken from q in the first round, i.e. p^+ and then q^- fire in N^{env}; analogously the other rounds can be treated.

The second consequence is that the $t_{r..}^i$, $i = 1, 2$, must not fire; hence, just before e.g. the first σ occurs, the token on s_r^1 is old and there cannot be a token on z or an old token on x. Thus, N^{env} sees the refusal set $\{x^1, z^0\}$ at the end of the first and $\{z^1\}$ at the end of the second round.

For the last round, the reason that t_{rx0}^3 must not fire, is different: if it fires in v, i.e. at the latest in the third round, then we cannot have a time step before the 0-transition t_2^4 fires, which would satisfy the test.

In consequence, N^{env} must perform the pr-trace w to fail the test $(U, 3)$, and the above analysis should also show how to fail the test, when N^{env} performs w.   □

Observe that a faster system has less pr-traces, i.e. such a trace is a witness for slow behaviour, it is something 'bad' due to the refusal information it contains.

Also observe that the above proof is somewhat modular, i.e. it can be used to a large part for treating subcases. First, let us point out the following: we have given a characterization for a testing-based preorder for all testable nets; but it is not a priori clear that this result still holds if we restrict the class of nets under consideration – e.g. to the restricted setting where each interface place is *either* an input place with no arc entering *or* an output place with no arc leaving. The reason is that the test nets we used in the proof have IO-places and do not belong to the restricted class.

Luckily, they almost belong to the restricted class: if e.g. a p^+-transition occurs in U then p is an input place of the testable nets, which do not have arcs to p, and there will be no p^--, p^0- or p^1-transitions; hence, we can make p an output place of U with no arcs leaving p. Therefore, our result also holds in this type-restricted case.

Also, the proof can be adapted if we only consider nets without 0-transitions (and without 1-transitions with empty preset): in this case, we would not allow p^0's in refusal sets when defining pr-sequences and -traces; thus, U would have no 0-transitions – the only exception being t_2^4 in the above example-U; this represents a special treatment of the last refusal set, and this is only necessary since we might have time stops. Since these cannot occur in this second restricted class, also t_2^4 can be avoided (details omitted) and the result carries over.

Lastly, U is safe on its non-interface places. (s_2^4 is an exception, if the last refusal set has more than 1 element; but in this case, we could multiply s_2^4 and t_2^4 suitably.) Thus, our construction should also be an essential step for the treatment of safe nets, but here additional work needs to be done; see Section 6.

Theorem 8 quite immediately implies:

Theorem 9. $\sqsupseteq$ *is a precongruence w.r.t. relabelling and in- and output hiding.*

An important question is now how we show that one net is faster than another. If nets N^{env} were bounded, there would also be only finitely many reachable timed markings; the above characterization would allow to decide $\sqsupseteq$ by a static check plus a decision of regular-language-inclusion. Since N^{env} is generally unbounded, it does not seem feasible to decide $\sqsupseteq$. Still, one can treat many examples (even possibly infinite classes of examples) using simulations:

Definition 10. For nets N_1 and N_2, a relation $\mathcal{S}$ between some timed markings of N_1^{env} and some of N_2^{env} is a *(forward) simulation* from N_1 to N_2 if the following hold:

1. $(TM_{N_1}, TM_{N_2}) \in \mathcal{S}$
2. If $(TM_1, TM_2) \in \mathcal{S}$ and $TM_1[t\rangle_{pr} TM_1'$ or $TM_1[X\rangle_{pr} TM_1'$, then for some TM_2' with $(TM_1', TM_2') \in \mathcal{S}$ we have $TM_2[\hat{t}\rangle\rangle_{pr} TM_2'$ or $TM_2[X\rangle\rangle_{pr} TM_2'$, where $\hat{t}$ is t for $t \in T_2^{env} \setminus T_2$ and the empty word λ otherwise. Observe that these moves from TM_2 to TM_2' may involve several transitions of N_2. $\square$

The following theorem is straightforward; compare e.g. [LV95] for a similar result and a survey on the use of simulations; note that a simulation does not have to exist in each case where $N_1 \sqsupseteq N_2$.

Theorem 11. *If there exists a simulation from N_1 to N_2, then $N_1 \sqsupseteq N_2$.*

5   Examples

Our first two examples are more technical in nature; the first discusses aspects of asynchronous communication and timing. N_1 and N_2 in Figure 5 both have the pr-trace $p_2^- p_1^-$, showing that with asynchronous communication messages can be taken 'in the wrong order'.

In fact, $N_1 \sqsupseteq N_2$ (and vice versa), since the outputs must appear in the first round. The formal proof is maybe not that obvious: a suitable simulation relates reachable timed markings TM_1 of N_1 and TM_2 of N_2, if they are the initial timed markings or if : $M_1(s_1) = 0$, $M_1(p_1) = M_2(p_1)$, $M_1(p_2) + M_1(s_2) = M_2(p_2)$, the old tokens are the same on the interface places and M_2 is 0 on the other places. When N_1 fires its first transition, N_2 fires both its transitions, giving a related timed marking; then both can perform p_1^-, and before or after N_1 fires its second transition, bringing both nets to essentially the same timed marking. Only then refusal sets can occur.

Figure 5

If we change all transitions to 1-transitions (getting N_1' and N_2'), there could be enough time between appearance of the tokens to observe their order: $p_1^-\{p_2^0\} \in PR(N_1') \setminus PR(N_2')$.

The second example demonstrates the effect of interface typing. Also for N_1 and N_2 in Figure 6, we have $N_1 \sqsupseteq N_2$ and vice versa. (The dead transition in N_1 ensures that the nets satisfy the static condition of Theorem 8.) In N_1, tokens on s and on p_2 behave the same: they can get old, and then at the latest they move to the output. Thus, a simulation relates timed markings, if the sum of (old) tokens on s and p_2 in N_1 is the number of (old) tokens on p_2 in N_2, and they are equal on the other places. The situation would be completely different, if p_2 were an IO-place; in this case N_2 would have additional functionality as witnessed by $p_1^+p_2^- \in PR(N_2) \setminus PR(N_1)$.

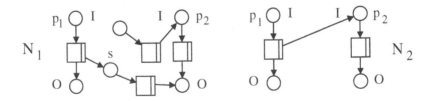

Figure 6

Now we will discuss a series of examples involving the production lines from Section 2. Generally, assume that N_1 has a transition t with a dot, i.e. with a 'private' loop place l, and we split t into two transitions as shown in Figure 7; i.e. t_1 inherits l and the preset of t, and t_2 the postset of t except l.

Then, $N_1 \sqsupseteq N_2$: this can be shown with a simulation that relates timed markings TM_1 of N_1 and TM_2 of N_2, if TM_1 is componentwise the restriction of TM_2 to S_1 and $M_2(s) = 0$; N_2 simulates each firing of t by firing $t_1 t_2$. As an application, we have $PL1 \sqsupseteq PL2$ and also e.g. $PL12 := PL1 \;|||\; PL2 \sqsupseteq PL2 \;|||\;$ $PL2 =: PL22$ by 3. In fact, $PL1$ is strictly faster than $PL2$, as witnessed by $p_1^+ \emptyset \{p_2^0\} \in PR(PL2) \setminus PR(PL1)$, which means that only with $PL1$ an order will certainly result in a product before time 2.

Figure 7

We will now show that $PL22 \sqsupseteq PL2$, denoting the loop place of t_1 by l and giving a dash to the items of the second copy of $PL2$ in $PL22$. (I.e. this second copy has a transition t_1' with a loop place l' etc.) A respective simulation relates TM_{22} and TM_2 if they are equal on the interface, $M_{22}(s) + M_{22}(s') = M_2(s)$, $M^{old}{}_{22}(s) + M^{old}{}_{22}(s') = M^{old}{}_2(s)$, $M_{22}(l) = M_{22}(l') = 1 = M_2(l)$ and $\min(M^{old}{}_{22}(l), M^{old}{}_{22}(l')) = M^{old}{}_2(l)$.

It is clear that e.g. t_1 or t_1' in $PL22$ is simulated by t_1 in $PL2$, and that, after firing them, one of l and l' in $PL22$ does not have an old token, which also holds for l in $PL2$, and this is okay since the minimum above is 0. A bit tricky is the occurrence of a refusal set in $PL22$, where the interface transitions are no problem since the interface places are marked the same way in both nets. Let us just consider t_1; since t_1 and t_1' in $PL22$ are not enabled by old tokens, p_1 does not carry old tokens, or otherwise at least one of l and l' does not carry old tokens; in either case, p_1 or l does not carry old tokens in $PL2$, which therefore can perform a refusal set as far as t_1 is concerned.

Again, $PL22$ is strictly faster – as witnessed by $p_1^+ p_1^+ \emptyset \emptyset p_2^- \{p_2^0\}$, where the tokens on p_1 are old after the first $\emptyset$, but all of them have to leave p_1 before the second $\emptyset$ only in $PL22$.

We close by comparing $PL12$ and $PL1$. One might expect the former to be faster, since the additional production line should help somehow; but this is in fact not true: an order might be processed by this slower line giving rise to the slower behaviour $p_1^+ \emptyset \{p_2^0\} \in PR(PL12) \setminus PR(PL1)$.

On the other hand, $PL2$ does help indeed if there is a rush of orders. If $PL1$ gets four orders, the tokens will be old in the second round; so in this and the third and fourth round, at least one token will be moved to p_2. If these three tokens are removed, the fourth refusal set occurs at a time when p_2 is empty; hence, $p_1^+ p_1^+ p_1^+ p_1^+ \emptyset \emptyset \emptyset p_2^- p_2^- p_2^- \{p_2^0\} \in PR(PL1)$. But this pr-trace is not in $PR(PL12)$: here, in the second round at least two tokens will be moved from p_1 and the remaining ones in the third; thus, in the fourth round there will be some tokens on p_2 and some old tokens on s, which have to be moved; after taking three tokens, p_2 is still not empty at the fourth refusal set. We conclude that *PL12* and *PL1* are incomparable.

Since all phenomena encountered in the above examples had an intuitive explanation, these examples demonstrate that our efficiency preorder makes sense. It should also have become clear, that statements about $\sqsupseteq$ as presented above can only be treated using some characterization – as we have provided it.

6   Conclusion and Work in Progress

For nets as models of asynchronous systems, we have presented a parallel composition $|||$ that merges places and have introduced a typing of such interface places. This composition was used for a testing scenario which gives rise to a faster-than relation for components that communicate asynchronously. (This is in analogy with previous studies for the case of synchronous communication.) We have characterized this faster-than relation and used the characterization to compare the efficiency of some examples.

The most important next step will be to adapt the presented approach to the study of safe nets, and in Augsburg work on this is in progress along the following lines. Since safe nets can be build from components that are not safe in other environments, one possibility is to require that N_1 is a faster implementation of N_2 only if $N_1 ||| U$ is safe whenever $N_2 ||| U$ is; this would lead to considering an additional set of sequences that make a component unsafe – compare $U(N)$ in [Vog92, Section 3.3.1].

Another possibility is to enforce safety on the interface by coupling each interface place with a complement place, and to change the definition of $|||$ accordingly. Then, it is much more reasonable to require the components to be safe in all environments, which makes the *PR*-semantics of a net the language of a finite automaton, so inclusion would certainly be decidable. In fact, wrapping a place-bordered net N into a standard environment in N^{env} wraps the place-oriented problem into a transition-oriented one; thus, it should be possible to check *PR*-inclusion with our tool FASTASY, which was designed for the case of synchronous communication; see [BV98] for results regarding the MUTEX-problem obtained with this tool.

We have convinced ourselves that 0-transitions allow to express solutions to the MUTEX-problem in the cases of asynchronous and synchronous communication. This is remarkable, since [Vog97] relates weakly fair behaviour of ordinary safe nets (which disregards any timing) to timed behaviour of nets with only 1-transitions and shows that in any case the MUTEX-problem cannot be solved – unless one extends nets e.g. with so-called read arcs. We will study how read arcs are related to 0-transitions in general, prove formally how the MUTEX-problem can be solved with 0-transitions and compare in particular MUTEX-solutions in a setting with asynchronous communication.

References

[BDE93]   E. Best, R. Devillers, and J. Esparza. General Refinement and Recursion for the Box Calculus. In P. Enjalbert, A. Finkel, and K. W. Wagner, editors, *STACS'93*, Lect. Notes Comp. Sci. 665, 130–140. Springer, 1993.

[Bih98]   E. Bihler. Effizienzvergleich bei verteilten Systemen – eine Theorie und ein Werkzeug. Diplomarbeit an der Uni. Augsburg, 1998.

[BV98]   E. Bihler and W. Vogler. Efficiency of token-passing MUTEX-solutions – some experiments. In J. Desel et al., editors, *Applications and Theory of Petri Nets 1998*, Lect. Notes Comp. Sci. 1420, 185–204. Springer, 1998.

[Che91]   G. Chehaibar. Replacement of open interface subnets and stable state transformation equivalence. In *Proc. 12th Int. Conf. Applications and Theory of Petri Nets, Gjern*, 390–409, 1991.

[CZ91]   R. Cleaveland and A. Zwarico. A theory of testing for real-time. In *Proc. 6th Symp. on Logic in Computer Science*, pages 110–119. IEEE Computer Society Press, 1991.

[dBKPR91]   F.S. de Boer, J.N. Kok, C. Palamidessi, and J.J.M.M. Rutten. The failure of failures in a paradigm for asynchronous communication. In J.C.M. Baeten and J.F. Groote, editors, *CONCUR '91*, Lect. Notes Comp. Sci. 527, 111–126. Springer, 1991.

[dBZ99]   F.S. de Boer and G. Zavaratto. Generic process algebras for asynchronous communication. In J.C.M. Baeten and S. Mauw, editors, *CONCUR '99*, Lect. Notes Comp. Sci. 1664, 226–241. Springer, 1999.

[DNH84]   R. De Nicola and M.C.B. Hennessy. Testing equivalence for processes. *Theoret. Comput. Sci.*, 34:83–133, 1984.

[Gom96]   D. Gomm. *Modellierung und Analyse verzögerungs-unabhängiger Schaltungen mit Petrinetzen*. PhD thesis, Techn. Univ. München, Dieter Bertz Verlag, 1996.

[Kin97]   E. Kindler. A compositional partial order semantics for Petri net components. In P. Azema et al., editors, *Applications and Theory of Petri Nets 1997*, Lect. Notes Comp. Sci. 1248, 235–252. Springer, 1997.

[LV95]   N. Lynch and F. Vaandrager. Forward and backward simulations I: Untimed systems. *Information and Computation*, 121:214–233, 1995.

[Lyn96]   N. Lynch. *Distributed Algorithms*. Morgan Kaufmann Publishers, San Francisco, 1996.

[NC96]   V. Natarajan and R. Cleaveland. An algebraic theory of process efficiency. In *11th Ann. Symp. Logic in Computer Science (LICS '96)*, 63–72. IEEE, 1996.

[PF77]   G. Peterson and M. Fischer. Economical solutions for the critical section problem in a distributed system. In *9th ACM Symp. Theory of Computing*, pages 91–97, 1977.

[Pop91]   L. Popova. On time Petri nets. *J. Inform. Process. Cybern. EIK*, 27:227–244, 1991.

[SDdSS94]   P. Senac, M. Diaz, and P. de Saqui-Sannes. Toward a formal specification of multimedia synchronization scenarios. *Ann. of telecommunications*, 49:297–314, 1994.

[Val94]   A. Valmari. Compositional Analysis with Place-Ordered Subnets. In R. Valette, editor, *Application and Theory of Petri Nets '94*, Lect. Notes Comp. Sci. 815, 531–547. Springer, 1994.

[Vog92]   W. Vogler. *Modular Construction and Partial Order Semantics of Petri Nets*. Lect. Notes Comp. Sci. 625. Springer, 1992.

[Vog95]   W. Vogler. Faster asynchronous systems. In I. Lee and S. Smolka, editors, *CONCUR 95*, Lect. Notes Comp. Sci. 962, 299–312. Springer, 1995. Full version as Report Nr. 317, Inst. f. Mathematik, Univ. Augsburg, 1995.

[Vog97]   W. Vogler. Efficiency of asynchronous systems and read arcs in Petri nets. In P. Degano, R. Gorrieri, and A. Marchetti-Spaccamela, editors, *ICALP 97*, Lect. Notes Comp. Sci. 1256, 538–548. Springer, 1997.

CASCADE: A Tool Kernel Supporting a Comprehensive Design Method for Asynchronous Controllers

Jochen Beister[1], Gernot Eckstein[2], and Ralf Wollowski[1]

[1] Department of Electrical Engineering, University of Kaiserslautern
P.O. Box 3049, D-67653 Kaiserslautern, Germany
{beister,wollo}@rhrk.uni-kl.de
[2] Infineon Technologies, P.O. Box 801760, D-81617 Munich, Germany
Gernot.Eckstein@infineon.com

Abstract. CASCADE is a tool kernel that supports the synthesis of asynchronous controllers. It uses a generalized STG (an interpreted Petri net) as a unified design entry and allows the designer to choose between several appropriate design methods. It then transforms the initial specification into the design entry required by the chosen style, and interfaces with existing synthesis tools (*petrify* for SI circuits, *3D* for XBM synthesis). By decomposition, certain problems involving output concurrency and MOC behaviour are made XBM-feasible.

1 Introduction

CASCADE is a hardware design tool that supports a comprehensive Petri net based design method for asynchronous controllers. Currently, a designer wishing to synthesize a controller consisting of one or several communicating asynchronous circuits is faced with the problem of first having to choose an appropriate design style and then formulating the design problem in that style's particular specification scheme (which in general is not a Petri net). A better approach would

1. start from a unified design entry using a specification scheme capable of expressing every known kind of asynchronous controller behaviour, and
2. <u>then</u> decide upon the <u>appropriate</u> synthesis method.

Asynchronous circuits are event-driven. It would be adequate, therefore, to specify the required input-output behaviour from a <u>causal</u> point of view. A causal specification scheme already in use is the signal transition graph (STG) [8]. STGs are interpreted place-transition Petri nets where the firing of a transition represents the occurrence of a rising(+) or falling(-) edge of the binary signal with which it is labelled. Black transitions are used for input signals, white ones for output signals. STGs express causal dependence, independence and exclusion (choice, conflict) between signal edges. However, conventional STGs are unable to express certain kinds of asynchronous behaviour such as pseudo-causalities,

M. Nielsen, D. Simpson (Eds.): ICATPN 2000, LNCS 1825, pp. 445–454, 2000.

causal linkage, biased concurrency, and race causality. These shortcomings have been overcome by the introduction of the generalized STG (gSTG) [5]. CAS-CADE supports the full modelling power of the gSTG. The additional net elements needed, such as unlabelled and tc-labelled read and inhibitor arcs, have been incorporated into the Petri net editor PED [6] that outputs net data for further processing by CASCADE. This meets our first demand. To meet the second demand (choice of design style), the net data can be preprocessed and handed over to existing synthesis tools (Fig. 1). CASCADE supports speed-independent (SI) synthesis with *petrify* [3] and extended-burst-mode (XBM) synthesis with *3D*. Support of hazard-tolerant synthesis [7] is currently being incorporated (dotted arc in Fig. 1).

STG data can be directly handed over to *petrify* (which includes a feasibility checker) using a format converter. Interfacing to *3D* is not possible directly because *3D* starts from an XBM machine (XBMM), an FSM-like specification which, if it exists, guarantees implementability [2]. However, CASCADE can derive a primitive flow table (PFT) from the gSTG, check it for XBM feasibility, and, if positive, transform it into an XBMM. Certain forms of output concurrency, not implementable by a single XBMM [2], are treated by a parallel decomposi-

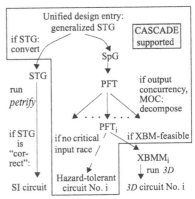

Fig. 1. Comprehensive design method

tion algorithm for PFTs [4] as part of CASCADE's transformation procedure. Multiple-output-change (MOC) behaviour (in the sense of [1], where a single change of the input state causes a sequence of output-state changes), which is forbidden in a single XBMM, but may realized by a set of interacting *3D* circuits, can also be handled using CASCADE, as shown in Sect. 3.1. This enables designers to implement systematically a larger class of behaviour than ever berfore.

2   Program Structure and Use of CASCADE

CASCADE consists of several modules which can be invoked individually via a simple graphical user interface (GUI).

2.1   Program Structure of CASCADE

The graphical Petri net editor PED [1] is used for the design entry (Fig.2). It offers designers the full variety of net elements needed to create a gSTG. CASCADE

---

[1] PED is by courtesy of Monika Heiner, Cottbus Technical University, and is available from http://www-dssz.informatik.tu-cottbus.de

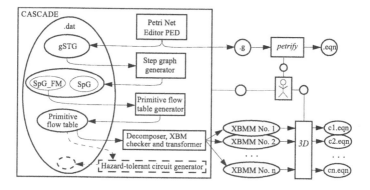

Fig. 2. Current structure of CASCADE

includes converters to generate net data either in the format needed by *petrify* [3] or in its own internal format (.dat). In the latter case, a .dat-file is created that contains the entire net information and will be supplemented in the steps of the design process. The first design step generates either the step graph (SpG) under the tc-firing rule [5] or the fundamental mode step graph (SpG_FM) [4]. Data describing the SpG or the SpG_FM are written into the .dat-file. Fig. 3 shows an example (explanatory comments have been added): the right-hand gSTG from Fig. 6.b is designed with the help of PED (Fig. 3.a), the net data are put into one data block of the .dat-file (Fig. 3.b), and the step graph data into another (Fig. 3.c, cf. Fig. 6.c). Then the PFT can be generated, and a data block describing it (Fig. 3.d) is added to the .dat-file. Finally, these data are read by the program module that checks for XBM feasibility (Fig. 2) and - if feasible - transformed into one or several XBM machines (Fig. 4.a) each of which can be synthesized by the *3D* tool (Fig. 2, Fig. 4.b).

2.2   Using CASCADE

The GUI offers a browser with a file selection push button linked to a text editor. There is one menu for each of the four module programs shown in Fig. 2 (cf. Fig. 5). Having invoked the gSTG module, which in turn invokes PED, the designer creates a gSTG. He then proceeds with the step graph menu that contains a file selection dialogue to convert the output file generated by PED into CASCADE's internal format (.dat). It also contains the parameter selection dialogue (Fig. 5) where the user can choose which firing rule to apply. The FM firing rule must be chosen for synthesis because the PFT generator requires the SpG_FM [4]. While generating either the SpG or the SpG_FM, the step graph generator checks whether rising and falling transitions alternate for each signal. It also checks the single value-set requirement (each marking must be associated with only one set of signal levels). Optionally, the step graph generator aborts if this requirement is violated (Fig. 5). The net data and the signal levels associated with the individual markings are written into the .dat-file (Fig. 3.c). The next

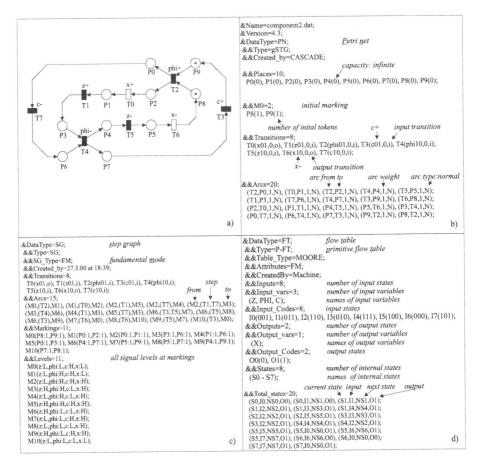

Fig. 3. Example net and related data blocks in the .dat-file
(author's comments in italics not by CASCADE)

design step starts with the PFT generation menu where the user can let the PFT module generate a primitive Moore flow table from the last step graph in the input file. A list of the PFT's total states is written into the .dat-file (Fig. 3.d). Finally, the XBM transformation menu may be used to invoke a program module that decomposes the PFT, if necessary, into parallel component flow tables without output concurrency, eliminates redundant input variables, checks whether all components can be transformed into XBM machines and, if so, performs the transformation. The XBM-specification for each component machine is written into a separate file in the format required by the *3D* tool (Fig. 4.a). This can then be used to synthesize the logic circuit equations (Fig. 4.b).

input	z	0			
input	phi	0	} *initial values*	$x = z + phi + zzz00$	
output	x	0			

current state	*next state*	*input burst*	*output burst*	
0	1	phi+	| x+	
1	2	phi- z+	|	
2	0	z-	| x-	

a)

$zzz00 = phi + z'\ zzz00z$

b)

Fig. 4. The XBM machine for the example of Fig. 3 and equations as derived by *3D*

3 Examples

3.1 Treatment of Multiple-Output-Change (MOC) Behaviour

The extended-burst-mode methodology restricts the set of permissible design problems to those where every output burst must be preceded by a non-empty input burst. This means that MOC behaviour is not implementable by a single XBM machine. But it can be synthesized as a set of component machines where some output signals of each component can be input signals of others; implementation results in a system of interacting circuits. These must be obtained by non-parallel decomposition, because the input signals of the components now are both primary inputs and primary outputs of the overall circuit. CASCADE has no module for non-parallel decomposition. However, MOC behaviour can be trivially recognized at the gSTG level (e.g. Fig. 6.a) and, using PED, it is very easy to generate copies of the original gSTG each of which specifies the interface behaviour of a component with only single-output-change behaviour. For example, the MOC behaviour between x and z (x+ → z+ and z- → x-) in Fig. 6.a is treated by letting x be generated by one component and z by another (Fig. 6.b); then z is made an input to the first component, and x to the second. By applying the FM firing rule, the component SpG_FMs are generated (Fig. 6.c). Every signal is still included in every component's interface, but CASCADE removes redundant input signals before starting the XBM transformation. Input signal c is found to be redundant [4] for both component XBM machines and removed from their interfaces (Fig. 6.d, e, f). The CASCADE representations of the right-hand side of Fig. 6.b to e are shown in Fig. 3 and Fig. 4.a, and the *3D* representation of the right-hand component of Fig. 6.f in Fig. 4.b.

3.2 Controlling the Cardinality of Input Bursts

Part of the transformation of a component PFT into an XBMM is finding the input bursts. The transformation algorithm constructs maximal input bursts because the gSTG does not show how much time elapses between two causally dependent input edges [4]. But maximal input bursts create two problems. The degree of concurrency is increased, making the circuit more complex than if the

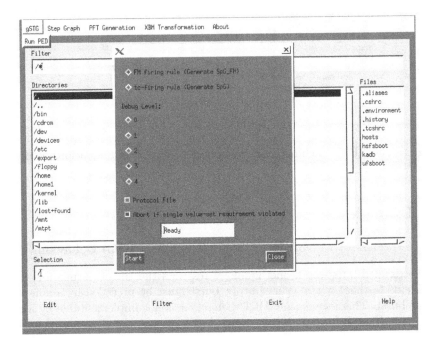

Fig. 5. CASCADE with popped up step graph menu

input edges were known to be far enough apart. And the larger the input bursts, the greater the danger of violating the distinguishability constraint when passing through a state with several successors in the primitive state transition graph. Already when creating the design entry (the gSTG), the user ought to be allowed to specify that time between two successive input edges is sufficiently long for the circuit to stabilize. This can be achieved by using time-stamped places [7] marked with T (e.g. Fig. 6.b). Another way to stop the collection of edges for an input burst is to insert dummy output signals into the gSTG that have to be removed from the XBMM before synthesis is started. Using CASCADE, the designer may introduce these additional signals by PED. Fig. 7 demonstrates this by a section of the benchmark sbuf-send-pkt2 that has been made free of MOC behaviour as discussed in Sect. 3.1. Note that input signals sbufsendpkty2 and sendline are redundant and therefore automatically removed by CASCADE in the decomposition step.

3.3   Processing a Three-Way Arbiter

With the conventional STG, certain kinds of asynchronous behaviour such as critical input races can be modelled only incorrectly or not at all. These short-comings have been overcome by the introduction of the generalized STG (gSTG) [5]. Fig. 8.a shows an example: a three-way arbiter [5], edited with PED. Two tc-labelled read arcs are needed to specify the symmetric and deterministic race

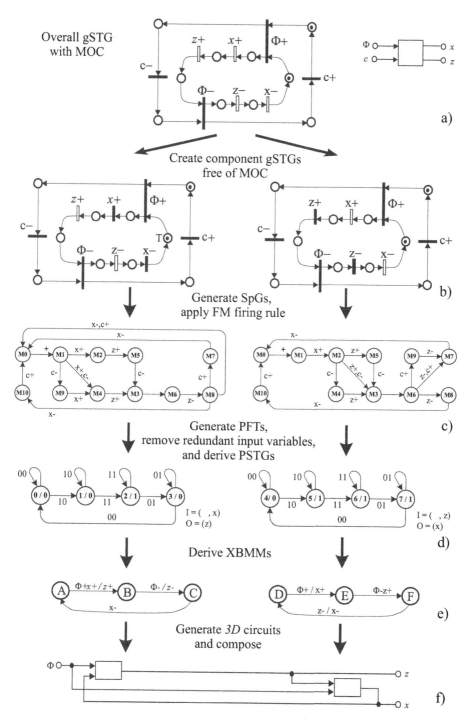

Fig. 6. Example that deals with MOC behaviour [7]

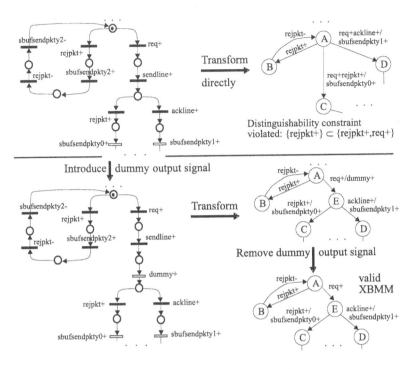

Fig. 7. Signal insertion example

between R1+ and R2+. PED supports the complete expressiveness of the gSTG model. From it, CASCADE generates the step graph (SpG) of the three-way arbiter (Fig. 8.b) by using the tc-firing rule. Note that M3 in Fig. 8.b cannot be reached by any firing sequence consisting of single edges, but only by the simultaneous occurrence of R1+ and R2+. This can only be expressed in a step graph, but not by a reachability graph. But specifying this simultaneous occurrence is essential for the deterministic three-way arbiter: it is to signal a tie (T+) if and only if it recognizes R1+ and R2+ as simultaneous. for every gSTG.It may also generate an SpG_FM from which a PFT may be derived. Unfortunately, critical input race behaviour cannot be correctly implemented by a logic circuit. Therefore, after CASCADE has generated the PFT, arbiter extraction by PFT decomposition is called for. The deterministic part may then be synthesized by *petrify* or, after XBM transformation of the PFT, by *3D* or by hazard-tolerant synthesis [7]. The indeterministic part, not synthesizable by CASCADE, has to be delegated to a separate circuit containing microelectronic devices designed not to show anomalous behaviour at their outputs (e.g. mutual-exclusion elements).

3.4   Results

Among other examples, CASCADE has been applied to the meat_examples subset of benchmarks (Table 1). Those with an asterisk occur in different versions.

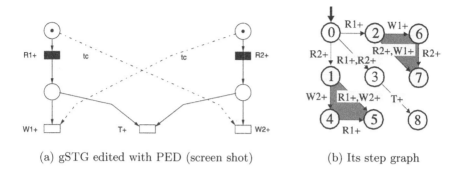

(a) gSTG edited with PED (screen shot)   (b) Its step graph

Fig. 8. Race part of a three-way arbiter

We took the SIS-versions, most of which have MOC behaviour and hence are not XBM feasible without decomposition. The remaining benchmarks from the meat subset: pe-send-ifc, sbuf-send-ctl, and sbuf-send-pkt2 are not listed in the table, but can be solved by the burst-limiting method outlined in Sect. 3.2. Benchmarks stn1 and hazard are from other subsets, and the rest of the examples in Table 1 are our own. Some benchmarks are still not XBM feasible, even if CASCADE's decomposition methods are applied to them, e.g. because an irrelevant edge input may change twice during an input burst, or no compulsory edge exists. Such examples can be solved by circuits with direct feedback and under the assumption of pure bounded wire delays, using the hazard-tolerant design style introduced in [7] (hazard pulses are tolerated on internal feedback lines, not on outputs!).

4   Conclusion and Future Work

CASCADE is a new tool kernel that supports a comprehensive design method for asynchronous controllers. The graphical Petri net editor PED is used for specifying a unique design entry, the gSTG, as the causal model of the required behaviour of the circuit at its input-output interface. Then the designer can choose which design style to apply: currently either SI (supported by an interface to *petrify*) or XBM synthesis. CASCADE is not an XBM synthesis tool itself. It supports other XBM synthesis tools, mainly *3D*, by transforming the gSTG into one or - in certain cases of output concurrency and MOC behaviour - several XBM machines, if this is possible. CASCADE can also derive step graphs and primitive flow tables for gSTGs, specifying critical input race behaviour. Future work will deal with arbiter extraction at the gSTG level (cf. Sect. 3.3). We are about to expand CASCADE to a complete synthesis tool (Fig. 1) by integrating hazard-tolerant synthesis [7]. The designer will then be able to solve all problems that do not involve critical input races under the assumption of bounded wire delays with one tool.

Circuit	signals I/O	XBM feasible	XBM feasible after decomposition	signals after MOC decomposition (I/O)	signals after parallel decomposition (I/O)	states of XBM machine(s)
stn1	2/3	no	yes	2/2 + 3/1	-	4 + 4
alloc-outbound*	4/5	no	yes	6/3 + 4/2	-	8 + 4
mp-forward-pkt*	3/5	no	yes	4/4 + 2/1	-	4 + 3
nak-paa*	4/6	no	yes	5/5 + 4/1	-	6 + 2
pe-rcv-ifc*	4/7	no	no	-	-	-
ram-read-sbuf*	5/6	no	yes	6/5 + 3/1	-	7 + 3
rcv-setup*	3/2	yes	-	-	-	6
sbuf-ram-write*	5/7	no	yes	4/5 + 5/2	-	6 + 4
sbuf-read-ctl*	3/5	no	yes	4/3 + 3/2	-	5 + 3
sendr-done*	2/2	no	yes	2/1 + 2/1	-	2 + 3
hazard	2/2	yes	-	-	-	4
async99	3/3	no	yes	-	3/2 + 3/1	4 + 4
c_sg	4/1	yes	-	-	2/1[a]	2
c_sg1	4/2	no	no	-	-	-
c_sg2	4/2	no	yes	-	2/1 + 2/1	3 + 3
dcko2	3/1	yes	-	-	2/1[a]	3
five	6/5	no	yes	-	2/1 + 2/1 + 2/1 + 2/1 + 1/1	3 + 3 + 3 + 3 + 2

a.   Only redundant input signals were removed

Table 1. Experimental Results

References

1. Unger, S.H.: Asynchronous Sequential Switching Circuits. R.E. Krieger, reprint 1983 (original edition 1969)
2. Yun, K.Y.: Synthesis of Asynchronous Controllers for Heterogeneous Systems. PhD thesis, Stanford University (1994)
3. Cortadella, J.: Petrify: A tutorial for the designer of asynchronous circuits. Available as part of the petrify tool package from: http://www.lsi.upc.es/~jordic/petrify.
4. Beister, J., Eckstein, G., Wollowski, R.: From STG to Extended-Burst-Mode Machines. In: Proc. of the 5th Int. Symp. on Advanced Research in Asynchronous Circuits and Systems, Barcelona (April 1999). IEEE Computer Society Press.
5. Wollowski, R., Beister, J.: Comprehensive Causal Specification of Asynchronous Controller and Arbiter Behaviour. In: Yakovlev, A., Gomes, L., Lavagno, L. (eds.): Hardware Design and Petri Nets. Kluwer Academic Publishers, Boston (2000) 3-32
6. Tiedemann, R.: Dokumentation PED Version 4.3 (Benutzerleitfaden). Technical Report, Cottbus Technical University (June 1997)
7. Eckstein, G.: Logischer Entwurf hasardtoleranter asynchroner Schaltwerksverbünde (Logical design of hazard-tolerant communicating asynchronous circuits). Submitted dissertation, University of Kaiserslautern (to appear in 2000)
8. Kondratyev, A., Kishinevsky, M., Yakovlev, A.: Hazard-free implementation of speed-independent circuits. IEEE Transactions on Computer-Aided Design of Integrated Circuits and Systems, Volume 17 (September 1998)

Ex*Spect* 6.4
An Executable Specification Tool for Hierarchical Colored Petri Nets

Wil M. P. van der Aalst[1], Poul J. N. de Crom[2], Roy R. H. M. J. Goverde[2],
Kees M. van Hee[1,2], Wout J. Hofman[2], Hajo A. Reijers[1,2], Robert A. van der Toorn[1,2]

[1]Eindhoven University of Technology,
Department of Mathematics and Computing Science,
P.O. Box 513, NL-5600 MB, Eindhoven, The Netherlands
{wsinwa, wsinhee, hreijers, rvdtoorn}@win.tue.nl

[2]Deloitte & Touche Bakkenist,
Department of Information and Communication Technology,
P.O. Box 23103, NL-1100 DP, Amsterdam ZO, The Netherlands
{pdcrom, rgoverde, kvhee, whofman, hreijers, rvdtoorn}@bakkenist.nl

Abstract. Ten years ago Ex*Spect* became available on the market. Since then a lot of modeling and simulation projects in logistics, workflow and electronic commerce have been performed using Ex*Spect*. In the past ten years the *heart* of Ex*Spect*, the simulation engine, has never been changed: it still executes models of hierarchical, timed, colored Petri nets with priorities. Over the years new features have been introduced based on user requests. Three extensions dominate the new functionality of Ex*Spect*. The first is 'ease of use' in simulating and carrying out quantitative analysis of workflows. The second is to view Message Sequence Charts for electronic commerce applications using Ex*Spect*. The last is the integration of Ex*Spect* and applications; i.e., to use Ex*Spect* to handle the *flow of control* for other applications.

1 Introduction

Ex*Spect* is a simulation and animation tool for hierarchical timed colored Petri nets with priorities [9]. The first version of Ex*Spect* was launched ten years ago and is described in [10]. Since then the tool has been used both in the academic and in the business world. Ex*Spect* should be compared for instance with tools such as Design/CPN, CPN-AMI and Great SPN (see [17] for pointers). Ex*Spect* proved to be particularly successful in the fields of workflow management, logistics, and electronic commerce. The deployment of the tool in workflow and electronic commerce projects required a number of new features. We will discuss the three most important new functionalities of Ex*Spect*.

The first feature has been developed to meet consultant needs simulating business processes with Ex*Spect*. In workflow management the use of workflow management systems (WFMSs) to support business processes (also called workflows) is widespread. Until now these WFMSs have poor simulation and verification facilities [1]. For organizations it is, however, extremely important to verify and simulate new

M. Nielsen, D. Simpson (Eds.): ICATPN 2000, LNCS 1825, pp. 455-464, 2000.

business processes before implementing them. Ex*Spect* does not have verification options like, for instance, Woflan [16], but Ex*Spect*-simulations are useful for various reasons.

- The *graphical animation of a workflow* is a good way to provide feedback to the owners of the workflow, i.e., employees of an organization. For instance, the sequence of tasks can be checked for correctness.

- The *quantitative analysis* of a workflow process gives insight into waiting times, service times, throughput times, occupation levels and queue lengths. Moreover resource allocation scenarios may be carried out in order to optimize the distribution of resources over processes and tasks.

Quantitative analysis of workflows is facilitated by a large number of statistical functions (all listed in the Ex*Spect* help file [6]) and a large library of *workflow building blocks*. Building blocks are predefined Petri net structures with a particular function in a workflow, for instance the creation of a case or the choice between two paths according to a certain mathematical distribution. A case study in which these qualities of Ex*Spect* are exploited is described in [14]. However in practice constructing a workflow simulation model in Ex*Spect* to perform quantitative analysis was often too expensive (in terms of costs and time) and could only be carried out by someone with a deep knowledge of the Ex*Spect* workflow library. To achieve 'ease of use' it was very fruitful to develop translators. Translators are applications that are able to read a workflow process definition created in tools solely dedicated for workflow design and translate this definition to an executable Ex*Spect* specification, using building blocks from the workflow library. This resulted in the applications CO2EX and Pros*pect*. These applications are *interfaces* (i.e., links) between respectively the tools COSA® [15] and Ex*Spect* and PROTOS [13] and Ex*Spect*.

The second feature has been developed to meet customer requirements for specifying message exchange between information systems to support electronic commerce. Important application areas in this respect are the business-to-business and business-to-administration communication [2]. The integration of business processes of a large number of organizations is the central issue. Organizations, also called *actors*, exchange information to support their business processes. Modeling the interaction between these actors in Ex*Spect* can be done easily because of the expressive power of Ex*Spect*. However, it appeared to be a problem to present complicated Ex*Spect* models to persons without a Petri net background. It is easier to provide them with interaction scenarios derived from an Ex*Spect* model. This motivated the extension of the Ex*Spect* Dashboard (see Section 2) with the option to generate Message Sequence Charts [5] during a simulation run of Ex*Spect*. Note that these Message Sequence Charts correspond to Sequence Diagrams in UML [3].

The third feature is the 'componentization' of the Ex*Spect* engine. There is an ongoing trend in software engineering to separate the specific functionality of applications from 'standard' functionality and make a software component of the separated functionality. This has been done with Ex*Spect* as well. The Ex*Spect* engine has been separated from the other parts of the Ex*Spect* application and has been 'componentized' according to the Microsoft COM standard [12]. The new component is called the *ExSpect Server*. The Ex*Spect* Server is particularly useful in applications with a complicated flow of control. Moreover customers require more often not only an Ex*Spect* simulation of the system specification but also an Ex*Spect* prototype of the system that is able to operate in the environment of the system in mind. To meet such demands it is necessary to have a tool that can be integrated in various applications. With Microsoft COM the Ex*Spect* Server can be integrated in Visual Basic, JAVA and C++ based applications.

The structure of this paper is as follows. Section 2 discusses the architecture of Ex*Spect* 6.4. Functionality that has not been presented in previous publications will be described briefly; for known functionality references will be given. Section 3 presents the first feature, the interface to other tools. Section 4 describes the option to generate Message Sequence Charts. In Section 5, the Ex*Spect* Server is explained. Section 6 formulates conclusions and indicates plans for future development.

2 Architecture of Ex*Spect*

Fig. 1 depicts the architecture of Ex*Spect*. We will now describe the modules of this architecture.

- The heart of Ex*Spect* is the *ExSpect Server (engine)*. It is able to execute the token-game for models that are expressed in the Ex*Spect* language. The Ex*Spect* language is a typed functional language close to a subset of the Z-*language* defined in [9]. It is based on hierarchical timed colored Petri nets with priorities. The elementary transitions in the Ex*Spect* language have a logical relation that describes their consumption-production behavior. A precise description of the language can be found in [8] and [10].

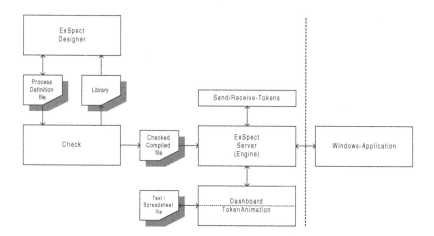

Fig. 1. The architecture of ExSpect 6.4

- The *ExSpect Designer* provides a user interface to create new Ex*Spect* models and to update existing ones. Creating and installing system, processor, function and type components results in an Ex*Spect* model. A detailed description of these components and the construction process is given in [7].

- Once Ex*Spect* systems are defined and installed, it is possible to perform a syntactical check of these systems by activating the module *Check*. Errors are reported and the user can improve the model until it is correct.
- If a closed system, i.e., a system with no input-, output-, or store-pins [6, 9], is correct, the Ex*Spect* Engine can simulate it. To display simulation results from the Ex*Spect* Engine and to communicate with the engine, the *Dashboard/TokenAnimation* module is used. This module is activated from the Ex*Spect* Designer module. The Dashboard/TokenAnimation module consists of two parts: the dashboard window and the token animation windows (available for each system).
  - The *Dashboard* window is used to monitor and edit the contents of channels (i.e., places) during a simulation run. A user can place several types of DashBoard Objects (DBOs) on the dashboard. Every DBO either shows a graphical representation of the contents of a channel or allows the user to add tokens to a channel. Fig. 2 depicts all Dashboard objects. The newest DBO is the representation of Message Sequence Charts. The functionality of DBOs is described in the Ex*Spect* help file [6].

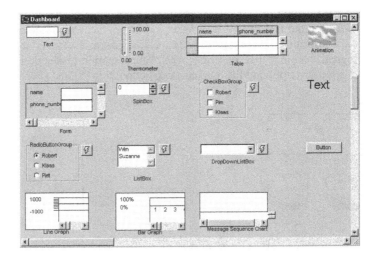

Fig. 2. Dashboard objects

- The *Token Animation* of a system is displayed in one or more Animation-windows. Starting point of an animation is the main system, and from that point on animations of sub-systems can be opened. Each animation-window displays a system in the same way it is displayed by the Designer, but without the editing facilities. When a system is being simulated, the movement of the tokens on connectors (i.e., arcs) is animated graphically. The value of a token residing in a place can be inspected and a token can be added or deleted from the Petri-net during simulation. All DBOs can be displayed in the animation-window as well. Fig. 3 is an example of an animation window.

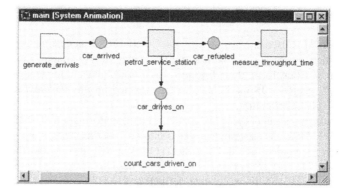

Fig. 3. System animation window

- An important advantage of Ex*Spect* is the development of a number of *libraries* for various types of problems that have been analyzed over the past 10 years. Libraries are available for modeling workflow, logistics, electronic commerce and administrative processes. These libraries can be used in the Ex*Spect* Designer. It is easy to create new libraries; only specifications that are checked may be used as a library.
- Once an Ex*Spect* file has been created, it is saved as a *Process Definition file*.
- The Ex*Spect* engine uses a *Checked and Compiled* version of the Process Definition file.
- Import and export files that can be used by the Dashboard/TokenAnimation model are flat *Text or Spreadsheet files*. The content of these files can be read before and written after a simulation.
- *Send/Receive-Tokens* is an application that is able to put token values from a file in an Ex*Spect* channel and it is able to get a token value from an Ex*Spect* channel and put it in a file. The precise functionality is described in the Ex*Spect* help file [6]. Windows applications use Send/Receive-Tokens to communicate with Ex*Spect*. The file-interface offered by Send/Receive-Tokens is an alternative communication channel for the one offered by the ExSpect Server.
- The *Windows-application* is not really part of the Ex*Spect* application. It signifies a COM-application that can communicate with the Ex*Spect* Engine in process.

3   Interfaces with Other Tools

Interfaces with other tools give Ex*Spect* the 'ease of use' in simulating and carrying out quantitative analysis of workflows. Two interfaces are discussed: the PROTOS-Ex*Spect* and the COSA®-Ex*Spect* interface.

3.1   Interface between PROTOS and Ex*Spect*

PROTOS [13] is used for business process modeling. It is designed to enable users to describe their business process in an intuitive way. It has no underlying semantics, or at least a weaker form than Ex*Spect*. If a user takes into account a number of restrictions, a PROTOS model can be transformed into an Ex*Spect* model by the so-called Pros*pect* application.

There are two aspects that play a role in this transformation. The first is the way in which PROTOS processes are translated into Ex*Spect* processes and the second is the definition of statistical functions in PROTOS. We will first discuss the way PROTOS processes relate to Ex*Spect* processes. Fig. 4 depicts three PROTOS processes. The first illustrates how a process can be defined in PROTOS: transitions, called *activities* in PROTOS, are connected directly to other activities. The second and the third processes illustrate that the interpretation of the first process may cause confusion. Does the first process correspond to the second or the third process? The Pros*pect* application always assumes that places should be inserted on all arcs between two activities. (Situation 2 in Fig. 4.)

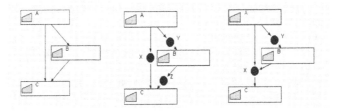

Fig. 4. Three PROTOS processes

PROTOS offers dedicated windows to add simulation settings to a process definition. Simulation settings of four types are introduced in a PROTOS model 1) per activity, 2) per role, 3) for a process model and 4) per choice-construction:

1) For each activity the necessary number of persons to execute the activity has to be put in. The relative frequency of an activity related to other activities has to be entered and service times have to be entered in the form of a probability distribution. Optionally a priority and the cost of an activity can be inserted.

2) In PROTOS each activity is connected to a role. For each process definition the number of resources available in that role have to be entered.

3) For each process, settings concerning reliability have to be entered as well as an arrival pattern. The reliability information of the simulation includes: the number of cases per batch, the number of batches, the number of sub-runs, the length of a sub-run, the number of start-runs. An arrival pattern is given by a probability distribution; often this is the negative exponential distribution.

4) The choice construction is the construction in which a choice is based on the value of the data element. The information that is added specifies the conditions to be used.

If the process definition extended with simulation data is correct then after the transformation by Prospect the result can directly be simulated in ExSpect.

3.2   Interface between COSA® and Ex*Spect*

A similar simulation model as obtained with PROTOS, can be obtained from the workflow management system COSA® [15]. Also a number of simulation settings has to be added. Since COSA® is a Petri net based tool, the translation is straightforward. The application that translates a COSA® model to an Ex*Spect* model is called CO2EX.

4   Generate Message Sequence Charts During Simulation

During simulation Ex*Spect* can generate Message Sequence Charts (MSCs) [5]. We illustrate this feature by means of the Transit process, which has been modeled in Ex*Spect* in a design project for European Commission. Transit is a European customs process. It controls the communication between organizations that exchange messages about a transit declaration of goods between member states of the European Union. With the model, particular interaction scenarios were generated. Fig. 5 shows a number of Ex*Spect* systems. 'NCTS' represent roles of organizations. Fig. 5 also shows an MSC that corresponds with these roles. The MSCs are particularly useful to evaluate interaction scenarios between roles. MSCs are derived from an Ex*Spect* system by monitoring the communication channels between roles. In case of the example an internal channel in the system 'network' is monitored.

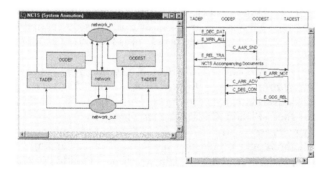

Fig. 5. Actor system and Message Sequence Chart

Each message token in a message channel is monitored: the sender and recipient of the token are registered. This results in an arrow in the MSC at the right-hand side of Fig. 5. In the Ex*Spect* help file [6] a technical description of the DBO MSC is given.

UML (Unified Modeling Language), see [3] has been accepted as standard modeling language, where designers often have to conform to. UML Sequence Diagrams (SDs) model the interaction between objects. A designer can interactively construct SDs with tools. However, (s)he cannot verify these SDs with other models like state charts or activity diagrams. Because an MSC is similar to a SD and because an MSC represents one particular trace of the dynamic behavior of a system, Ex*Spect* can be used to validate these diagrams with the integrated system description in Ex*Spect*.

5   Ex*Spect* as a COM-Component

Recently, the Ex*Spect* engine has been isolated from the other parts of Ex*Spect* and made a COM component [12] with a well-defined interface. This offers the opportunity to develop applications with "Ex*Spect*-inside". In applications of this type Ex*Spect* handles the control flow while the user interface and the integration with data can be handled by the window application (called the *client*). Microsoft COM components can be integrated in a number of programming environments. Some of the well known are Visual Basic, Java and C++.

A client can interact with the Ex*Spect* COM server through methods and events of the server's Application Programming Interface (API). The API intervenes in the process of the engine. A client calls methods and the server triggers events as illustrated in Fig. 6.

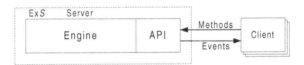

Fig. 6. Interaction between the ExSpect server and clients (windows applications)

The process of invoking methods and receiving events is asynchronous. After a client invokes a method it expects an event as a response. A period of time passes before the client is reached. Meanwhile the client is blocked to avoid synchronization problems.

The methods of the API are: *AddBreakpoint, ConsumeToken, Continue, GetTime, GetTokenValue, HaltAfterSteps, HaltOnTime, Idle, Init, LoadState, Pause, ProduceToken, RemoveBreakpoint, SaveState, TracePlace* and *UntracePlace*.

The events of the API are: *Consume, Continue, FireEnd, FireStart, Initialised, Paused, PlaceType, Present* and *Produce*.

With these methods and events and a thorough understanding of the protocol between the Ex*Spect* Server and its Clients, it is possible to obtain the same monitoring and control options as one would have with the ExSpect Simulator and Dashboard. Presently, a tutorial on the Ex*Spect* Server is developed that describes all methods, events and protocols in detail.

Fig. 7 shows an MSC, which is an example of an ExSpect Client - Server protocol. This protocol illustrates the following. To start client-server interaction a client calls the method *Init*. As a result a client receives an event *PlaceType* for every place in the system for which the "show in simulation" option is selected [6] and it also receives an event *Initialised* from the COM server. The event *PlaceType* provides information concerning the type of a place while the event *Initialised* indicates if the initialization has succeeded or failed. After initialization a client can start with the simulation by calling method *Continue*. A client will receive event *Continued* as an acknowledgement. In case of more than one client, the COM server will broadcast event *Continued* to all clients. To interrupt the simulation a client has to call method *Pause*. Event *Paused* will be sent by the COM server. In this case there will also be a broadcast to all clients if there is more than one client connected to the COM server.

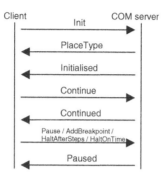

Fig. 7. ExSpect Client - Server protocol

6   Conclusions and Future Developments

This paper demonstrates a number of new features of ExSpect. In the past ten years the high level Petri-net concept appeared to be successful in solving a number of modeling problems in workflow, logistics and electronic commerce. Users requested changes of other aspects, such as user-friendliness, better integration with other tools and the capability to create prototypes rather then specifications of systems.

Ex*Spect* addressed these requests by:
-    improving the user interface and creating interfaces to other tools,
-    offering the possibility to view aspects of a system, such as MSCs,
-    "componentizing" the Ex*Spect* engine and thereby offering the possibility to create user dedicated interfaces for Ex*Spect*.

In the future the tool will be extended with other architectural views, such as a component view [11]. Also an attempt will be made to add workflow verification options to Ex*Spect* [16].

References

1. van der Aalst, W.M.P.. The Application of Petri Nets to Workflow Management. *The Journal of Circuits, Systems and Computers*, 8(1):21-66, (1998).
2. van der Aalst, W.M.P.. Process-oriented Architectures for Electronic Commerce and Interorganizational Workflow. *Information Systems*, 24(8):639-671, (2000).

3. Booch, G., Rumbaugh, J., Jacobson, I.: The Unified Modeling Language User Guide. Addison Wesley (1998).
4. de Crom, P.J.N.: The design of a new graphical user interface for Ex*Spect*. Final report of the postgraduate program Software Technology. ISBN 90-5282-566-1. Stan Ackermans Instituut, Eindhoven (1995).
5. CCITT. CCITT Recommendation Z.120: Message Sequence Chart (MSC92). Technical report, CCITT, Geneva (1992).
6. Deloitte & Touche Bakkenist. Ex*Spect*. Product Management Ex*Spect* P.O. Box 23103, 1100 DP Amsterdam, The Netherlands. http//www.exspect.com/ (1999).
7. Deloitte & Touche Bakkenist. Ex*Spect* 6 User Manual. Product Management Ex*Spect* P.O. Box 23103, 1100 DP Amsterdam, The Netherlands (1997).
8. Deloitte & Touche Bakkenist. Ex*Spect* Language Tutorial (rel 6.0). Product Management Ex*Spect* P.O. Box 23103, 1100 DP Amsterdam, The Netherlands (1997).
9. van Hee, K.M. Information Systems Engineering: a Formal Approach. Cambridge University Press (1994).
10. van Hee, K.M., Somers, L.J., Voorhoeve, M. Executable Specifications for Distributed Information Systems, in E.D. Falkenberg, P. Lindgreen (eds.), Information System Concepts: an in-depth analysis, North Holland, pages 139 - 156 (1989).
11. van Hee, K.M., van der Toorn, R.A., van der Woude, J., Verkoulen, P.: A Framework for Component Based Software Architectures. In W.M.P. van der Aalst, J. Desel, and R. Kaschek, editors, *Software Architecture Business Process Management (SABPM'99)*, pages 1-20, Heidelberg, Germany, June 1999, Forschungsbericht Nr. 390, University of Karlsruhe, Insitut AIFB, Karlsruhe, Germany (1999).
12. Microsoft. Component Object Model (COM). http://www.microsoft.com/com/.
13. Pallas Athena. PROTOS User Manual. Pallas Athena BV, Plasmolen, The Netherlands (1997).
14. Reijers, H.A. and Van der Aalst, W.M.P.: Short-term Simulation: Bridging the gap between operational control and decision making. In Proceedings of the IASTED International Conference on Modeling and Simulation 1999, May 5-8, Philadelphia. ACTA Press, Pittsburg PA, pages 417 - 421 (1999).
15. Software-Ley. COSA® User Manual. Software-Ley GmbH, Pullheim, Germany (1996).
16. SMIS group, Department of Computing Science, Eindhoven University of Technology. Woflan. http://www.win.tue.nl/~woflan.
17. Department of Computing Science, University of Aarhus. Tools on the Web. http://www.daimi.aau.dk/PetriNets/tools/.

LoLA
A Low Level Analyser

Karsten Schmidt

Humboldt–Universität zu Berlin
kschmidt@informatik.hu--berlin.de

Abstract. With LoLA, we put recently developed state space oriented algorithms to other tool developers disposal. Providing a simple interface was a major design goal such that it is as easy as possible to integrate LoLA into tools of different application domains. LoLA supports place/transition nets. Implemented verification techniques cover standard properties (liveness, reversibility, boundedness, reachability, dead transitions, deadlocks, home states) as well as satisfiability of state predicates and CTL model checking. For satisfiability, both exhaustive search and heuristically goal oriented system execution are supported. For state space reduction, LoLA features symmetries, stubborn sets, and coverability graphs.

1 General Remarks

The development of LoLA started in the end of 1998 as an experimental implementation of several state space reduction techniques. Meanwhile, a data stream oriented interface has been added that allows other tools to exchange data with LoLA.

In principle, LoLA can run stand-alone (with a text editor for creating input). However, interface design was guided by the idea to run LoLA as a background service of another tool, invisible to the user. Therefore, LoLA does not have a graphical interface nor does it have any features that support modelling (such as hierarchies, modules, or version control).

We see the following advantages of a pure verification-only tool running in the background of a domain-specific tool:

- Developers of application oriented tools do not need to implement general-purpose verification techniques by their own;
- it is easier to keep pace with changes such as further improvements, optimisations, or new techniques (several verification techniques share the same pattern of communication: input system description and specification of the property to be verified, output verification result yes/no/don't know and — if available — a witness or counterexample execution);
- publications of verification techniques do not always report all details, possible shortcuts and optimisations of the presented techniques;

M. Nielsen, D. Simpson (Eds.): ICATPN 2000, LNCS 1825, pp. 465–474, 2000.

– it is easier to maintain parallel execution of different techniques (or the same technique with different parameters) in competition — this is particularly interesting in the face of well known discrepancies between worst case complexity and "typical case" run time ("typical" usually depends on the verification technique).

Example. In 1994, we reported an algorithm that is able to calculate symmetries of place/transition nets [4] and implemented the technique in the tool INA[1](Humboldt-Universität zu Berlin). Other tool developers implemented this algorithm, too, and reported some optimisations they had found. It turned out that most of these optimisations were covered by our own INA implementation as well (though not mentioned in our report). Thus, double work was done concerning both implementation and further tuning.

In 1997 and 1999, we published significant improvements to the symmetry algorithm, both concerning run time and memory efficiency. Had we distributed our technique in LoLA style already in 1994, it would be easy for the above mentioned developers to participate in our improvements just by downloading a new LoLA version. Without a service like LoLA, they either need to keep the old, less powerful algorithm or have to implement the new technique again.

Integrating the existing INA implementation into other tools (as done with the PEP tool (University of Hildesheim)) turned out to be a costly task since INA's interface was first and foremost designed for interactive use which is difficult to be simulated by machines. For creating LoLA, providing simple patterns for communication with other tools was consequently a major criterion.

We would like to notice that there are several tools that are prepared to integrate external services or actually do so. Among these tools are PEP (University Hildesheim/University Oldenburg), CPN-AMI (Université Paris VI), and the Petri Net Kernel (Humboldt–Universität zu Berlin).

As we understand LoLA as a service for other tools, LoLA itself does not have features that are specific for particular application domains. Input consists of plain and unstructured place/transition nets and parameters to the verification task specified at compile time (for instance, the name of a transition to be checked for liveness). Output consists of the answer to the verification query (a value ranging between "yes", "no", and "no answer") and, if existing, a witness or counterexample execution proving the given answer. On demand, additional output such as the system symmetries or the produced state space can be generated, too.

Benefits from LoLA's integration depend on the question whether there are verification problems that LoLA is able to solve and which are important for

---

[1] references to all tools mentioned in this paper can be found through the Petri net database that is accessible from the Petri net home page http://www.daimi.au.dk/PetriNets/

the host tool's application domain. LoLA supports the exploration of deadlocks (dead states), liveness, boundedness, reversibility, home states, reachability of states or state properties, and properties that are expressible in a simple branching time temporal logic.

Additionally, it is crucial whether LoLA can cope with typical system sizes of that domain. For the latter condition, it is difficult to give any prediction since the power of most state space reduction techniques implemented in LoLA varies from "amazing" to "almost none", depending on the particular system under verification. We have tested nets with about 100 nodes where LoLA failed for some verification task while we have seen other nets with 90000 nodes where LoLA was successful.

2 How to Use LoLA

LoLA distribution contains a configuration file. By editing this file, the user can customise the verification technique he or she wants to use. Customisation consists of fixing the property to be analysed, selecting the reduction techniques to be applied, choosing between breadth-first and depth-first strategy (use of breadth-first strategy is crucial for preserving shortest paths). Furthermore, capacities of certain data structures can be customised. Fig. 1 shows a part of the configuration file.

Having fixed a configuration, a compiler run generates an executable program. The resulting program reads a place/transition net description from a file or data stream, computes (if specified) the net symmetries, generates a (reduced) reachability graph and outputs the answer to the specified analysis query. The amount of output information (target state, witness path, state space, computed symmetries) can be controlled by command line options. Additionally, statistical information (such as number of states and edges of the generated graph, number of computed symmetries) is reported.

LoLA expects net descriptions in programming language style (see figure 2). In similar style, additional information about the verification task can be passed to LoLA. Each kind of LoLA output is preceded by a similar keyword identifying it as witness path, final state or whatever. Beyond passing net and task dependent parameters to LoLA and scanning LoLA results, no communication is necessary to control LoLA. Input and output languages follow regular (CHOMSKY type 3) grammars documented in the LoLA manual.

LoLA regularly produces messages in order to reveal its own progress. Frequency of these messages is customisable.

The style of using LoLA is partly comparable with INA and partly with PROD. In both INA and LoLA, the net description can be passed through a data file (though INA's language is much harder to read). However, in INA the task to be executed must be chosen through an interactive dialogue which is rather case sensitive and is responsible of most integration problems with INA. Additionally, pre–compiled configuration of the verification task has the advantage that code not relevant to the chosen task is simply not present in

```
//#define COVER
#define STUBBORN
#define SYMMETRY

#define GRAPH depth_first
//#define GRAPH breadth_first


//#define REACHABILITY
//#define MODELCHECKING
//#define BOUNDEDPLACE
//#define BOUNDEDNET
#define DEADTRANSITION
//#define REVERSIBILITY
//#define HOME
//#define FINDPATH
//#define FULL
```

Fig. 1. Part of LoLA's configuration file. Lines starting with // are comments and thus considered not present. In the sketched version, the resulting program would expect a net and a name of a transition t. Then it would start constructing a symmetrically and stubborn reduced reachability graph using depth first strategy. Both the used symmetry group and the version of stubborn sets would assure that death of t is preserved by the reduction. As soon as a state enabling t is found, the program returns with a path from the initial marking to that state. Due to depth first exploration this path is not necessarily a shortest one. If there is no state enabling t, the program returns after exhaustive exploration of the reduced graph.

the executable version. In INA, there is a lot of statements like IF strategy = depth_first .... The evaluation of such statements costs only little, but inside a frequently executed loop this time sums up to a significant amount. This is one of the reasons why LoLA constantly outperforms INA on comparable tasks.

In PROD, both verification task and net description are subject to the compiler run. This way, additional speed-ups can be achieved. On the other hand, changes in the net require new compiler runs. Thus, an integrating tool needs to control the generation process of the PROD executable program during it's own run.

Concerning LoLA, we think that pre-compiled versions for at most a handful of configurations (of 350 KByte each) are sufficient for any host tool such that at run time of a host tool neither compilation (as in PROD) nor simulating an interactive session (as for INA) are necessary at the time the host tool is running.

```
PLACE eating0, eating1, ( ..... )
      thinking2, thinking3, thinking4;

MARKING thinking0 : 1, thinking1 : 1, ( ..... ), fork4 : 1;

TRANSITION releaseleft0
CONSUME eating0: 1;
PRODUCE hasright0: 1, fork0: 1;

( ......)

TRANSITION takeright4
CONSUME hasleft4: 1, fork0: 1;
PRODUCE eating4: 1;
```

Fig. 2. Sketch of a LoLA input file for a dining philosophers net

3   State Space Exploration Techniques in LoLA

LoLA is able to compute the symmetries of a place/transition net automatically using the method presented in [5] and to calculate a symmetrically reduced reachability or coverability graph. For the integration of the symmetries into reachability graph generation, the user can choose between three different strategies. This choice allows the user to use the optimal method for a particular case (advantages and disadvantages of the strategies are discussed in [6]). LoLA always uses the largest possible symmetry group (which leads to the most condensed state space) that is capable of preserving the property to be verified. For model checking and home state verification, symmetries are switched off.

With or without symmetries, stubborn set techniques [9,10]) can be applied. This way, reduced state spaces that preserve a particular property can be generated. The list of analysable properties includes reachability (of a full marking or a state predicate), boundedness (of the net or a particular place), dead transitions, home states, and reversibility. If possible, this list will be extended. Symmetry and stubborn set methods can be applied jointly. The user can switch between different graph exploration strategies (depth first, breadth first). When using the latter strategy, shortest paths can be calculated for reachability related properties (including reachability of predicates and dead transitions). LoLA always selects a version of stubborn sets that preserves the property under verification (for standard properties, the versions are discussed in [8]).

A model checker based on the ALMC algorithm [11] is implemented and can be used with stubborn set reduction. Thereby, AG and EF operators are bound to stubborn set reduction as described in [7] while the remaining operators rely on the reduction technique of [2].

The size of systems that can be handled by LoLA depends on the reduction power of the symmetry and stubborn set techniques for the particular system and verification problem. Especially stubborn set reduction power is rather difficult

```
# lola ph500.net -S
2500 Places
2000 Transitions
 computing symmetries...
499 generators in 1 groups for 500 symmetries found.
0 dead branches entered during calculation.
>>>>> 1499 States, 2496Edges, 1499 Hash table entries
>>>>> 1 States, 0Edges, 1 Hash table entries
not reversible: no return to m0 from reported state
STATE
hasleft0 : 1,
hasleft1 : 1,
(...)
hasleft499 : 1
#
```

Fig. 3. LoLA execution in a configuration that checks reversibility using symmetrically and stubborn reduced graph. `ph500.net` is a file containing a net for a 500 dining philosophers system. Option -S signals that a witness state be reported. LoLA reports the size of the net and the number of found symmetries. In pass 1 of the reversibility check, LoLA calculates 1499 states for finding representatives of all terminal strongly connected components. In pass 2, it checks for theses representatives whether it is possible to return to the initial marking. This check fails for the reported ("witness") marking which is the well known dead state (all philosophers have taken the fork to their left).

to predict. LoLA has successfully analysed the existence of a deadlock for the 1000 dining philosophers net (9000 nodes, 3^{1000} states) using a symmetrically and stubborn reduced graph of about 3000 nodes. On the other hand, it has failed for some much smaller nets due to the too large number of states.

4   Attracted Simulation

LoLA can be used for testing whether a state satisfying a given state predicate is reachable. Testing is based on simulation runs where only the simulation path but not the intermediate states are stored. Selection of the transition to be fired in a state is based on a heuristics that is based on stubborn set techniques and hash table hit statistics. In several cases, the heuristics proved its capability to attract simulation towards a state satisfying the given predicate (if such a state is reachable). For instance, LoLA has found a satisfying state for some system and some property during the first simulation run though there was only one such state among 3^{10000} which was several 10000 transition occurrences away from the initial state.

If simulation runs into a deadlock or exceeds a user defined limit of path length, LoLA resets to the initial state and starts new runs until the program is terminated.

If a state satisfying the predicate is found, LoLA can output a path leading to that state. In contrast to witness paths arising from state space exploration, this path may contain cycles. Nevertheless, according to our experience, the witness path produced by attracted simulation tends to be rather short.

Attracted simulation is between 20 times (net with less then 100 nodes) and 1000 times (nets with several 1000 nodes) faster than traditional state space exploration (concerning the number of transition occurrences per time unit) and much more memory–efficient (only initial and current state as well as the simulation path need to be stored). Attracted simulation provides only semi decision of the tested property. However, its possible answers (yes, reachable/don't know) are dual to linear programming techniques based on the state equation (no, not reachable/don't know).

5   Implementation Details

LoLA's architecture is rather simple. It consists of a parser for reading net description and verification information, a state space depository, and two search procedures — one for depth-first, one for breadth first search. Depth first search is enhanced by Tarjan's algorithm to detect strongly connected components in the state space on the fly. Depending on the user configuration, there are compiler directives that include or exclude specific code for applying selected reduction techniques and for instantiating search to the configured verification task.

5.1   State Space Management

For storing states, LoLA uses a hybrid data structure. The first state is stored as a simple vector. The more states one includes, the more the structure converges to a decision tree. Atop of this structure there is a hash table, i.e. every hash class is stored as a separate tree. Parameters of the hash function are randomly chosen. However, if symmetries are applied, the parameters are adapted such that that the hash function becomes symmetry respecting.

With this structure of the marking depository, containment of a marking can be decided in linear time. Furthermore, tree like structures have advantages when symmetries are used [6]. For the array kind of storage, the user can control the number of bits to represent the token count of a place. This way, LoLA can be customised for safe nets such that a 32 bit word can store the marking of 32 places. Graph information is separated from the search structure. A reachability graph node stores information about successor and parent states, dfs and min numbers (for detection of strongly connected components), a link into the marking depository, and (conditionally) information related to the analysis query.

5.2   Firing

Firing transitions is the core activity in state space generation. Thus, is it particularly important to spend as few as possible run time for this task.

We apply the single transition firing rule. We have minimised the number of enabledness checks. After having fired a transition, only transitions where enabledness could have changed are re–investigated. If, for example, transition $t1$ is fired in the situation depicted in figure 4, only for $t2$ and $t3$ an enabledness check would be forced while the remaining transitions retain their values from the previous state. Backtracking is organised by firing transitions backwards.

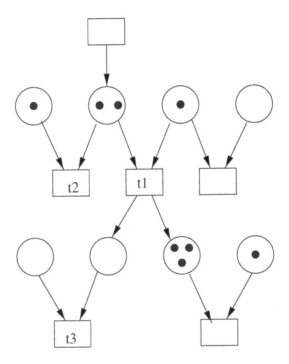

Fig. 4. Enabledness check in LoLA

5.3   Reduction Techniques

All reduction techniques are implemented as modifications to LoLA's central search procedures.

Stubborn set reduction is implemented as a modification of the routine that produces the list of transitions to be fired at a state. Stubborn sets are calculated using a closure operation on a property–dependent starting set. Reversibility and home states are checked using a two–phased procedure. In the first phase, representatives of all terminal strongly connected components of the reachability graph are computed using terminal component preserving stubborn sets. In the second phase, the property is verified by checking mutual reachability of these representatives using reachability preserving stubborn sets. For terminal component preserving stubborn sets as well as deadlock preserving stubborn sets, LoLA calculates minimal stubborn sets with respect to set inclusion.

Symmetric reduction influences the search of markings in the depository. Symmetry calculation and symmetry search is implemented exactly as described in [5,6]. If symmetries are used, LoLA's hash function will be symmetry respecting.

For coverability graph generation, a backward search for a covered state is performed. That slows down generation speed significantly. LoLA produces coverability graphs in the KARP–MILLER–style [3] rather than in the FINKEL–style [1]. The latter would require to remove already calculated states whereas the data structure of the marking depository supports only search and insertion as efficient operations.

Since all reduction techniques concern different parts of reachability graph generation procedure, they can be switched on or off independently. However, LoLA does not allow the application of reduction techniques that do not preserve the chosen target property.

6   Performance

On a 400MHz Pentium II PC (LINUX), reachability graph generation speed (without reduction) ranges between 25000 states/sec (net with 20 places) and 700 states/sec (net with 5000 places). With stubborn reduction, we still obtain up to 16000 states/sec (small net) or 400 states/sec (large net). Attracted simulation reaches about 500000 transition occurrences per second independently of the net size.

Symmetry calculation for a net with a few thousand nodes requires usually 3 to 5 minutes. The speed of producing a symmetrically reduced graph varies significantly, depending on the applied integration technique and the structure of the symmetry group.

For more detailed results concerning run–time and space consumptions when using reduction techniques, we refer the reader to the quoted papers presenting these techniques. The data reported there were all achieved using LoLA.

7   Availability

Under WWW page
>   http://www.informatik.hu-berlin.de/~kschmidt/lola.html

a free download as well as an online manual can be found.

LoLA runs on virtually all UNIX platforms (we have tested SOLARIS, LINUX and SUN-OS) as well as under WINDOWS-NT (using CYGNUS). If other platforms provide a C++ compiler and the GNU compiler generation tools (`bison` and `flex`), we do not expect problems for using LoLA there.

8   Future Work

We are going to enrich LoLA by more verification techniques, particularly in the area of efficient semi-decision procedures. Then, we plan to provide a platform that allows competition parallel run of several instances of LoLA (in different configurations) or even distributed state space generation in workstation or PC networks.

References

1. A. Finkel. A minimal coverability graph for Petri nets. *Proc. 11th International Conference on Application and Theory of Petri nets*, pages 1–21, 1990.
2. R. Gerth, R. Kuiper, D. Peled, and W. Penczek. A partial order approach to branching time logic model checking. *3rd Israel Symp. on the Theory of Computing and Systems, IEEE 1995*, pages 130–140, 1995.
3. R. M. Karp and R. E. Miller. Parallel programm schemata. *Journ. Computer and System Sciences 4*, pages 147–195, 1969.
4. K. Schmidt. Symmetries of Petri nets. *Informatik-Bericht* 33, Humboldt-Universität zu Berlin, 1994.
5. K. Schmidt. How to calculate symmetries of Petri nets. *Acta Informatica* 36, pages 545–590, 2000.
6. K. Schmidt. Integrating low level symmetries into reachability analysis. *Proc. 6th International Conference Tools and Algorithms for the Construction and Analysis of Systems*, LNCS 1785, pages 315–330, 2000.
7. K. Schmidt. Stubborn sets for model checking the EF/AG fragment of CTL. to appear in FUNDAMENTA INFORMATICAE, 2000.
8. K. Schmidt. Stubborn sets for standard properties. *20th International Conference on Application and Theory of Petri nets*, LNCS 1639, pages 46–65, 1999.
9. A. Valmari. Error detection by reduced reachability graph generation. *Proc. 9th European Workshop on Application and Theory of Petri Nets*, 1988.
10. A. Valmari. State of the art report: Stubborn sets. *Petri net Newsletter 46*, pages 6–14, 1994.
11. B. Vergauwen and J. Lewi. A linear local model checking algorithm for ctl. *Proc. CONCUR, LNCS 715*, pages 447–461, 1993.

Woflan 2.0
A Petri-Net-Based Workflow Diagnosis Tool

Eric Verbeek and Wil M.P. van der Aalst

Department of Technology Management, Eindhoven University of Technology,
the Netherlands
{H.M.W.Verbeek, W.M.P.v.d.Aalst}@tm.tue.nl

Abstract. Workflow management technology promises a flexible solution facilitating the easy creation of new business processes and modification of existing
ones. Unfortunately, most of today's workflow products allow for erroneous processes to be put in production: these products lack proper verification mechanisms
in their process-definition tools for the created or modified processes. This paper
presents the workflow diagnosis tool Woflan, which fills this gap. Using Petri-net
based techniques, Woflan diagnoses process definitions before they are put into
production. These process definitions can be imported from commercial workflow products. Furthermore, Woflan guides the modeler of a workflow process
definition towards finding and correcting possible errors.

1 Introduction

Today's workflow management systems are ill suited to dealing with frequent changes:
there are hardly any checks to assure some minimal level of correctness on the process
[Aal98, AH00]. Even a simple change like adding a task can cause serious problems like
deadlock or livelock. As a result, an erroneous process definition may be taken into production as a workflow, causing dramatic problems for the organization. Therefore, it is
important to verify the correctness of a process definition before it becomes operational.
The role of verification becomes even more important as many enterprises are making
Total Quality Management (TQM) one of their focal points. For example, an ISO 9000
certification and compliance forces companies to document business processes and to
meet self-imposed quality goals [IC96]. Clearly, verification of these process definitions
can be used to ensure certain levels of quality.

The tool Woflan was built in response to the need for a workflow verification tool.
Right from the start, three requirements have been imposed on the tool:

1. Woflan should be independent of the process definition tool used by the modeler.
2. Woflan should be able to handle complex process definitions (up to hundreds of
   tasks).
3. Woflan should give the modeler to-the-point diagnostic information to find and
   repair errors.

Based on these requirements, we decided to use Petri nets because they are a universal modeling language with a solid mathematical foundation, are close to diagramming

M. Nielsen, D. Simpson (Eds.): ICATPN 2000, LNCS 1825, pp. 475–484, 2000.

techniques used in practice, and have efficient analysis techniques that are already available. The primary goal of the Woflan tool is to verify a process definition, i.e., to check whether a process definition is a workflow process definition that satisfies the so-called soundness property [Aal97, Aal98].

Workflow process definition A process definition is called a *workflow* process definition if it has a single start condition (indicating the arrival of a new case), a single end condition (indicating the completion of a case) and if all tasks contribute to completing a newly-arrived case.

Soundness property A workflow process definition is called sound if it is always possible to complete a case (i.e., if it is always possible to reach the end condition), if completion is always proper (i.e., if no references to the case are left behind when it reaches the end condition), and if every task can be executed in some way.

In case the process definition is not a workflow process definition that satisfies the soundness property, Woflan's diagnostic information guides the developer towards finding and correcting the errors.

First, we explain the terminology used in this paper. Second, we describe the architecture of the tool. Third, we discuss the properties used by the tool to decide whether it is a sound workflow process definition. Fourth, we introduce the diagnosis process that helps the developer in finding and correcting the errors. Fifth, we discuss a number of import- and export filters from third party (WFMS, BPR) tools that increase Woflan's usefulness, using a diagnosis process definition as example. Sixth, we diagnose and correct the flawed diagnosis process definition. Last, we conclude with conclusions and future work.

2   Terminology

The terminology used in this paper is based on the terminology used by the WfMC [WFM96]. However, to avoid confusion within the Petri-net community, we use the term *condition* instead of *transition* to describe places. Table 1 shows the mapping from the workflow terms [WFM96] used in this paper to Petri-net related terms. For some non-standard terms a brief explanation is given.

Short-Circuited Net In [Aal97] it has been shown that a workflow net is sound if and only if that net extended with an extra transition (called EXTENSION in Woflan) from the sink place(s) to the source place(s) is bounded and live. This extended net is called the short-circuited net.

In the remainder of this paper, we want to avoid mentioning the short-circuited net over and over again. For this reason, some properties of a process definition are defined (see Table 1) on the short-circuited P/T net, while others are defined on the original P/T net.

Restricted Coverability Graph A restricted coverability graph (RCG) is a coverability graph (CG) except for the fact that infinite states are not expanded during construction of the RCG. Like a CG, an RCG is not uniquely defined if the net is unbounded. If no infinite states exist, an RCG equals the occurrence graph (OG) [VBA99].

Workflow	Petri net
Process definition	P/T net
Workflow process definition	Workflow (WF) net [Aal97, Aal98]
Condition	Place
Task	Transition
Start condition	Source place
End condition	Sink place
Useless task or condition	Strongly unconnected nodes in the short-circuited net
Thread of control	S-component in short-circuited net projected to places
Uniform invariant	P-invariant in short-circuited net containing only weights 0 and 1
Weighted invariant	P-invariant in the short-circuited net containing only semi-positive weights
Proper condition	Bounded place in minimal coverability graph (MCG, [Fin93]) of short-circuited net
Improper scenario	Unbounded sequence in restricted coverability graph (RCG, [VBA99]) of short-circuited net
Live task	Live transition in RCG of short-circuited net
Dead task	Dead transition in MCG of short-circuited net
Deadlock scenario	Non-live sequence in RCG of short-circuited net
Confusion	Non-free-choice cluster [DE95] in short-circuited net
AND-OR mismatch	TP-handle [EN94] in short-circuited net
OR-AND mismatch	PT-handle [EN94] in short-circuited net

Table 1. Mapping from workflow terms to Petri-net terms

Sequence A sequence is a firing sequence of minimal length (e.g., paths in the (R)CG) such that states with a given property become unavoidable (fairness etc. assumed). It is minimal in the sense that up to the last-but-one transition in the sequence (e.g., the last-but-one state in the path) the property is avoidable: the last transition in the sequence (e.g., the last edge in the RCG) makes the property unavoidable.

3   Architecture

The core of Woflan consists of Petri-net-based analysis routines. Using these routines, Woflan can verify the soundness of a given process definition. This soundness property is the minimal requirement any workflow process definition should satisfy. Because soundness is equivalent to the boundedness and liveness of the short-circuited WF net [Aal97], it can be verified using standard Petri-net techniques. Although it is possible to verify the soundness property for many process definitions in polynomial time, Woflan uses the general approach by constructing a minimal and/or restricted coverability graph. The diagnosis of the process definition is also partly based on these constructed CG's. The Woflan tool contains a number of modules:

1. One GUI module (`wofapp`),
2. One analysis module (`wofdll`) for loading, verifying and diagnosing process definitions, and

3. Three conversion modules (`scr2tpn`, `wil2tpn`, and `gwd2tpn`) for process definitions from commercial products (Cosa [SL98], Meteor [SKM], resp. Staffware [Sta97].

The GUI- and conversion modules are implemented in the main executable (called `wofapp.exe`). To support the use of Woflan as a back-end tool, the analysis module is implemented in a separate DLL (`wofdll.dll`).

4   Properties

Soundness of a workflow process definition is equivalent to that definition being proper and live, i.e., all conditions must be proper and all tasks must be live. Therefore, to decide soundness, Woflan computes whether all conditions are proper and all tasks are live. Preceding these two properties, Woflan has to decide whether or not the process definition is indeed a workflow process definition.

4.1   Workflow

The definition of a workflow process definition is straightforward: it should be a process definition with exactly one start condition, exactly one end condition, and no useless tasks or conditions. Because these properties are of a structural nature and do not require the construction of an MCG, RCG or OG, they are relative easy to check.

4.2   Properness

Properness of conditions can be decided using the conventional method, i.e., by generating the net's MCG etc. However, because of its complexity, we would like to avoid this if possible. Fortunately, there are alternatives that are less expensive from a computational point of view: all conditions covered by threads of control, uniform invariants or weighted invariants are proper. Because a thread of control is also a uniform invariant and a uniform invariant is also a weighted invariant, we have ordered these alternatives from more desirable to less desirable.

Threads of Control From the workflow point of view, threads of control are very desirable. A workflow case typically consists of a number of documents. Each document has its own route through the workflow. A thread of control coincides with such a document route. So, if threads of control cover a workflow process definition, then each workflow case can be split into a number of documents such that each condition can be linked to some (possibly all) of these documents. If there is not such a cover, the uncovered conditions cannot be linked to any document. As a result, for a workflow process definition to be sound, both confusions and mismatches have to be present [VBA99]. Apparently, these constructions are vital to 'cure' the net from these uncovered conditions. This soundness-related property is called *interim soundness*: a process definition containing uncovered conditions is called interim sound if and only if it contains confusions and mismatches.

Diagnostic Properties If a definition contains improper conditions, Woflan computes some additional properties that can help finding and correcting the properness problem: AND-OR mismatches (they endanger properness), confusions, and improper scenarios.

4.3   Liveness

Suppose we have a process definition that can be covered by invariants (i.e., that contains no improper conditions), that can not be covered by threads of control, and that has either no confusions or no mismatches. For such a definition we can conclude that it is unsound, i.e., it contains non-live tasks.

Likewise, suppose we have a process definition containing no improper conditions and for which we have detected substates during the construction of the MCG. A reachable markings M_1 is a substate of another reachable marking M_2 iff $M_1 < M_2$. At this point, we can conclude that from M_1 the extra task EXTENSION is dead.

Otherwise, Woflan has no method yet to decide liveness without generating the OG. Note that generating this OG is only possible if the process definition contains no improper conditions.

Diagnostic Properties If a definition contains non-live tasks, Woflan computes some additional properties that can help finding and correcting the liveness problem: OR-AND mismatches (they endanger liveness), confusions, dead tasks, and deadlock scenarios.

5   Diagnosis

Based on practical experiences with earlier versions of Woflan we have developed a method for detecting errors in a workflow process. This method is supported by Woflan 2.0 and uses the diagnosis process shown in Figure 1. The diagnosis process can either be executed in-succession or step-by-step. In the latter case, dialogs are used to communicate with the user. First, we give the diagnosis process. Second, we explain one of the dialogs in detail, using the diagnosis process as shown before as example. Last, we explain the general view in detail, using again the diagnosis process itself. Please note that Figure 1 is used both as a meta-model describing the functionality of Woflan *and* as a concrete example of a workflow process.

5.1   Process

Because of the fact that we need the OG (and therefore a workflow process definition containing no improper conditions) for deciding liveness, it is obvious to decide it only if the workflow process definition has been proven to contain only proper conditions. Because the workflow properties are easy to check, Woflan starts with them. As a result, the main steps in the diagnostic process are:

1. Decide workflow,
2. Decide properness, and
3. Decide liveness.

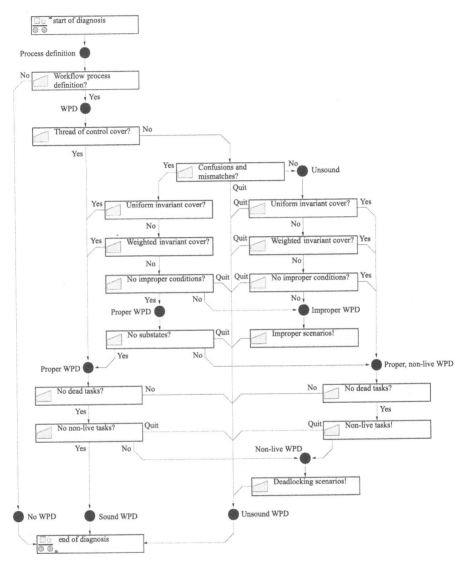

Fig. 1. Diagnosis process, modeled using Protos [Pal97]

If some step fails, there is not much use in continuing with the next step. The modeler of the process definition first has to correct the errors present.

Figure 1 shows a graphical representation of the diagnosis process definition. Note that in some cases the diagnosis process may be continued when unsoundness of the process definition has been detected. The reason for this is to collect more diagnostic information.

5.2   Dialogs

The diagnosis process uses a series of dialogs to guide the user step-by-step through the process. Depending on the diagnostic results, either a next dialog is presented or the process is finished (end of diagnosis). Properties that are likely to be of interest to the modeler are automatically unfolded. As running example, we take the diagnosis process definition as shown in Figure 1 and show the dialog concerning the thread of control cover.

Thread of Control Cover?   The dialog as shown in Figure 2 shows that, using previous dialogs, Woflan has concluded that the diagnosis process definition is a workflow process definition, but that no threads of control exist. As a result, the 20 conditions of the diagnosis process definition are listed.

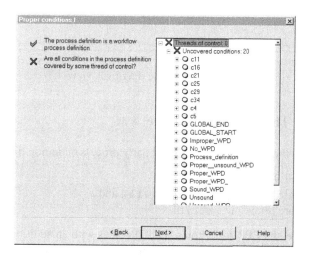

Fig. 2. Example "Thread of control cover?" dialog

5.3   Diagnosis View

The diagnosis view shows all properties of the process definition in a tree-like manner. At the root, the name of the process definition file is shown. This root node has two child nodes: the upper for the diagnosis results, the lower for the diagnostic properties. The diagnosis results node shows in brief the results on the main properties (workflow, safeness, liveness, soundness). The diagnostic properties node combines the diagnostic information from all dialogs.

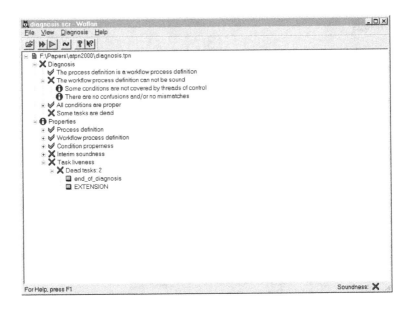

Fig. 3. Example diagnosis view

6   Links to Third-Party Software

Woflan embeds three filters to import third party process definition files:

1. For Cosa [SL98] script files (*.scr),
2. For Meteor [SKM] workflow files (*.wil), and
3. For Staffware [Sta97] (*.xfr) files.

Furthermore, the BPR tool Protos [Pal97] comes with an additional Woflan export filter, which uses Cosa script files as an intermediate format. Each embedded filter shows the results (which could be error messages) of the import process in a dialog. For readability's sake the comments are colored gray, the keywords green, and error messages red.

The dialog as shown in Figure 4 results from the diagnosis process definition (see Figure 1) that was designed using Protos, exported to Woflan (i.e., to a Cosa script file) and imported by Woflan's Cosa import filter. Note that the Cosa import filter automatically added a start condition (GLOBAL_START), an end condition (GLOBAL_END), and a token in the start condition representing a newly-arrived case.

7   Example

Apparently, the diagnosis process definition from Figure 1 is unsound. In this case, particularly the dead tasks are of interest. Note that the short-circuiting task EXTENSION is dead. As a result, all tasks are non-live. The task EXTENSION can only be dead if task end_of_diagnosis is dead, which is dead because it acts as an AND-join instead

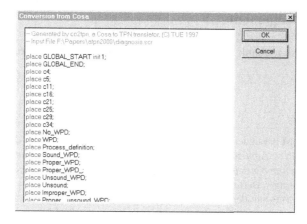

Fig. 4. Example import filter dialog

of an OR-join. Protos' export filter and Woflan's import filter allow to have this property changed in Protos, both filters can handle a task which acts as an OR-join (split) instead of as an AND-join (split). After changing this property in Protos, exporting it to Woflan and importing the resulting Cosa script file, the diagnosis process appears to be a sound workflow process definition. Although this error seems trivial, taking a workflow with such a flawed process definition into production will result in much irritation and agony: prevention is better than cure. Also note that real world examples are not as straightforward as the workflow process definition shown in Figure 1.

8 Conclusions and Future Work

For several commercial WFMS/BPR products, Woflan can be used to verify a process definition, checking both syntactic (cf. Section 4.1) and behavioral (cf. Sections 4.2 and 4.3) properties. By using Woflan and its state-of-the-art techniques it is possible to prevent that an unsound workflow is taken into production.

In the nearby future we hope to extend Woflan with two more features: transition invariants and visualization.

A *sound* workflow process definition is covered by non-negative transition invariants. If the process definition is safe, a task that is not covered by these invariants cannot be live. In a next version of Woflan we hope to use this property to avoid the use of the OG, if possible.

We also would like to visualize the diagnosis results in some intuitive way, using Petri nets of course. If possible, we even want to visualize these results in the WFMS/BPR tool the modeler is using. To support this use of Woflan as a back-end tool, we separated the analysis techniques from the rest of the tool.

Woflan can be downloaded from [VA].

9   Acknowledgements

The authors wish to thank Twan Basten and Hajo Reijers for their fruitful remarks and suggestions.

References

[Aal97]   W.M.P. van der Aalst. Verification of Workflow Nets. In P. Azéma and G. Balbo, editors, *Application and Theory of Petri Nets 1997, Proceedings*, volume 1248 of *Lecture Notes in Computer Science*, pages 407–426, Toulouse, France, June 1997. Springer, Berlin, Germany, 1997.

[Aal98]   W.M.P. van der Aalst. The Application of Petri Nets to Workflow Management. *The Journal of Circuits, Systems and Computers*, 8(1):21–66, 1998.

[AH00]   W.M.P. van der Aalst and A.H.M. ter Hofstede. Verification of Workflow Task Structures: A Petri-net-based Approach. *Information Systems*, To appear, 2000.

[DE95]   J. Desel and J. Esparza. *Free Choice Petri Nets*, volume 40 of *Cambridge Tracts in Theoretical Computer Science*. Cambridge University Press, Cambridge, UK, 1995.

[EN94]   J. Esparza and M. Nielsen. Decidability Issues for Petri Nets - a Survey. *Journal of Information Processing and Cybernetics*, 30(3):210–242, 1994.

[Fin93]   A. Finkel. The Minimal Coverability Graph for Petri Nets. In G. Rozenberg, editor, *Advances in Petri Nets 1993*, volume 674 of *Lecture Notes in Computer Science*, pages 210–243. Springer, Berlin, Germany, 1993.

[IC96]   R.R.A. Issa and R.F. Cox. Using Process Modeling and Workflow Integration to gain (ISO 9000) Certification in Construction. In *CIB W89 Beijing International Conference on Construction, Modernization, and Education*, Beijing, China, 1996.

[Pal97]   Pallas Athena. *Protos User Manual*. Pallas Athena BV, Plasmolen, The Netherlands, 1997.

[SKM]   A. Sheth, K. Kochut, and J. Miller. Large Scale Distributed Information Systems (LSDIS) laboratory, METEOR project page. http://lsdis.cs.uga.edu/proj/meteor/meteor.html.

[SL98]   Software-Ley. *COSA User Manual*. Software-Ley GmbH, Pullheim, Germany, 1998.

[Sta97]   Staffware. *Staffware 97 / GWD User Manual*. Staffware Plc, Berkshire, UK, 1997.

[VA]   H.M.W. Verbeek and W.M.P. van der Aalst. Woflan Home Page. http://www.win.tue.nl/~woflan.

[VBA99]   H.M.W. Verbeek, T. Basten, and W.M.P. van der Aalst. Diagnosing Workflow Processes using Woflan. Computing Science Report 99/02, Eindhoven University of Technology, Eindhoven, The Netherlands, 1999.

[WFM96]   WFMC. Workflow Management Coalition Terminology and Glossary (WFMC-TC-1011). Technical report, Workflow Management Coalition, Brussels, 1996.

Author Index

Lecture Notes in Computer Science

For information about Vols. 1–1750
please contact your bookseller or Springer-Verlag

Vol. 1792: E. Lamma, P. Mello (Eds.), AI*IA 99: Advances in Artificial Intelligence. Proceedings, 1999. XI, 392 pages. 2000. (Subseries LNAI).

Vol. 1793: O. Cairo, L.E. Sucar, F.J. Cantu (Eds.), MICAI 2000: Advances in Artificial Intelligence. Proceedings, 2000. XIV, 750 pages. 2000. (Subseries LNAI).

Vol. 1795: J. Sventek, G. Coulson (Eds.), Middleware 2000. Proceedings, 2000. XI, 436 pages. 2000.

Vol. 1794: H. Kirchner, C. Ringeissen (Eds.), Frontiers of Combining Systems. Proceedings, 2000. X, 291 pages. 2000. (Subseries LNAI).

Vol. 1796: B. Christianson, B. Crispo, J.A. Malcolm, M. Roe (Eds.), Security Protocols. Proceedings, 1999. XII, 229 pages. 2000.

Vol. 1800: J. Rolim et al. (Eds.), Parallel and Distributed Processing. Proceedings, 2000. XXIII, 1311 pages. 2000.

Vol. 1801: J. Miller, A. Thompson, P. Thomson, T.C. Fogarty (Eds.), Evolvable Systems: From Biology to Hardware. Proceedings, 2000. X, 286 pages. 2000.

Vol. 1802: R. Poli, W. Banzhaf, W.B. Langdon, J. Miller, P. Nordin, T.C. Fogarty (Eds.), Genetic Programming. Proceedings, 2000. X, 361 pages. 2000.

Vol. 1803: S. Cagnoni et al. (Eds.), Real-World Applications and Evolutionary Computing. Proceedings, 2000. XII, 396 pages. 2000.

Vol. 1805: T. Terano, H. Liu, A.L.P. Chen (Eds.), Knowledge Discovery and Data Mining. Proceedings, 2000. XIV, 460 pages. 2000. (Subseries LNAI).

Vol. 1806: W. van der Aalst, J. Desel, A. Oberweis (Eds.), Business Process Management. VIII, 391 pages. 2000.

Vol. 1807: B. Preneel (Ed.), Advances in Cryptology – EUROCRYPT 2000. Proceedings, 2000. XVIII, 608 pages. 2000.

Vol. 1810: R.López de Mántaras, E. Plaza (Eds.), Machine Learning: ECML 2000. Proceedings, 2000. XII, 460 pages. 2000. (Subseries LNAI).

Vol. 1811: S.W. Lee, H.. Bülthoff, T. Poggio (Eds.), Biologically Motivated Computer Vision. Proceedings, 2000. XIV, 656 pages. 2000.

Vol. 1815: G. Pujolle, H. Perros, S. Fdida, U. Körner, I. Stavrakakis (Eds.), Networking 2000 – Broadband Communications, High Performance Networking, and Performance of Communication Networks. Proceedings, 2000. XX, 981 pages. 2000.

Vol. 1816: T. Rus (Ed.), Algebraic Methodology and Software Technology. Proceedings, 2000. XI, 545 pages. 2000.

Vol. 1817: A. Bossi (Ed.), Logic-Based Program Synthesis and Transformation. Proceedings, 1999. VIII, 313 pages. 2000.

Vol. 1818: C.G. Omidyar (Ed.), Mobile and Wireless Communications Networks. Proceedings, 2000. VIII, 187 pages. 2000.

Vol. 1819: W. Jonker (Ed.), Databases in Telecommunications. Proceedings, 1999. X, 208 pages. 2000.

Vol. 1821: R. Loganantharaj, G. Palm, M. Ali (Eds.), Intelligent Problem Solving. Proceedings, 2000. XVII, 751 pages. 2000. (Subseries LNAI).

Vol. 1822: H.H. Hamilton, Advances in Artificial Intelligence. Proceedings, 2000. XII, 450 pages. 2000. (Subseries LNAI).

Vol. 1823: M. Bubak, H. Afsarmanesh, R. Williams, B. Hertzberger (Eds.), High Performance Computing and Networking. Proceedings, 2000. XVIII, 719 pages. 2000.

Vol. 1824: J. Palsberg (Ed.), Static Analysis. Proceedings, 2000. VIII, 433 pages. 2000.

Vol. 1825: M. Nielsen, D. Simpson (Eds.), Application and Theory of Petri Nets 2000. Proceedings, 2000. XI, 485 pages. 2000.

Vol. 1830: P. Kropf, G. Babin, J. Plaice, H. Unger (Eds.), Distributed Communities on the Web. Proceedings, 2000. X, 203 pages. 2000.

Vol. 1831: D. McAllester (Ed.), Automated Deduction – CADE-17. Proceedings, 2000. XIII, 519 pages. 2000. (Subseries LNAI).

Vol. 1834: J.-C. Heudin (Ed.), Virtual Worlds. Proceedings, 2000. XI, 314 pages. 2000. (Subseries LNAI).

Vol. 1835: D. N. Christodoulakis (Ed.), Natural Language Processing – NLP 2000. Proceedings, 2000. XII, 438 pages. 2000. (Subseries LNAI).

Vol. 1838: W. Bosma (Ed.), Algorithmic Number Theory. Proceedings, 2000. IX, 615 pages. 2000.

Vol. 1839: G. Gauthier, C. Frasson, K. VanLehn (Eds.), Intelligent Tutoring Systems. Proceedings, 2000. XIX, 675 pages. 2000.

Vol. 1840: F. Bomarius, M. Oivo (Eds.), Product Focused Software Process Improvement. Proceedings, 2000. XI, 426 pages. 2000.

Vol. 1842: D. Vernon (Ed.), Computer Vision – ECCV 2000. Part I. Proceedings, 2000. XVIII, 953 pages. 2000.

Vol. 1843: D. Vernon (Ed.), Computer Vision – ECCV 2000. Part II. Proceedings, 2000. XVIII, 881 pages. 2000.

Vol. 1844: W.B. Frakes (Ed.), Software Reuse: Advances in Software Reusability. Proceedings, 2000. XI, 450 pages. 2000.

Vol. 1845: H.B. Keller, E. Plöderer (Eds.), Reliable Software Technologies Ada-Europe 2000. Proceedings, 2000. XIII, 304 pages. 2000.

Vol. 1846: H. Lu, A. Zhou (Eds.), Web-Age Information Management. Proceedings, 2000. XIII, 462 pages. 2000.

Vol. 1847: R. Dyckhoff (Ed.), Automated Reasoning with Analytic Tableaux and Related Methods. Proceedings, 2000. X, 441 pages. 2000. (Subseries LNAI).

Vol. 1848: R. Giancarlo, D. Sankoff (Eds.), Combinatorial Pattern Matching. Proceedings, 2000. XI, 423 pages. 2000.

Vol. 1849: C. Freksa, W. Brauer, C. Habel, K.F. Wender (Eds.), Spatial Cognition II. XI, 420 pages. 2000. (Subseries LNAI).

Vol. 1850: E. Bertino (Ed.), ECOOP 2000 – Object-Oriented Programming. Proceedings, 2000. XIII, 493 pages. 2000.

Vol. 1851: M.M. Halldórsson (Ed.), Algorithm Theory – SWAT 2000. Proceedings, 2000. XI, 564 pages. 2000.

Vol. 1857: J. Kittler, F. Roli (Eds.), Multiple Classifier Systems. Proceedings, 2000. XII, 404 pages. 2000.

Vol. 1860: M. Klusch, L. Kerschberg (Eds.), Cooperative Information Agents IV. Proceedings, 2000. XI, 285 pages. 2000. (Subseries LNAI).